ELECTRICAL AND ELECTRONIC DRAFTING

BRUCE A. RENTON,
Coordinator of Technical Subjects
The Ottawa Board of Education
Ottawa, Ontario

ELECTRICAL AND ELECTRONIC DRAFTING

Bruce A. Renton

HAYDEN BOOK COMPANY, INC., NEW YORK

Printed in Canada

Library of Congress Catalog Card Number: 70-155864

Illustrations, unless otherwise stated, by:
DAVID C. LUNN, *Toronto*

1 2 3 4 5 6 7 8 9 PRINTING
71 72 73 74 75 76 77 78 79 YEAR

PREFACE

MANY DRAFTSMEN WITH MINIMAL ELECTRICAL KNOWLEDGE BECOME EMBROILED in the intricacies of electrical and electronic drafting without having had any special training in this work. Observations made during years as a practising draftsman led to the conviction that a major emphasis should be placed on some formal instruction.

It became increasingly apparent as a drafting instructor in the teaching profession that the teaching of electrical and electronic drafting should be covered in a logical and systematic manner. The writing of this textbook is the outgrowth of this conviction and experience.

It is not possible to cover in one volume every area of this widely expanding field, nor to present every topic in depth. Therefore, this textbook is designed to present a broad cross section of circuits, symbols, and techniques relevant to this field by drafting examples in a manner that will enable the student to progress in a systematic learning sequence.

The Table of Contents lists the topics that are covered, and the Projects are intended to serve as a guide for the instructor to additional projects that will challenge the student's skill and knowledge. Unit 4 is included specifically to show the physical appearance of a component that is represented on the drawing by a symbol.

The writing of a technical textbook requires the help and cooperation of many people. Therefore, appreciation is extended to the professional people and the manufacturers who have been most generous with their assistance. My thanks go to Pat Rolland for her excellent typing of the manuscript, and to Mr. Chas. Yeo, Senior Instructor of Applied Electricity, and Technical Subjects Consultant, Ottawa Board of Education for the time he has spent in scrutinizing the manuscript and illustrations. The help of a good editor is necessary to bring a book to completion, and Ewart Davies has amply fulfilled that role.

B. A. R.

CONTENTS

UNIT 1

Introduction

THE EXTENT TO WHICH ELECTRICAL AND ELECTRONIC DEVICES ENTER OUR daily lives is multitudinous and indeed continues to grow at a rapid pace. In the process of manufacturing these devices much thought and 'paper work' is a prior requirement. And, when we make reference to this paper work, we mean the development of engineering data gained by research and analysis, and the engineering sketches and working drawings made by the electrical draftsman.

Purpose

The purpose of this text is to explain by printed word and illustration alike the proper forms that electrical and electronic working drawings must take in their diversity. It is not the intention of the writer to explain electrical nor electronic theory, although the necessity of having an intimate understanding of electrical theory and practice is an extremely important asset. If such is not the case the draftsman may never be able to cope with advanced work which in turn will limit his chances of advancement in this field.

Furthermore, it is not the intention of the writer to explain techniques of lettering, basic view presentation, nor any other practices common to technical drafting in detail. However, where delineation varies for purposes of identification, and wherever practices vary greatly from standard technical drafting procedures, explanations will be given.

The skills required

Already briefly mentioned is the weakness of those who make electrical drawings without having a solid foundation of electrical theory. Unfortunately this condition does exist in part and therefore harm is being done to two groups, those who employ and those who are employed. Whether this condition has been brought about by a deficiency of formal instruction in this field is difficult to say. However, opportunities for a student to develop his abilities in electrical drafting properly are vital to the rapidly expanding electrical and electronic technologies.

What then are the skills necessary of a good electrical draftsman?

First, he must have a sound background of electrical theory and practice upon which to build. If this is not the case, then his use to his employer will be limited to that of a junior draftsman or tracer.

Second, he must have the necessary skill of hand, and judgment of eye to execute drawings which are neat, pleasing to the eye, and correctly laid out.

Third, he must have the ability to construct drawings in a minimum amount of time relative to the complexity of the drawing. It is essential that the electrical drafting student fully appreciate this last skill, since the cost of producing a drawing is largely based on the drafting time consumed. However, in no way should the time element dominate the neatness and correctness demanded of good electrical or electronic drawings.

KNOWLEDGE TESTERS

1. What are the necessary prime requirements before electrical and electronic devices are manufactured?
2. Suggest a reason or reasons to support your answer to Question 1.
3. What limitations are imposed on a draftsman who does not have a good background of electrical theory?
4. List the skills necessary for a competent electrical draftsman.
5. Give at least two reasons why the element of time is of prime consideration in the production of electrical drawings.

UNIT 2

Types of Electrical and Electronic Drawings

Variety of electrical drawings

There is a wide variety of electrical drawings currently in use. And, it is perhaps because of their variety and ultimate use that the student draftsman has a partial if not total misconception of their true picture. To many students, an electrical drawing might conjure up the idea of a house wiring diagram. Fig. 2-1 illustrates the floor plan of a single family residence upon which is superimposed the electrical outlets, lamps, and switches. Most of the building dimensions and notes have been left off this drawing so that the electrical symbols will show clearly. Other students already interested in electronics would conceive of a solid state amplifier circuit diagram as being representative. Such a schematic drawing is shown in Fig. 2-2. To others still, the electrical transformer vault drawing shown in Fig. 2-3 would suitably define their concept of an electrical drawing. The truth of the matter is that they can all be broadly classified as electrical drawings but individually falling within limited areas or divisions.

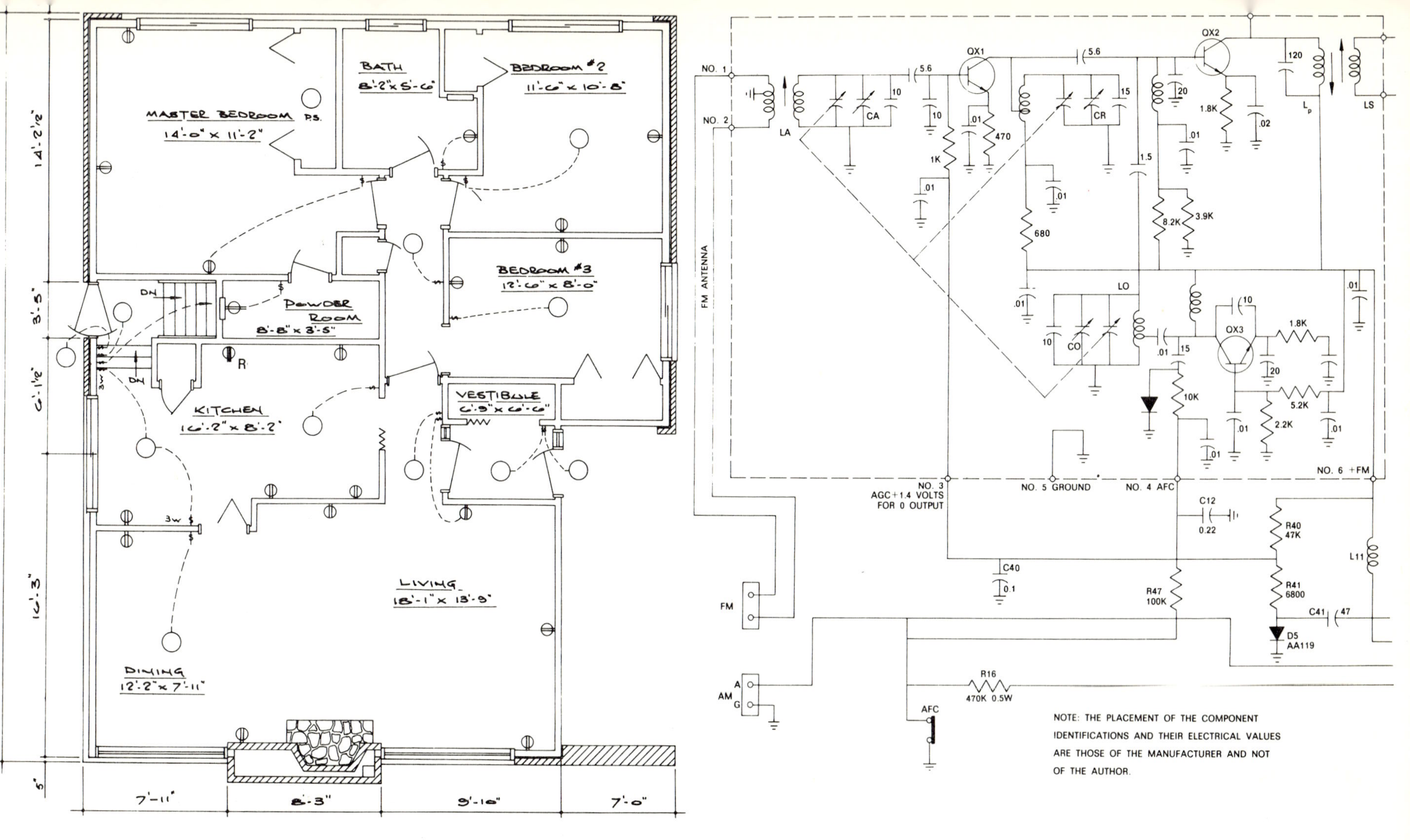

Fig. 2-1 A ground floor plan of a single-family residence showing the superimposed electrical system in the Chatelaine Magazine Award Winning Home. (Courtesy Minto Construction Co. Ltd.)

Fig. 2-2 A part of a schematic wiring diagram of a solid-state F.M. receiver. (Courtesy Clairtone Sound Corp'n Ltd.)

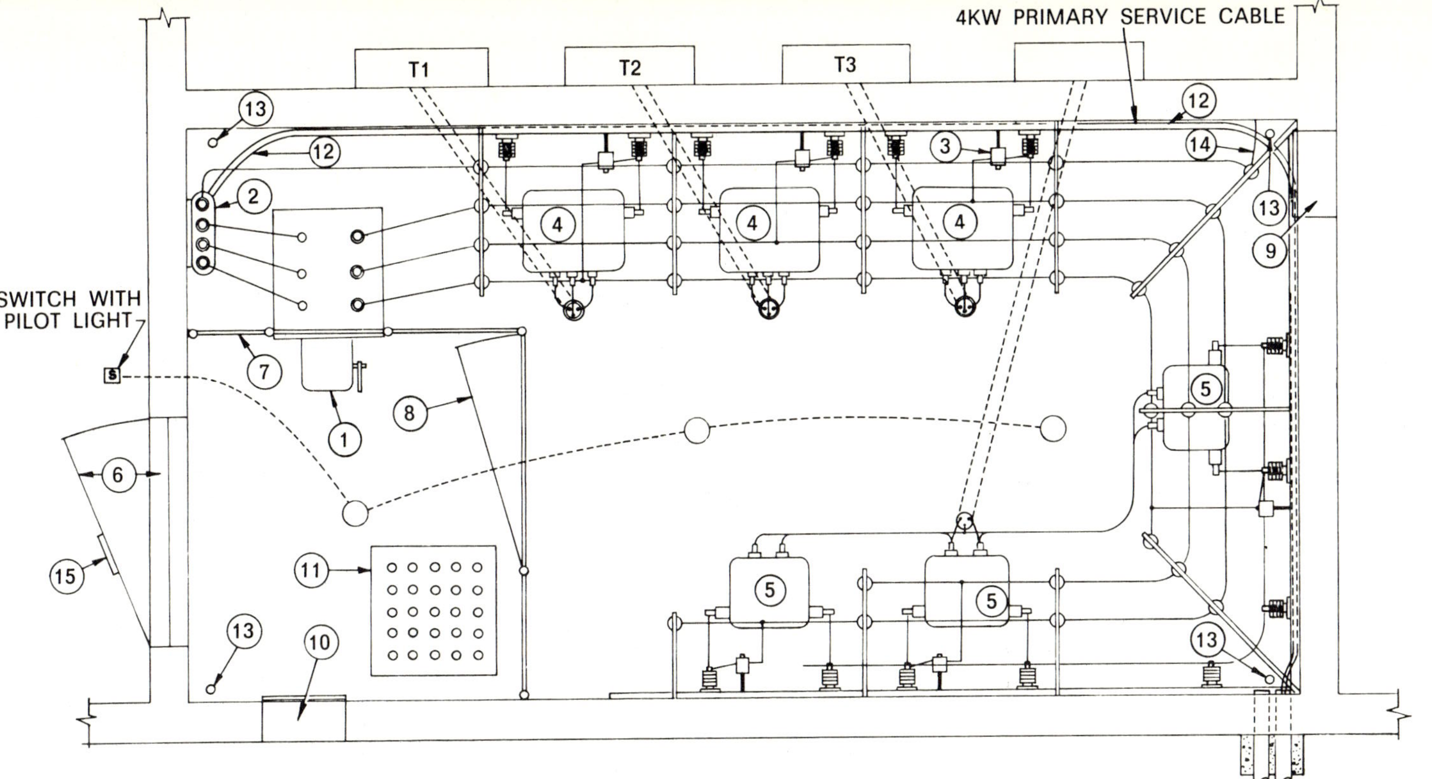

BILL OF MATERIALS

ITEM			
1	Oil circuit breaker shows operating data & interrupting capacity	9	Ventilation near ceiling
2	Pothead	10	Ventilation 12″ from floor
3	Fuse cutouts	11	Sump cover 3/8″ perforated steel plate
4	2400/120/240 V transformers	12	4 KV service cable
5	2400/600 V transformers	13	10′ x 3/4″ ground rods
6	Underwriter's approved class "A" fire door 4′ x 7′ with 4″ sill	14	2/10 connection between ground bus & neutral bus
7	7′ heavy screen guard	15	Weatherproof enameled sign "Danger high voltage" "Entry by Ottawa Hydro personnel only"
8	4′ gate		

Fig. 2-3 A plan view of an electrical transformer vault. (Courtesy Ottawa Hydro-electric Commission)

DRAWINGS USING STANDARD GRAPHIC SYMBOLS

Having said the above we can describe an electrical schematic drawing as one which conveys electrical information in graphic means by the use of standardized component symbols suitably interconnected together to perform some useful function. Fig. 2-4 illustrates this idea graphically.

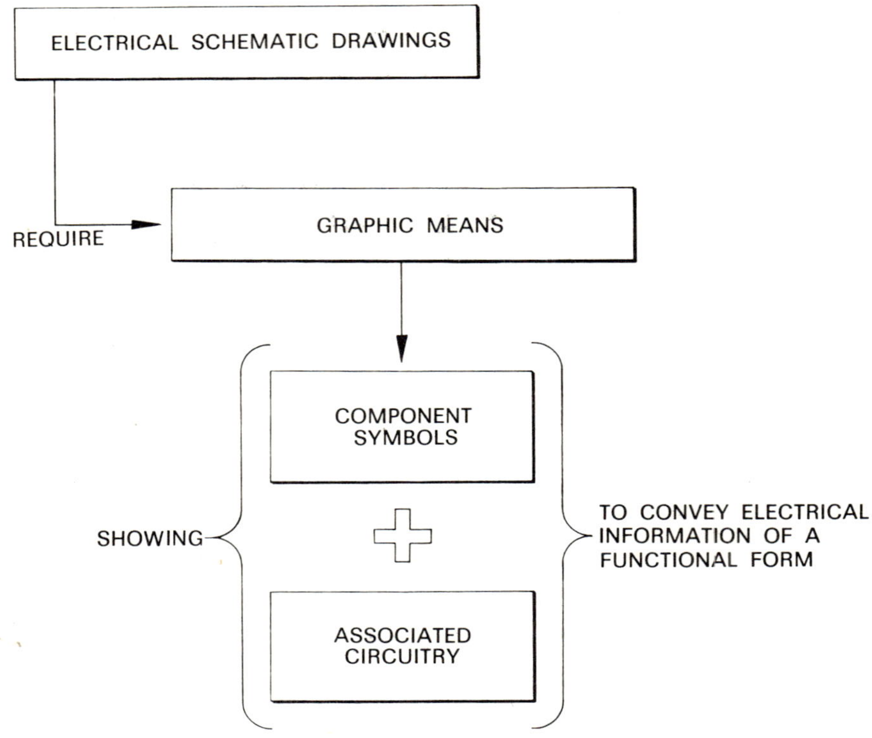

Fig. 2-4 A block diagram illustrating the requirements for an electrical schematic drawing

Schematic wiring diagram

This is a type of drawing frequently associated with electronic applications. Standardized symbols are used and point-to-point connections are illustrated. See Fig. 2-5(b).

It is also known as an elementary wiring diagram if it pertains to electrical control systems. The term **wiring diagram** is also referred to as a **connection diagram,** which is perhaps a more modern term. In any event they can be considered as interchangeable terms.

Pictorial wiring diagram

This is a type of drawing showing **three dimensional** views of the components of a functional circuit in their relative position of assembly. To this extent, even the runs of the individual leads are shown in their proper position.

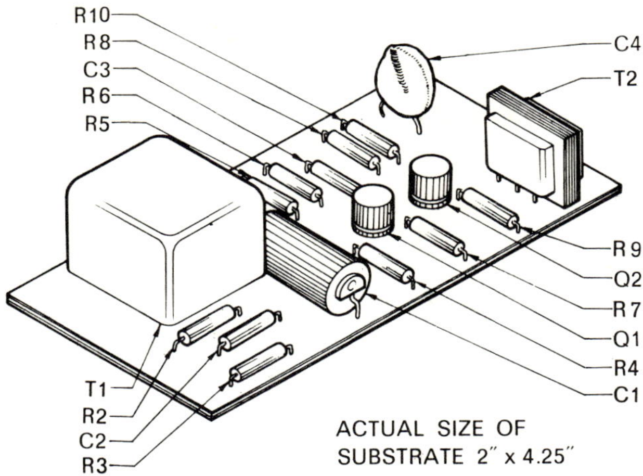

PICTORIAL DRAWING OF AN ACTUAL CIRCUIT BOARD AND COMPONENTS

FIG. 2-5(a)

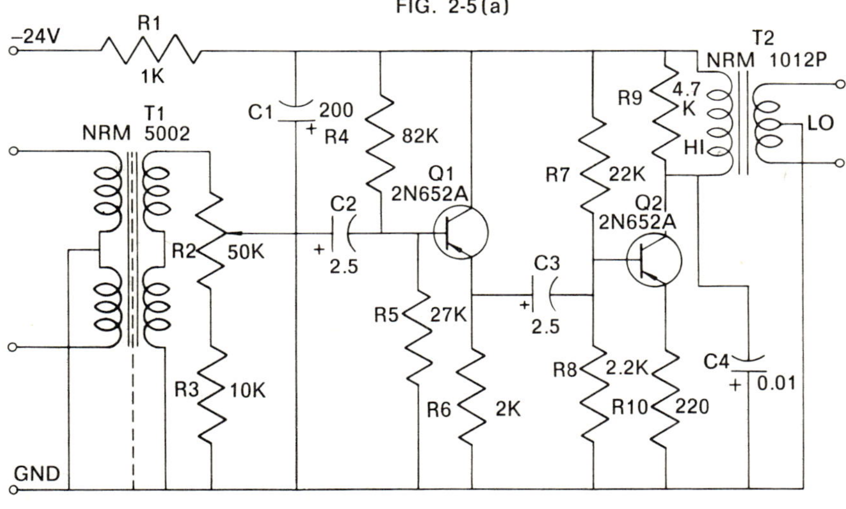

FIG. 2-5(b)

Fig. 2-5(b) A schematic wiring diagram of an operator's telephone amplifier. (Courtesy Northern Radio Mfg. Co.)

In the electronic kit assembling business this type of drawing has wide application. A typical drawing is shown in Fig. 2-6.

An adaptation of the basic pictorial wiring diagram shown in Fig. 2-6 is the type used in the electrical sections of automotive service manuals. While the components illustrated are not accurate representations of the actual parts they are not strictly symbols. No attempt is normally made to give the illustrations a three-dimensional aspect. Fig. 2-7 illustrates this type of pictorial wiring diagram.

Interconnection wiring diagram

Circuits in which individual components such as switches and circuit breakers, or other functional units, are joined together by individual leads terminating at screws or other patented connections are referred to as interconnec-

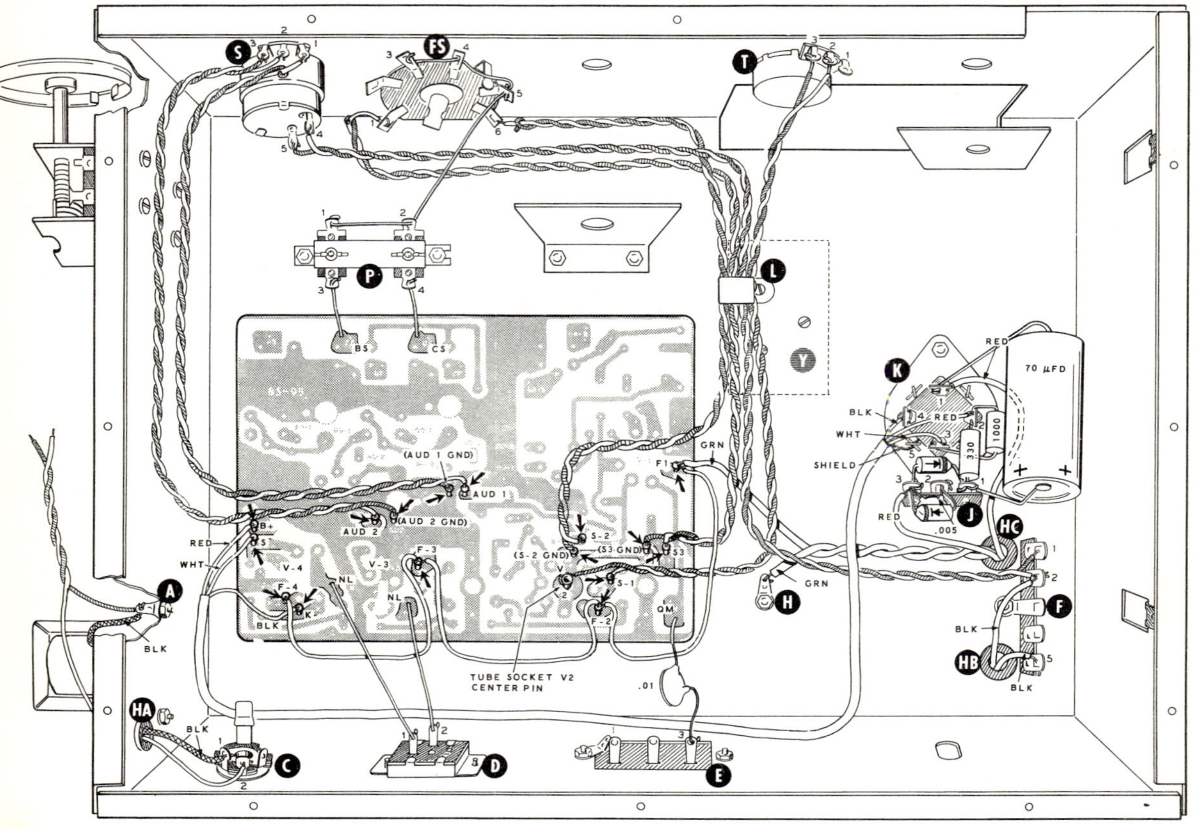

Fig. 2-6 One of a series of pictorial wiring diagrams used in the assembly of a radio receiver kit. (Courtesy Heath Company)

tion wiring diagrams. This is quite different from the wiring or connection diagrams used for electronic circuitry, in which many individual components are soldered together to form a functional circuit. In contrast to the highway type of wiring diagram, corners are drawn with sharp 90° bends. A simplified interconnection wiring diagram is shown in Fig. 2-8.

Highway wiring diagram

This type of connection wiring diagram has the superficial appearance of a cabling diagram but in actual fact it is a point-to-point type of diagram. Curved corners are used, and main lines (highways) are drawn from which branch lines fan out to the various components. In a sense it performs the function of an interconnection diagram, since it is used in circuits where terminations are usually screw connected, or other than soldered. It is important that a logical method of circuit identification be used, otherwise, it would be impossible to trace a circuit. This aspect is recognized in Fig. 2-9 which illustrates a highway or trunk-line wiring diagram.

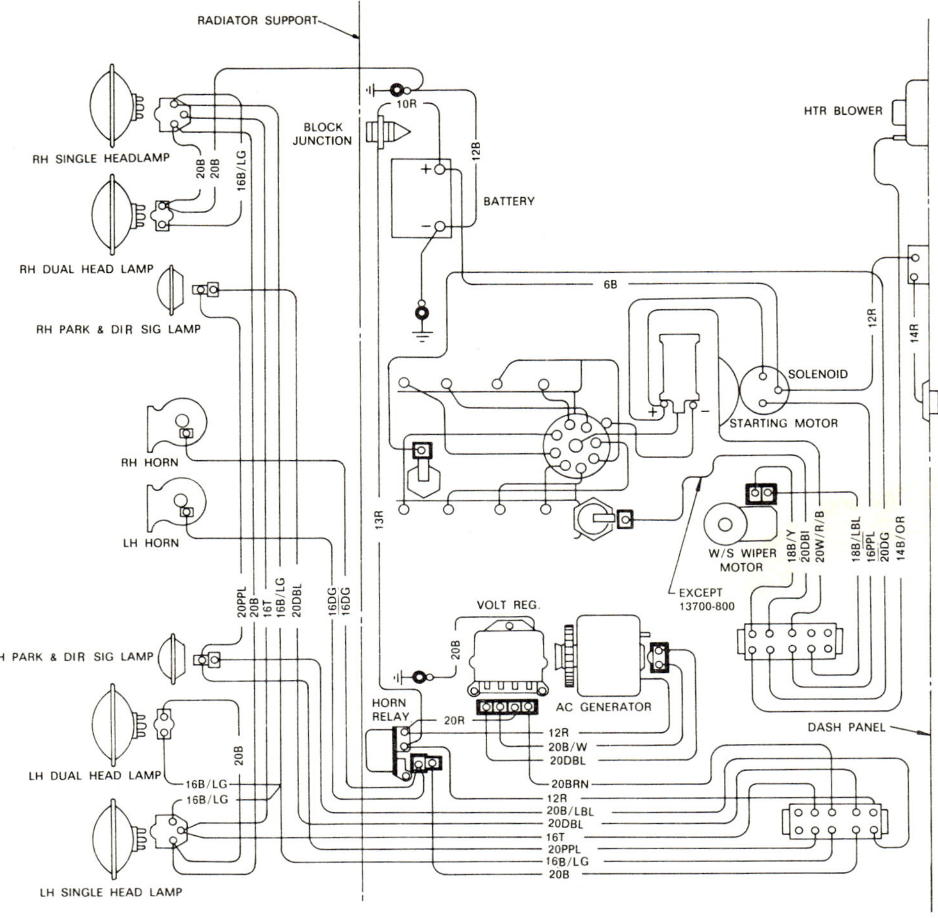

Fig. 2-7 A specialized type of electrical pictorial wiring diagram used in the automotive industry. (Courtesy General Motors Company)

Airline or baseline wiring diagram

Fig. 2-10 illustrates an additional type of connection diagram that is known as an airline or baseline wiring diagram. Note that the major artery (airline) which is shown drawn in a horizontal sense, but could be made in a vertical one, is used to tie together graphically the circuits of all the electronic components. **However, this line is strictly imaginary.** Hence, a system of circuit identification is mandatory.

In highly complex electronic circuits this type of diagram is more easily read than the conventional schematic wiring diagram which shows every lead or conductor travelling from point-to-point connections. Extremely large electronic circuits that are shown by an airline diagram may be subdivided into strips or columns and drawn in a side-by-side fashion. An interconnecting feeder line is used to join these strips or columns, and it may actually represent more than one conductor.

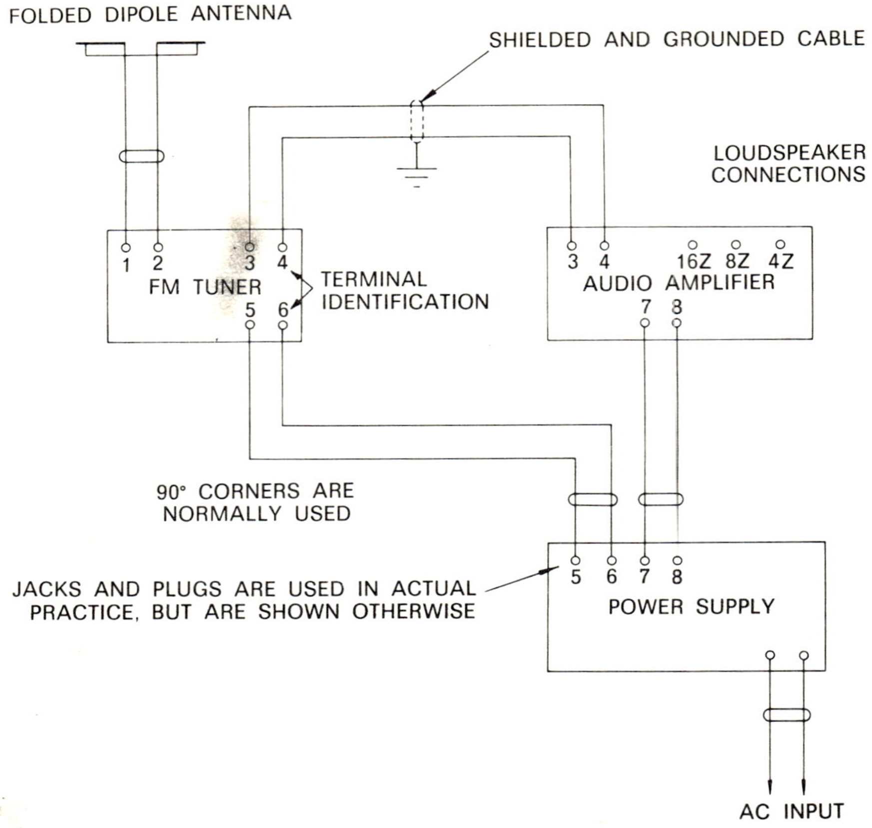

Fig. 2-8 A simplified interconnection wiring diagram. This type of wiring diagram shows the external connections between major functional units. Leads are drawn directly from connection to connection and must be identified suitably. No internal connections are shown

Cabling diagram

A cable is essentially a number of insulated conductors enclosed by either a plastic or metal sheath. When a drawing illustrates major electrical components connected by multiconductor cables, it is referred to as a cabling diagram. Fig. 2-11 shows this type of drawing.

In the United States of America, the term **cabling diagram** is seldom used, and generally does not show the external connections at a functional unit. In Canada, this type of wiring diagram is now much more frequently known as an **interconnection wiring diagram,** but in appearance it is highly suggestive of the **highway wiring diagram.** It is obvious that some ambiguity in terminology exists; but it is the author's point of view that the terms cabling, interconnection, and highway, as they refer to electrical wiring diagrams, all have a common meaning.

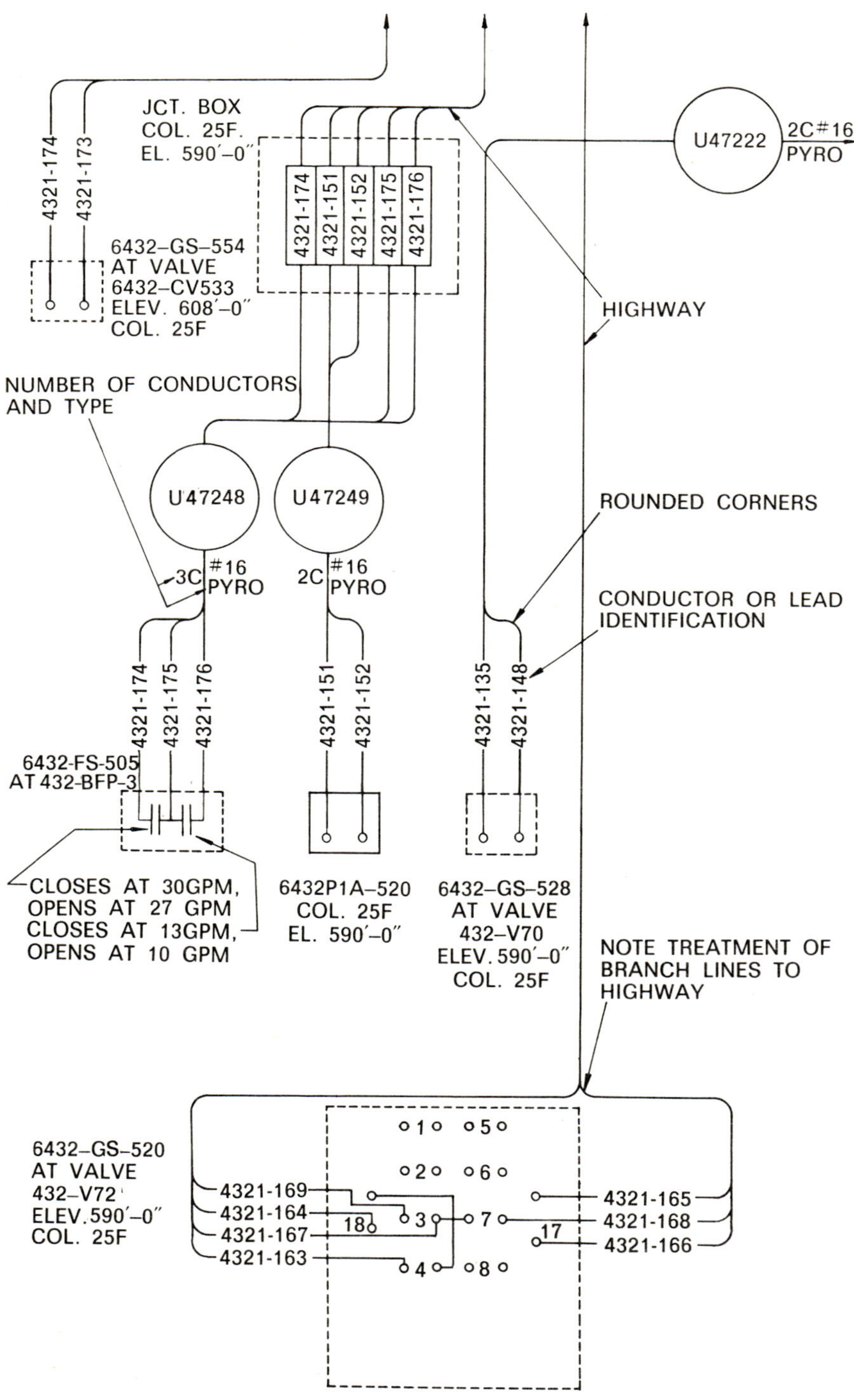

Fig. 2-9 A highway type connection diagram. This is a part of the wiring for certain field connection junction boxes of the main feed water system for a turbine unit of a nuclear power station at Douglas Point, Ontario. (Courtesy Ontario Hydro)

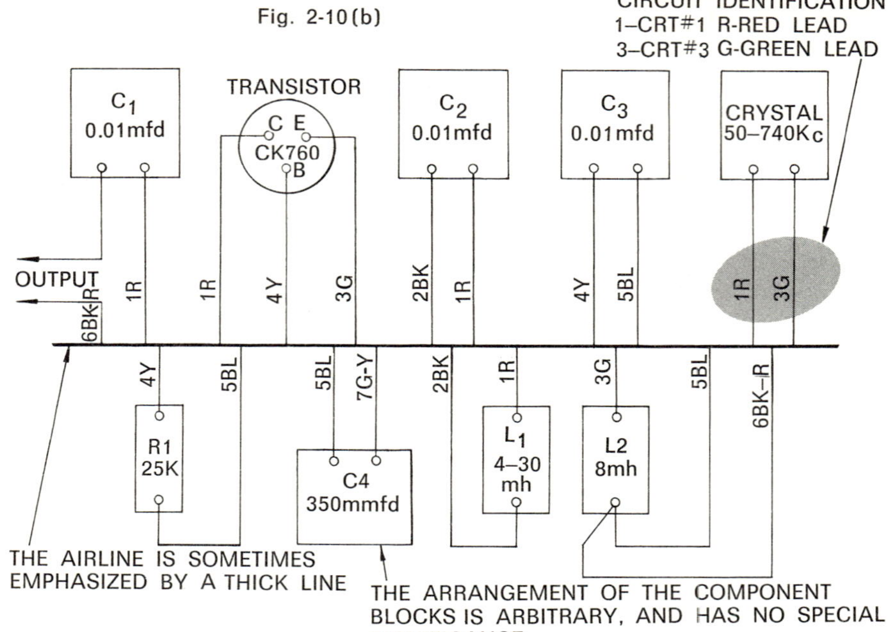

Fig. 2-10(a)

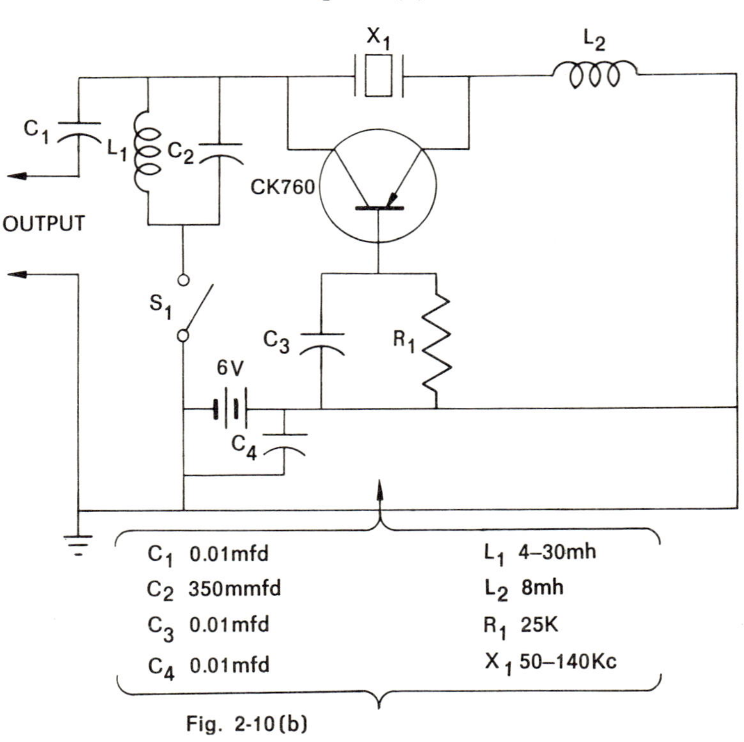

Fig. 2-10(b)

Fig. 2-10(a) *An airline connection wiring diagram of a part of a frequency standard circuit*

(b) *The equivalent schematic wiring diagram of the circuit shown in Fig. 2-10(a)*

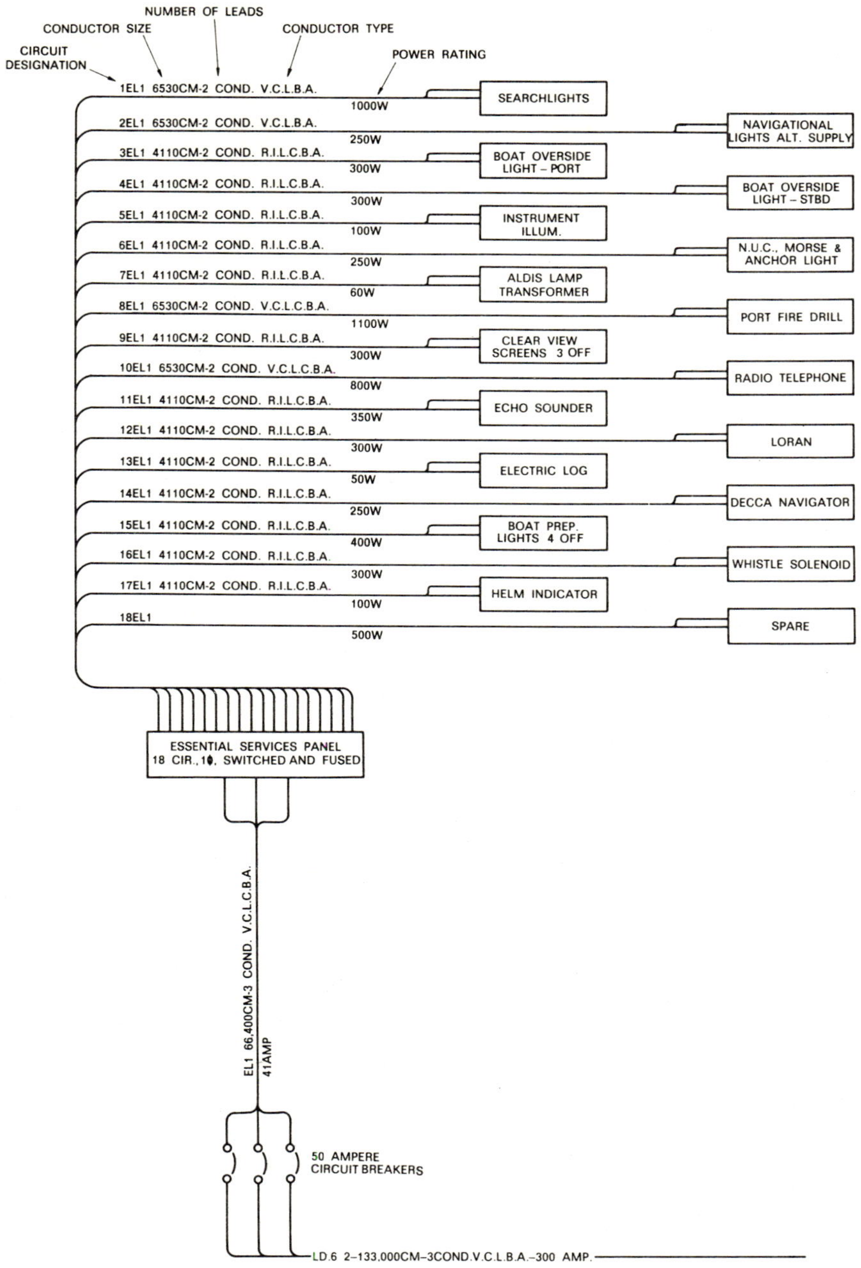

NUMBER OF LEADS
CONDUCTOR SIZE
CONDUCTOR TYPE
CIRCUIT DESIGNATION
POWER RATING

1EL1 6530CM-2 COND. V.C.L.B.A. 1000W SEARCHLIGHTS
2EL1 6530CM-2 COND. V.C.L.B.A. 250W NAVIGATIONAL LIGHTS ALT. SUPPLY
3EL1 4110CM-2 COND. R.I.L.C.B.A. 300W BOAT OVERSIDE LIGHT – PORT
4EL1 4110CM-2 COND. R.I.L.C.B.A. 300W BOAT OVERSIDE LIGHT – STBD
5EL1 4110CM-2 COND. R.I.L.C.B.A. 100W INSTRUMENT ILLUM.
6EL1 4110CM-2 COND. R.I.L.C.B.A. 250W N.U.C., MORSE & ANCHOR LIGHT
7EL1 4110CM-2 COND. R.I.L.C.B.A. 60W ALDIS LAMP TRANSFORMER
8EL1 6530CM-2 COND. V.C.L.C.B.A. 1100W PORT FIRE DRILL
9EL1 4110CM-2 COND. R.I.L.C.B.A. 300W CLEAR VIEW SCREENS 3 OFF
10EL1 6530CM-2 COND. V.C.L.C.B.A. 800W RADIO TELEPHONE
11EL1 4110CM-2 COND. R.I.L.C.B.A. 350W ECHO SOUNDER
12EL1 4110CM-2 COND. R.I.L.C.B.A. 300W LORAN
13EL1 4110CM-2 COND. R.I.L.C.B.A. 50W ELECTRIC LOG
14EL1 4110CM-2 COND. R.I.L.C.B.A. 250W DECCA NAVIGATOR
15EL1 4110CM-2 COND. R.I.L.C.B.A. 400W BOAT PREP. LIGHTS 4 OFF
16EL1 4110CM-2 COND. R.I.L.C.B.A. 300W WHISTLE SOLENOID
17EL1 4110CM-2 COND. R.I.L.C.B.A. 100W HELM INDICATOR
18EL1 500W SPARE

ESSENTIAL SERVICES PANEL
18 CIR., 1φ, SWITCHED AND FUSED

EL1 66,400CM-3 COND. V.C.L.C.B.A.
41AMP

50 AMPERE CIRCUIT BREAKERS

LD.6 2–133,000CM–3COND.V.C.L.B.A.–300 AMP.

Fig. 2-11 *A cabling diagram. Note the rounded corners where the cable direction changes. The method of displaying information regarding a specific cable is not standardized, and it varies according to the manufacturer. (Courtesy Canadian Vickers Shipyards Ltd.)*

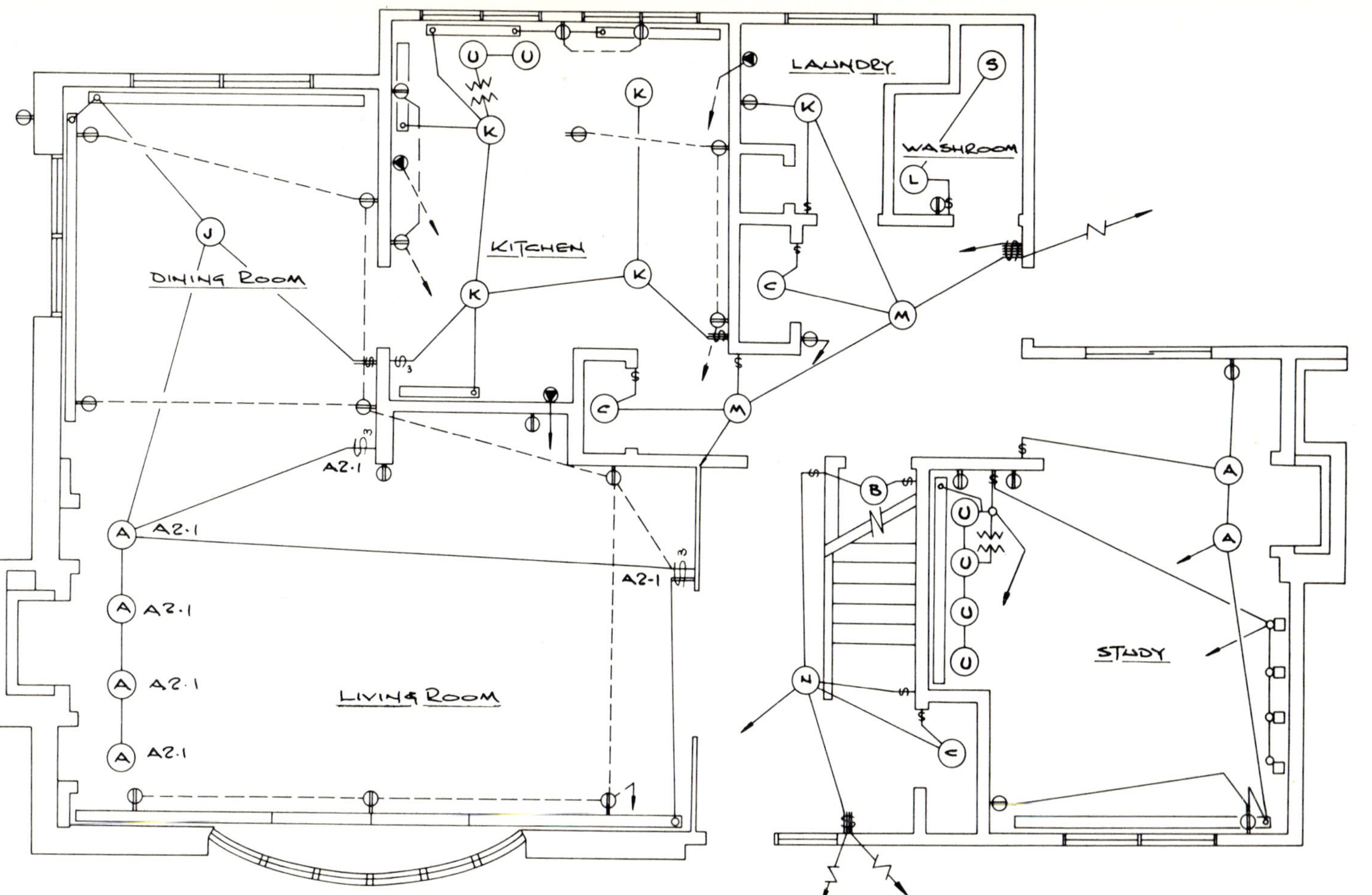

Fig. 2-12 A ground floor plan of a custom designed home showing receptacles, light outlets, switches, and wiring connections. (Courtesy J. L. Richards & Associates, Ottawa, Ontario)

Pictorial cabling diagram

In a similar manner to the pictorial wiring diagram, a cabling diagram can be represented in pictorial fashion. This method has some application in shipboard wiring where, due to the extreme length of the ship, and the number of deck levels, it is advantageous to show the lay of the cabling in pictorial fashion.

Residential building wiring diagram

This type of drawing utilizes standard symbols drawn within the scaled plan view of a house. Unlike an electronic wiring diagram it does not indicate the point-to-point soldered or joined connections, nor does it concur with the nature of a true cabling diagram. Nevertheless, the wiring diagram for a house is essentially a cabling diagram, since at the least each line joining components represents a cable containing two insulated conductors plus a ground wire. A typical example is shown in Fig. 2-12.

The letters within the symbols on the drawing of Fig. 2-12 refer to the various types of the electrical components, and the letter and number combinations beside the symbols refer to the switch or switches to which the component is wired. For example, the two A2.1 switches shown control the four A lights. The original drawing shows the control and reference number for each component in the house, but only the one combination is shown on this drawing for the sake of clarity.

Commercial building wiring diagram

In the design of multi-storey commercial and residential apartment buildings, the electrical and associated services are shown on separate sheets for each level or storey of the building, or at the least a typical floor layout is shown. This is necessary because such things as lighting and power circuits, and T.V. antenna outlets become quite complex with their multiplicity of numbers. Again, standardized symbols are used, although the number and variety of them will be more than in a typical residential wiring diagram. An individual floor layout is shown in Fig. 2-13.

Specialized electrical wiring diagrams

In addition to the foregoing types of drawings, a variety of others exist which are peculiar to their own industry. These are in the fields of aviation, and more broadly, aero-space, and naval construction. While drawings in these areas may follow under some of the more basic types previously mentioned such as cabling or schematic wiring diagrams, they are likely to use some component symbols unfamiliar to the student of electrical drafting. Nevertheless, they constitute an important field in electrical drafting. Fig. 2-14 illustrates an example of an aircraft wiring diagram. This drawing, when completed fully, includes specification references for all of the electrical and electronic components. A few of these references are listed here and are related to the numbered components in the illustration.

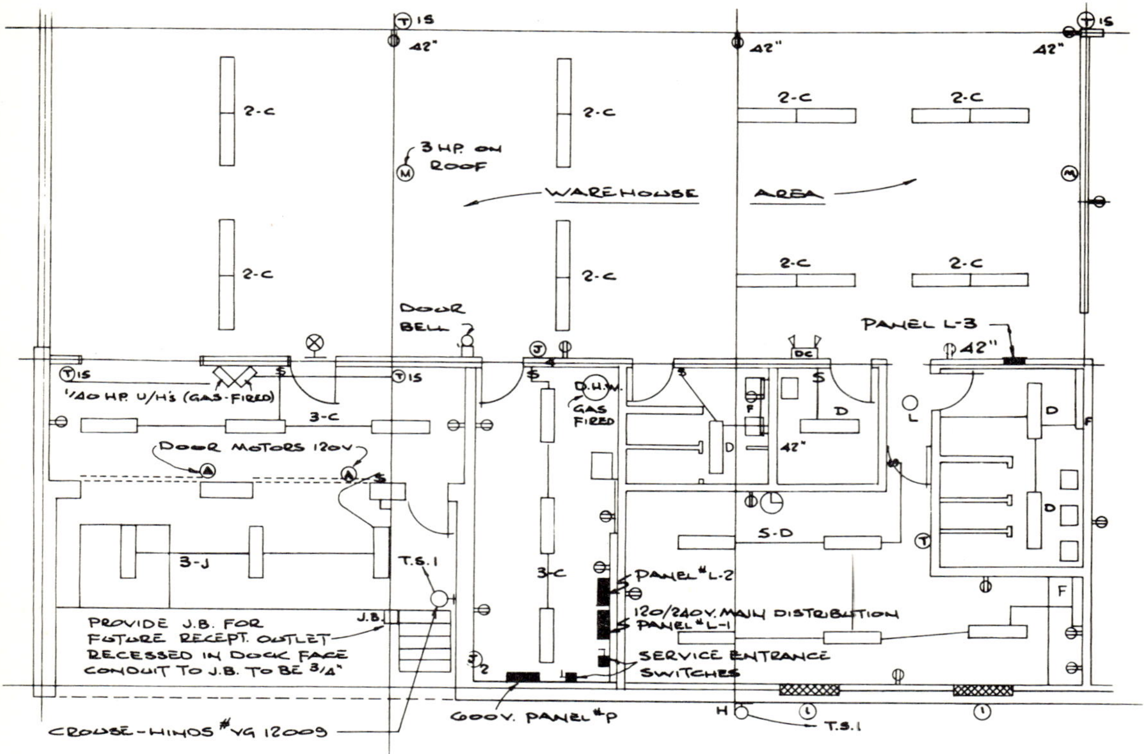

Fig. 2-13 *A type of commercial building having both business offices and warehouse facilities. The Grand and Toy Building, Ottawa, Ontario. (Courtesy Balharrie, Helmer and Associates, Ottawa, Ontario)*

Block or flow diagrams

In addition to drawings using standardized symbols there are also drawings made in block fashion to show the relationship of the major units within an electrical system. As illustrations or examples one could show in block form the essential units in a radar system, an electrical power generating station, or an industrial process requiring electrically operated automated machinery. In Fig. 2-15 a block diagram of a radar transmitter-receiver is shown. An additional illustration of a block diagram is shown in Fig. 2-16. Note that the only conventional symbols used are for the antenna and the loudspeaker. The reason for this variation is that these two units are not constructed from a number of integral components such as resistors, capacitors, and transistors. The purpose of this type of drawing is for preliminary engineering evaluation in the first stages of planning. It will also have value in engineering sales brochures and also as a medium for explaining a system to the layman.

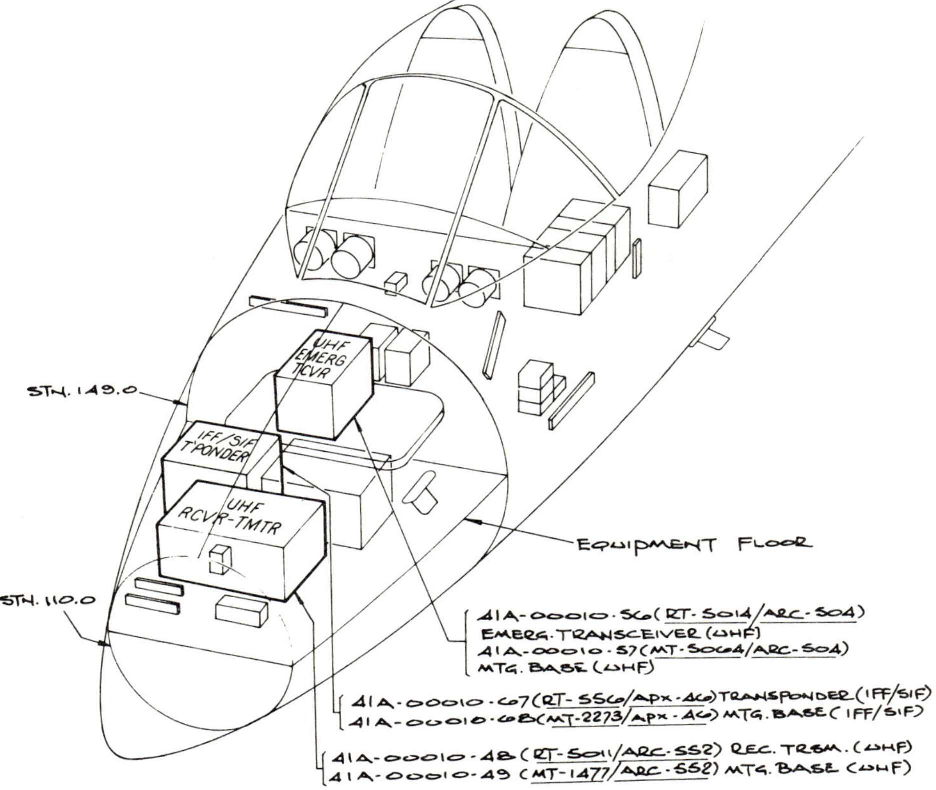

NOTES:

(1) THE ABBREVIATION STN REFERS TO STATION OR DATUM, IN INCHES. GENERALLY THE MOST EXTREME FORWARD POSITION ON THE FUSELAGE WOULD BE STN-100.

(2) REFERENCE TO ALL THE PIECES OF ELECTRICAL EQUIPMENT HAS NOT BEEN DONE IN THE INTERESTS OF CLARITY.

41A – 42100 INSTALL — ELECTRONIC EQUIPMENT — NOSE SECT

1. 41A – 00010-56 (RT-5014/RAC-504) EMERG. — TRANSRECEIVER (UHF)
   41A – 00010-57 (MT-5064/RAC-504) MTG. BASE (UHF)
2. 41A – 00010-67 (RT-556/APX-46) TRANSPONDER (IFF/SIF)
   41A – 00010-68 (MT2237/APX-46) MTG. BASE (IFF/SIF)
3. 41A – 00010-48 (RT-5011/ARC-552) REL-TRSM (UHF)
   41A – 00010-49 (MT-1477/ARC-552) MTG. BASE (UHF)

Fig. 2-14 An aircraft electrical diagram

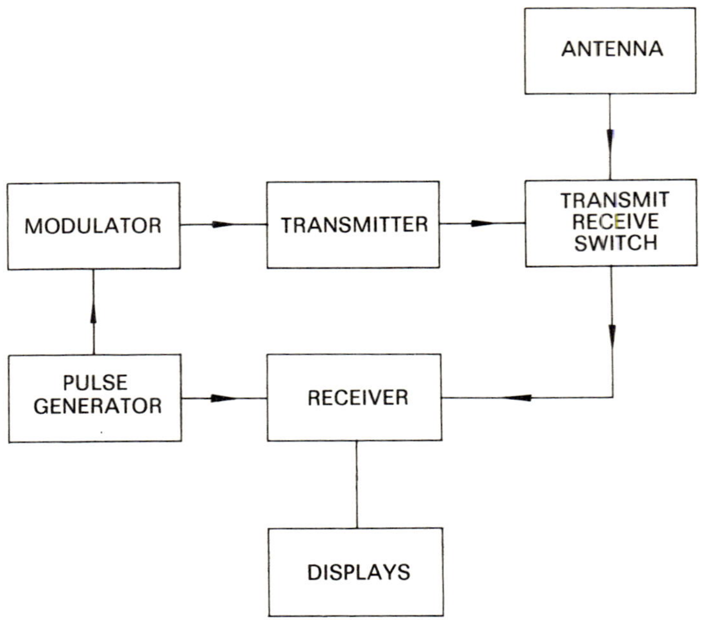

Fig. 2-15 A block diagram of a radar transmitter/receiver

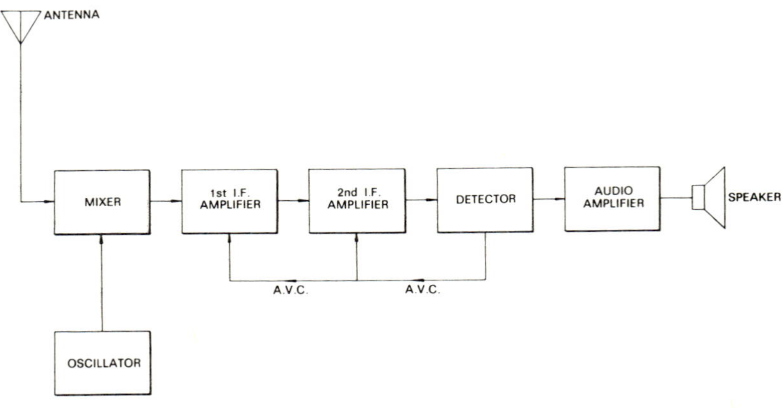

Fig. 2-16 A block or flow diagram

Electromechanical parts drawings

Within the area of electrical drafting a further subdivision exists. This is the realm of drafting associated with the design and drawing of electromechanical parts. To illustrate this concept how would one classify the drawing of a bus bar, a single-pole single-throw switch, a sheet metal radio receiver chassis, or the parts of a transformer? Certainly they perform operations necessary to the overall function of an electrical circuit, yet they are mechanical in nature. Representative examples of electromechanical parts are shown in Figs. 2-17, 2-18, 2-19, and 2-20.

Would one expect an electrical draftsman to have the capacity to undertake both types of drawings? The answer would probably take this form. A draftsman who has a high level of mechanical reasoning and a thorough knowledge of electrical theory and practice could very well accommodate both situations, but in actual practice this is unlikely to be the case. With some reservations, it is probably safe to say that good mechanical draftsmen are not the most competent in the field of electrical drafting. Conversely, those people who know their way in and out of electrical circuitry may fall short of the mechanical reasoning skills of the former group.

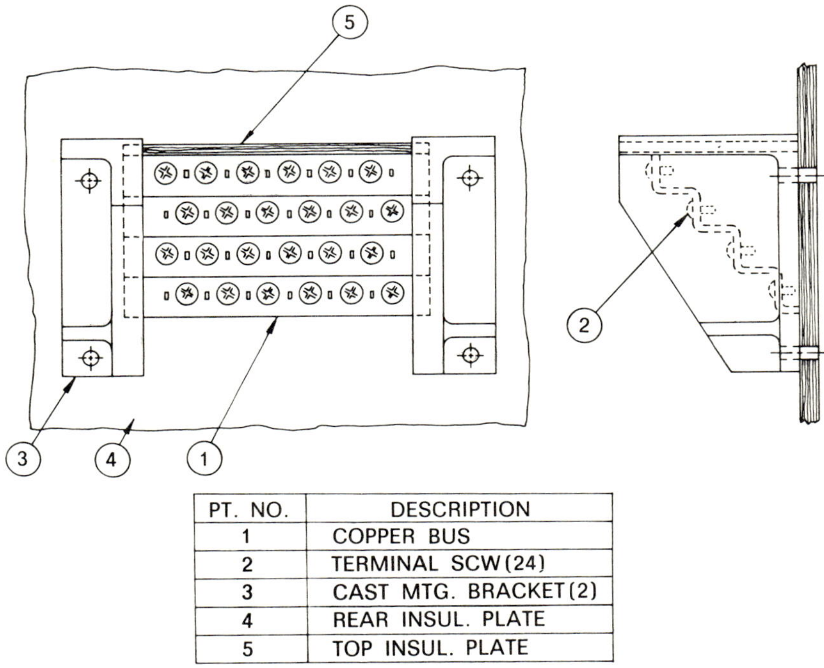

PT. NO.	DESCRIPTION
1	COPPER BUS
2	TERMINAL SCW(24)
3	CAST MTG. BRACKET(2)
4	REAR INSUL. PLATE
5	TOP INSUL. PLATE

Fig. 2-17 Orthographic views of a neutral wire bus for a combination load centre having a 110/220 V.A.C. 100A service

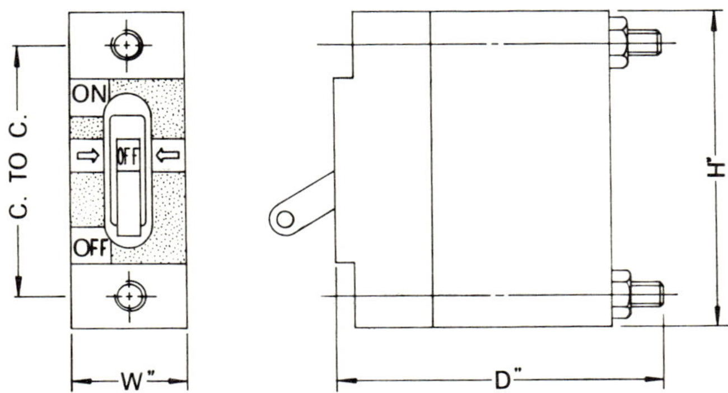

Fig. 2-18 Orthographic views of a S.P.S.T. circuit breaker switch with a 120 V.A.C. 1A rating

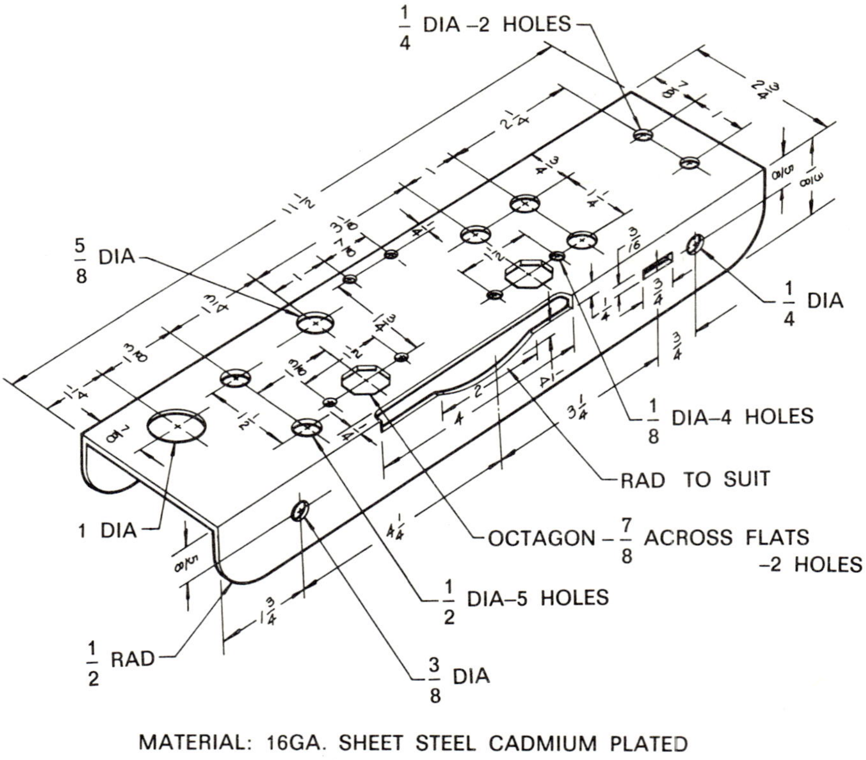

$\frac{1}{4}$ DIA -2 HOLES

$\frac{5}{8}$ DIA

$\frac{1}{4}$ DIA

$\frac{1}{8}$ DIA-4 HOLES

RAD TO SUIT

1 DIA

OCTAGON - $\frac{7}{8}$ ACROSS FLATS -2 HOLES

$\frac{1}{2}$ DIA-5 HOLES

$\frac{1}{2}$ RAD

$\frac{3}{8}$ DIA

MATERIAL: 16GA. SHEET STEEL CADMIUM PLATED

Fig. 2-19 A pictorial view of a chassis punched for the components of a superheterodyne radio

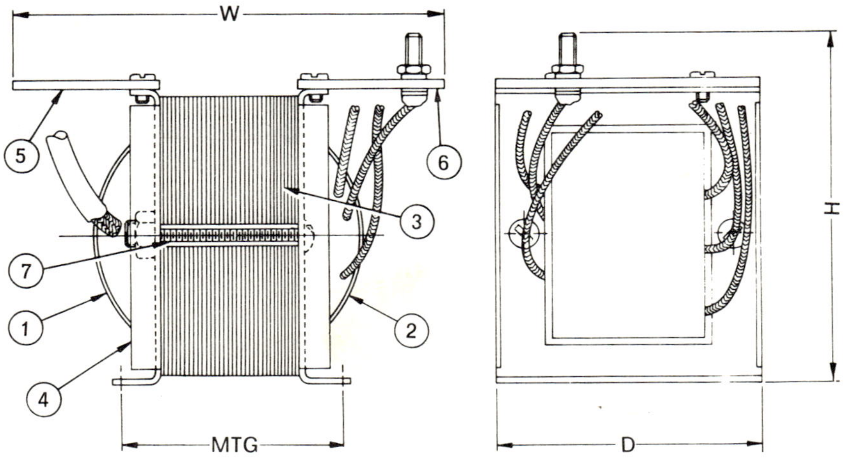

Fig. 2-20 Orthographic views of a step-up transformer

KNOWLEDGE TESTERS

1. Describe the purpose of block diagrams.
2. What form of symbolism if any, is used in a block diagram?
3. Why would individual drawings likely be used for the different levels of a 30-storey building?
4. List 10 different types of drawings that could be labelled as electrical.
5. What is the main difference between the drawings illustrated in Figs. 2-1 and 2-2, other than the fact that one pertains to house wiring and the other to electronics wiring?
6. List and draw the different symbols shown in Figs. 2-1 and 2-2.
7. Refer to the book glossary of terms or other sources and define the terms bus bar, D.P.S.T. switch, transformer.
8. What purpose do the arrowheads in Figs. 2-16 perform?
9. Since it is obvious that the receiver circuit could not be wired from the diagram in Fig. 2-16, what purpose will the diagram serve?
10. Compare the terms 'conductor' and 'cable'.
11. In electronic wiring diagrams, soldered connections may be made along any conductor (consistent with the circuit design). What can you infer from this drafting technique?
12. It has been pointed out that scaled plan views are used in residential wiring diagrams. How does this affect the laying out of the wiring?
13. A suggestion has been made that pictorial wiring diagrams have wide application in the electronic kit assembling business. Why is this likely so?
14. Pictorial cabling diagrams are frequently used in naval construction. Give your ideas as to why this form of presentation would be advantageous.
15. If individual drawings are used for each level of a 30-storey commercial building, suggest an important feature that the draftsman must consistently carry out in order that the drawings are meaningful.
16. Write a brief technical report on the 'qualifications' of a good electrical draftsman.
17. What other term is used in place of 'schematic wiring diagram', if it pertains to electrical control systems?
18. What variation of a pictorial wiring diagram is used in the electrical sections of automotive service manuals?
19. How does an interconnection wiring diagram differ from an elementary wiring diagram?
20. Describe what is meant by a 'highway' in referring to a highway wiring diagram.
21. What type of corners are used in a highway wiring diagram?
22. What special precaution must be understood with regard to an airline wiring diagram?
23. Why is circuit identification vital in an airline wiring diagram?

PROJECTS

1. Design a block diagram of a typical electrically ignited, forced hot air, thermostatically and humidity controlled oil heated furnace.
2. Make a block diagram showing the major components in an automotive electrical system.
3. Make a freehand sketch of the electrical system of your electrical laboratory at school. The sketch should show outlets, lighting switches, etc. Redraw the sketch using drafting instruments.
4. Make a list of at least **10** different electromechanical parts. Describe their purpose as simply as possible in technical terms and make freehand pictorial sketches of them.

UNIT 3

Specialized Drafting Equipment

Templates

In order to carry out the construction of diagrams in a minimum amount of time, a variety of templates can be used. Some templates are exclusively designed for drawing electronic schematics. Others may have a wider range of application in the electrical field.

Note that the templates shown in Figs. 3-1 and 3-2 are used with the pencil point or pen nib (fountain pen type, see Fig. 3-3) penetrating directly through the symbol opening. The symbol openings in the plastic templates are bevelled back slightly on one side. This is the side that is placed upwards.

Templates are normally placed against the edge of a parallel rule or drafting machine for alignment purposes. An additional technique is to place a gridded paper underneath the tracing vellum for locating the position of the symbol relative to the remainder of the circuit lines.

Another device for drawing symbols mechanically is shown in Fig. 3-4. In this instrument the symbols are engraved in a plastic strip which is used in conjunction with a specially designed nib holder. The use of this device is limited to ink drawings. Furthermore, devices of this type require a considerable amount of practice before results of a professional nature are obtained. However, they do have numerous advantages. The uniformity of symbol size, shape, and line thickness are important considerations. Also, the design of the template tends to eliminate smudging. Frequently, junior draftsmen and draftswomen become highly proficient in the use of these devices and spend a major part of their employed time in their operation.

Nevertheless, although these devices are of simple construction, it is necessary to use them carefully in order to produce high quality drawings.

Transfer symbols

In the past few years materials have been developed in the printing industry which eliminate the time spent in drawing component symbols. Initially large size component symbols are drawn accurately. These are photographically reduced to a size commensurate to that used in electrical drafting, and are

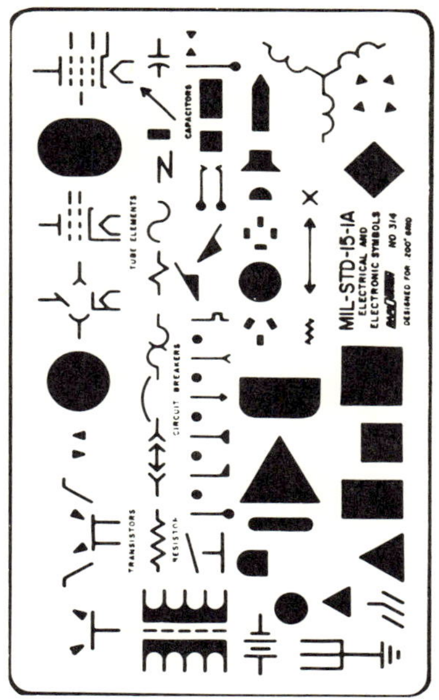

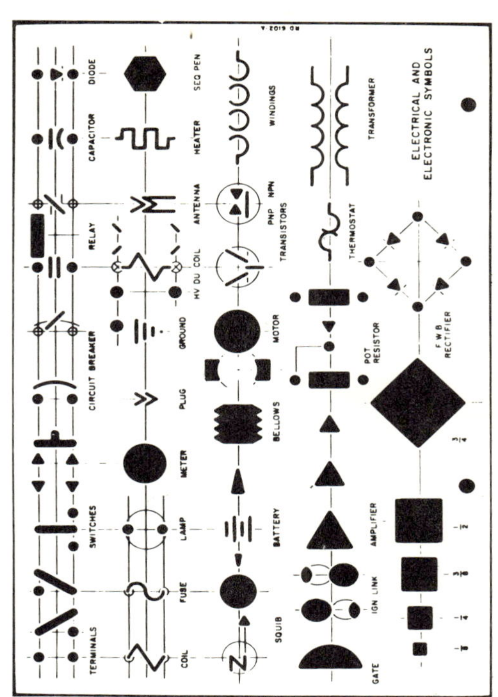

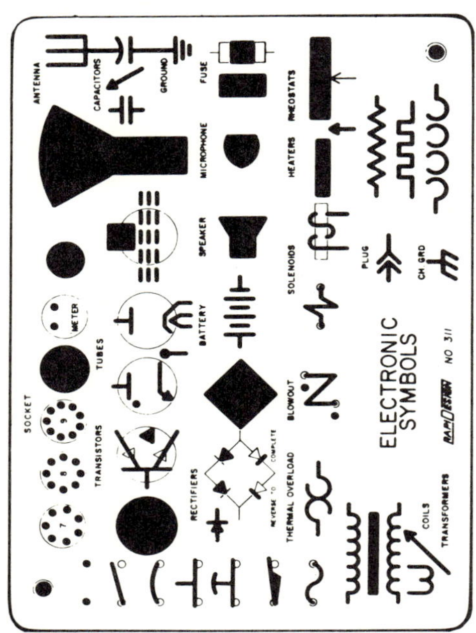

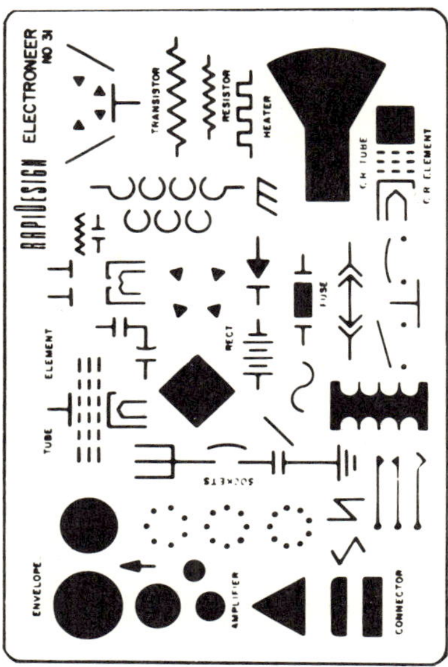

Figs. 3-1 and 3-2 Electrical-electronic symbol templates. (Courtesy RapiDesign, Inc., Burbank, California)

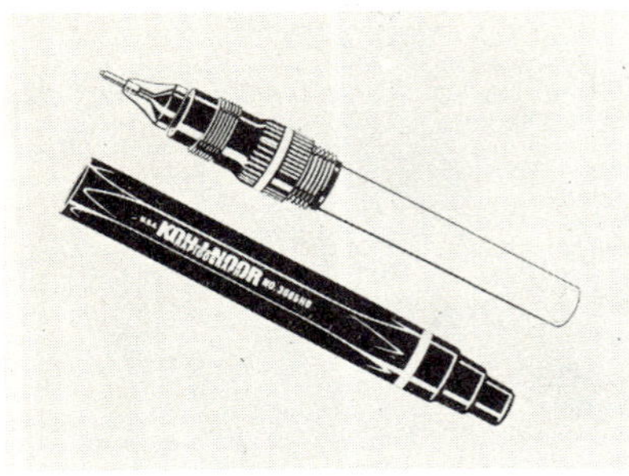

Fig. 3-3 A type of a technical pen. (Courtesy Koh-i-noor (Canada) Ltd., Montreal, P.Q.)

printed in black ink on thin translucent material. Transferring the symbol to the drawing under construction is done by rubbing the face of the transfer medium that does NOT have the symbol on it with a hard surface, such as the handle of a pen. The result is a well defined, accurately shaped symbol image. There are some drawbacks. Care must be taken after application, otherwise the transferred image will be disturbed and possibly broken or distorted. Also, care must be taken in positioning and aligning the transfer material before the actual transferring is carried out. While the cost may be initially high for sheets of transfer material, the time saved and the fact that a less skilled person can build up a drawing by such a means becomes an important asset.

Press-on symbols

Another medium manufactured in a similar way to transfer symbols, but utilizing the adhesiveness of a wax film on its reverse side, is press-on symbols. The use of the press-on medium is not as great as the transfer medium because of two factors. First, the application of the press-on requires that a part of the sheet must be cut, and second, the range of different symbols manufactured as press-on medium does not lend itself especially to electrical drafting, but has a greater application in machine and cartographic work. Samples of commercially available press-on and transfer media are shown in Figs. 3-5 and 3-6. Fig. 3-5(a) shows a few of the 200 Letratone shading patterns that are available. These are printed on heat resistant adhesive film, and include mechanical tints, dots, light and heavy lines, squares, representations, and graduations. Fig. 3-5(b) shows how the film is applied to a drawing.

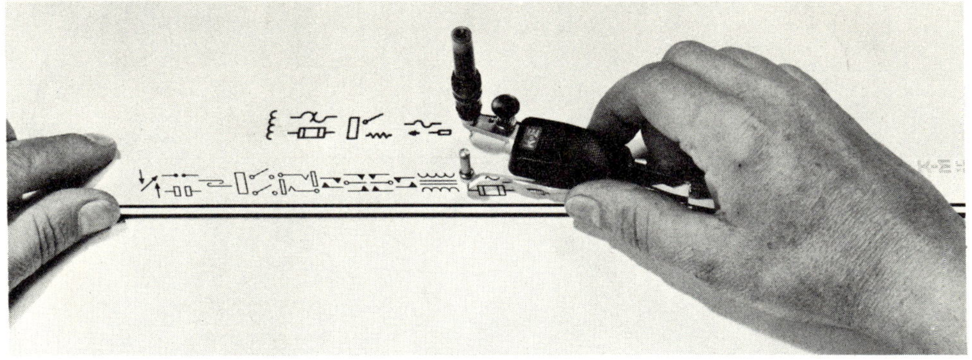

Fig. 3-4 *The mechanical method of drawing electrical component symbols (Courtesy Keuffel and Esser, Hoboken, New Jersey)*

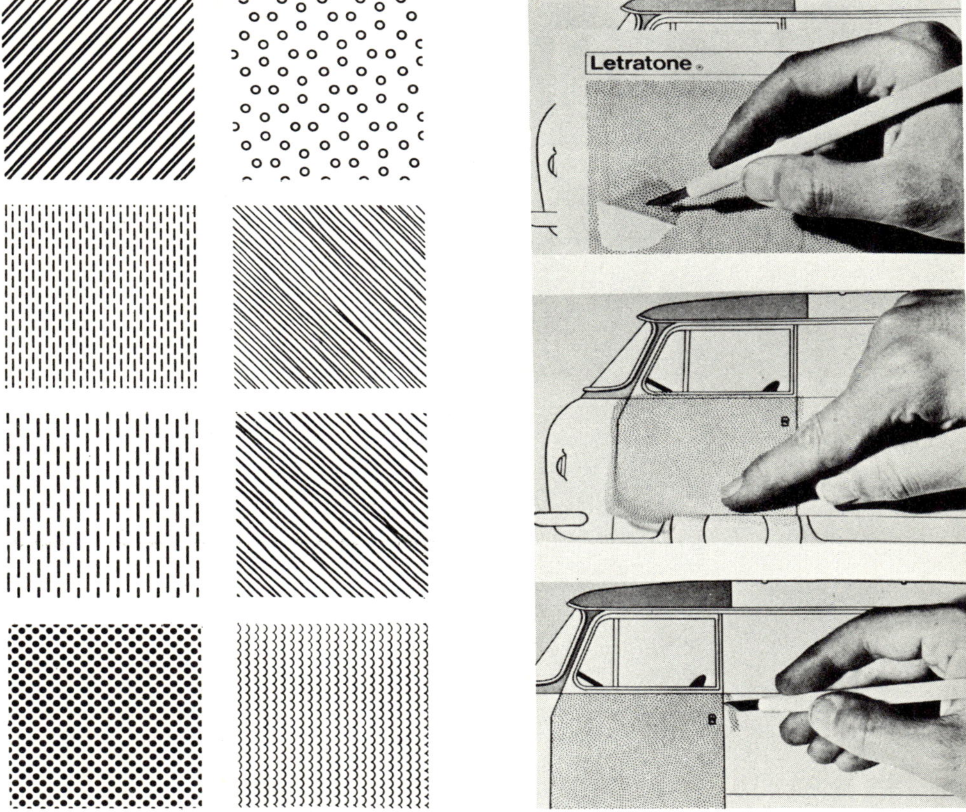

Fig. 3-5(a)

Fig. 3-5(b)

Fig. 3-5(a) *Examples of press-on shading symbols (b) The method of applying shading film (Courtesy Letraset Canada Limited)*

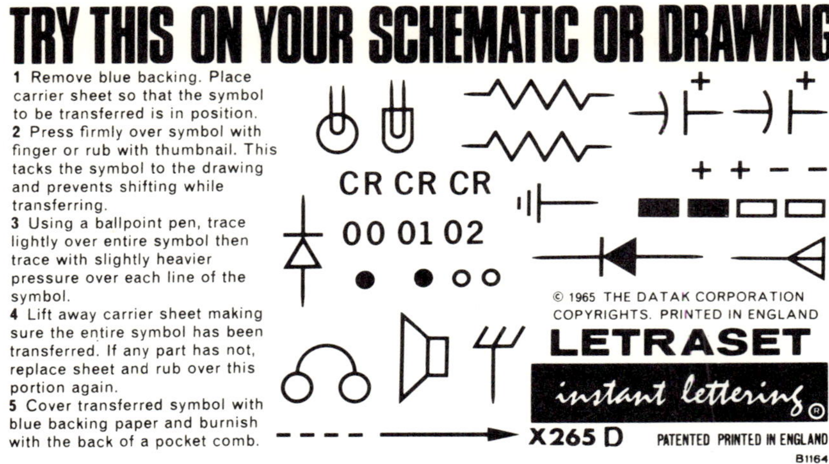

Fig. 3-6 Transfer symbols and letters. (Courtesy Letraset Canada Limited)

Tape

Another medium which now has widespread use in preparing the large-scale drawings necessary in the preliminary stages of making printed circuits is black plastic tape of various widths. To facilitate its application the tape is pre-coated with pressure-sensitive adhesive.

The tape is manufactured in conveniently sized rolls in a simple plastic dispenser. The advantages of the tape are uniformity of width, density of blackness, and speed of application. Since these drawings are photo-reduced to a much smaller size, sharply defined and consistently wide lines result.

THE USE OF CONVENTIONAL DEVICES

Instrument drawing of resistances

A 30°-60° triangle can be used for drawing resistances. After two parallel guide lines are lightly drawn, the triangle is placed against a vertical or horizontal straight edge in an alternating manner as shown in Fig. 3-7.

Instrument drawing of coils

Coils which represent transformers or chokes can be effectively drawn by the use of a compass. The method of construction is shown in Fig. 3-8. While this is a slow procedure, it is perhaps the safest way for ink work. If the finished drawing is to be in either pencil or ink, the necessity of drawing the coil lightly at first is mandatory in order to produce correctly formed loops.

Nonstandard symbols

With the inherent high costs of drafting and the necessity of maintaining uniformity of symbol shape and size, the use of templates should be stressed. However, on occasions, nonstandard symbols, and standard shaped symbols larger than those existing in templates may be required. If such is the case the necessity of using drafting instruments other than templates will be obligatory.

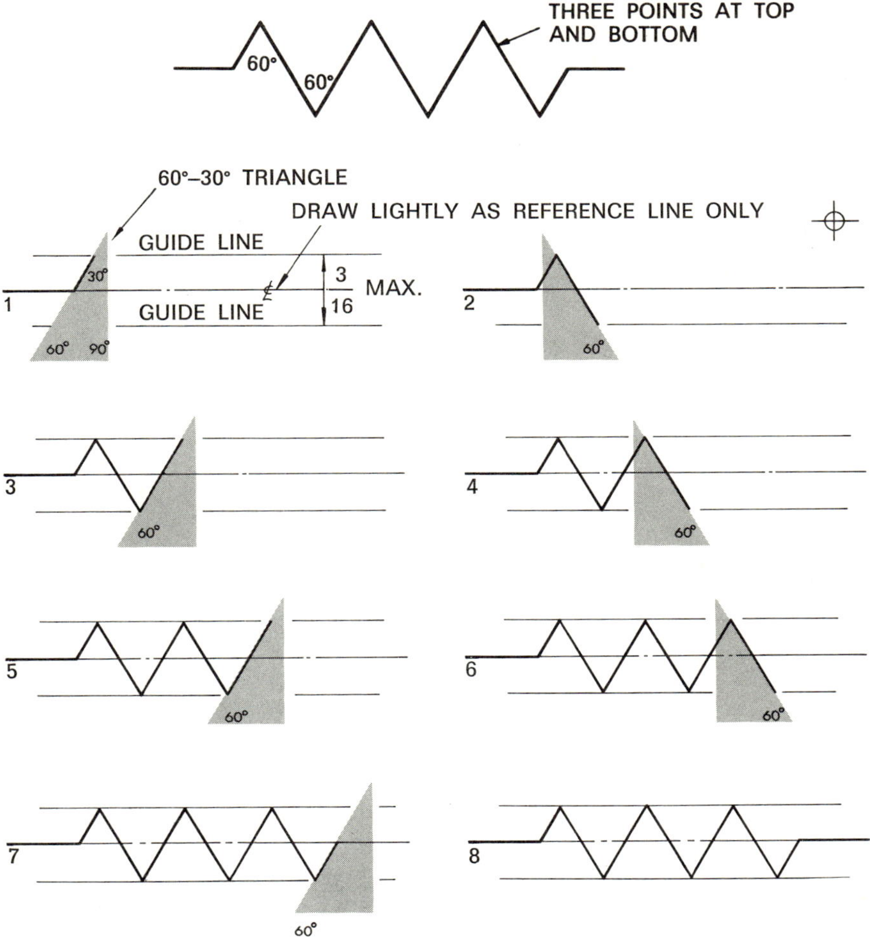

Fig. 3-7 The instrument drawing of resistances

Determining sheet size

In the process of laying out an electrical wiring diagram, some basic steps must be taken to assure optimum results. First of all, a sheet size must be chosen suitable for the complexity and size of the circuit to be drawn. More simply expressed, it is more desirable to complete a drawing with some blank space remaining, then to approach the completion of a drawing without space to do so.

The dimensions shown in Fig. 3-9 are the overall size of the blank sheet. The sizes are based on 'A' format dimensions. To obtain 'B' size, for example, the smallest dimension, 8½″, is doubled, and the largest dimension, 11″, is carried over. To obtain 'C', 'D', or 'E' sizes, the same procedure is used and the

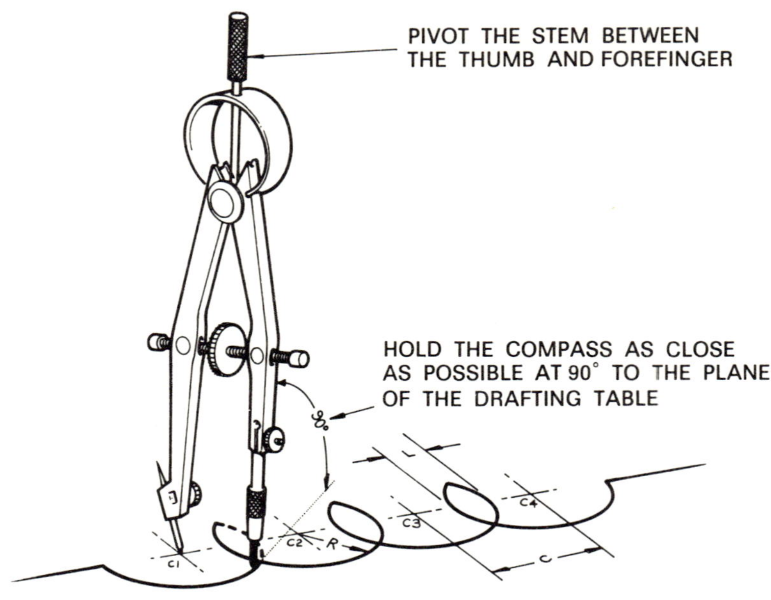

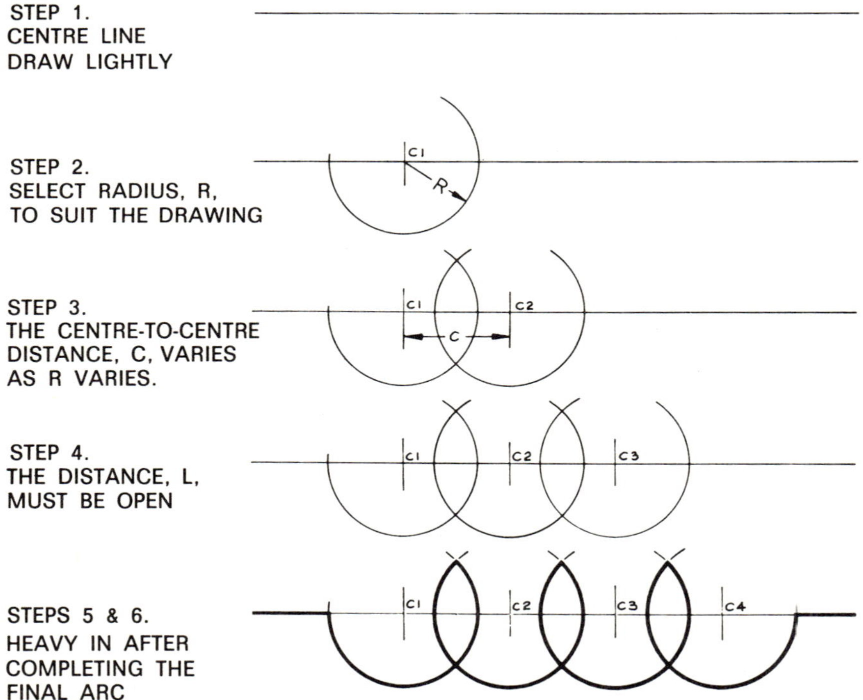

NOTE. CGSB ELECTRICAL SYMBOL FOR COILS INDICATES THAT THE STANDARD SYMBOL SHALL NOT SHOW MORE THAN FOUR LOOPS

Fig. 3-8 The instrument drawing of coils

SMALLEST PRECEDING DIMENSION DOUBLED

LARGEST PRECEDING DIMENSION CARRIED OVER

LETTER DESIGNATION	$8\frac{1}{2}$ x 11 FORMAT	9 x 12 FORMAT
A	$8\frac{1}{2}$ x 11	9 x 12
B	11 x 17	12 x 18
C	17 x 22	18 x 24
D	22 x 34	24 x 36
E	34 x 44	36 x 48

Fig. 3-9(a) A chart showing the dimensions of standard sheet sizes

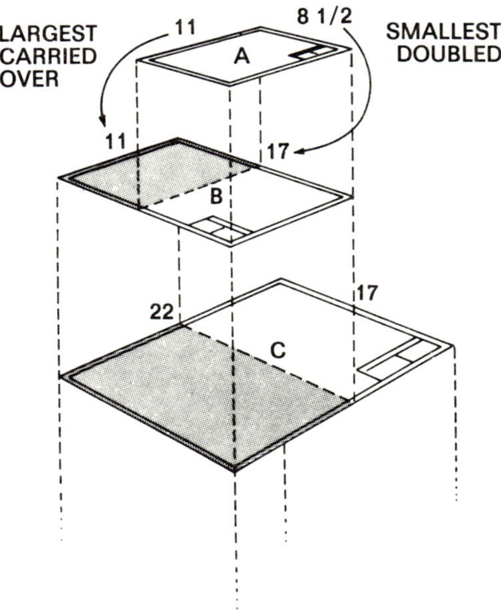

Fig. 3-9(b) A pictorial presentation of sheet sizes

reference format is based on the immediate preceding sheet size. A nonstandard sheet size may be required on occasion, and this is cut from a roll of paper. Consideration must be given to border and title block size when determining what overall sheet size is required. Generally borders are 1/2″ in width, although this may decrease or increase somewhat depending on extremes of sheet size. An additional point regarding borders is that some draftsmen draw an additional border 1/2″ from the true border. This extra border is used as a reference for cutting purposes when prints are made.

Title blocks

Title block formats **are not standardized** and vary from engineering firm to firm. See Figs. 3-10, 3-11, and 3-12. Although they are not standardized, they can be classified into four types. First, in a small firm, the draftsman might simply establish a format and draw them for each tracing. Second, a rubber stamped title block could be used. Third, a printed, press-on title block which would provide uniformity could be used. It should be noted that the press-on

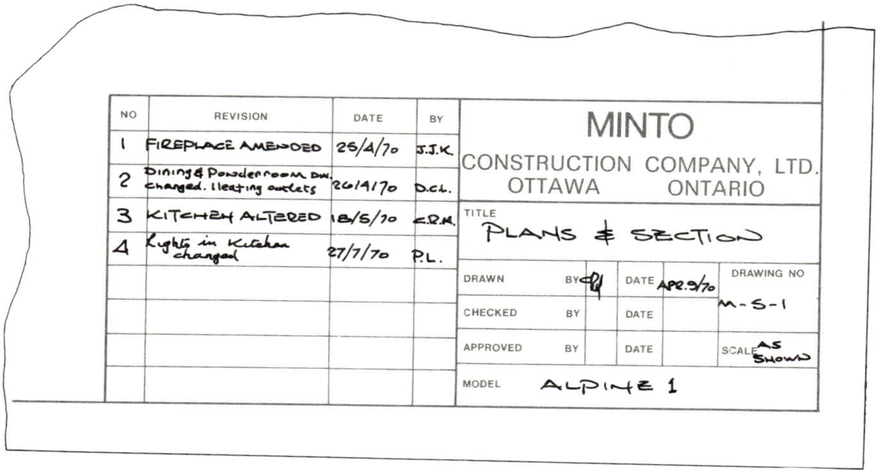

Fig. 3-10 A title block produced by the use of a rubber stamp

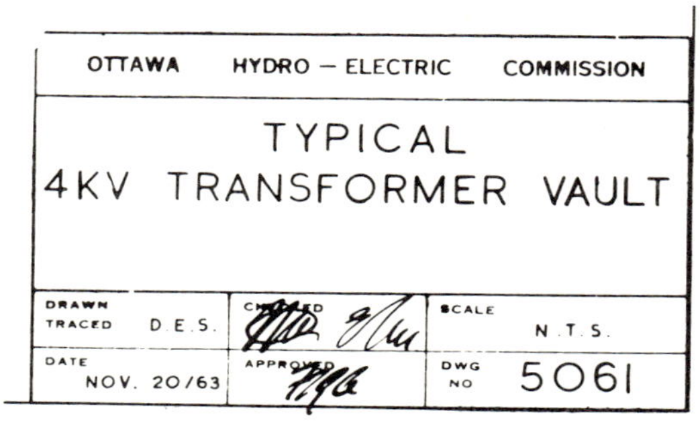

Fig. 3-11 A title block produced by using a mechanical lettering device. (Courtesy Ontario Hydro)

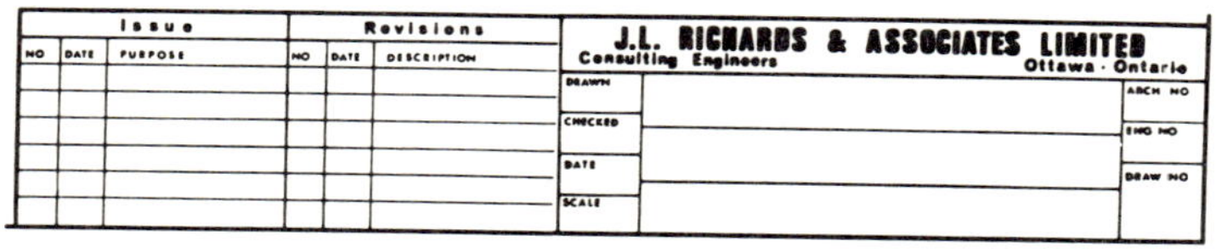

Fig. 3-12 A title block printed integrally on a standard size drafting sheet. (Courtesy J. L. Richards & Associates, Ottawa, Ontario)

title blocks are attached face up to the reverse side of the tracing. This is necessary in order that hand lettered, or machine lettered information can be placed on the tracing. Fourth, printed sheets which include title blocks, borders, and revision columns can be obtained. While the fourth type is more costly in initial outlay, this cost may be offset by the drafting time saved. Also, uniformity of sheet size, border size, and title block format, is constantly maintained.

DELINEATION TECHNIQUES

First, all of the symbol outlines for the major components should be drawn lightly. Their location should be established with regard to the wiring or cabling interconnecting them. This approach is EXTREMELY IMPORTANT. If this is not done and a **piecemeal** approach to the drafting problem follows, a slipshod drawing will likely result with an accompanying shortage of space to complete it.

By studying the rough or preliminary sketch, patterns of leads will become evident. The trick here, then, is to draw in lightly groups of leads which run parallel to each other and have no **crossovers**. Eventually it is probable that crossovers will be required; but if it is possible, **reroute** leads to eliminate as many crossovers as possible. In current practice the methods of using crossovers are shown in Fig. 3-13.

When a large proportion of the conductors or cables have been tied in with their related components, part of the line work could be safely darkened. However, the beginner must exercise precaution here and not heavy-in the drawing bit by bit as it is roughed in.

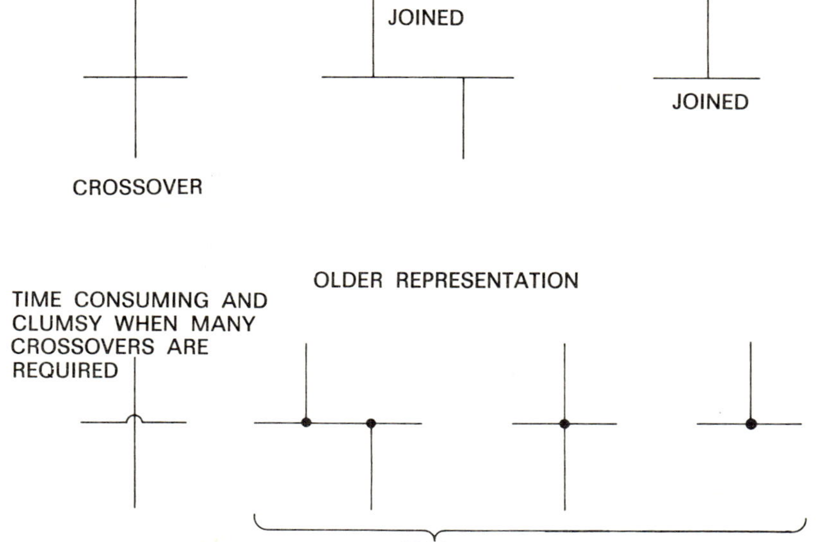

Fig. 3-13 The method of showing conductor crossovers

Weight of line

Lines of varying **weight** or thickness are important because they quickly point out those things which must be brought to the readers' attention. Lines for major component symbols should be drawn thicker for this reason. In cabling diagrams, the technique shown in Fig. 3-14 using varying line thickness is used sometimes.

Blackness of line

In addition, drawings executed on tracing paper or vellum must be sufficiently black to produce first-rate copies made by either blueprinting or diazo processes. The diazo process produces a copy known as a whiteprint. Surprisingly enough, thousands of dollars are wasted annually on poor quality engineering prints or copies, which in many cases can be related to poor tracings, in other words, tracings not black enough.

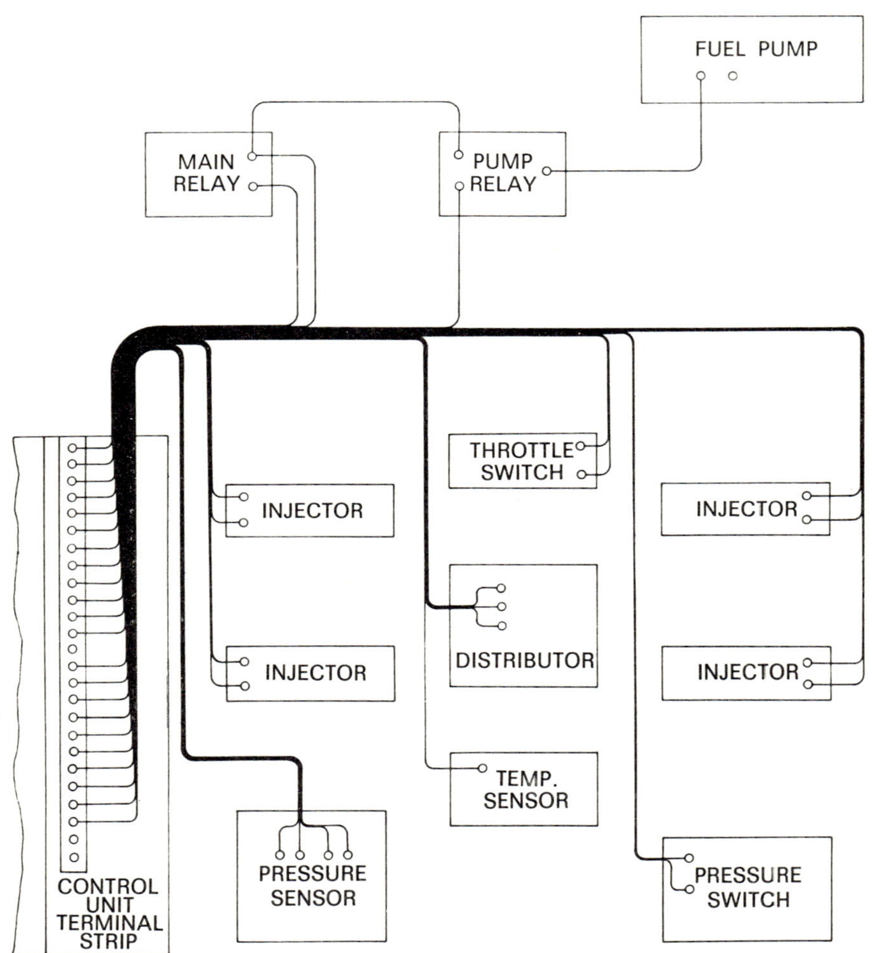

Fig. 3-14 A cabling diagram of a VW electronic fuel injection system. Note the progressively varying weight or thickness of line

Type of line

Lines are drawn normally solid or not broken. However, in order to indicate electrostatic shielding about a component, they will be shown as broken lines. Also, broken lines are used to indicate mechanical linkage between the sections of a ganged, variable tuning capacitor. Fig. 3-15 illustrates electrostatic shielding about a component. Other variations of delineation may be used for other specific purposes, but these will be dealt with in detail in Unit 4.

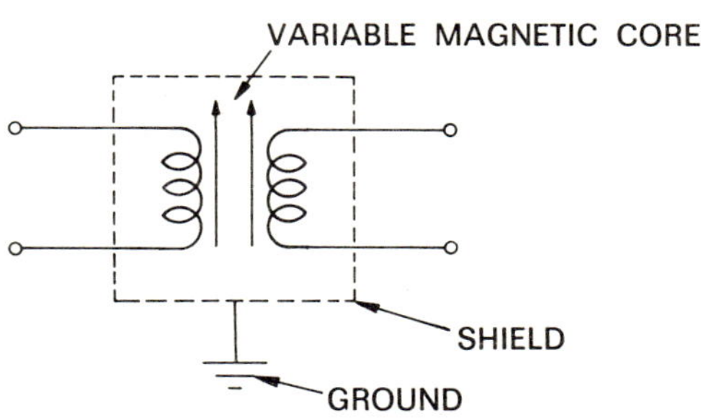

Fig. 3-15(a) An electronic symbol illustrating the use of broken lines to indicate a R.F. Shield surrounding an I.F. transformer

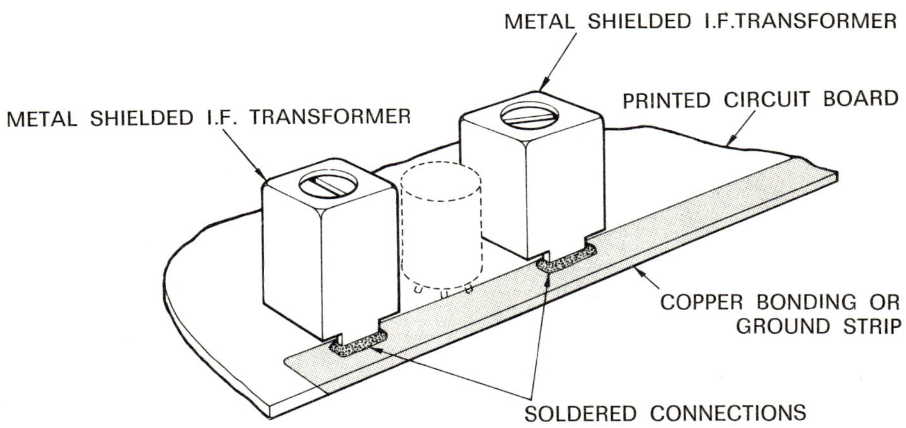

PART OF A 6-TRANSISTOR POCKET MODEL B.C. BAND RECEIVER,
SHOWING SHIELDED AND GROUNDED I.F. TRANSFORMERS

Fig. 3-15(b) A pictorial drawing showing a R.F. shielded I.F. transformer

LETTERING

The use of freehand lettering

What proper emphasis to place on lettering is difficult to decide. With so many and varied lettering aids now available, the stress which was applied to this art some years back perhaps now has lessened. Of course, this is not to suggest that quality lettering is not required anymore, but rather it is questioning the validity of expertise in freehand lettering.

Fig. 3-16(a) A special typewriter for placing data on engineering tracings

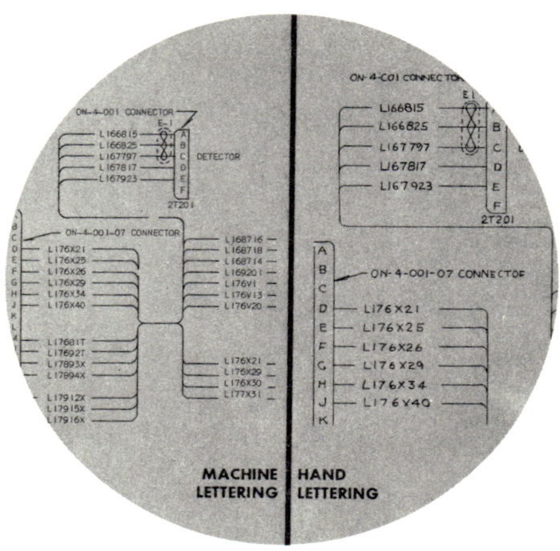

Fig. 3-16(b) A comparison of machine and hand lettering to illustrate the uniformity of shape and size of the lettering done by the typewritten method. (Courtesy VariTyper Corporation, Hanover, New Jersey)

Again, we must weigh our concepts against the use of lettering on a drawing and the purpose of the drawing.

Title blocks could well be done by mechanical means, or by the use of press-ons, and lettering used for descriptive and informative reasons on the body of the drawing could be done freehand.

Such devices as a special typewriter can be used to place all verbal information on the drawing. Fig. 3-16(a) shows the above mentioned device.

Furthermore, drawings used for internal use only, those used for specific clients, and those used for widespread publication such as a manual or brochure would obviously all require different approaches to lettering standards and excellence. Nevertheless, neatness, uniformity, and CORRECT SPELLING cannot be underrated.

Lettering style

Gothic vertical or slant style lettering should be used since its lack of ornateness lends itself well to engineering drawings. However, on an individual drawing the use of both slant and vertical style lettering should be avoided.

Lettering size

Letter size will vary with usage on a specific drawing. Title block numbers and names can be larger. On the body of the drawing lettering should be approximately 3/32" to 1/8" high. Guide lines lightly drawn must be used in order that the height will remain constant. Various guide line templates or devices are available when many lines are required, as for example, in lettering **notes**. See Fig. 3-17.

The correct shape and proportion of Gothic capitals, both vertical and slant, are shown in Fig. 3-18.

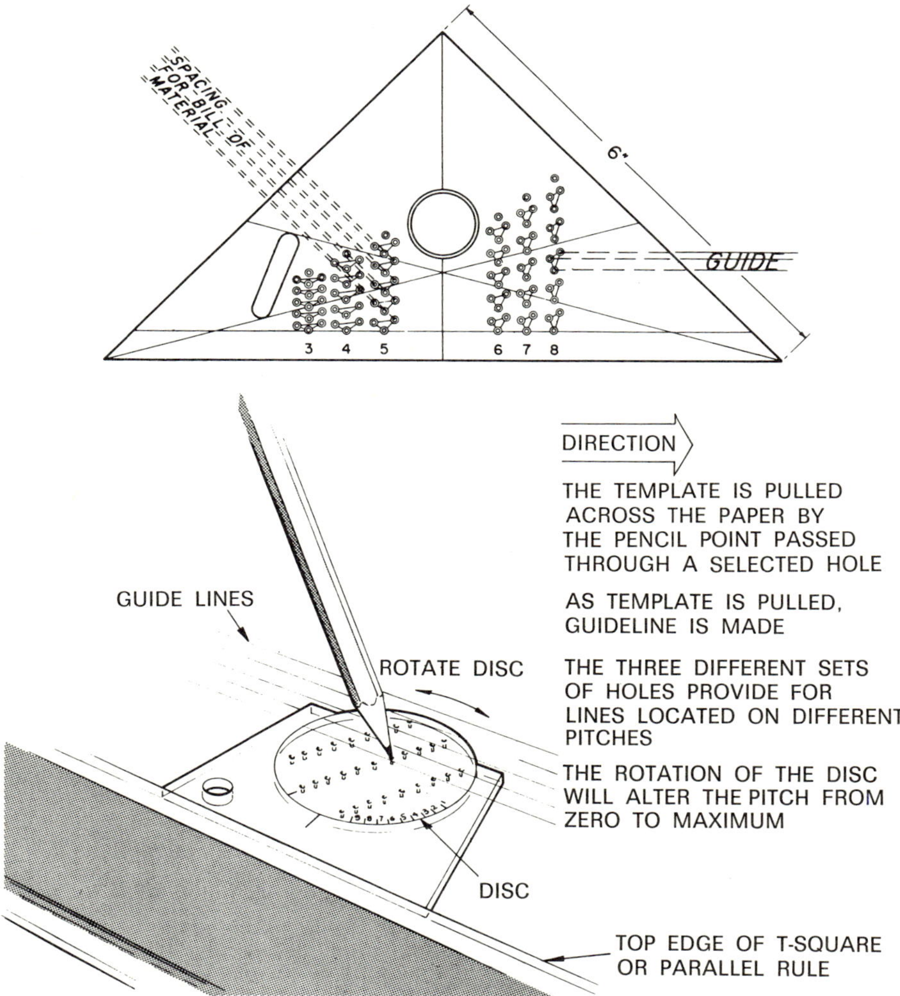

Fig. 3-17 Plastic guideline templates

SLANT	VERTICAL	*EXAMPLE*
A	A	ATOM
B	B	BEAT
C	C	CELL
D	D	DETECTOR
E	E	ELECTRICAL
F	F	F-M
G	G	GENERATOR
H	H	HYSTERESIS
I	I	ION
J	J	JOULE
K	K	KILOWATT
L	L	LASER
M	M	MICRO
N	N	NEUTRON
O	O	OHM
P	P	PEAKED
Q	Q	QUARTZ
R	R	RESISTOR
S	S	SEMICONDUCTOR
T	T	TETRODE
U	U	UNBIASED
V	V	VIDEO
W	W	WATT
X	X	X-RAY
Y	Y	YAGI
Z	Z	ZENER

0123456789	0123456789

Fig. 3-18 Gothic letters: slant and vertical style

KNOWLEDGE TESTERS

1. Why are electrical symbol templates used? Give at least three reasons.
2. List some of the problems encountered with the use of templates.
3. What type of template device is used in producing ink tracings?
4. How are press-on symbols or letters manufactured? Explain in simple terms.
5. Explain what precautions must be used when using this medium.
6. What are the advantages and disadvantages of press-on and transfer media?
7. Give the correct angle used when drawing resistance symbols.
8. How many points must be drawn?
9. Do some research on **printed circuits** and follow up with a brief but concise description.
10. Why are printed circuits drawn a number of times greater than full size?
11. Suggest a reason why drawing sizes are either multiples of 8½″ x 11″ or 9″ x 12″.

12. Why should the major component symbols be outlined first?
13. What is meant by the piecemeal approach when making an electrical drawing?
14. Why is an additional border line sometimes drawn?
15. Why is the system of crossovers and soldered connections shown thus

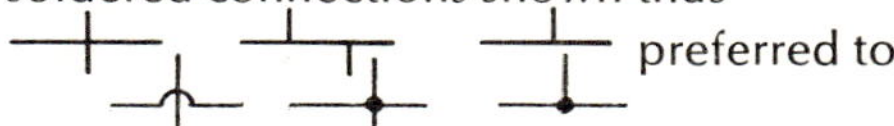

preferred to
16. Why must the lines on a tracing be an intense black?
17. List two purposes of using broken lines on electrical drawings.

PROJECTS

1. Design a title block. Rough it in lightly first in pencil and complete it in ink using a mechanical lettering device.
2. Without using the aid of templates, draw: (a) a circuit showing three resistances in series, and (b) a circuit showing three resistances in parallel.
3. Make a schematic drawing involving the use of symbols to illustrate a 1:4 step-up transformer with an alternating current input of 110 volts. What will be the output voltage?
4. Redraw the diagram, Fig. 3-19, eliminating as many crossover leads as is practical.
5. Design a simple circuit so that the operation of one S.P.D.T. switch turns on one lamp while at the same time it turns off another lamp in the same circuit. Assume that the lamp requires 110 volts of alternating current.

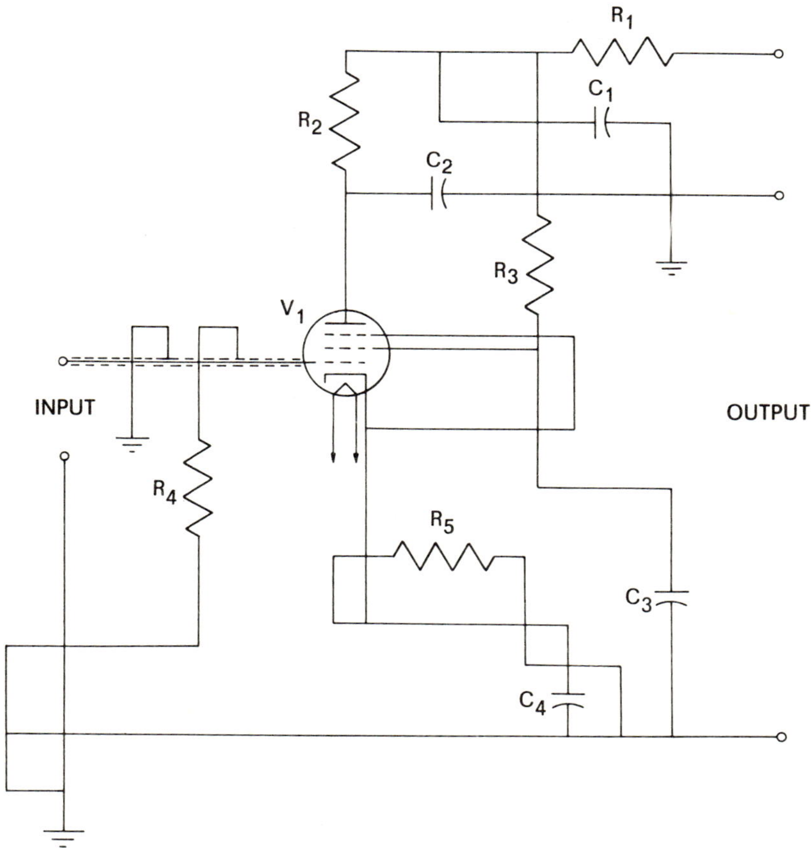

Fig. 3-19 A tube type preamplifier schematic wiring diagram

UNIT 4

Graphic Symbols

Purpose

Electrical and electronic symbols are graphic devices which rapidly convey to the reader of an electrical or electronic drawing, information regarding certain components. First of all, the symbols involved in a particular circuit represent specific components in a general manner. In addition, numerical values placed adjacent to the symbol indicate its representative component's size in terms of voltage, amperage, resistance, inductance, or capacitance. Also, terminal numbers or letters may be shown on some symbols, and manufacturers' part numbers or designations may be included. Figure 2-12 illustrates a portion of an electrical wiring diagram, and Fig. 2-5(b) shows a part of an electronic schematic to illustrate these points.

SYMBOL CHARACTERISTICS

Symbol size

The physical size of a symbol does not alter its electrical size, nor does it infer its importance in an electrical circuit. Within any given circuit diagram, symbol sizes should remain constant. If this is not possible, not more than two sizes are to be used.

Weight of symbol line

Line **weight** or thickness does not alter the meaning of a symbol, but it may be made thicker for particular emphasis.

Ornateness of symbol

All symbols must be drawn without any distracting artistic **ornateness**. In addition, they must be drawn clearly.

Symbol orientation

The **orientation** of a symbol does not affect its interpretation, and certainly not its electrical value. However, the manner in which symbols are shown in Unit 4, is the customary way of placing them. On occasions it may be necessary to rotate symbols 90° or even 180° from that shown, but this procedure is used as a final resort. In Fig. 4-1 examples of different symbol orientation are illustrated.

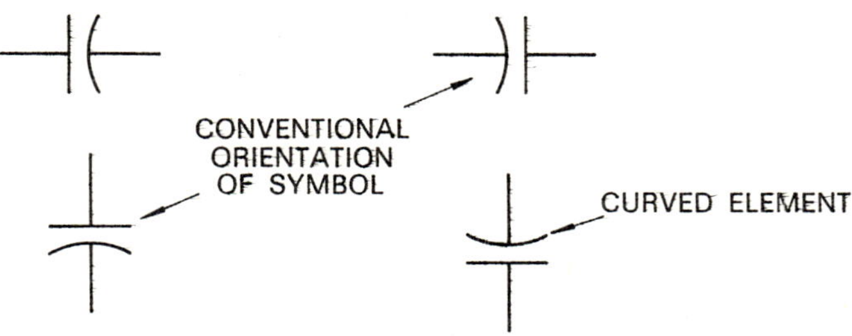

Fig. 4-1 *Four different ways of orienting a capacitor symbol. Since the curved element of the symbol represents the outside electrode in fixed capacitors, and the rotating element in variable ones, its position relative to a particular circuit will always be unique. Hence, it may be advantageous in terms of circuit clarity not to adhere to the conventional methods as shown*

Separation of symbol parts

It is sometimes necessary for the purpose of simplification to separate the parts of a symbol. Figure 4-2 shows part of a circuit involving the separation into two parts of a vacuum-tube symbol.

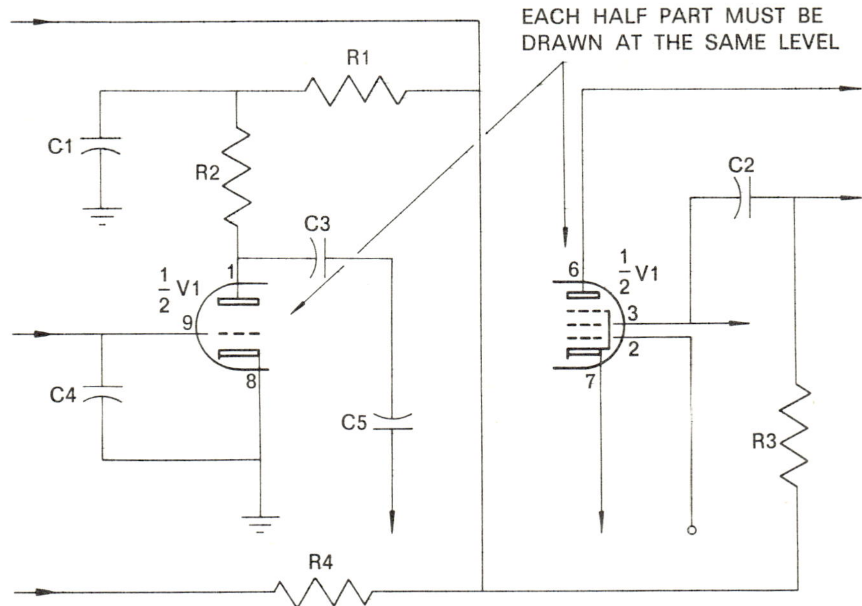

Fig. 4-2 *Part of a transceiver circuit showing the method of separating a vacuum-tube symbol into two parts.*

Leads connecting symbol

The angle at which a lead terminates at a symbol has no special significance. However, it is normal to draw them perpendicular to, or in line with, the symbol.

Lettering orientation about a symbol

All lettering is to be placed in a horizontal fashion as is shown in Fig. 4-3.

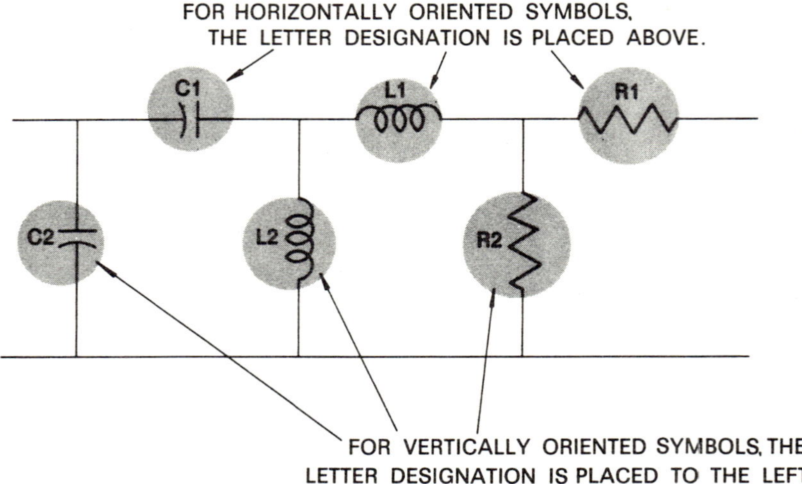

Fig. 4-3 A series-parallel circuit illustrating the correct letter orientation about graphic symbols

Placement of numerical values

If a symbol is horizontally oriented the value is placed below it. If a symbol is vertically oriented, the value is placed to the right of it. On occasions these rules may have to be violated because of space limitations. However, regardless of the placement position, the reader must never be left with any doubt in his mind. The resistances shown in Fig. 4-4 illustrate this method.

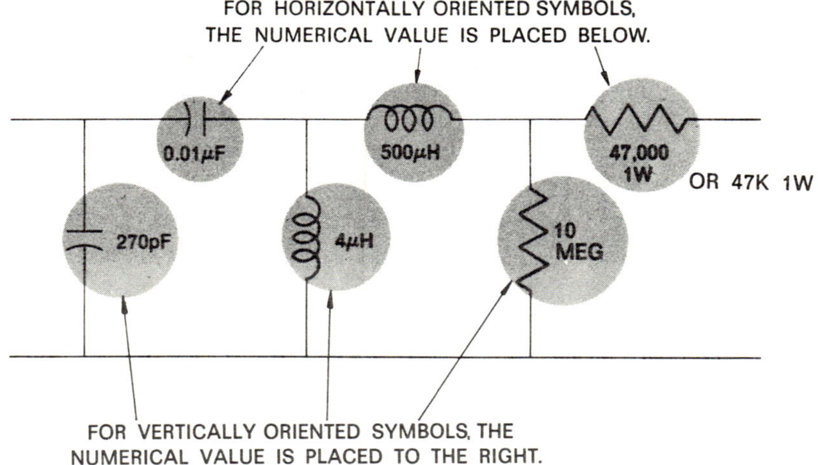

Fig. 4-4 A series-parallel circuit illustrating the correct placement of the numerical values for component symbols

Abbreviations for symbol component names

Special abbreviations using capital letters are required when designating different components. The table shown in Fig. 4-5 lists these abbreviations.

ABBREVIATION	COMPONENT NAME
C	Capacitor
CB	Circuit breaker
CR	Crystal diode
F	Fuse
I	Lamp
J	Jack
L	Inductance coil
M	Meter
PC	Printed circuit
R	Resistor
S	Switch
SP	Speaker
SR	Selenium rectifier
T	Transformer
V	Vacuum tube
Z	Capristor (resistor-capacitor circuit)

Fig. 4-5 Commonly used abbreviations for electrical and electronic components.

Placement of component name abbreviation

The component name abbreviation is to be placed immediately above its symbol, and in close proximity. See Fig. 4-6.

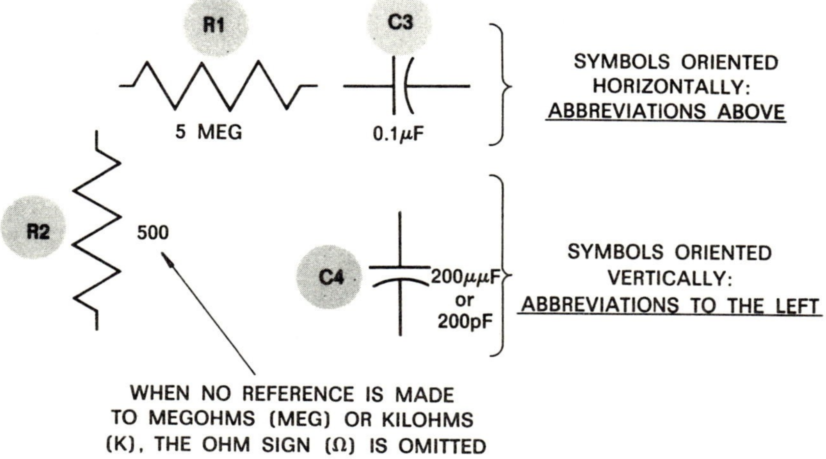

Fig. 4-6 The placement of component name abbreviations

Numerical sequencing of component symbols

Following the abbreviation for a component name, a number is placed in sequence. For example, if a circuit contains eleven resistors, they are referred

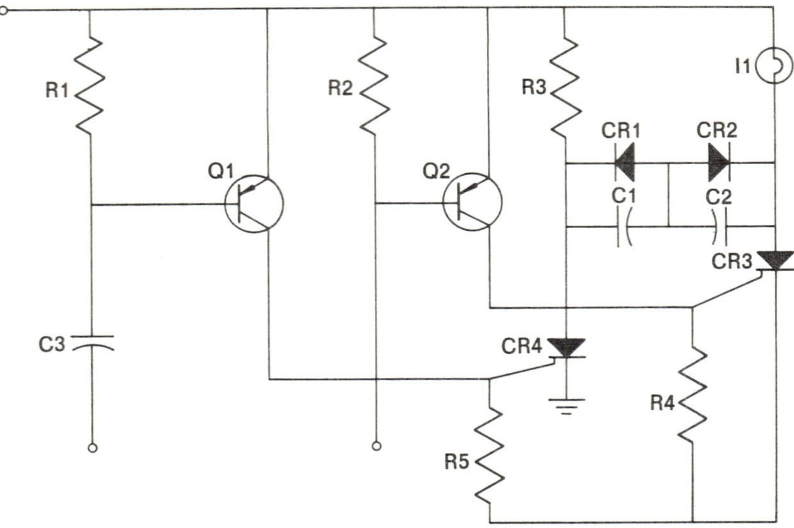

Fig. 4-7 The sequential numbering of component name abbreviations

to as R1, R2, and so on, up to R11. Other electrical components of the same kind are treated in a similar fashion. The illustration shown in Fig. 4-7 clearly represents the method of sequential numbering.

Organizations standardizing symbols

In Canada the **Canadian Government Specifications Board** (CGSB) located at the National Research Council is embodied to establish standardized electrical and electronic graphic symbols. The handbook issued by this organization is entitled 'Graphic Symbols, Designations and Abbreviations for use on Electrical and Electronic Schematic Diagrams — 33-GP-1'.

In the United States of America a standard known as USAS Y32.2 is approved by the United States of America Standards Institute. The sponsors of this standard are the American Society of Mechanical Engineers, and the Institute of Electrical and Electronic Engineers.

These symbols are divided into component groups and include the various types within a group. The components are numbered numerically, and are listed on the following pages in columnar form.

The first column in the book shows at the top the name of the component group, ALARMS. The number, name, and description of the first component in the group is shown next, e.g. 1. Annunciator.

The Canadian Government Specifications Board, CGSB, symbol is shown below and it is identified by (a) CGSB.

The United States of America standard symbol, identified by (b) USAS, is shown below the Canadian symbol. The note 'No equivalent symbol', means that there is no U.S. symbol.

A pictorial drawing of the symbol, identified by (c) Pictorial, is shown at the bottom of the column.

This column procedure is repeated to include all of the components.

INDEX

An index of these symbols is included at this point for quick reference.

The component groups are arranged alphabetically, and each particular type of component is listed under the group heading. The numbers correspond to the numbers of the components on the pages on which they are illustrated.

ALARMS

1. Annunciator alarm

This is a remotely operated communication device, capable of displaying visual signals from more than one place.

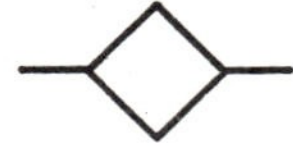

(a) CGSB

(b) USAS

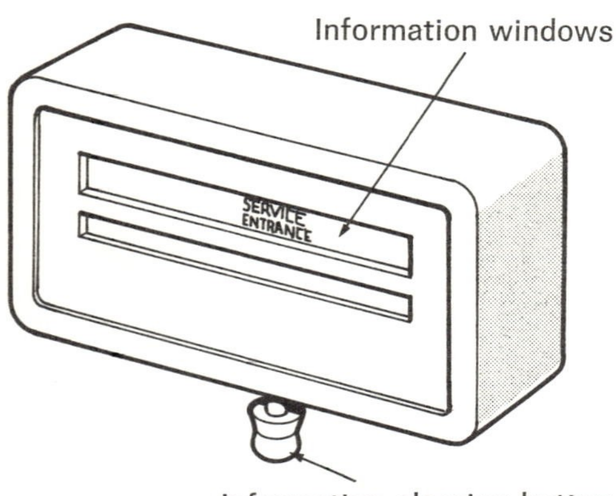

(c) Pictorial

2. Bell alarm

This is an audible signalling device utilizing a bell struck by a solenoid operated clapper, and it is remotely operated.

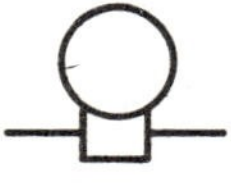

(a) CGSB

(b) USAS

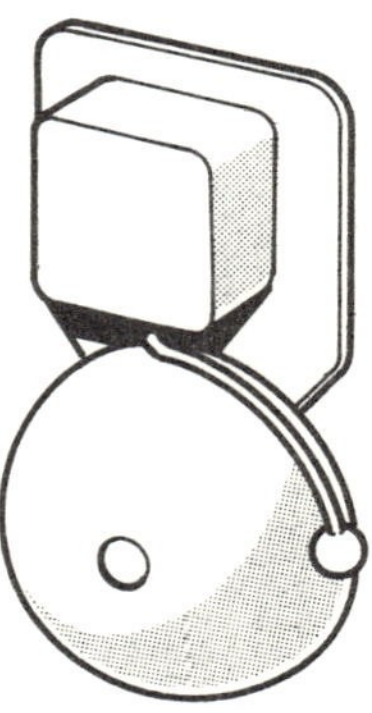

(c) Pictorial

3. Buzzer alarm
This is a signalling device operated remotely, similar to a bell, but having a muted tone.

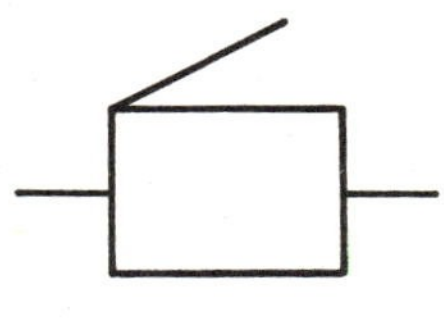

(a) CGSB

(b) USAS

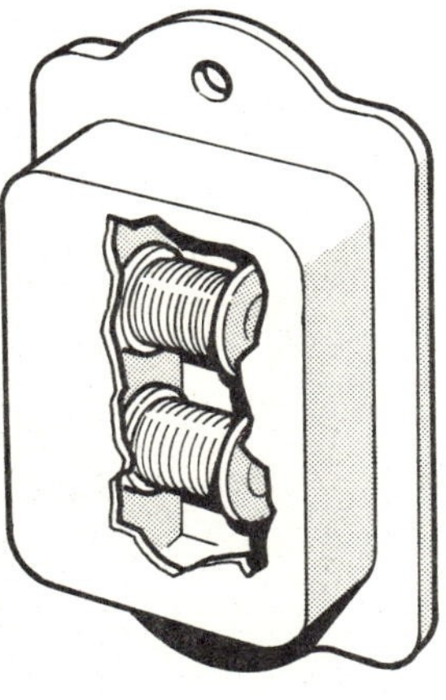

(c) Pictorial

4. Horn alarm
This is a signalling device operated remotely and having a harsh, loud tone. In an industrial situation where the background noise level is high, a signalling horn would find application.

(a) CGSB

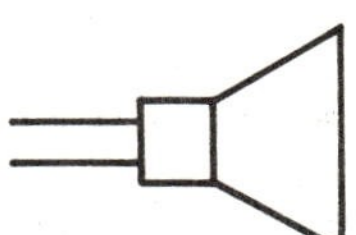

(b) USAS

(c) Pictorial

ANTENNAS

5. Antenna, general

This is a device used in conjunction with a receiver or transmitter in either initially receiving, or in the case of the transmitter, finally transmitting radio frequency energy.

(a) CGSB

The same as the Canadian

(b) USAS

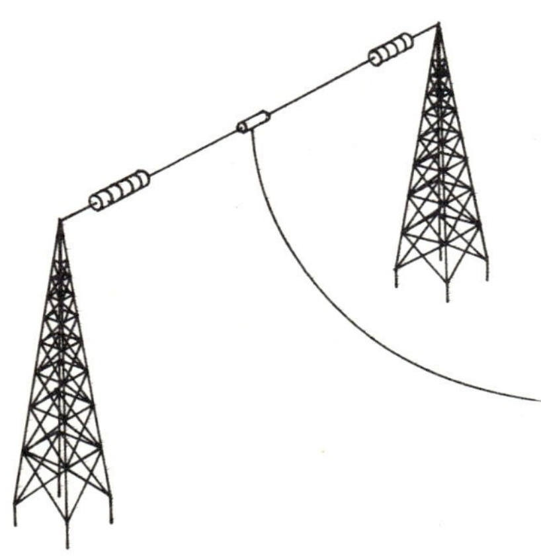

(c) Pictorial

6. Dipole antenna

This is a type of antenna containing two elements of equal length.

(a) CGSB

The same as the Canadian

(b) USAS

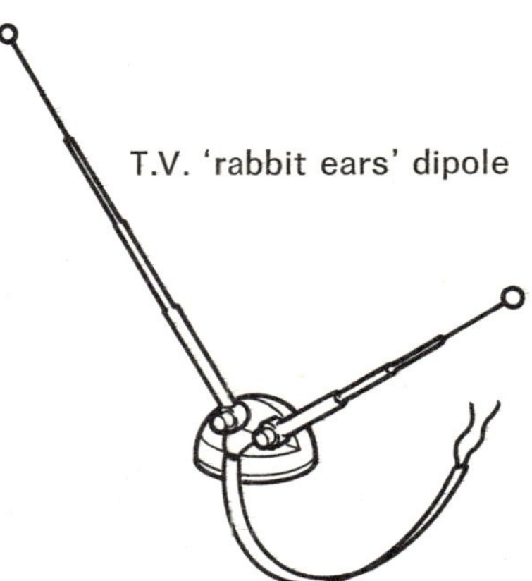

(c) Pictorial

7. Simple loop antenna
This is a type of antenna consisting of a number of turns of insulated wire wound in loop fashion and terminating with two connections.

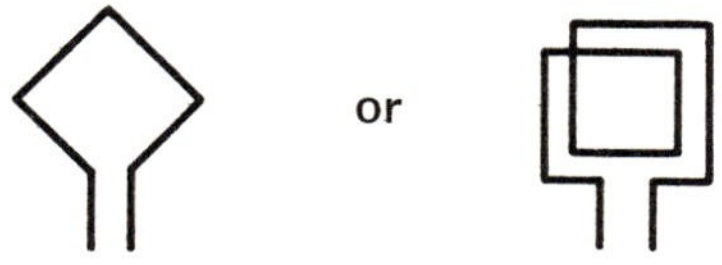

(a) CGSB

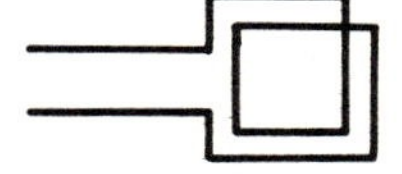

(b) USAS

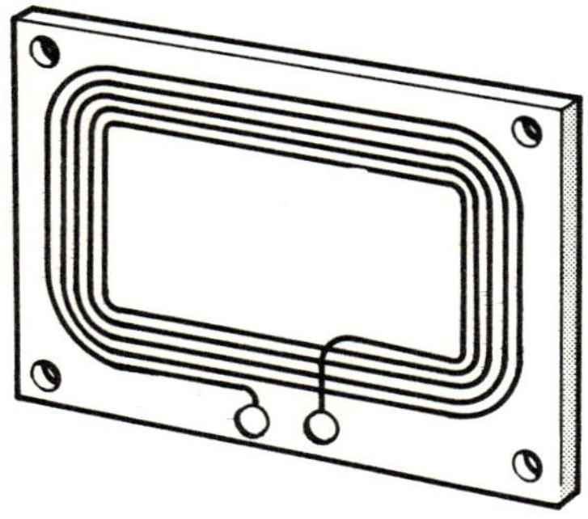

(c) Pictorial

BATTERIES

8. Single cell battery
This is an electrochemical device for producing direct current and having only one positive and one negative pole.

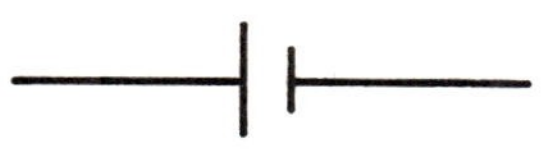

(a) CGSB

The same as the Canadian

(b) USAS

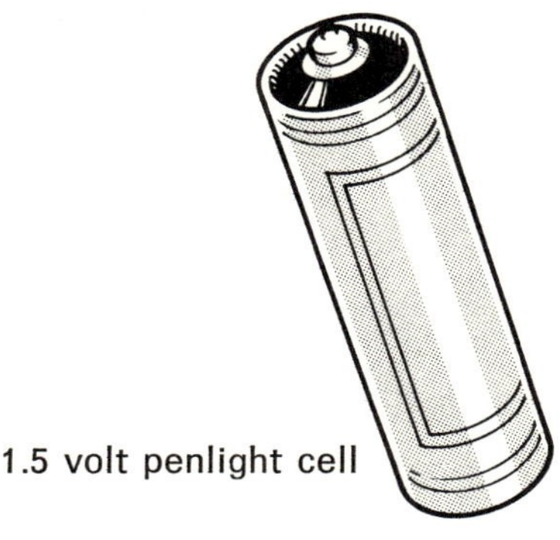

(c) Pictorial

9. Multiple cell battery

This is an electrochemical device for producing direct current and having more than one cell. The long line of the symbol is positive. If polarity and voltage are to be shown, the CGSB symbol is used.

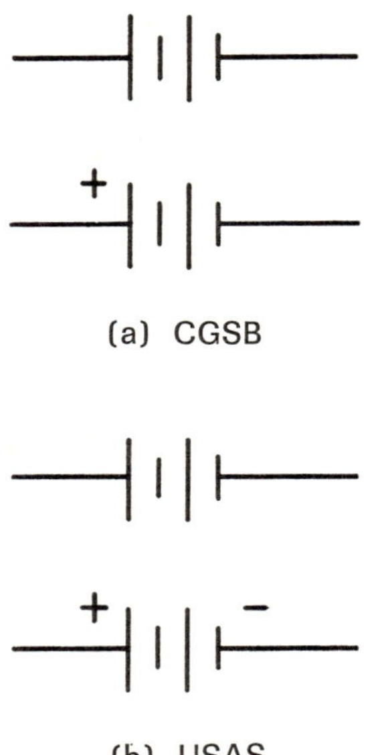

(a) CGSB

(b) USAS

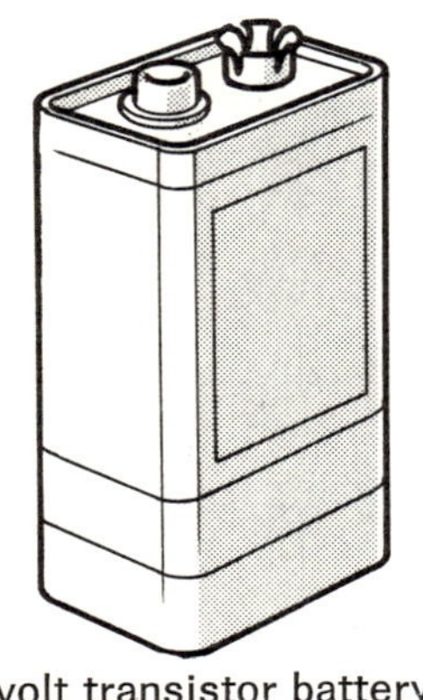

9 volt transistor battery

(c) Pictorial

10. BUS

This is a mechanical part frequently constructed from copper or brass which acts as a common attaching or bonding strip for a number of conductors.

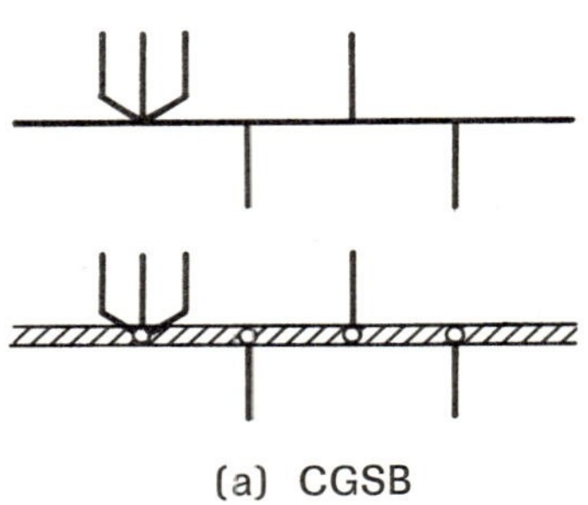

(a) CGSB

No equivalent symbol

(b) USAS

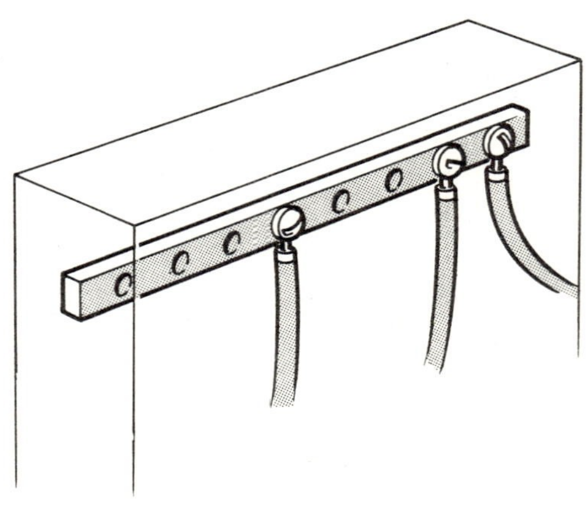

(c) Pictorial

CAPACITORS

11. Fixed capacitor

This is an electrical component capable of storing an electrical charge, and having one specific value of capacitance.

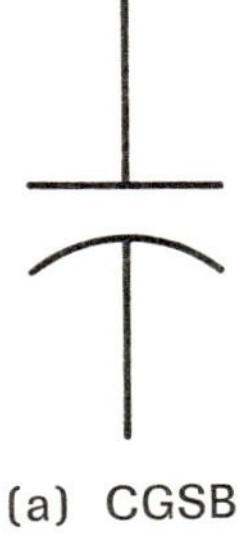

(a) CGSB

The same as the Canadian

(b) USAS

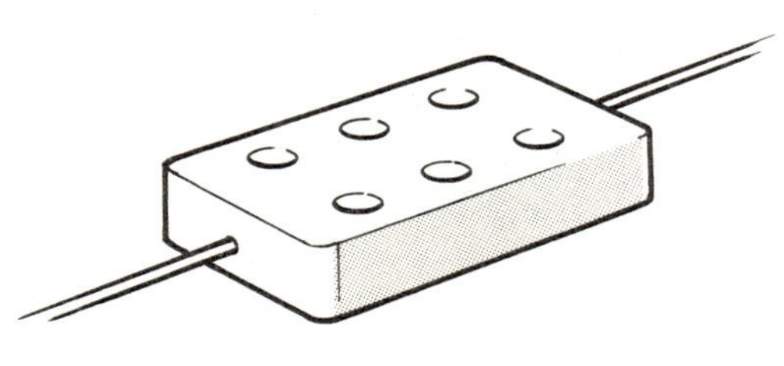

(c) Pictorial

12. Variable capacitor

This is a capacitor whose capacitance range can be varied between fixed limits.

(a) CGSB

The same as the Canadian

(b) USAS

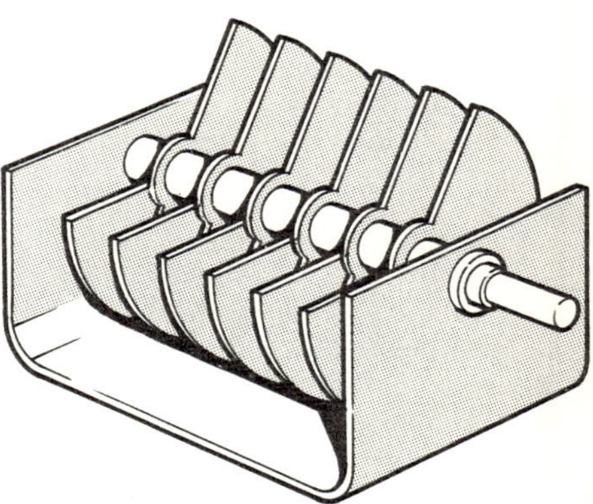

(c) Pictorial

13. Electrolytic capacitor
This is a capacitor of fixed value and specifically designed for use in power rectifying circuits.

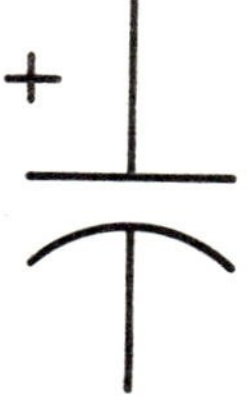

(a) CGSB

(b) USAS

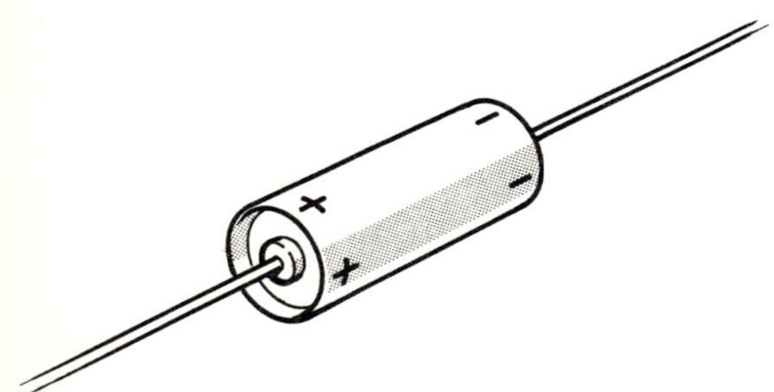

(c) Pictorial

CIRCUIT BREAKERS

14. Air circuit breaker
This is an electromechanical device designed to protect an electrical circuit from being overloaded, and in which the contacts of the breaker are air separated.

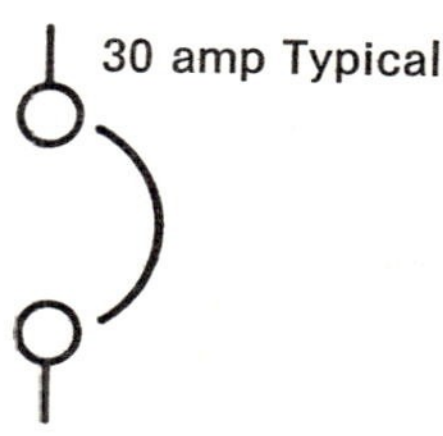

(a) CGSB

(b) USAS

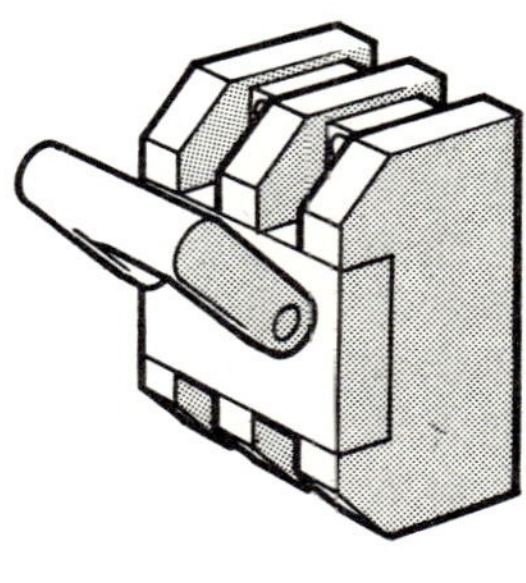

(c) Pictorial

15. Oil circuit breaker

This is a type of circuit breaker in which the contacts of the breaker are oil separated.

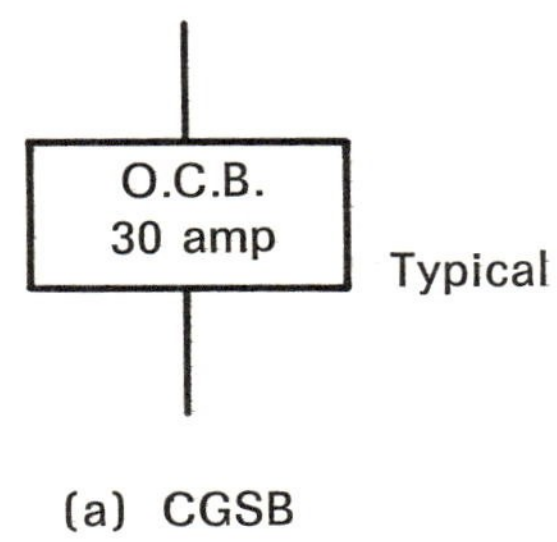

(a) CGSB

No equivalent symbol

(b) USAS

(c) Pictorial

16. COIL, IGNITION

This is a device for producing an electrical charge of high secondary voltage and low current from a primary source having a much lower voltage but higher current.

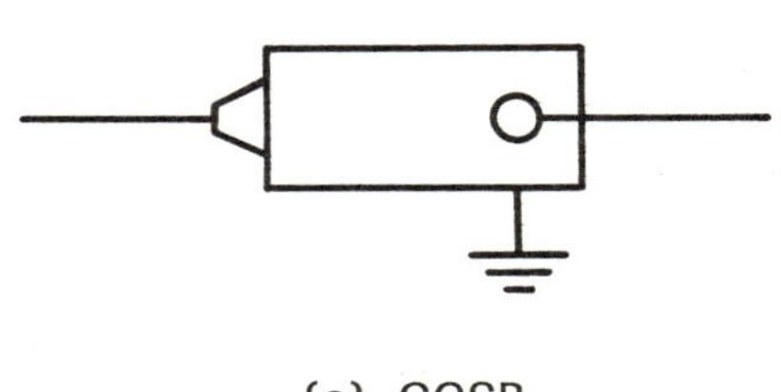

(a) CGSB

No equivalent symbol

(b) USAS

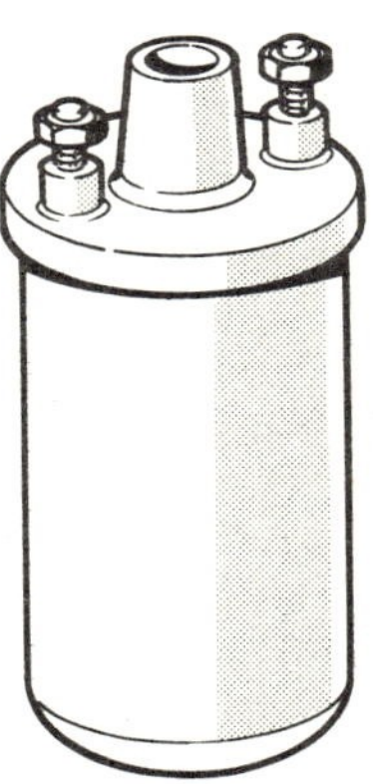

(c) Pictorial

CONNECTORS

17. Female contact part of connector

This is a mechanical device for joining two separate conductors with the provision that good electrical and mechanical contact is made, and that in conjunction with the male contact, it can be disconnected simply.

(a) CGSB

The same as the Canadian

(b) USAS

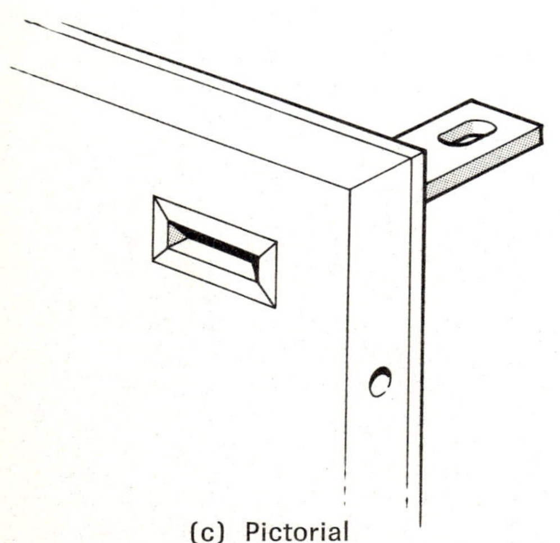

(c) Pictorial

18. Male contact part of connector

This is that part of a connector which is capable of entering a female contact.

(a) CGSB

The same as the Canadian

(b) USAS

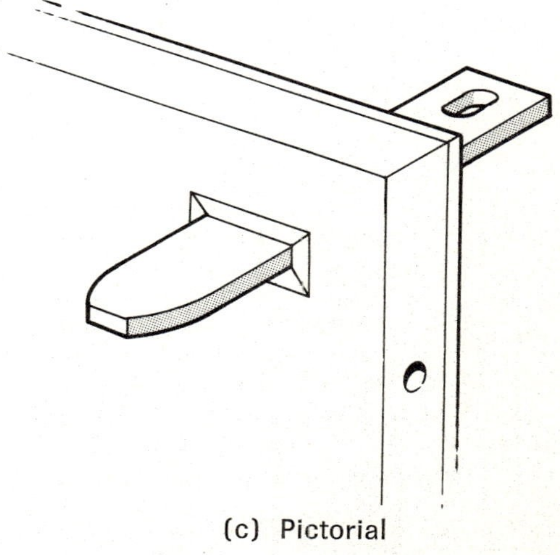

(c) Pictorial

19. Multiple connector
This is a connector used to join multi-conductor cables, or more than one single conductor.

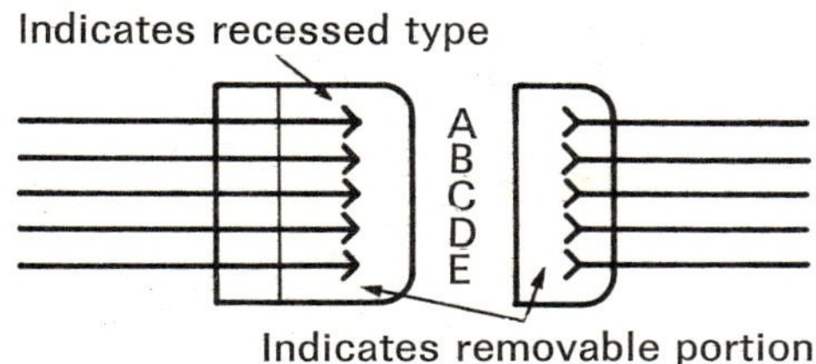

(a) CGSB

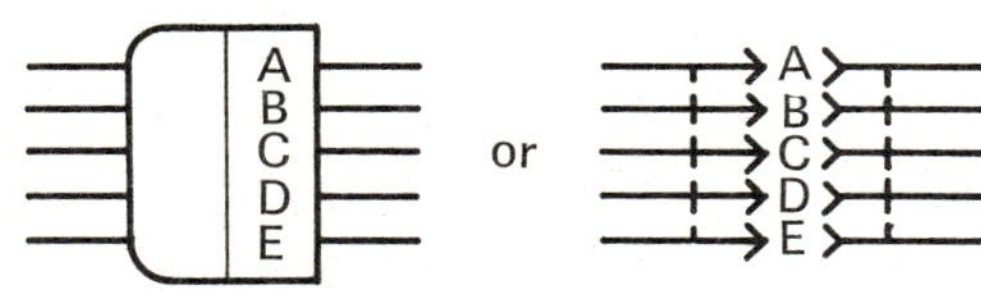

(b) USAS

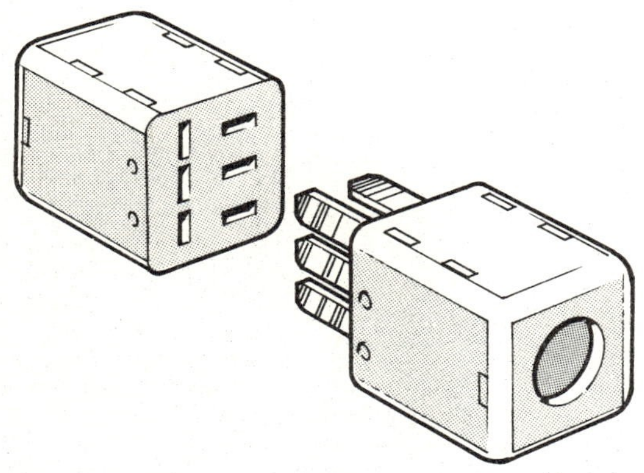

(c) Pictorial

20. Coaxial connector
This is a specific type of connector used in conjunction with coaxial cable.

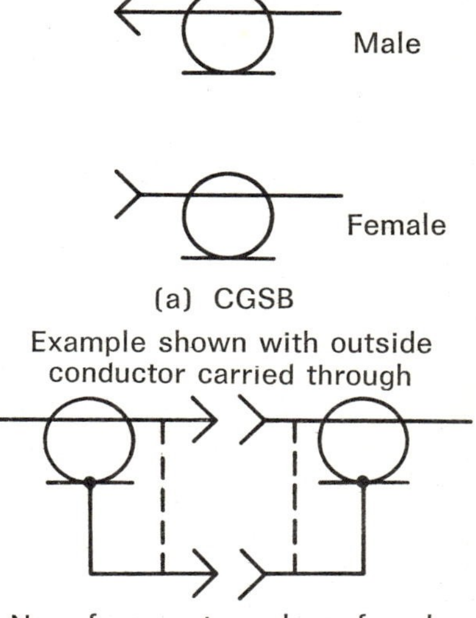

(b) USAS

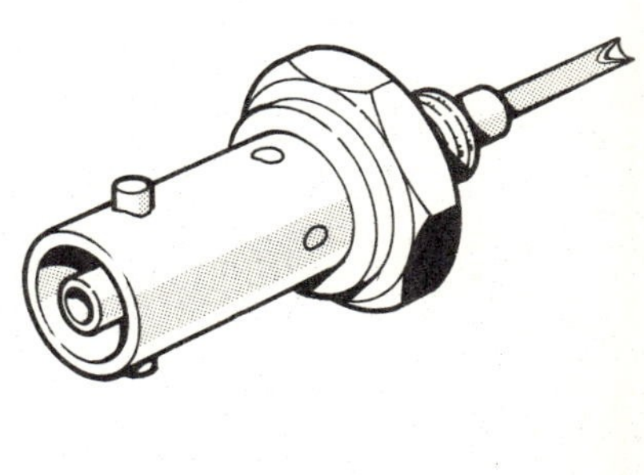

(c) Pictorial

21. Switchboard connector

This is a specific type of connector used on a telephone switchboard panel.

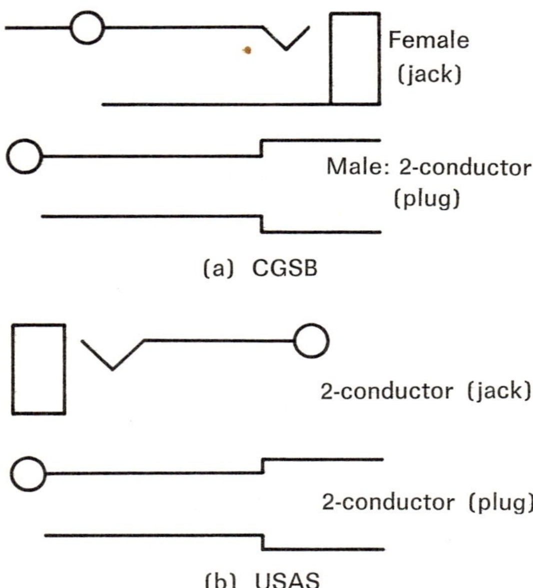

(a) CGSB

(b) USAS

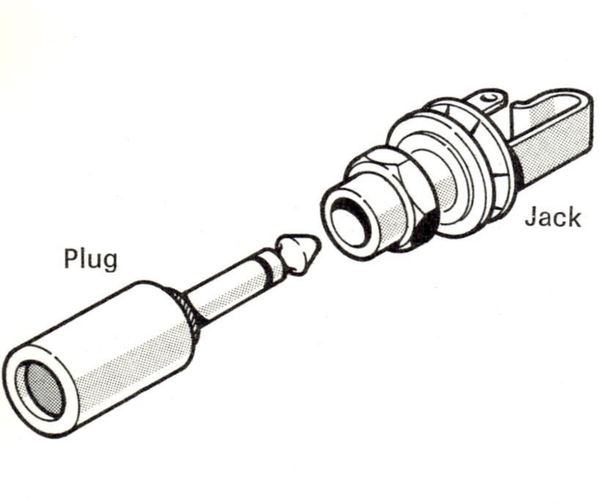

(c) Pictorial

CONTACTS

22. Normally-open power contact

This is the part of the power relay which permits the coil to receive electrical energy. In this case the contacts are normally open when the coil is in a nonenergized state.

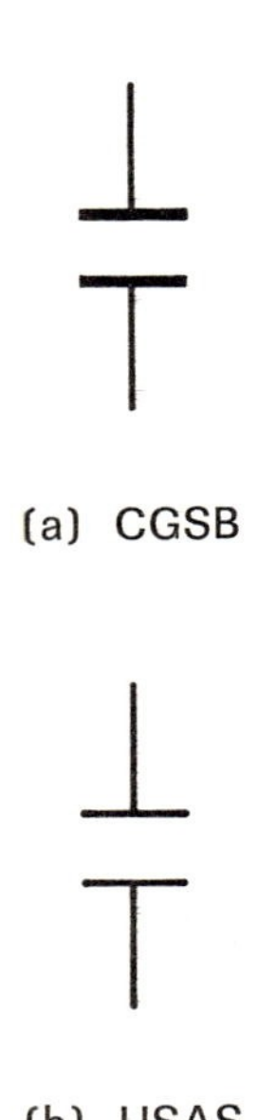

(a) CGSB

(b) USAS

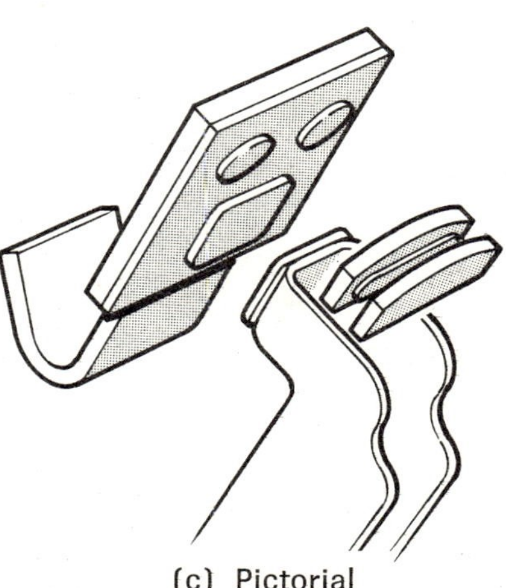

(c) Pictorial

23. Normally-closed power contact
This is a device similar to the one described for No. 22 except that the contacts are closed when the coil is in a nonenergized state.

(a) CGSB

(b) USAS

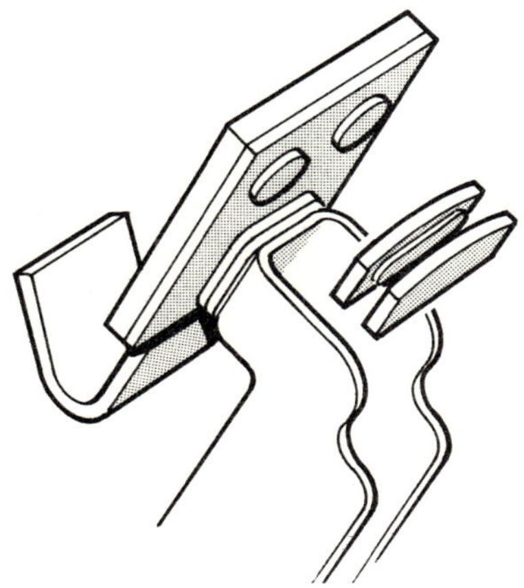

(c) Pictorial

24. Power switch moving contact
This is the moveable part or arm of a power switch

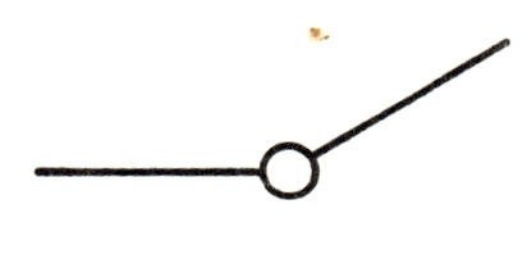

(a) CGSB

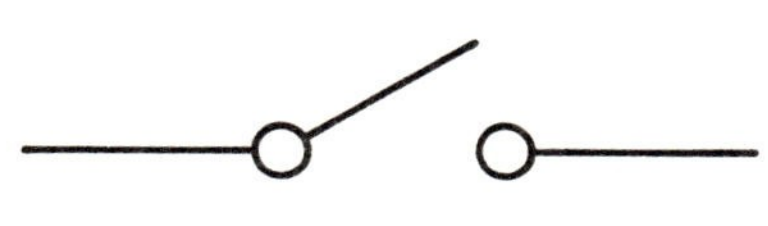

(b) USAS

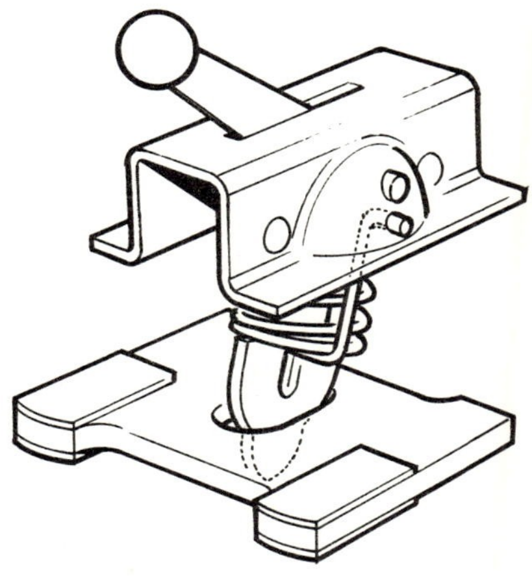

(c) Pictorial

25. Power switch fixed contact

This is the static or nonmoveable *part of a power switch.*

(a) CGSB

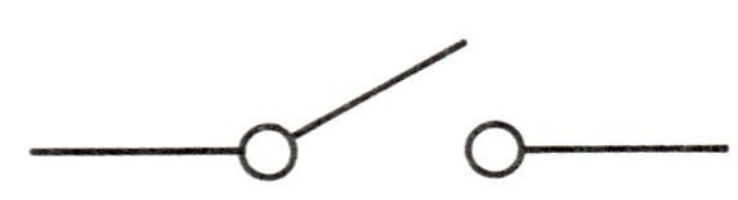

(b) USAS

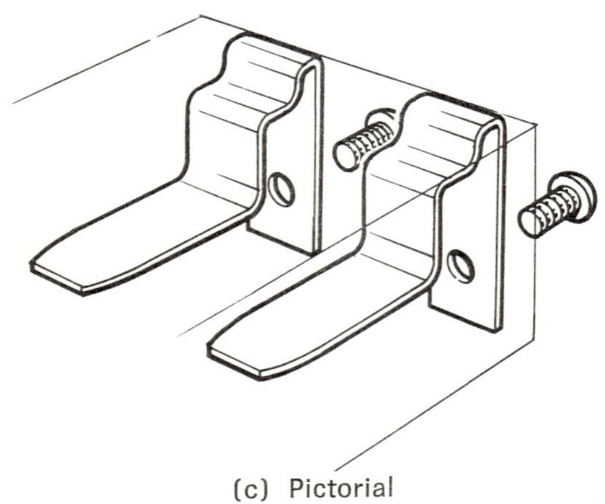

(c) Pictorial

26. Power switch, push-button type, moving contact

This is the moving part or plunger of a push-button power switch.

(a) CGSB

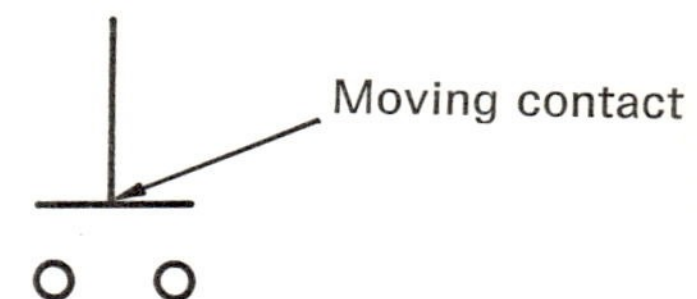

(b) USAS

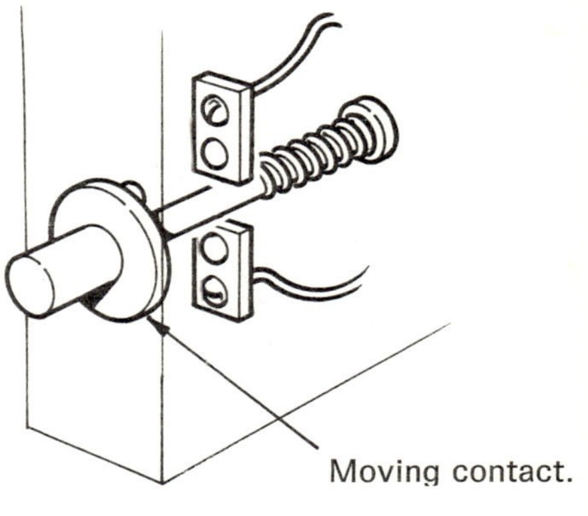

(c) Pictorial

27. Normally-open signal contact
It is that part of a relay, switch, or jack, which when closed causes the energization of the relay coil or other circuitry, and is normally open.

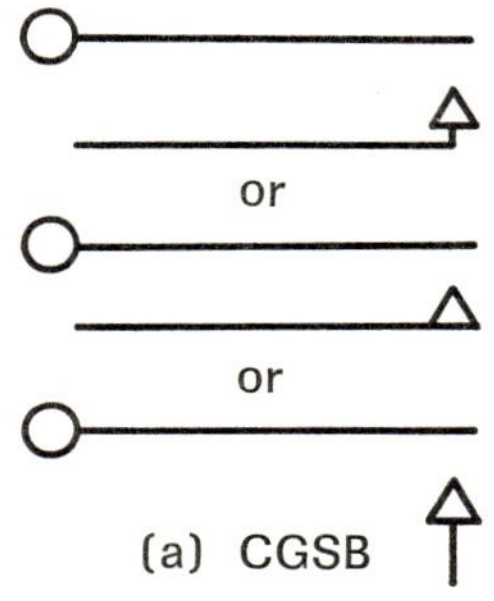

(a) CGSB

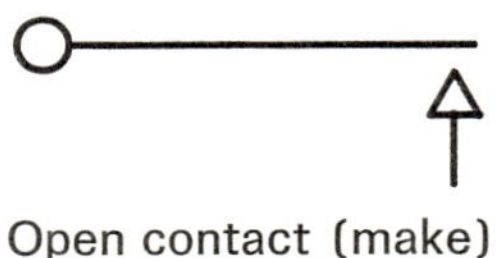

Open contact (make)

(b) USAS

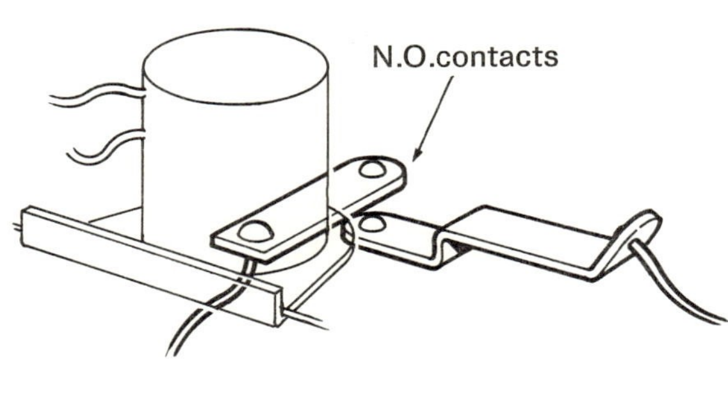

(c) Pictorial

28. Normally-closed signal contact
This is a type of signal contact that is normally closed when it is nonenergized, or when the other parts of a circuit associated with the contact are not energized.

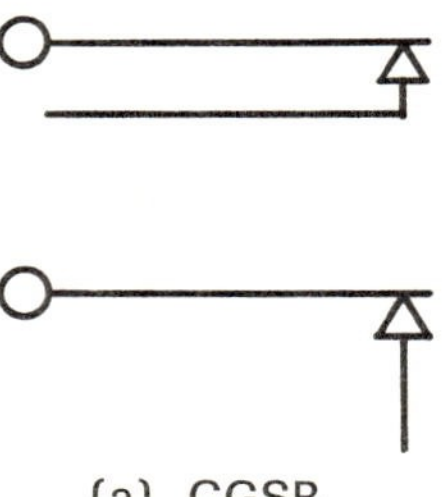

(a) CGSB

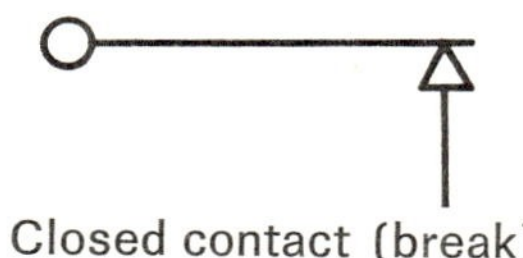

Closed contact (break)

(b) USAS

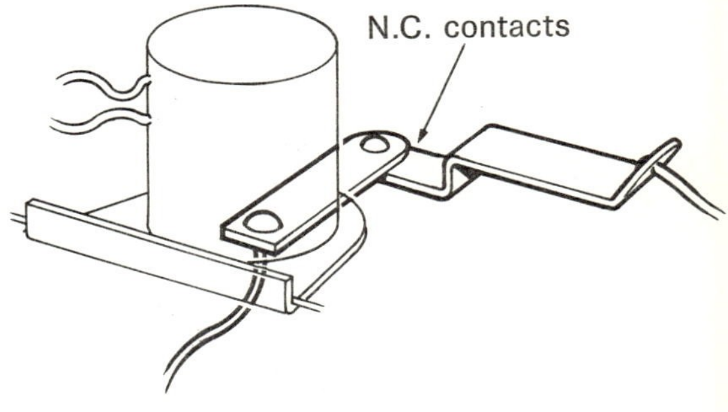

(c) Pictorial

CORES

29. Iron core

This is a type of choke or transformer core constructed from soft iron laminations, which can be either fixed or moveable. If it is moveable it is referred to as being variable.

30. Powdered iron core

This is a type of core constructed from powdered iron or ferroceramic material and can be either fixed or variable.

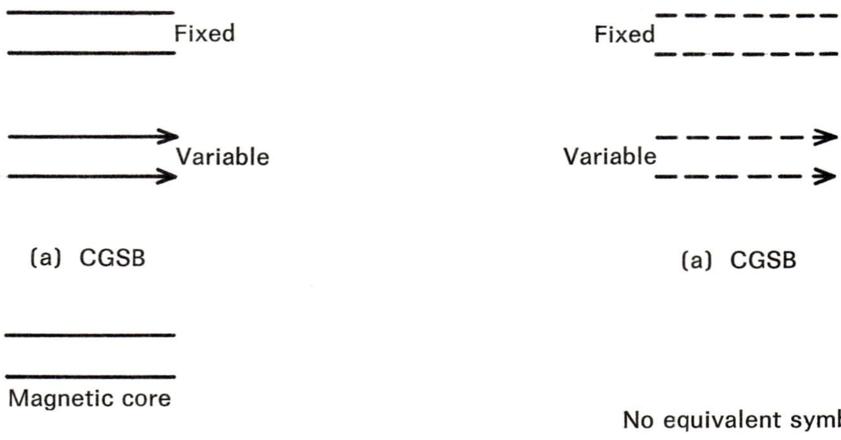

(a) CGSB

(a) CGSB

No equivalent symbol

No equivalent symbol

(b) USAS

(b) USAS

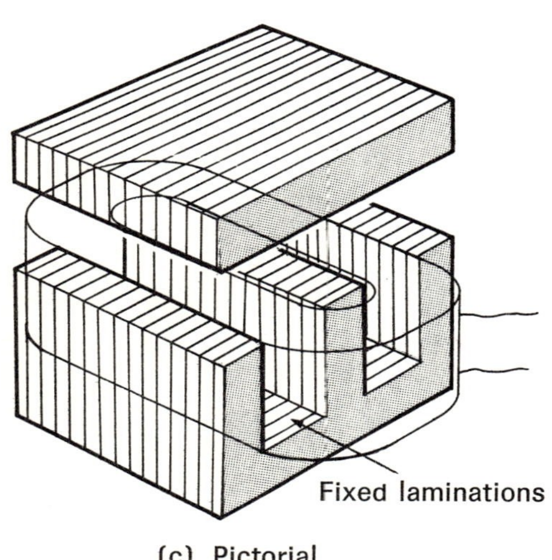

(c) Pictorial

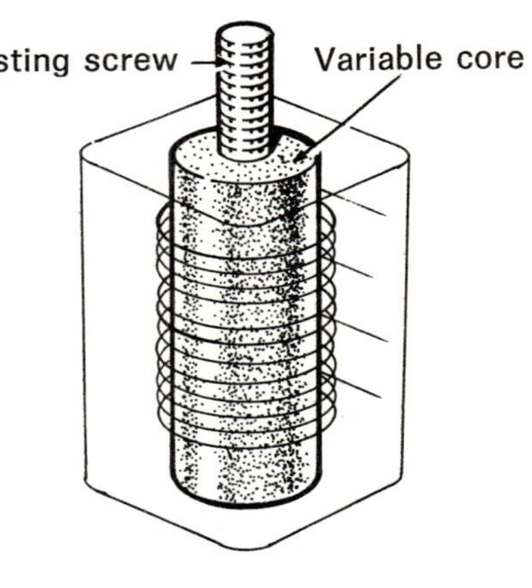

(c) Pictorial

31. CRYSTAL, PIEZOELECTRIC

This is an electrical device used in either microphones or record playing devices, and which functions by converting mechanical energy to electrical energy.

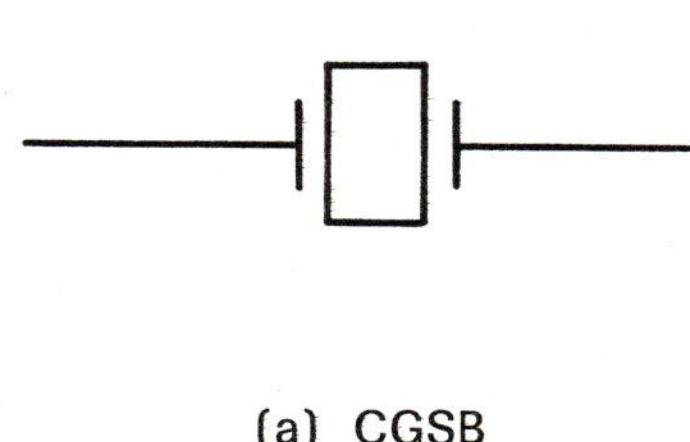

(a) CGSB

The same as the Canadian

(b) USAS

The type used in radio transmitter oscillator circuits

(c) Pictorial

GROUNDS

32. Earth ground

This is a device used to maintain an electrical path to ground for safety reasons and carried out practically by the use of grounding rods or by approved connecting clamps to a cold water piping system fabricated from copper tubing. In residential wiring systems this is carried out where the supply pipe first enters the residence.

(a) CGSB

The same as the Canadian

(b) USAS

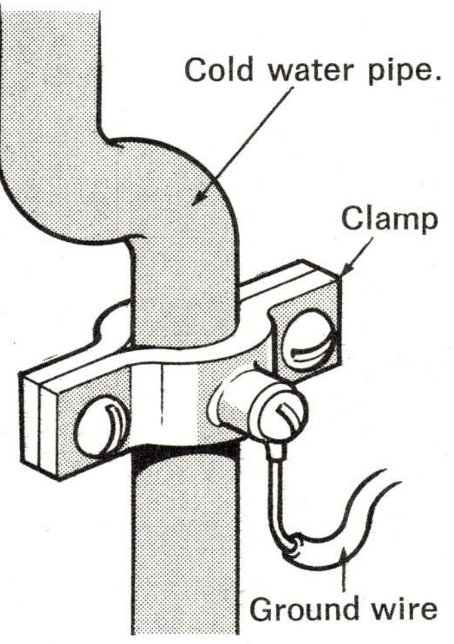

(c) Pictorial

33. Common earth ground
When more than one conductor is connected to a common earth ground, the CGSB symbol is used.

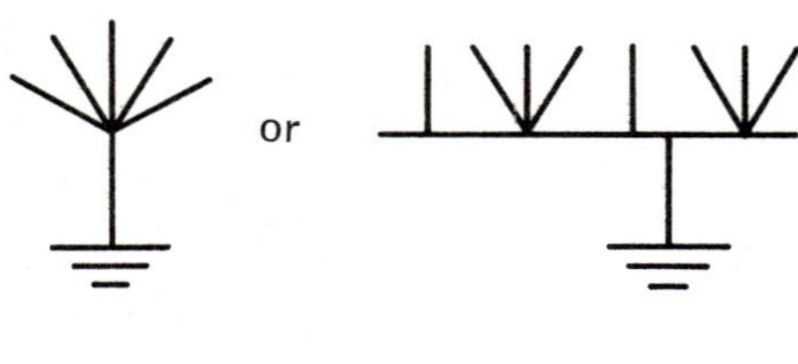

(a) CGSB

No symbol given, but 'Note' reads:
Common connections.
Conducting connections made to one another

(b) USAS

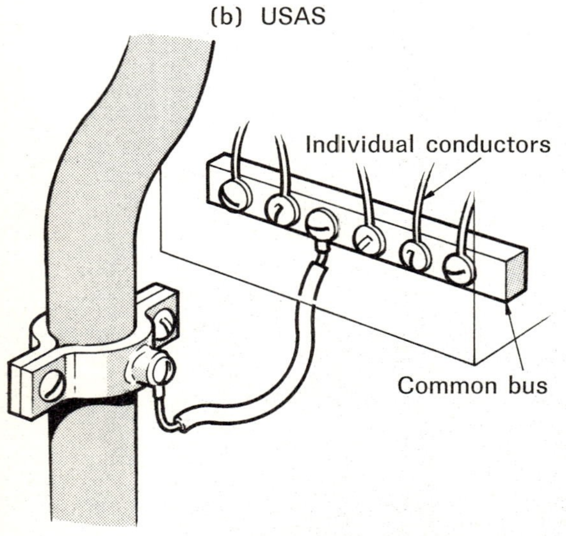

(c) Pictorial

34. Chassis ground
This is a method utilizing the metallic chassis upon which the electrical components are mounted, whereby certain parts of a circuit are commonly interconnected to the chassis without using interjoining leads.

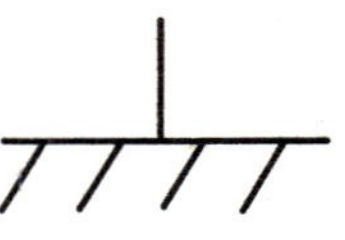

(a) CGSB

(b) USAS

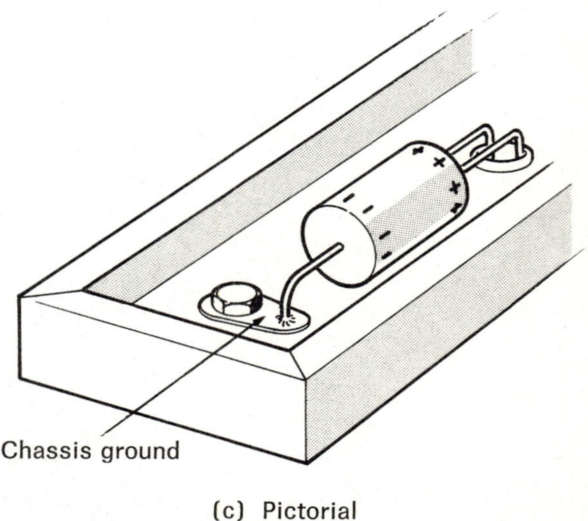

(c) Pictorial

35. HEATER

This is a specific type of resistance element designed and manufactured expressly for converting electrical energy into heat.

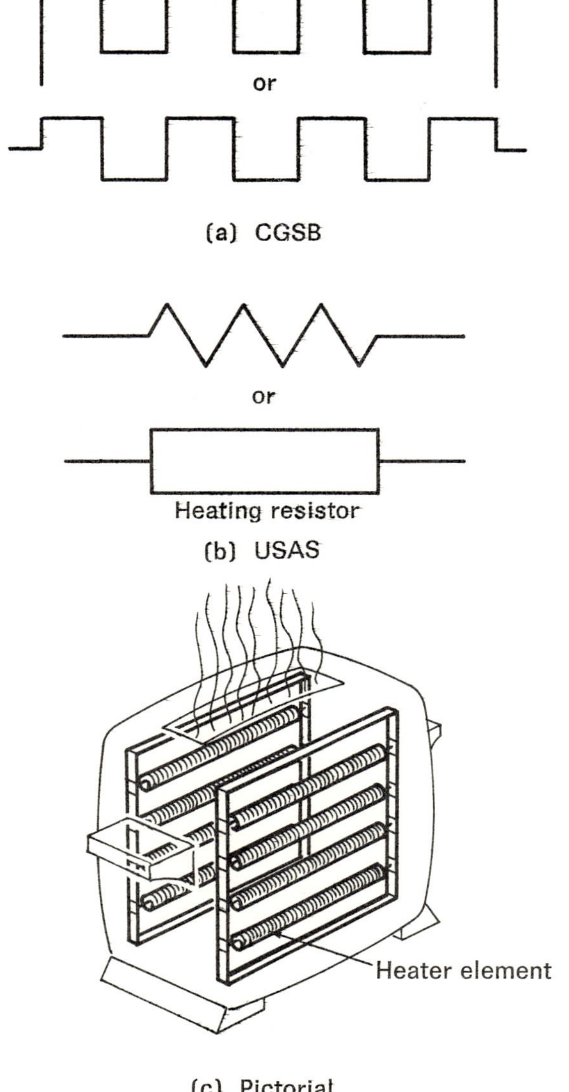

(a) CGSB

(b) USAS

(c) Pictorial

36. INSTRUMENTS AND METERS

These are devices for measuring and visually indicating certain electrical values within a circuit such as the electrical potential or the flow of current.

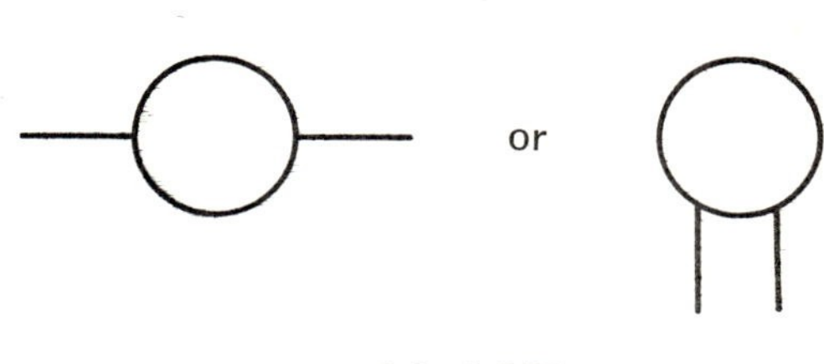

(a) CGSB

The same as the Canadian

(b) USAS

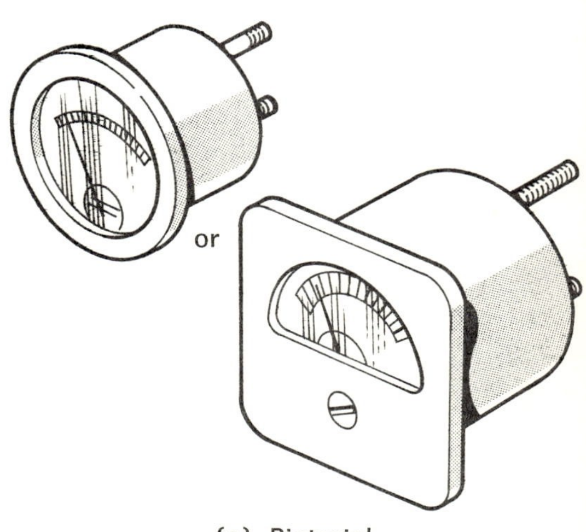

(c) Pictorial

37. Specifying type of meter

When a meter is used to measure a particular electrical characteristic such as the voltage, the approved abbreviation is included in the basic meter symbol to indicate the type of meter.

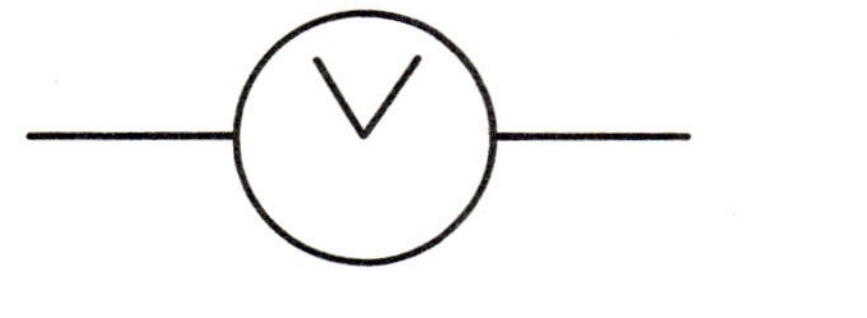

(a) CGSB

The same as the Canadian

(b) USAS

Not applicable

(c) Pictorial

38. Indicating full-scale deflection

If the full-scale deflection figure is required, it is shown in the symbol before the approved abbreviation.

(a) CGSB

No equivalent symbol

(b) USAS

Not applicable

(c) Pictorial

39. Alternating-current meter

If a meter is used in measuring an alternate current then the symbol '∼' is included in the basic meter symbol.

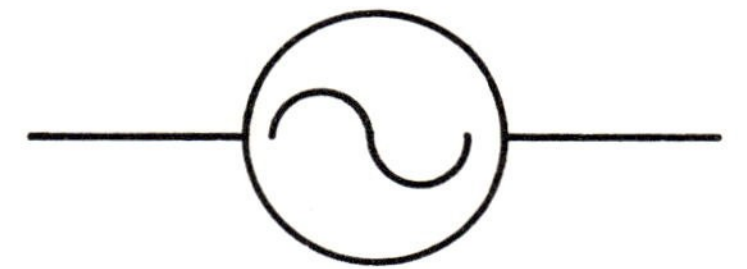

(a) CGSB

No equivalent symbol

(b) USAS

Not applicable

(c) Pictorial

40. Direct-current meter

If a meter is used in measuring a direct current, then the polarity of the instrument is indicated by a '+' symbol located adjacent to the positive lead.

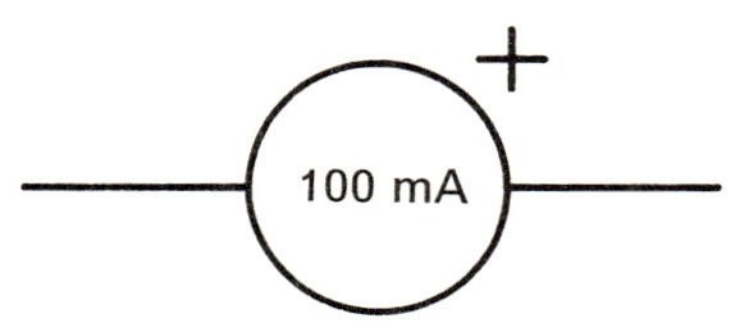

(a) CGSB

No equivalent symbol

(b) USAS

Not applicable

(c) Pictorial

KEY

41. Simple telegraph key

This is an electromechanical device used in conjunction with a transmitter to send Morse code signals over a wire circuit or by radio frequency transmission.

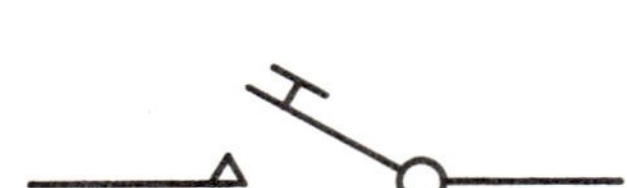

(a) CGSB

(b) USAS

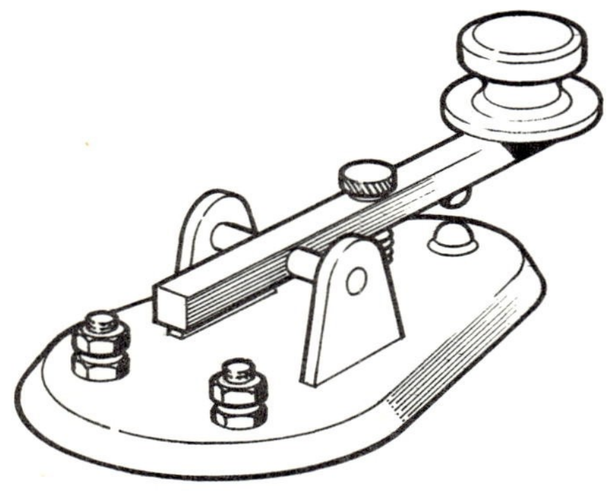

(c) Pictorial

LAMPS

42. Filament lamp

43. Filament lamp, coloured

This is a type of lamp used as pilot or warning lamp or for illuminating purposes, and operated by the principle of incandescence. If it is used as a coloured pilot lamp the specific colour is denoted below the symbol.

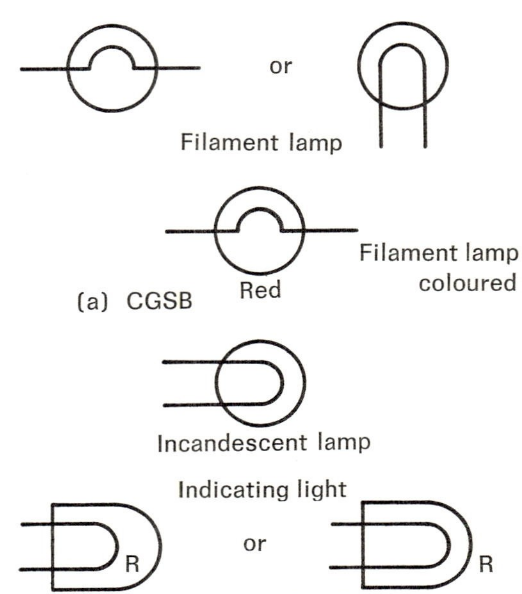

(a) CGSB

See U.S.A.S. Y 32-2-1967 for the list of colour abbreviations

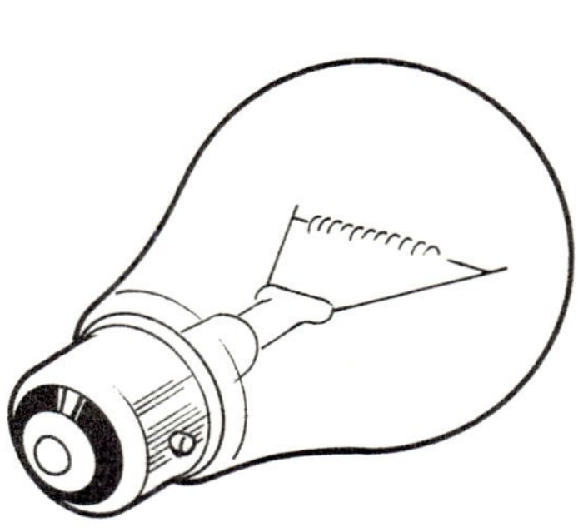

(c) Pictorial

44. Fluorescent lamp

This is a type of illuminating lamp having a special chemical coating on its interior walls which fluoresces to produce a whitish or daylight colour of light.

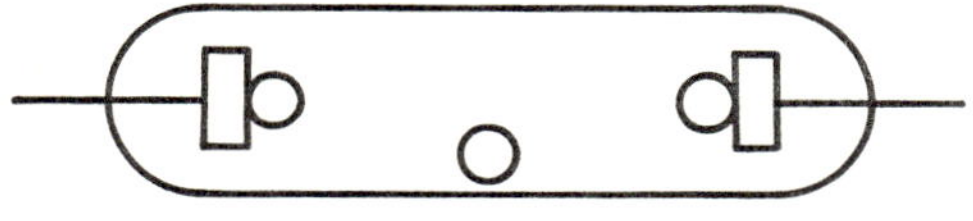

Fluorescent lamp

(a) CGSB

Fluorescent—2 terminal

(b) USAS

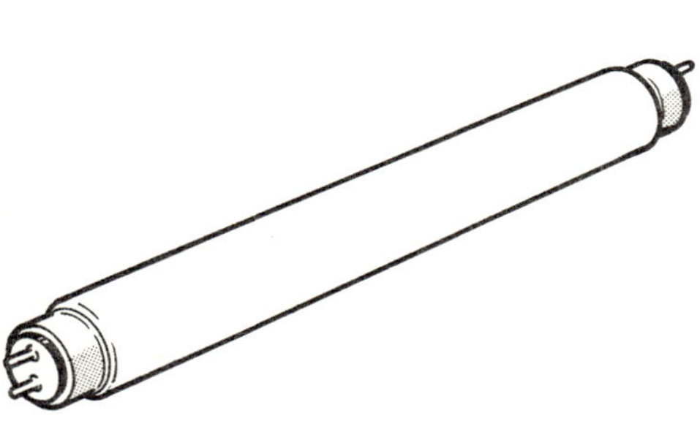

(c) Pictorial

MACHINES, ROTATING

45. Motor, basic symbol

This is an electromechanical device for converting electrical energy into mechanical energy.

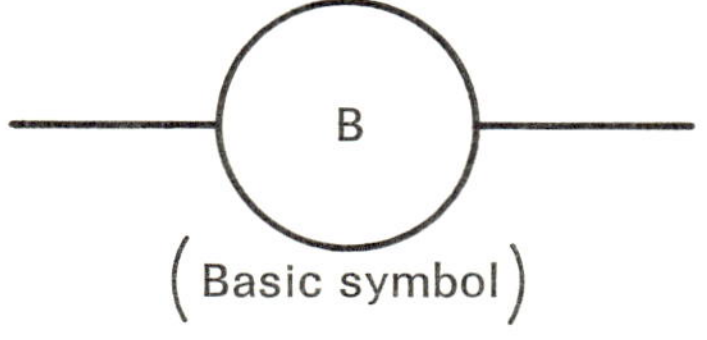

(Basic symbol)

(a) CGSB

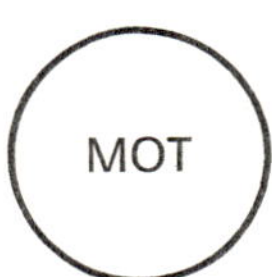

(b) USAS

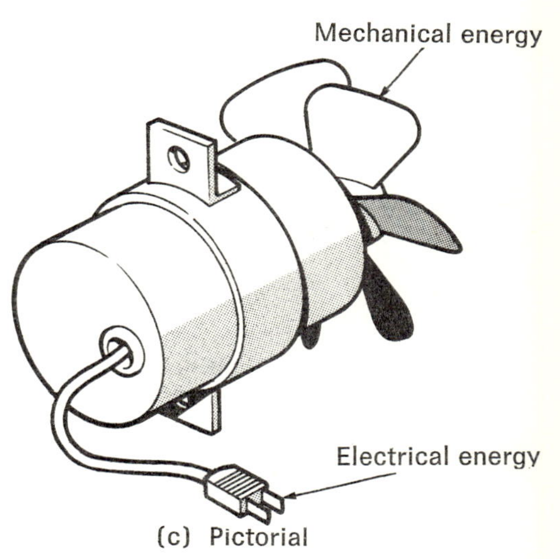

(c) Pictorial

46. Generator, basic symbol
This is an electromechanical device for converting mechanical energy into electrical energy.

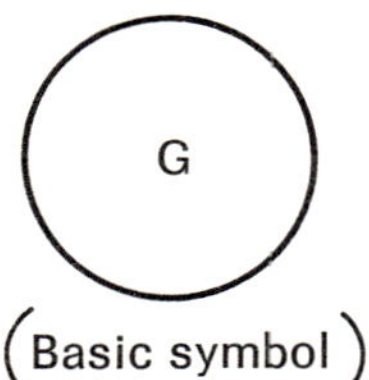

(Basic symbol)

(a) CGSB

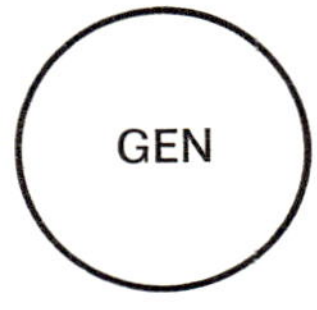

(b) USAS

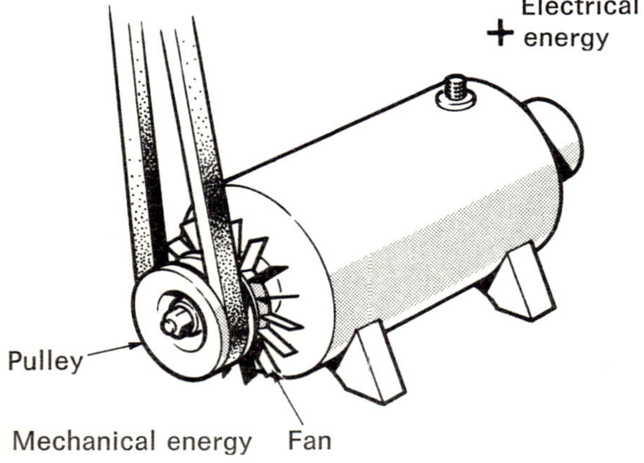

D.C. generator

(c) Pictorial

47. Motor generator
A motor generator is an electromechanical device consisting of an electric motor mechanically coupled to a generator, and normally manufactured as a complete set or unit.

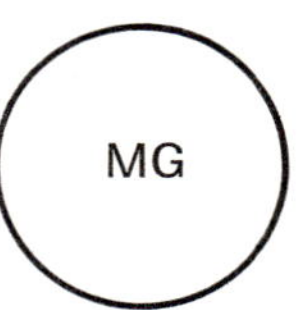

(a) CGSB

No equivalent symbol

(b) USAS

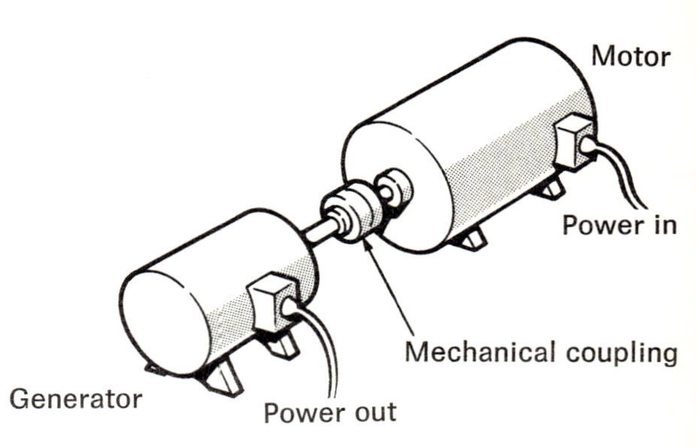

(c) Pictorial

FIELDS

48. Compensating or commutating field

A type of winding which is set into slots machined in the face of the pole pieces parallel to the armature slots is known as a compensating or commutating field winding. Current flowing in this winding is in opposition to that in the armature slots. Symbolically it is shown as a double open-looped coil.

(a) CGSB

The same as the Canadian

(b) USAS

Not applicable

(c) Pictorial

49. Series field

This is a type of field coil wired in series with the motor armature. Note that the symbol has three loops.

(a) CGSB

The same as the Canadian

(b) USAS

Not applicable

(c) Pictorial

50. Shunt field

This is a type of field coil wired in parallel with the motor armature. Note that the symbol has four loops.

(a) CGSB

The same as the Canadian

(b) USAS

Not applicable

(c) Pictorial

MOTORS, SPECIFIC

51. Series motor

When the field coil of a D.C. motor is connected in series between the source of power and one of the armature brushes, it is known as a series motor. This coil is wound from heavy, insulated copper wire.

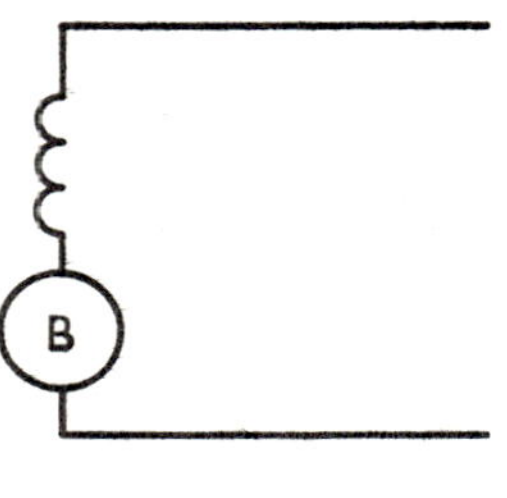

(a) CGSB

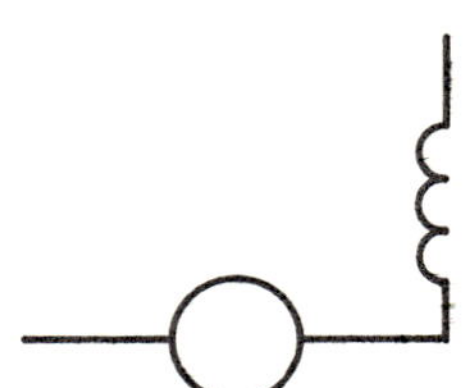

(b) USAS

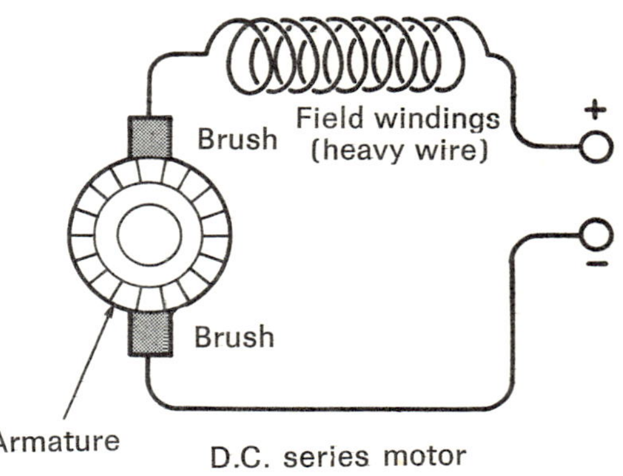

(c) Pictorial

52. Shunt motor

A shunt D.C. motor is one having a field coil wound from fine, insulated copper wire and connected across or in parallel to the armature brushes.

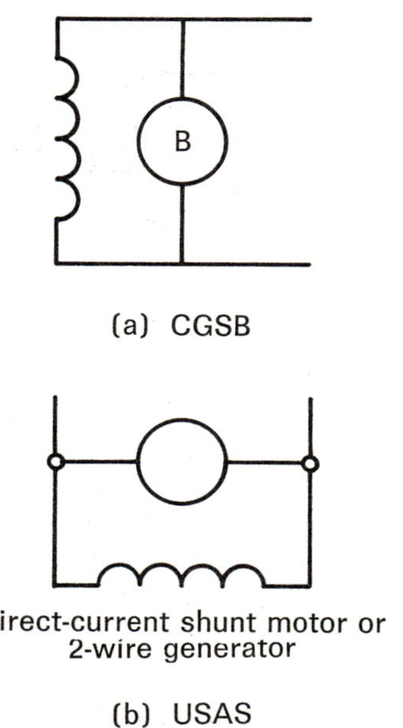

(a) CGSB

Direct-current shunt motor or 2-wire generator

(b) USAS

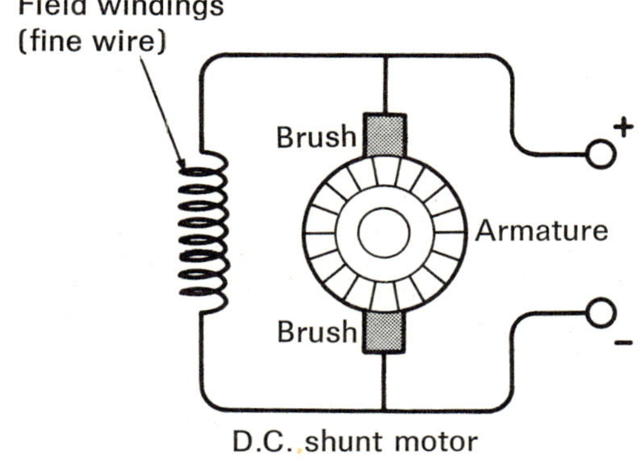

D.C. shunt motor

(c) Pictorial

53. Synchronous motor

This is a motor requiring a separate D.C. supply for its field coil, as well as a source of three-phase alternating current. Its special characteristic is that it runs at synchronous speed under varying load conditions.

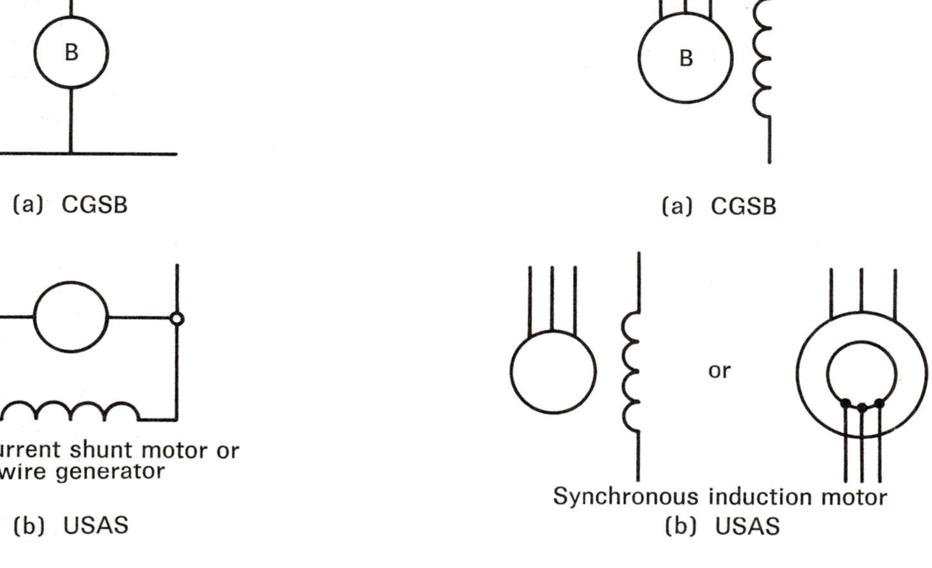

(a) CGSB

Synchronous induction motor

(b) USAS

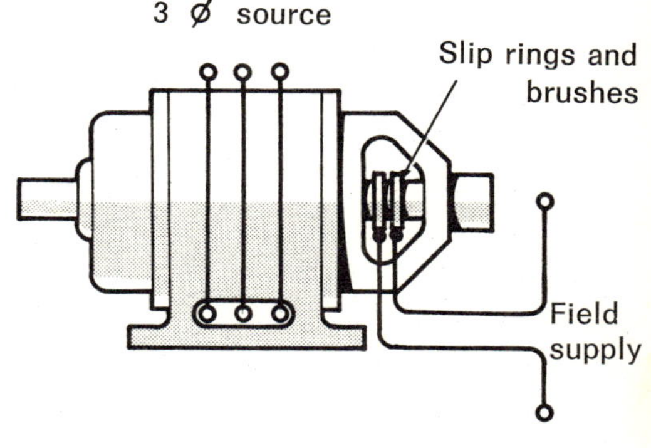

(c) Pictorial

54. Squirrel-cage induction motor
A motor of this type has a rotor resembling the shape of a squirrel cage. In addition it has no wound coils, slip rings, nor electrical connections from the rotor to the A.C. line. The rotor is formed of bars set into slots in it. These bars are in turn joined at the ends.

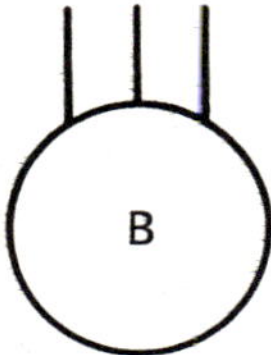

(a) CGSB

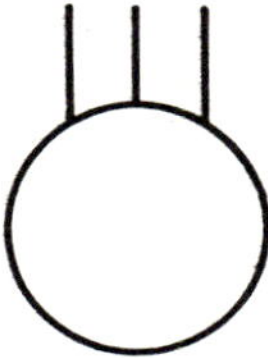

(b) USAS

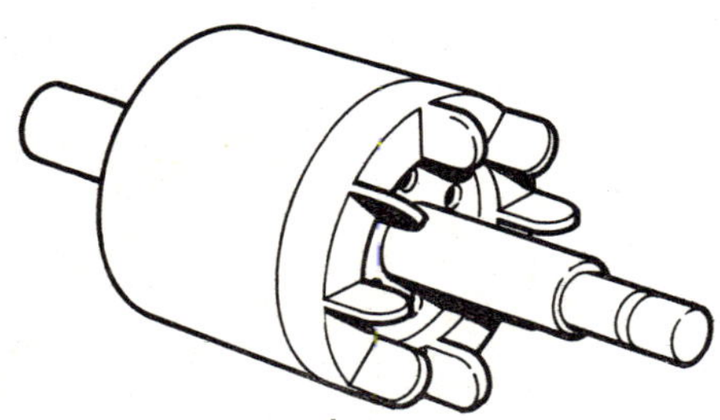

squirrel-cage rotor
1. No wire wound coils
2. No slip rings
3. No connection from rotor to line

(c) Pictorial

55. Wound-rotor inductor motor
The identification of this type of induction motor is made possible by the special design of its wound rotor. In addition special collector rings and brushes are used in conjunction with this rotor, as well as external resistors connected in series with each ring.

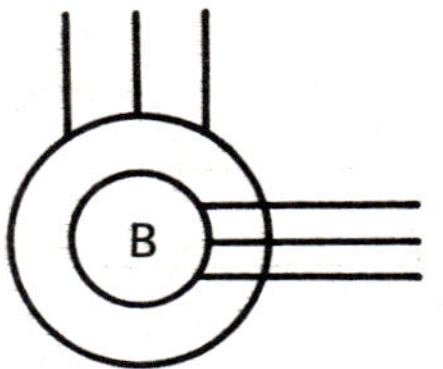

(a) CGSB

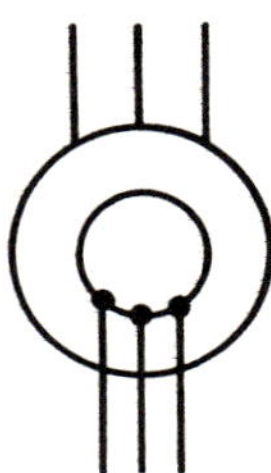

(b) USAS

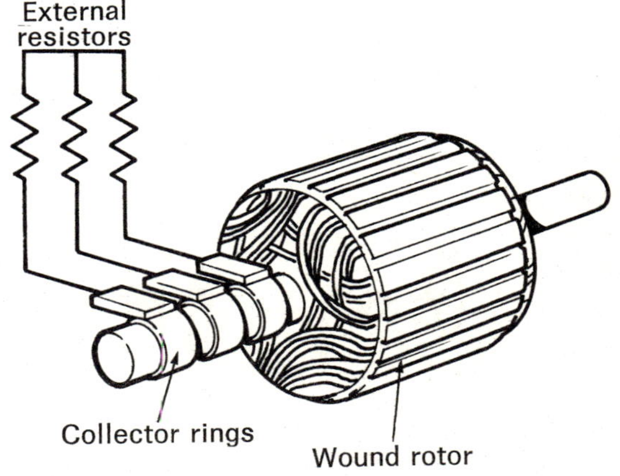

(c) Pictorial

MOTOR SYMBOLS DETAILED

56. Capacitor-start motor

This is a type of induction-run motor frequently used in refrigeration and compressor units, and which has good starting torque characteristics.

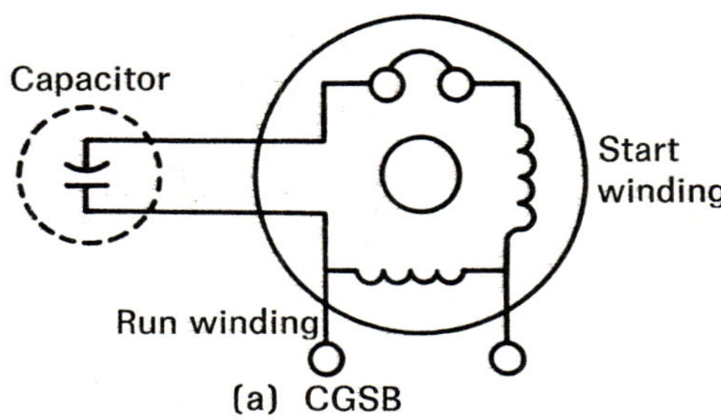

(a) CGSB

No equivalent symbol

(b) USAS

Centrifugal switch in motor end bell in capacitor-start motor

Capacitor

Induction type rotor

(c) Pictorial

57. Capacitor-run motor

The capacitor-run motor is similar to the motor illustrated by No. 56 (a) except that it does not have a centrifugal switch included in series between the capacitor and the starting winding.

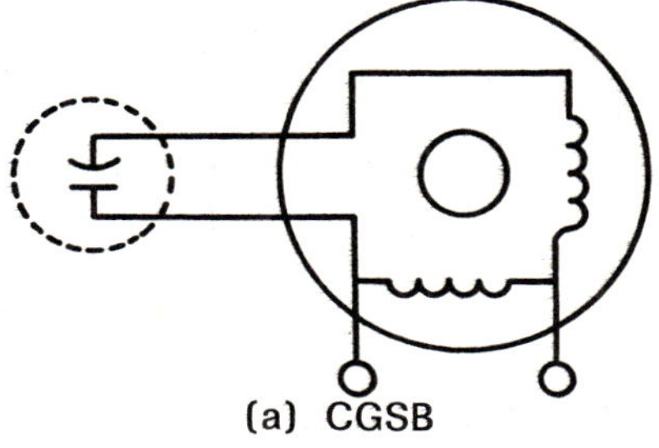

(a) CGSB

No equivalent symbol

(b) USAS

No centrifugal switch in capacitor-run motor.

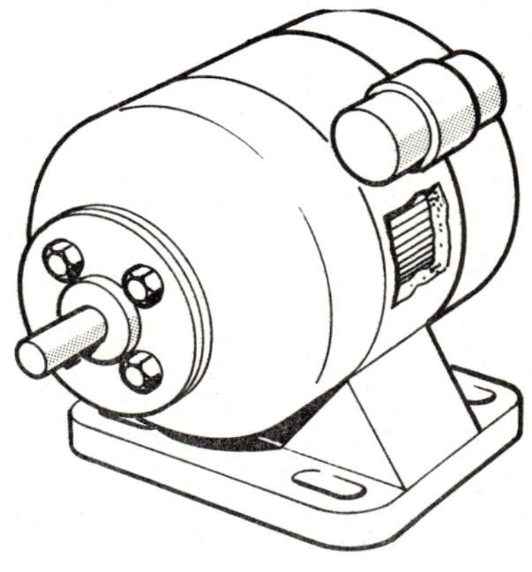

(c) Pictorial

58. Shaded-pole induction motor

This is a type of single-phase fractional horsepower motor having shading coils placed on one side of each stator pole. Its rotor is of the squirrel-cage type, and no centrifugal switch is required.

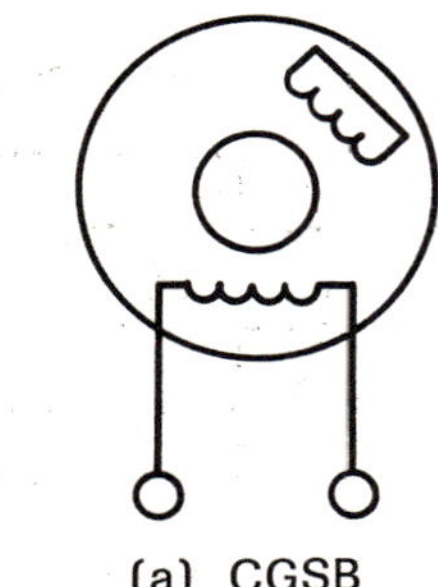

(a) CGSB

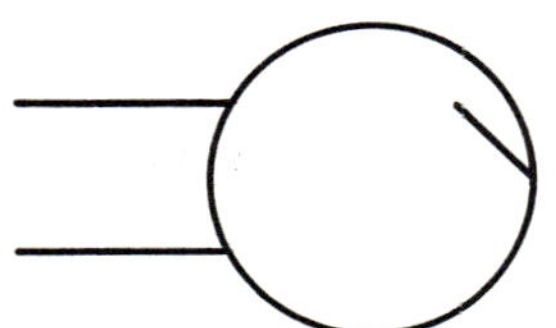

single-phase shaded-pole motor

(b) USAS

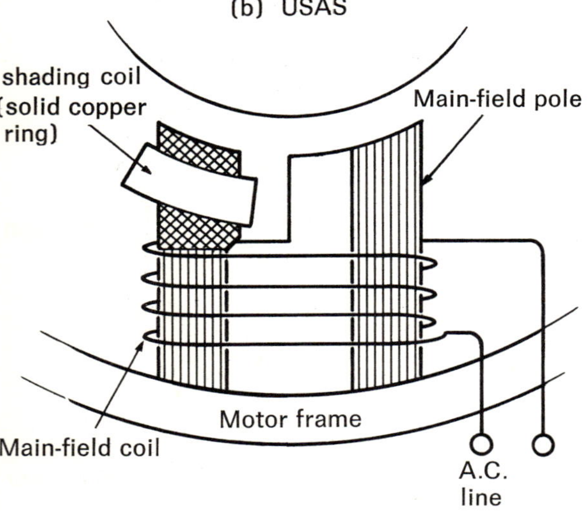

(c) Pictorial

59. Split-phase induction motor

A further kind of induction motor, known as a split-phase type is illustrated below. This single-phase motor requires a starting winding in series with the source of power, and a running winding across these power terminals. Physically, the motor has its running windings located outside of its starting windings, as is illustrated in No. 59(c). Starting torque is developed by phase displacement between its field windings.

(a) CGSB

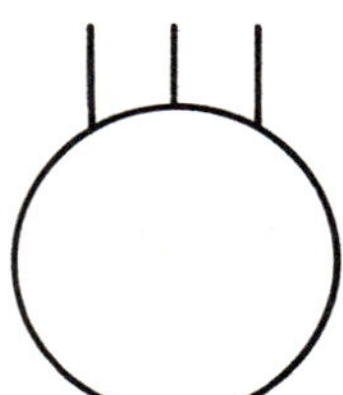

Split-phase induction motor

(b) USAS

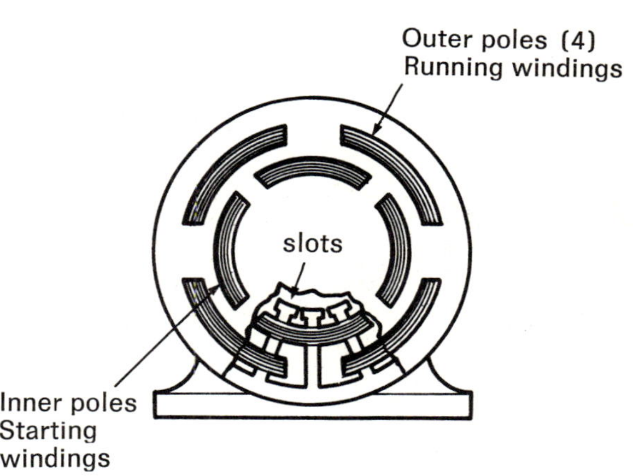

(c) Pictorial

60. MECHANICAL CONNECTION, LINKAGE

When mechanical parts or sections are connected together and operated as a unit, such as a two-section variable capacitor, the linkage is shown by a broken line consisting of alternate long and short dashes.

(a) CGSB

Mechanical connection

(b) USAS

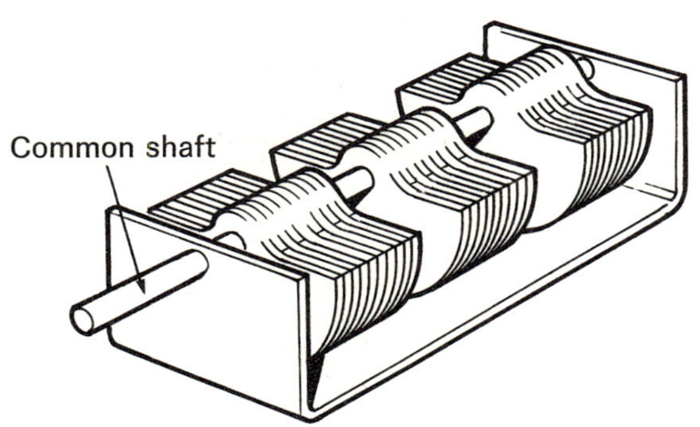

Ganged variable capacitor: 3 stage

(c) Pictorial

MICROPHONES

61. Moving coil or dynamic microphone

A dynamic microphone is a device which converts audible sound waves into a proportionately varying electrical current by generator action. This type of microphone is a voltage producing one.

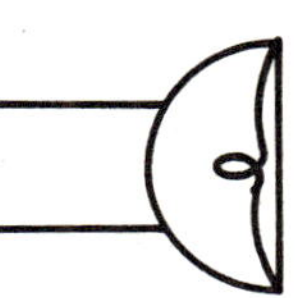

(a) CGSB

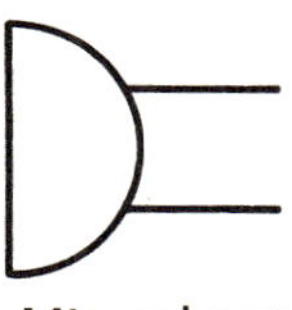

Microphone

(b) USAS

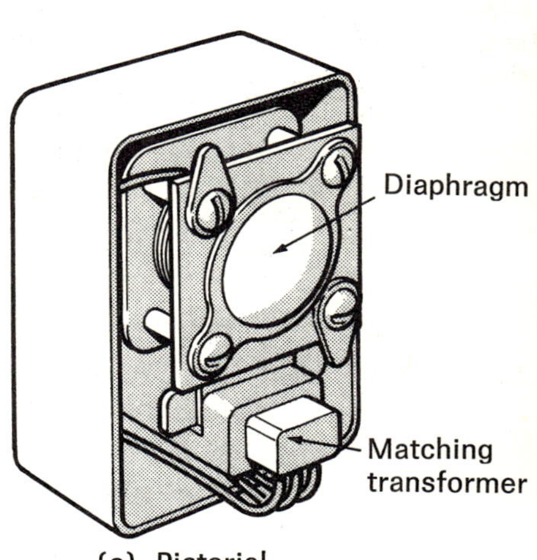

(c) Pictorial

62. Velocity or ribbon microphone
This is a microphone of a voltage generator type and it has a higher frequency response than a dynamic microphone.

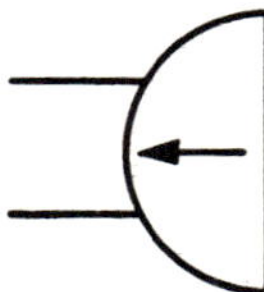

(a) CGSB

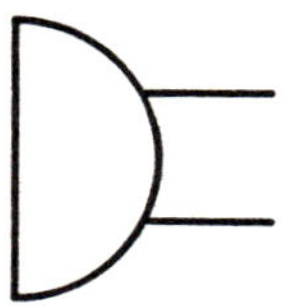

Microphone
(b) USAS

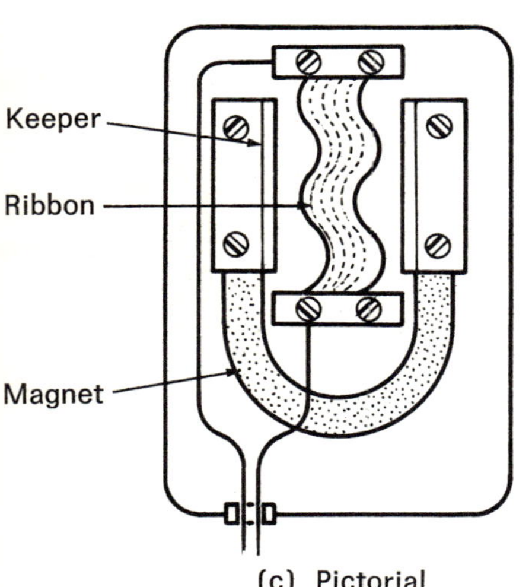

(c) Pictorial

63. Crystal microphone
This is a type of microphone utilizing the piezoelectric effect, in which the crystals used are rochelle salts.

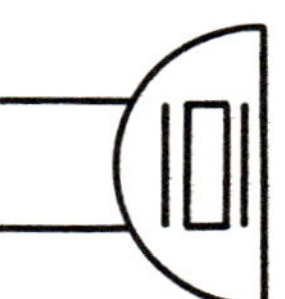

(a) CGSB

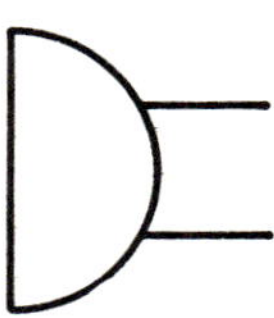

Microphone
(b) USAS

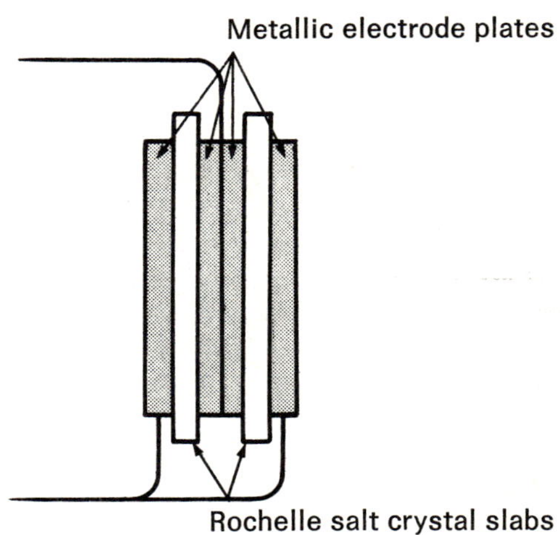

(c) Pictorial

64. Carbon, single or double-button microphone
This is a type of microphone utilizing a button or buttons of carbon granules, and it operates as a current-modulating instrument.

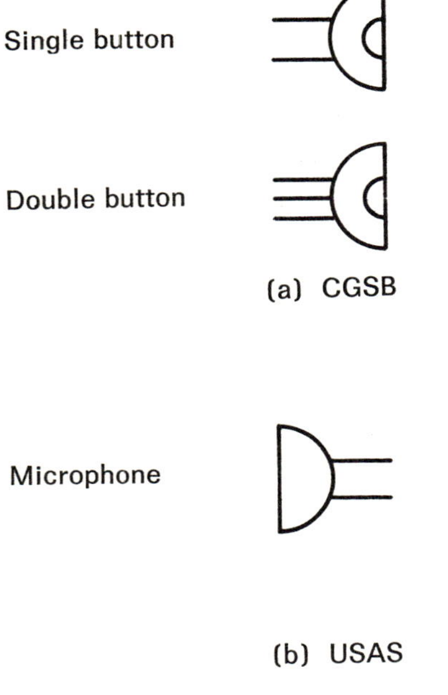

(a) CGSB

(b) USAS

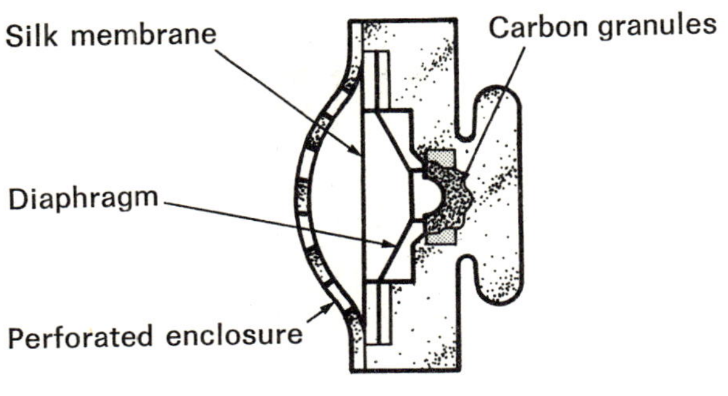

Single-button microphone

(c) Pictorial

PROTECTION DEVICES

65. Self-heated, normally-closed bimetal element
This is a type of protection device in which the bimetal element is self-heated by the overload current, and trips under this state, thus creating an open circuit. Upon cessation of the current flow, the strip cools down, flexes into a straight position, and closes the circuit.

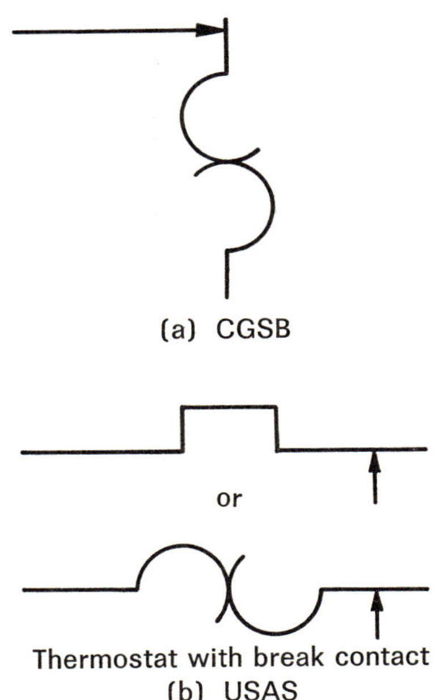

(a) CGSB

Thermostat with break contact
(b) USAS

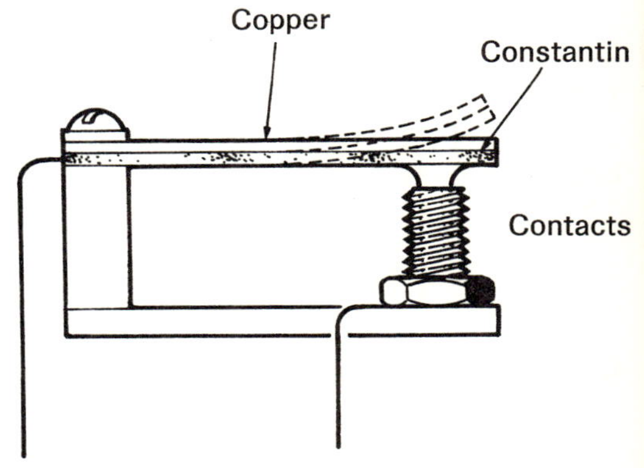

(c) Pictorial

66. Self-heated, hand reset, normally-closed bimetal element

This is a protection device which requires manual resetting to a closed position when it has been tripped by an overload current, but is similar to the protection device shown in No. 65(a) in all other respects.

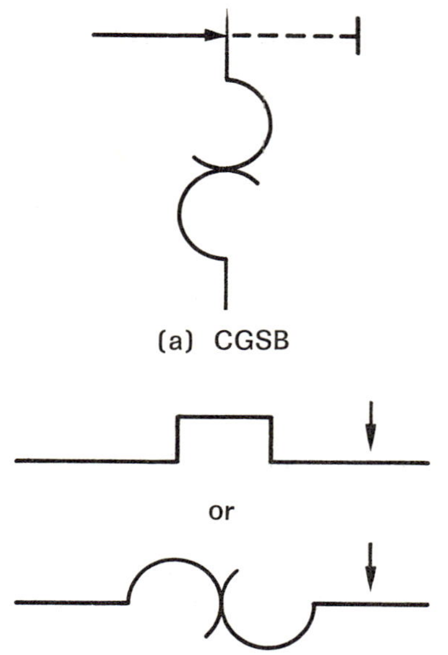

(a) CGSB

Thermostat with make contact (not hand reset)
(b) USAS

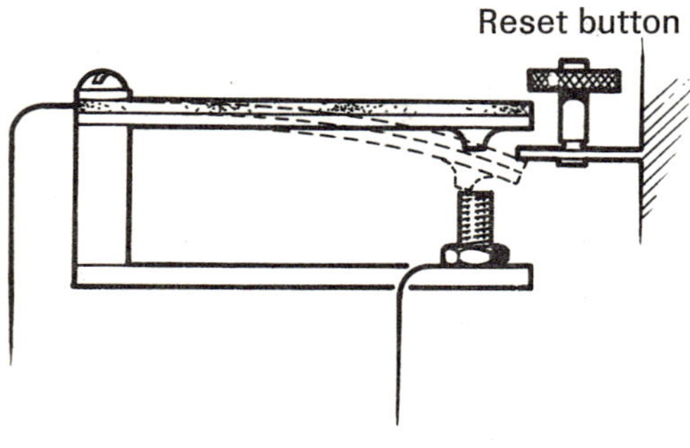

(c) Pictorial

67. Externally-heated bimetal element

This is a type of bimetal element protection not subject to operation by current overload but requiring external heat to trip it open.

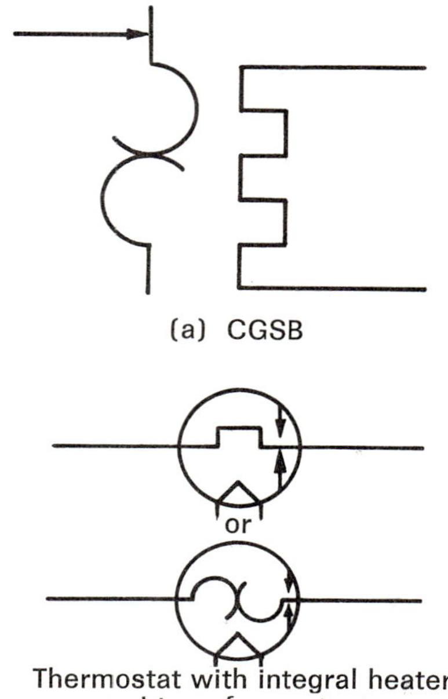

(a) CGSB

Thermostat with integral heater and transfer contacts
(b) USAS

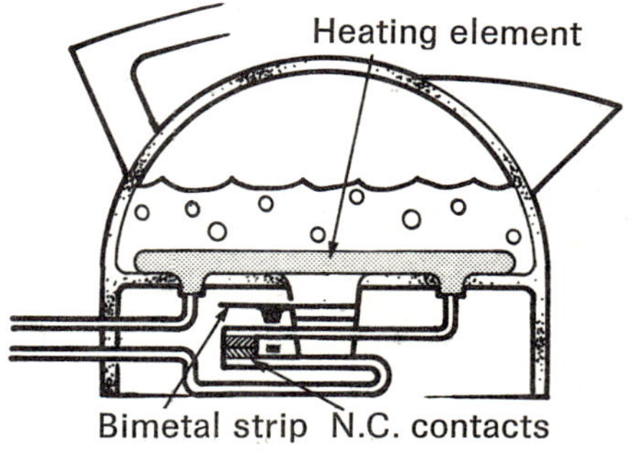

Electric kettle thermostat control

(c) Pictorial

68. Fuse
This is a protection device utilizing a metal of low specific melting point, which melts upon current overload, thereby opening the circuit. The individual current rating is placed adjacent to the CGSB symbol.

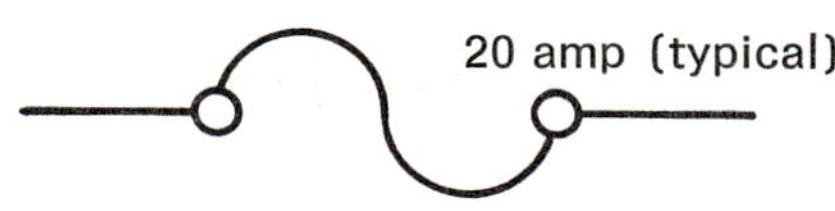

(a) CGSB

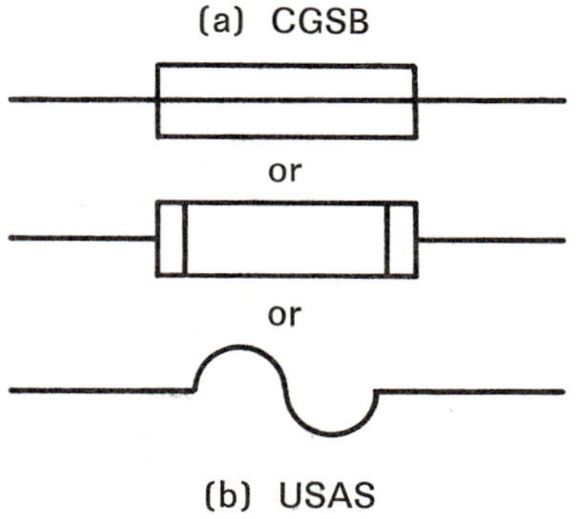

(b) USAS

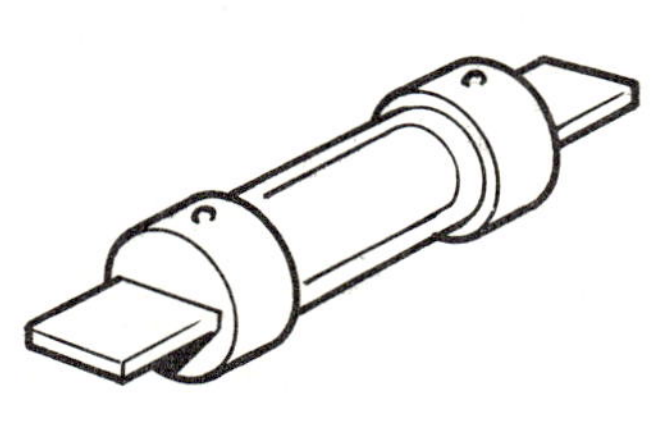

(c) Pictorial

RECEIVERS

69. Moving iron receiver
This is an electromechanical device which converts a pulsating direct current into audible sound waves by the motion of an iron diaphragm moving in a varying magnetic field.

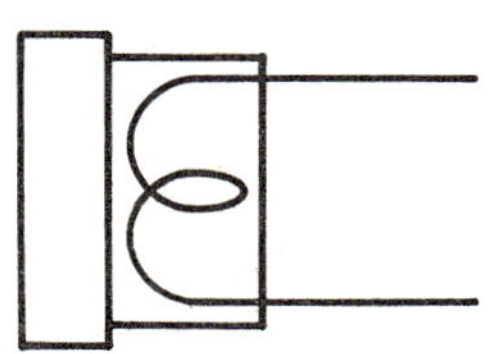

(a) CGSB

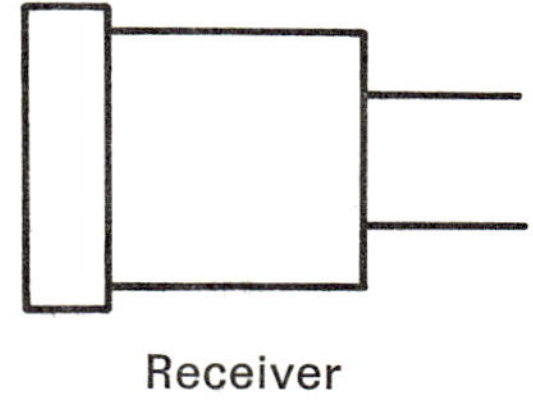

Receiver

(b) USAS

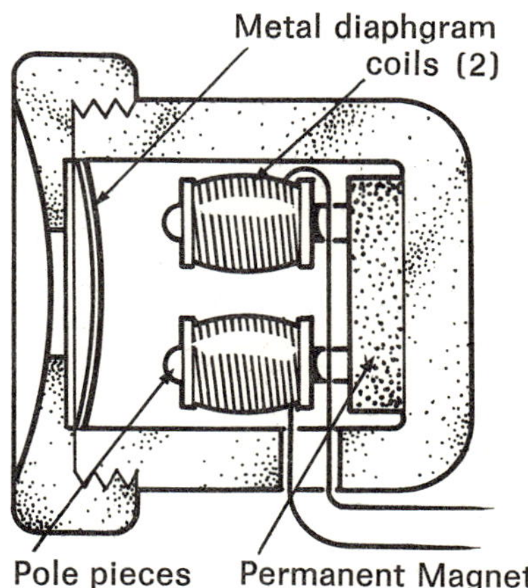

(c) Pictorial

70. Moving coil receiver
This is a receiver in which a moving coil is attached to a lightweight cone to produce audible sounds, when the coil is energized by a pulsating direct current.

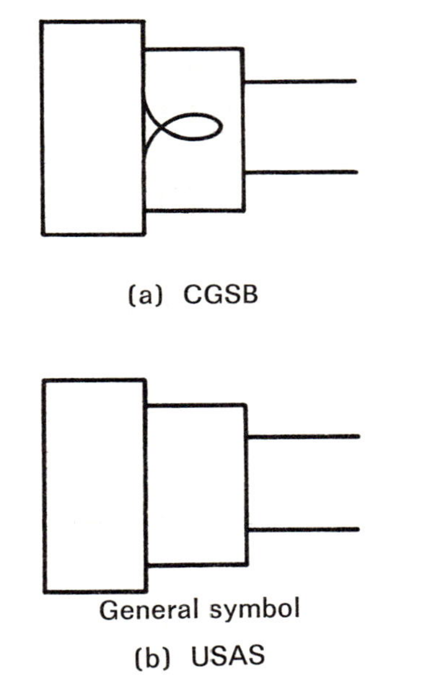

(a) CGSB

General symbol

(b) USAS

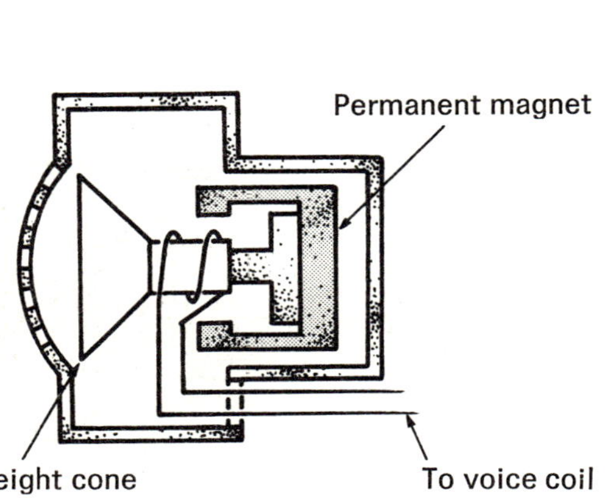

(c) Pictorial

71. Crystal receiver
This is a receiver operating by piezoelectric effect. Variations of the incoming pulsating direct current cause a proportionate deformation of the crystal, thereby creating audible sound waves.

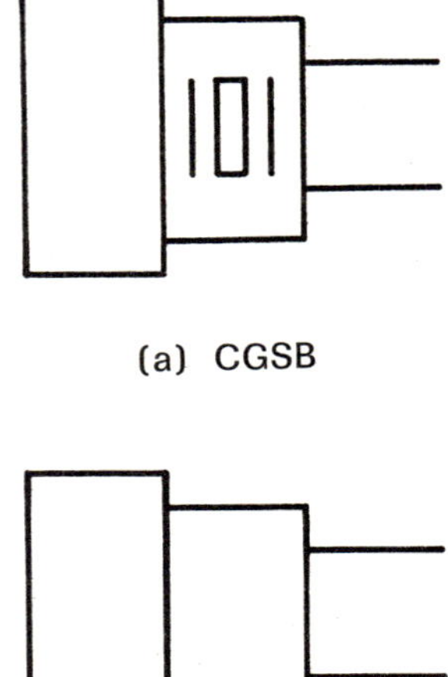

(a) CGSB

General symbol

(b) USAS

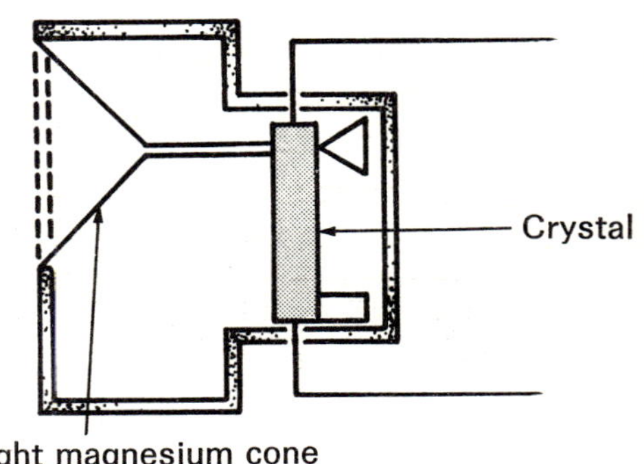

(c) Pictorial

RELAYS OR CONTACTORS

72. D.P.S.T. normally-open power type relay
This is a device for controlling a circuit remotely by energizing the coil of the relay and closing its double contacts to complete the circuit.

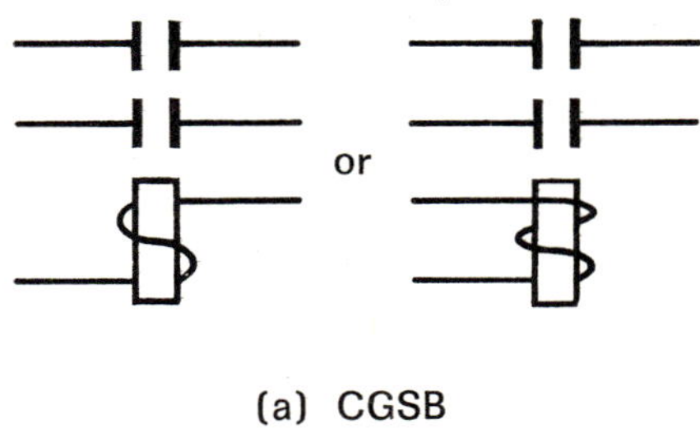

(a) CGSB

No symbol for complete relay
See contact and solenoid symbols
Nos. 22 and 91

(b) USAS

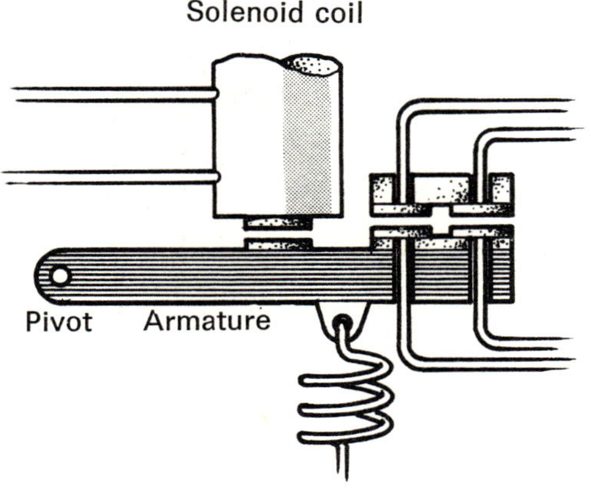

(c) Pictorial

73. D.P.S.T. normally-closed power type relay
This is a relay similar to the one described for No. 72(a) but having its contacts closed when the relay coil is in a nonenergized state.

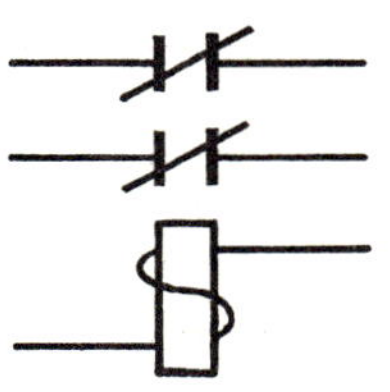

(a) CGSB

No symbol for complete relay.
See contact and solenoid symbols
Nos. 23 and 91

(b) USAS

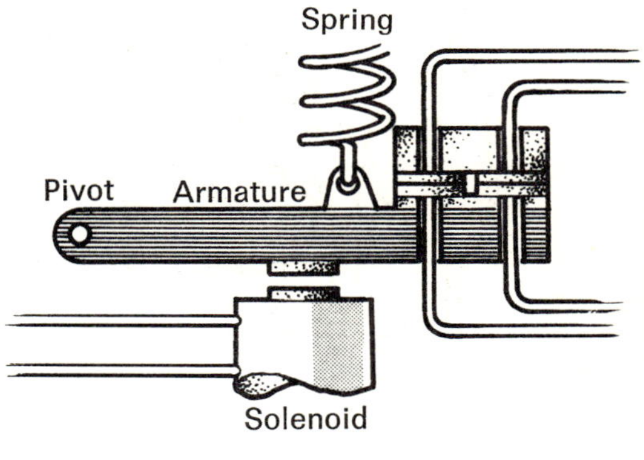

(c) Pictorial

74. S.P.S.T. power type relay
When a power relay has only one set of contacts it is known as a single-pole type and is graphically shown by the CGSB symbol in No. 74 (a).

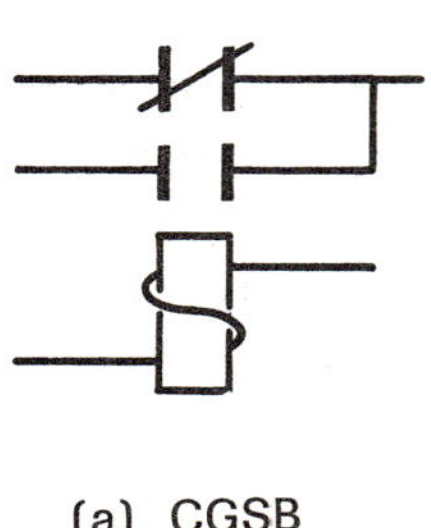

(a) CGSB

No symbol for complete relay.
See contact and solenoid symbols
Nos. 22, 23, and 91

(b) USAS

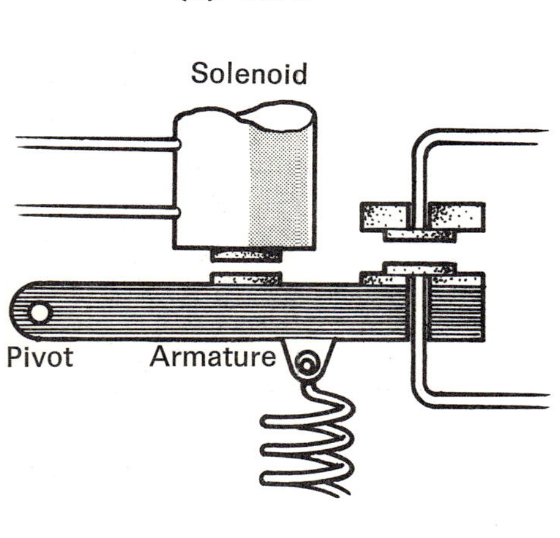

(c) Pictorial

75. Single-arm signal type relay
If a relay is used in a circuit to conduct an electrical current for signalling purposes, and is a single-arm type, the CGSB symbol is used.

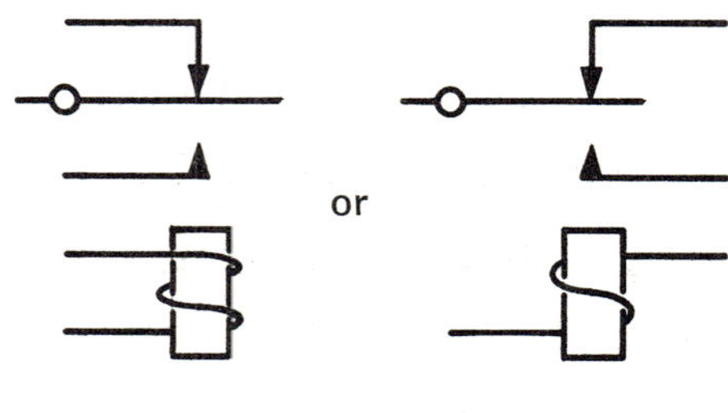

(a) CGSB

No symbol for complete relay.
See contact and solenoid symbols
Nos. 27, 28, and 91

(b) USAS

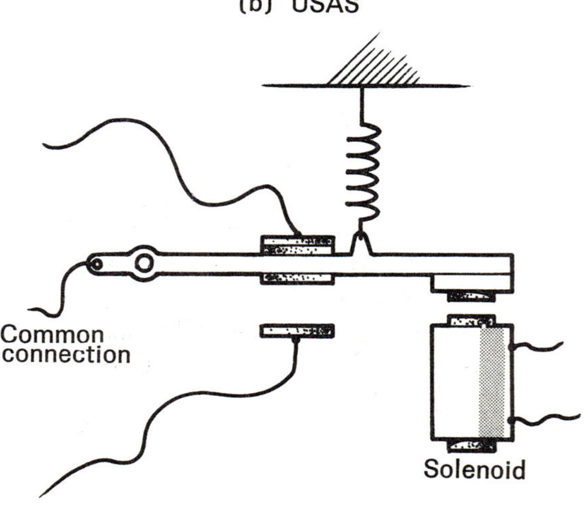

(c) Pictorial

REPRODUCERS OR RECORDERS

76. Crystal reproducer

A crystal reproducer is a device which utilizes the piezoelectric effect by converting mechanical vibrations into a pulsating direct current, which is subsequently amplified and converted into audible sound waves.

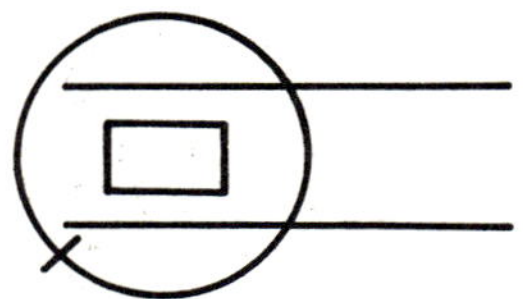

(a) CGSB

No equivalent symbol

(b) USAS

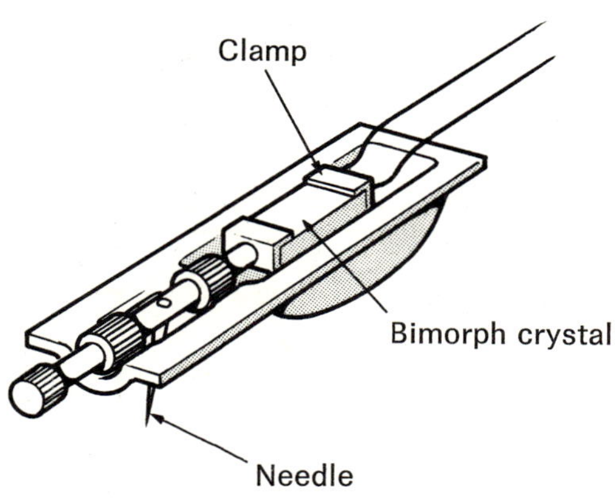

(c) Pictorial

77. Magnetic reproducer

This is a reproducing device utilizing magnetic field principles to produce audible sound waves.

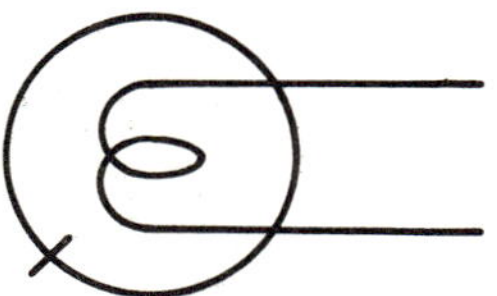

(a) CGSB

No equivalent symbol

(b) USAS

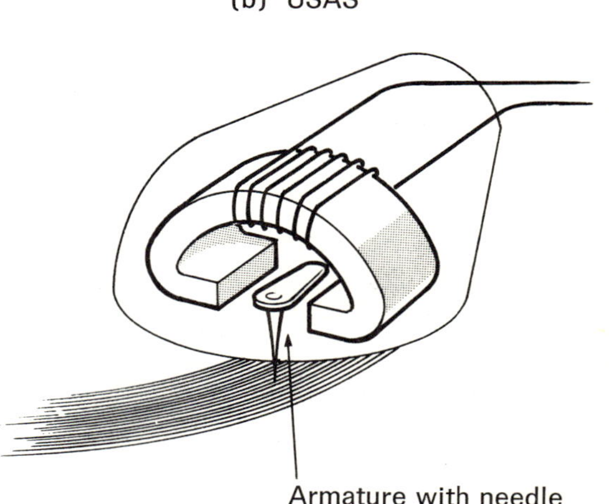

(c) Pictorial

78. Erasing head reproducer
When a reproducer has facility for electrically erasing a previously made tape, the CGSB symbol is used.

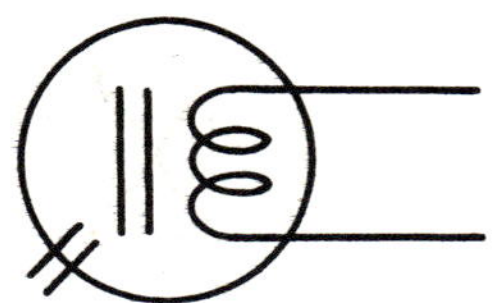
(a) CGSB

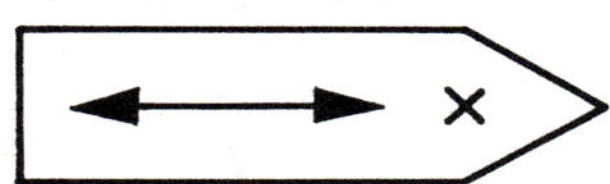
Writing, reading, and erasing pickup head

(b) USAS

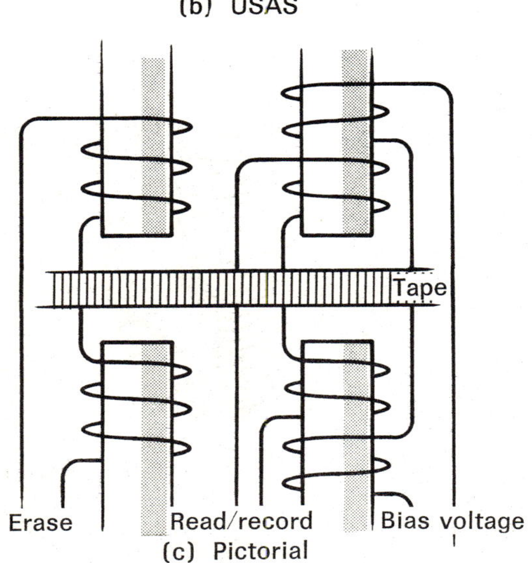

(c) Pictorial

79. Recorder or reproducer
The CGSB symbol illustrated is to be used for either a recording or reproducing head when no specified type is indicated.

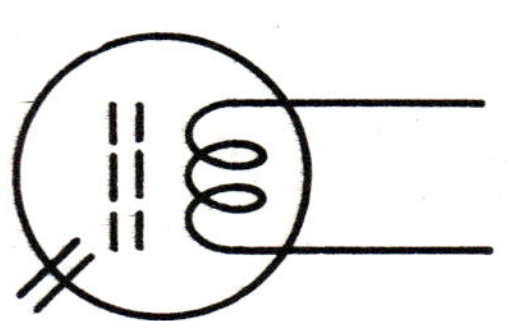
(a) CGSB

Sound recorder

Sound reproducer

(b) USAS

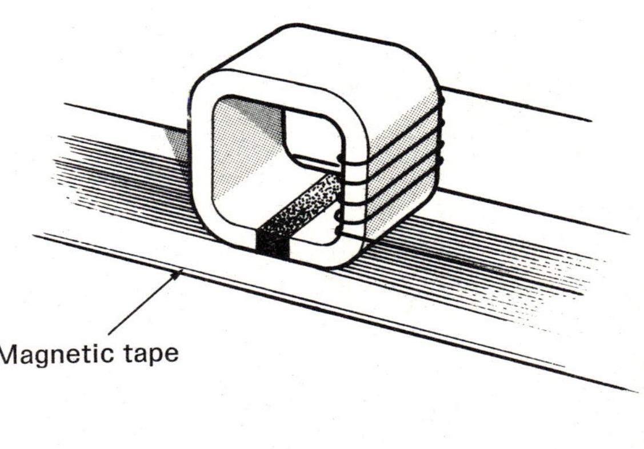

(c) Pictorial

RESISTORS

80. Fixed resistor

A fixed resistor is a device constructed from either special resistance wire or carbon, and has a specific resistance value, which cannot be mechanically altered.

(a) CGSB

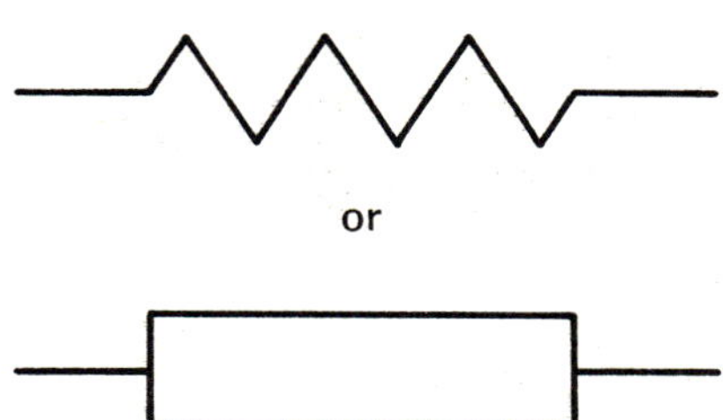

(b) USAS

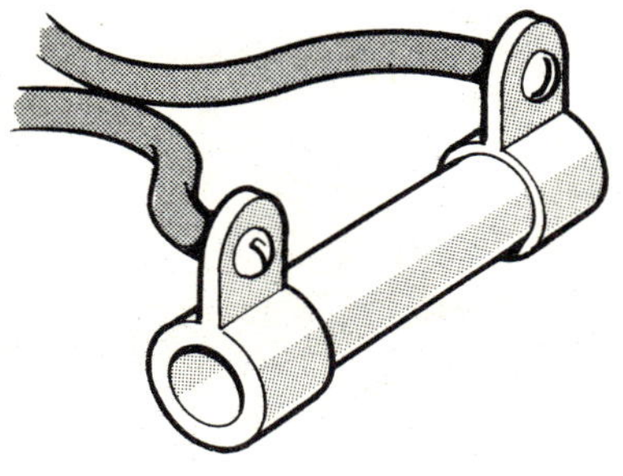

(c) Pictorial

81. Rheostat

This is a variable resistance used to control the current flow through a circuit.

(a) CGSB

Adjustable or continuously adjustable resistor

or

(b) USAS

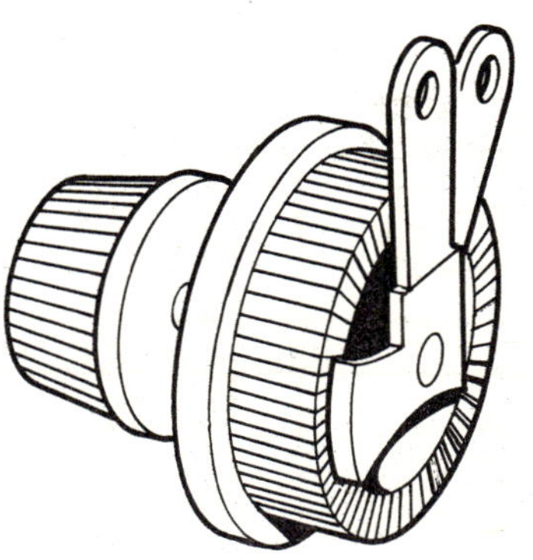

(c) Pictorial

82. Potentiometer

This is a variable type of resistance used in a circuit application when it is desired to select a portion of the total voltage impressed across the resistance.

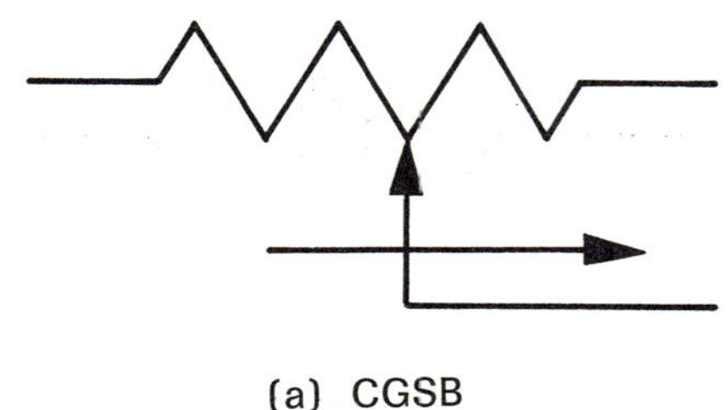

(a) CGSB

No equivalent symbol

(b) USAS

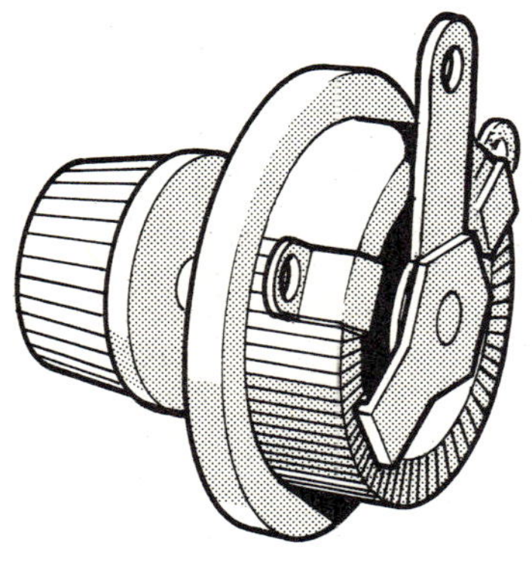

(c) Pictorial

83. Adjustable tap variable resistor

This is a resistance device having a tap which can be varied over the range of the resistor, by adjusting a screw holding the tap in place.

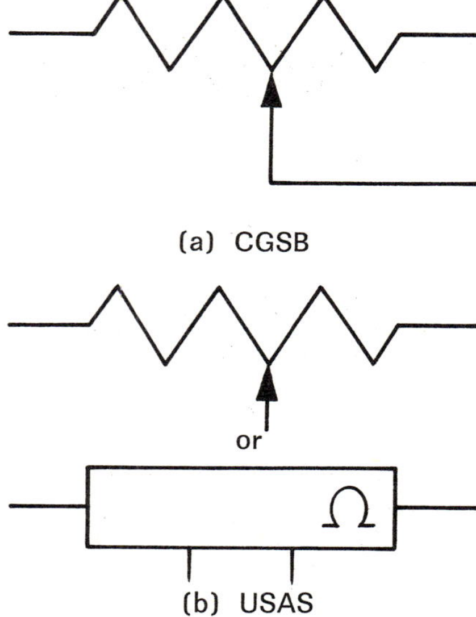

(a) CGSB

(b) USAS

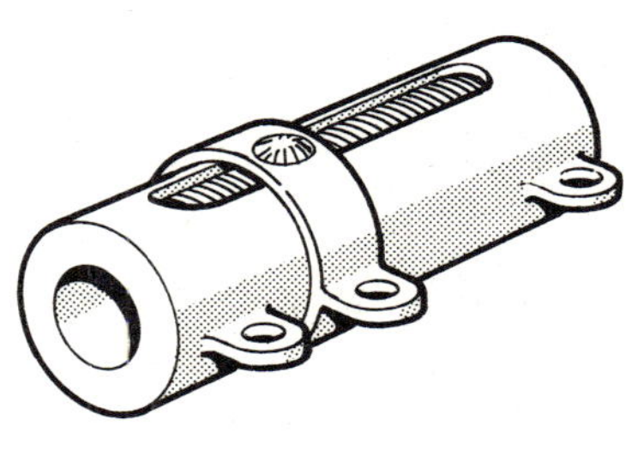

(c) Pictorial

84. Tapped variable resistor

A tapped resistance is a fixed resistor having a number of immovable taps, preset in position during manufacture.

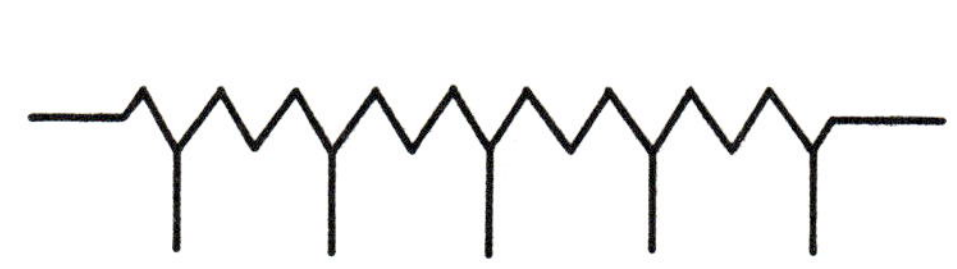

(a) CGSB

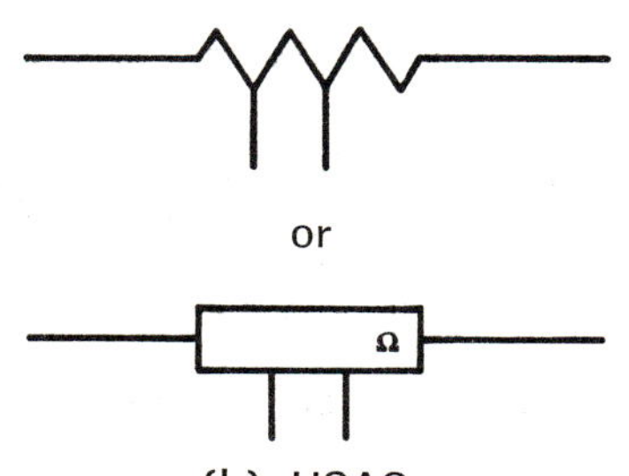

(b) USAS

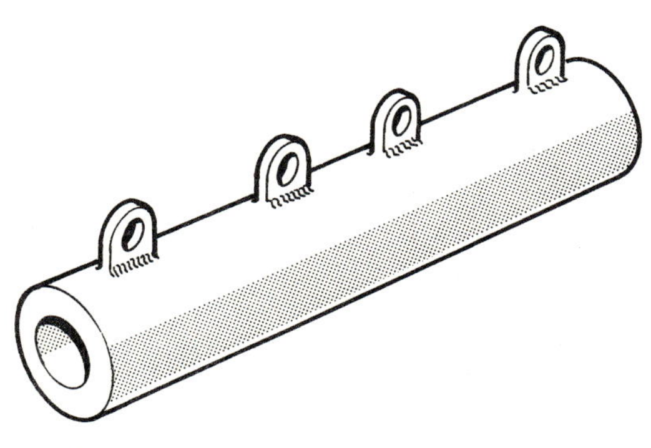

(c) Pictorial

SEMICONDUCTOR DEVICES

85. Diode rectifier, old symbol

This is a semiconductor device that is constructed from a number of selenium cells that are stacked together to perform the function of a power rectifier.

(a) CGSB

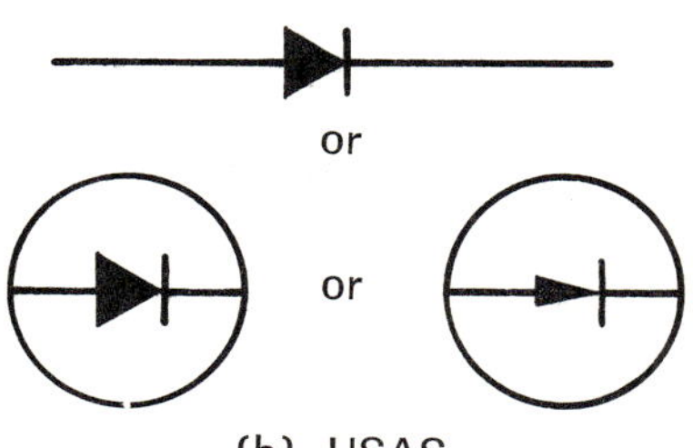

(b) USAS

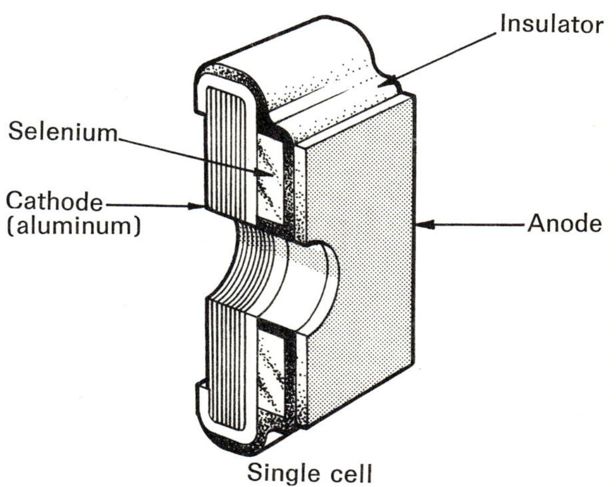

(c) Pictorial

86. Point diode, old symbol
This is a type of diode in which a 'catwhisker' of tungsten or gold comes in point contact with the germanium or silicon semiconductor wafer of the diode.

(a) CGSB

No equivalent symbol

(b) USAS

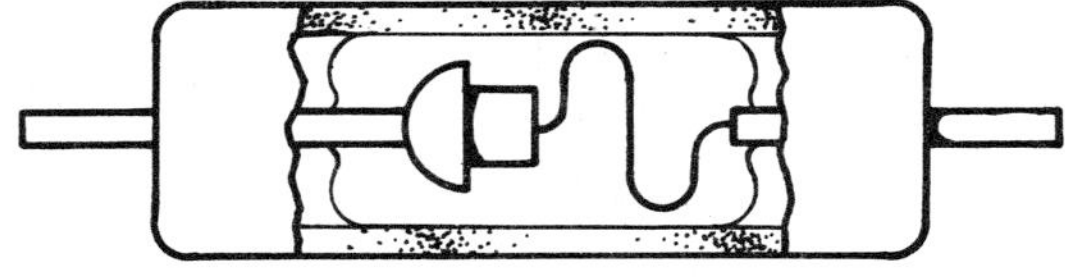

(c) Pictorial

87. Zener diode, old symbol
A zener diode is a semiconductor device used in waveshaping, voltage reference, and switching circuits. Its name is established from the term 'zener' or 'breakdown' voltage.

(a) CGSB

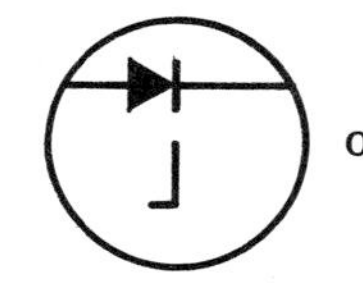

or

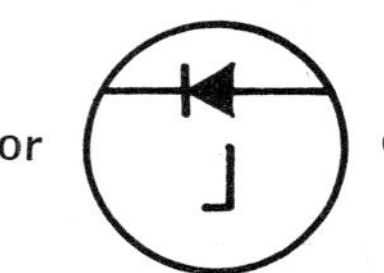

or

Unidirectional diode

(b) USAS

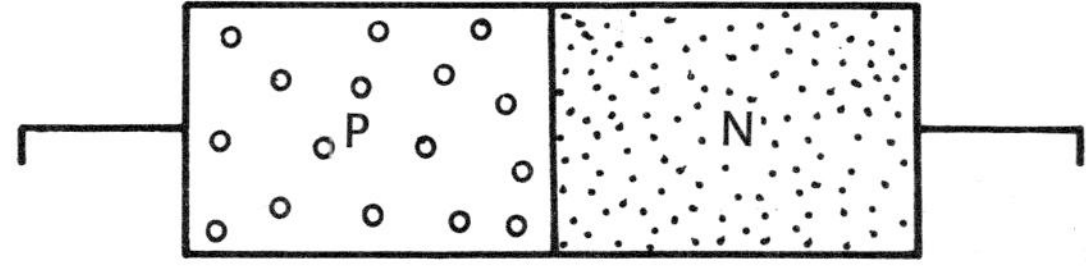

(c) Pictorial

88. Point photodiode, old symbol

This is a solid-state device employing a light-sensitive germanium wafer, and whisker contact. When light strikes the wafer, a very small unidirectional electrical current is produced.

(a) CGSB

No equivalent symbol

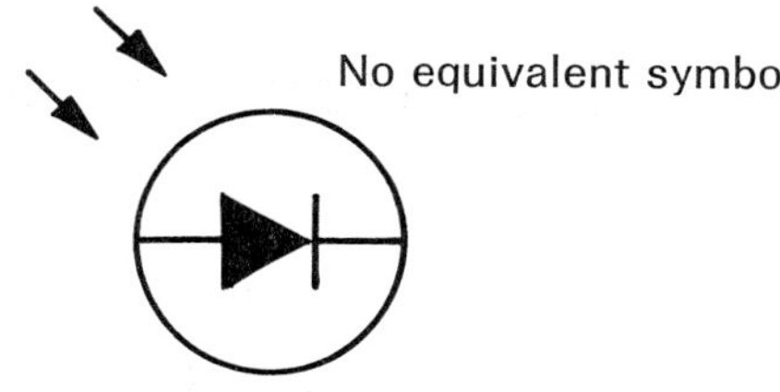

Photosensitive diode

(b) USAS

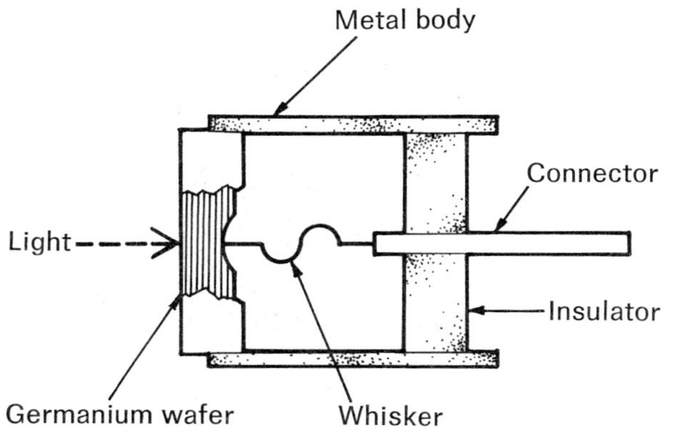

(c) Pictorial

89. Junction photodiode, old symbol

The device illustrated in No. 89(c) utilizes a special PN junction type diode. The action of light striking this photosensitive device produces a very small unidirectional electrical current.

(a) CGSB

No equivalent symbol

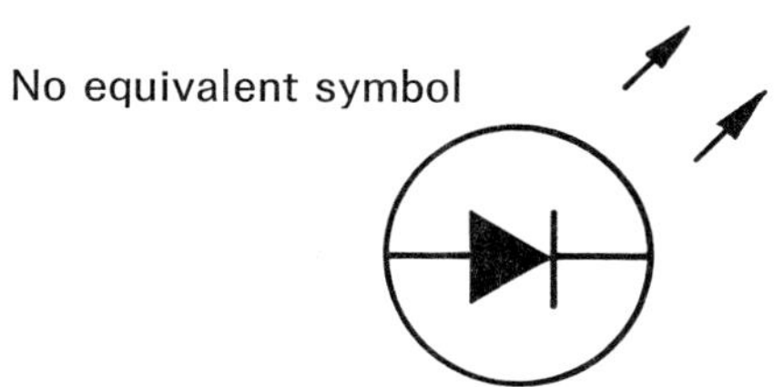

Photoemissive diode

(b) USAS

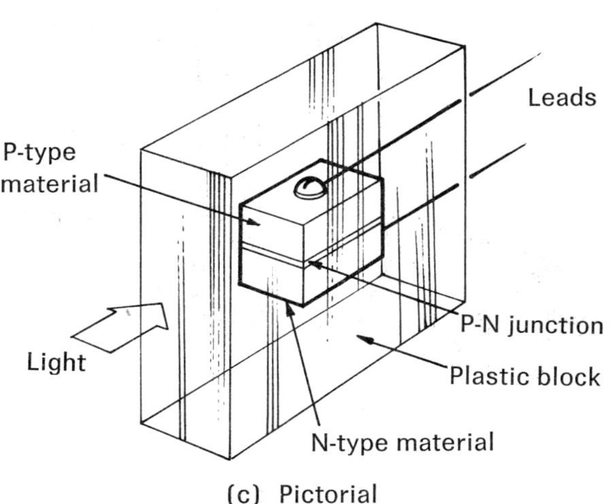

(c) Pictorial

90. SHIELDING

An electrostatic shield is a device consisting of a metallic cylindrical or cubic shape which surrounds a component to be shielded, and which has a connection to a chassis ground. In addition, an electrostatic shield about a conductor or cable takes the form of a braided metallic jacket which is also grounded.

(a) CGSB

(b) USAS

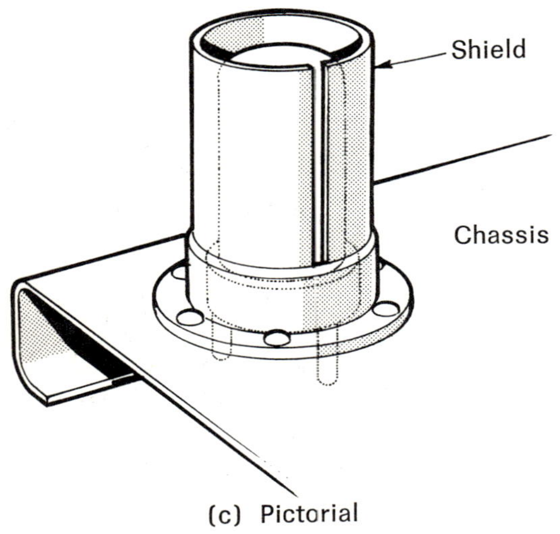

(c) Pictorial

91. SOLENOID

A solenoid is a device consisting of a number of turns of insulated wire, wrapped around a soft-iron bar, and which, when electrical current is passed through the wire, creates a magnetic field about the assembly.

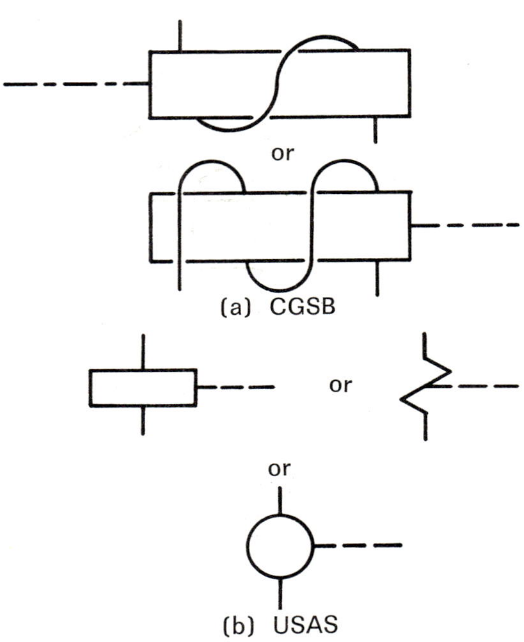

(a) CGSB

(b) USAS

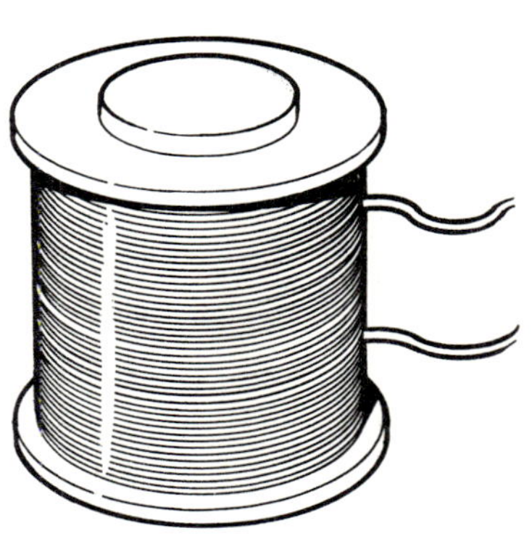

(c) Pictorial

SPEAKERS

92. Moving coil speaker

A moving coil or electrodynamic speaker is a device which converts a relatively weak audio input current into audible sound waves. To do this it utilizes a strong, independent magnetic field in which the voice coil is arranged to move about a cylindrical core of an electromagnet. Since the voice coil is attached to a lightweight cone, its action causes the cone to fluctuate and thus create audible sound waves.

93. Permanent-magnet coil speaker

This type of speaker has a powerful built-in permanent magnet, and hence requires no external source of electrical power to energize a field coil. In all other ways its construction is similar to an electrodynamic speaker.

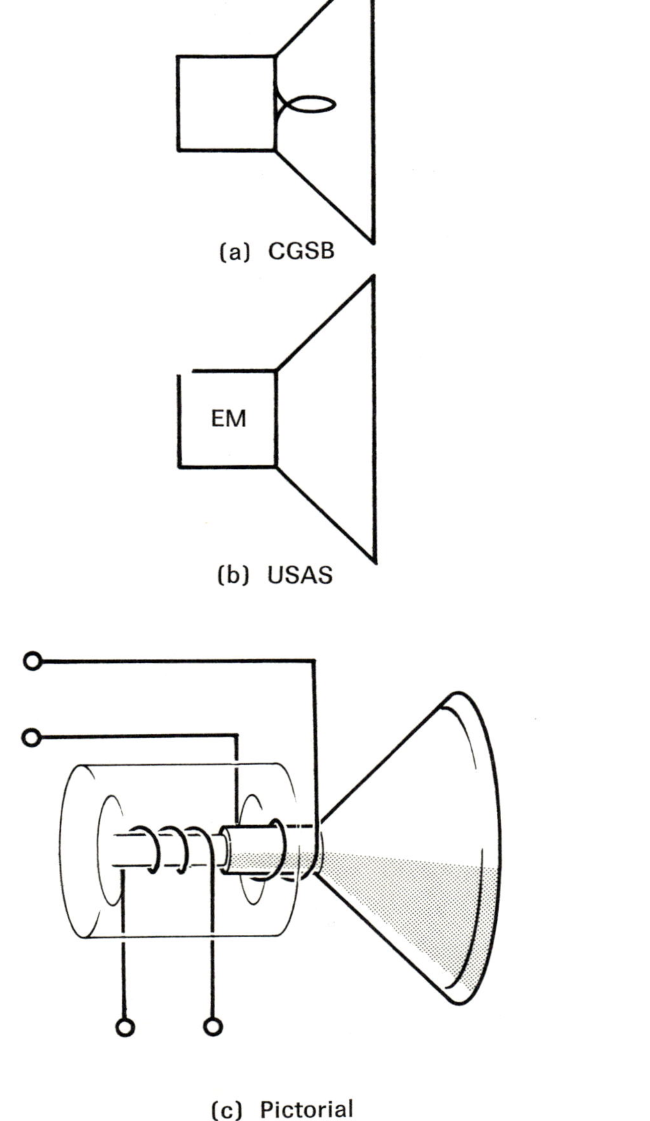

(a) CGSB

(b) USAS

(c) Pictorial

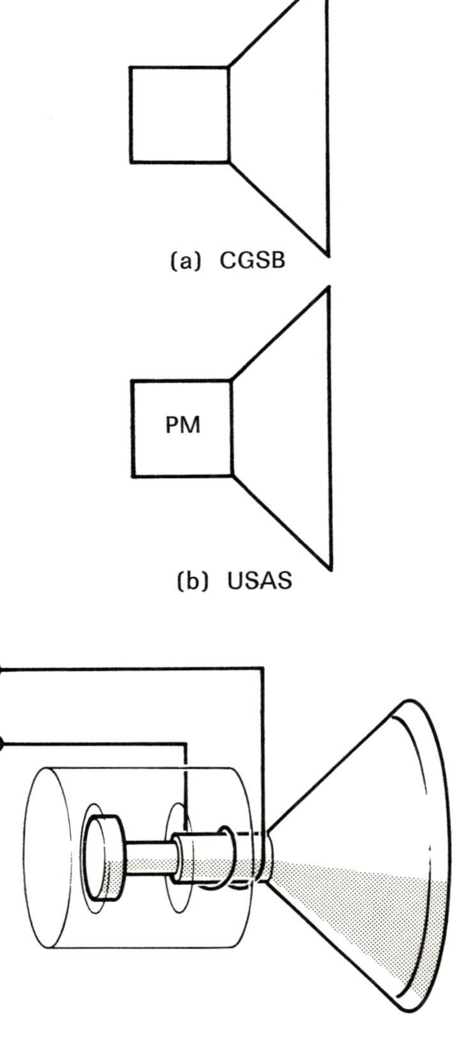

(a) CGSB

(b) USAS

(c) Pictorial

SWITCHES

94. S.P.S.T. knife or toggle, power type switch

This is an electromechanical device for opening or closing a circuit through a single conductor, and is manually operated.

95. D.P.D.T. knife or toggle, power type switch

A D.P.D.T. switch is an electromechanical device which can control two independent circuits depending upon what position the switch is placed in, and is manually operated.

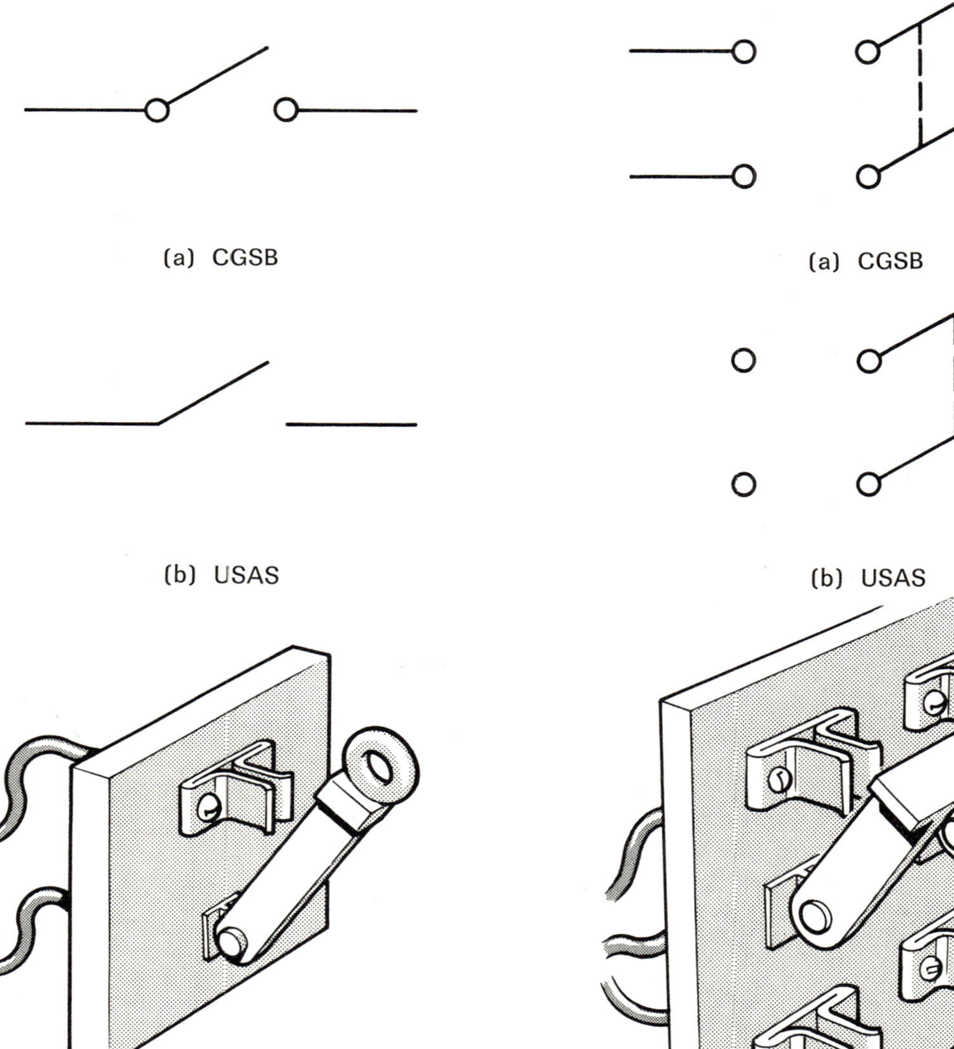

(a) CGSB

(b) USAS

(c) Pictorial

(a) CGSB

(b) USAS

(c) Pictorial

96. Normally-closed power push-button switch
A normally-closed power push-button switch is also a manually operated type. However, instead of moving a toggle or switch handle to control the switch mechanism, a push button is operated by thumb pressure. Also, this type of push-button switch is a normally-closed one.

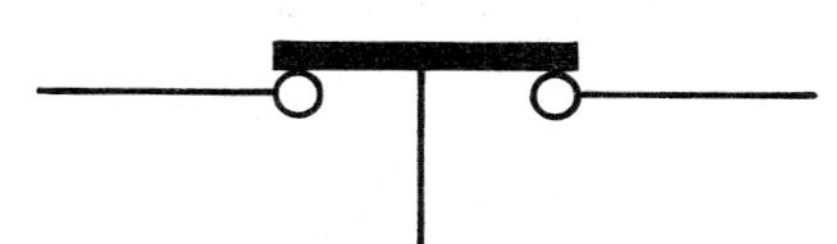

(a) CGSB

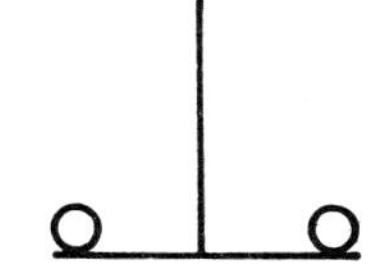

P.B. switch
Circuit opening (break)

(b) USAS

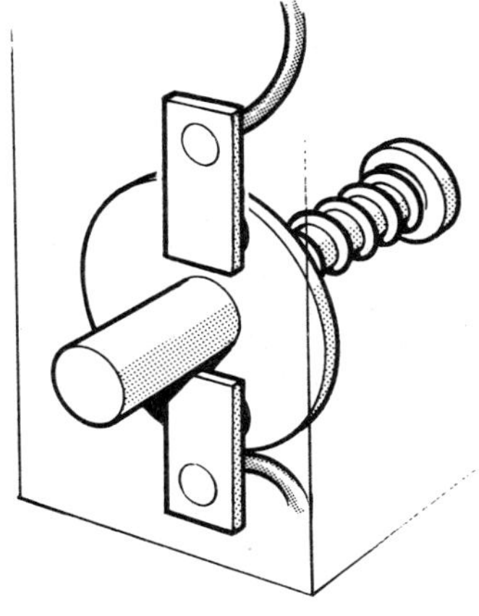

(c) Pictorial

97. Normally-open power push-button switch
This is a type of push-button switch similar to that described for No. 96 (a), but which is normally open until the switch is triggered.

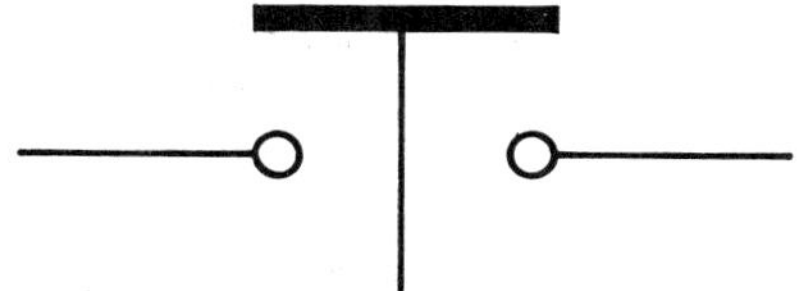

(a) CGSB

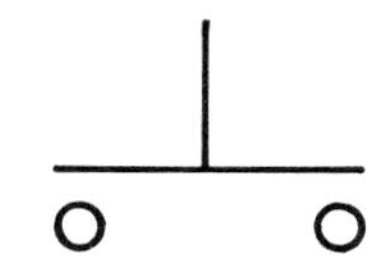

P.B. switch
Circuit closing (make)

(b) USAS

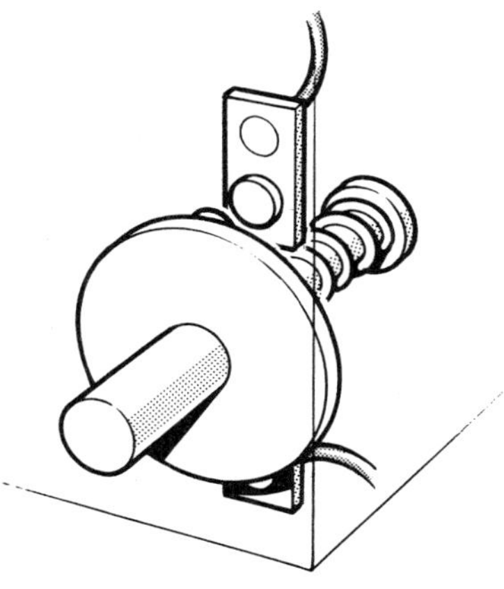

(c) Pictorial

98. Rotary selector switch
A rotary selector switch is a multiple contact switch having one, or more, dynamic contacts, operated in a radial fashion, by rotating a shaft to which it is attached, and which sweeps in a stepped fashion over a number of static contacts that are mounted in a circular pattern.

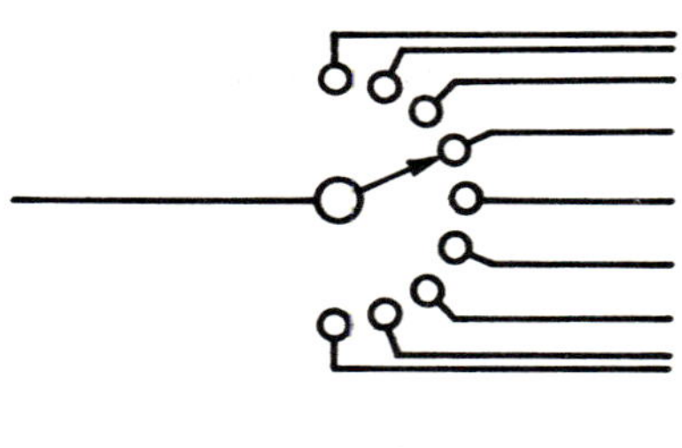

(a) CGSB

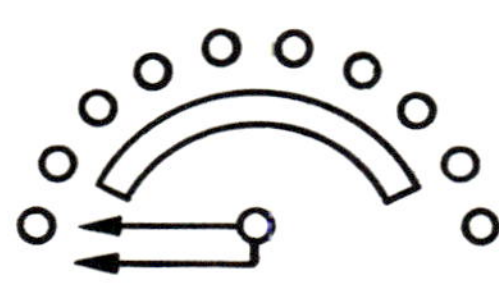

10 point selector switch with fixed segment
(b) USAS

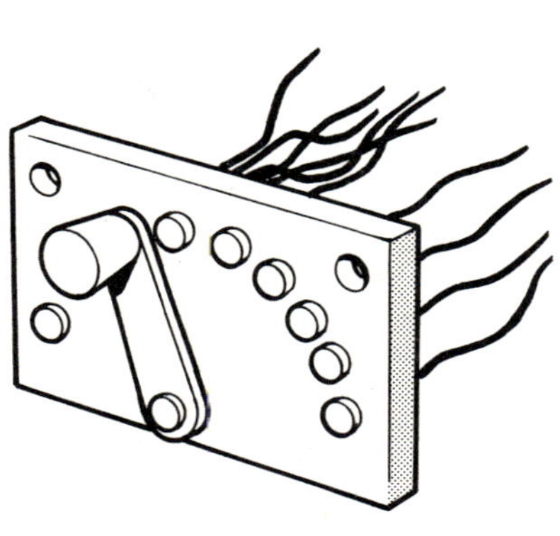

(c) Pictorial

99. Rotary wafer switch
A switch of this type has a number of fixed contacts fastened to a thin semicircular shaped phenolic laminate. The moving contact is secured to a centrally located revolving shaft.

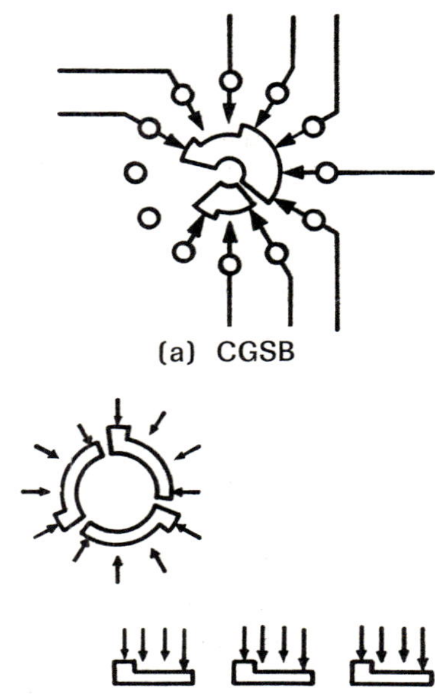

(a) CGSB

(b) USAS

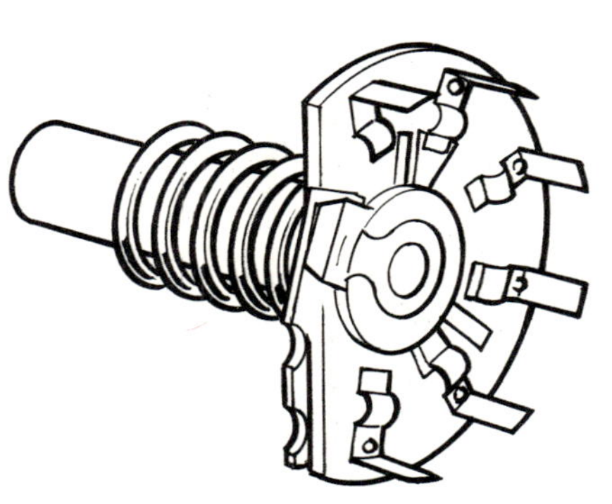

(c) Pictorial

SYNCHROS

100. Receiver, transmitter

These are motor like devices electrically connected to each other, whereby, when the shaft of the transmitter synchro is rotated, the shaft of the receiver synchro turns a corresponding amount in slave fashion.

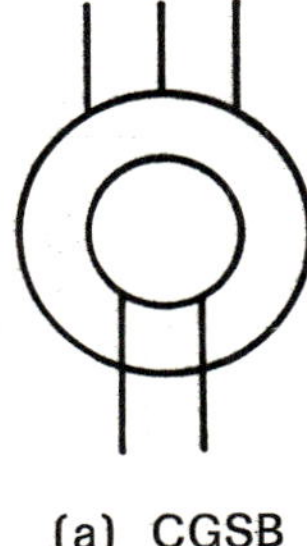

(a) CGSB

The same as the Canadian

(b) USAS

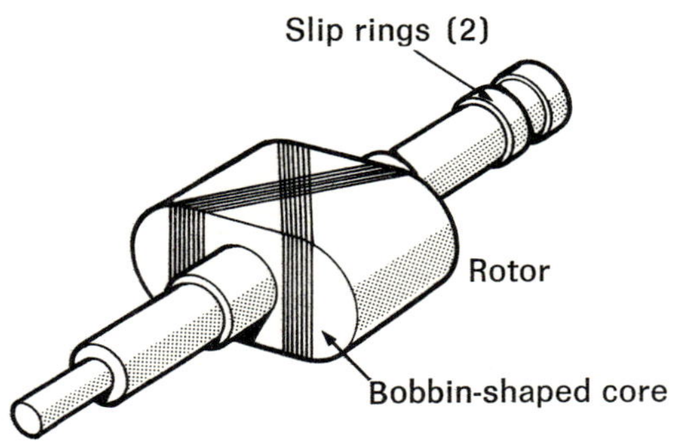

(c) Pictorial

101. Differential receiver, differential transmitter

Identification of this device is possible by internal inspection of its rotor. Whereas a standard synchro has a bobbin-shaped rotor, the differential has a cylindrically shaped one. In addition it has three slip rings, whereas the standard synchro has two.

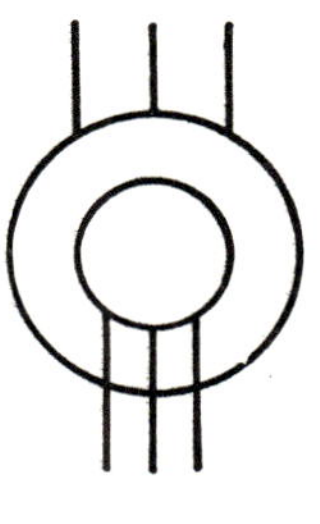

(a) CGSB

The same as the Canadian

(b) USAS

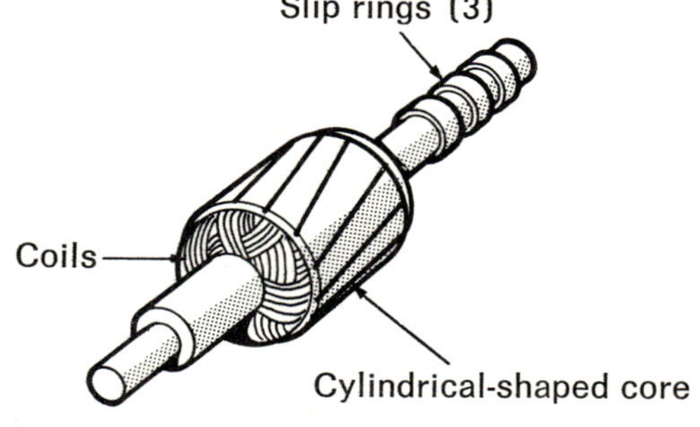

(c) Pictorial

TERMINALS

102. Binding post

When a number of conductors are terminated at an individual connection utilizing a screw clamp, the terminal is referred to as a binding post.

103. Terminal strip

When a number of conductors are to be connected to a variety of terminals, they are frequently done so to a device known as a terminal strip. The strip consists of a number of terminals mounted independently of each other on a length of material constructed from a nonconducting substance.

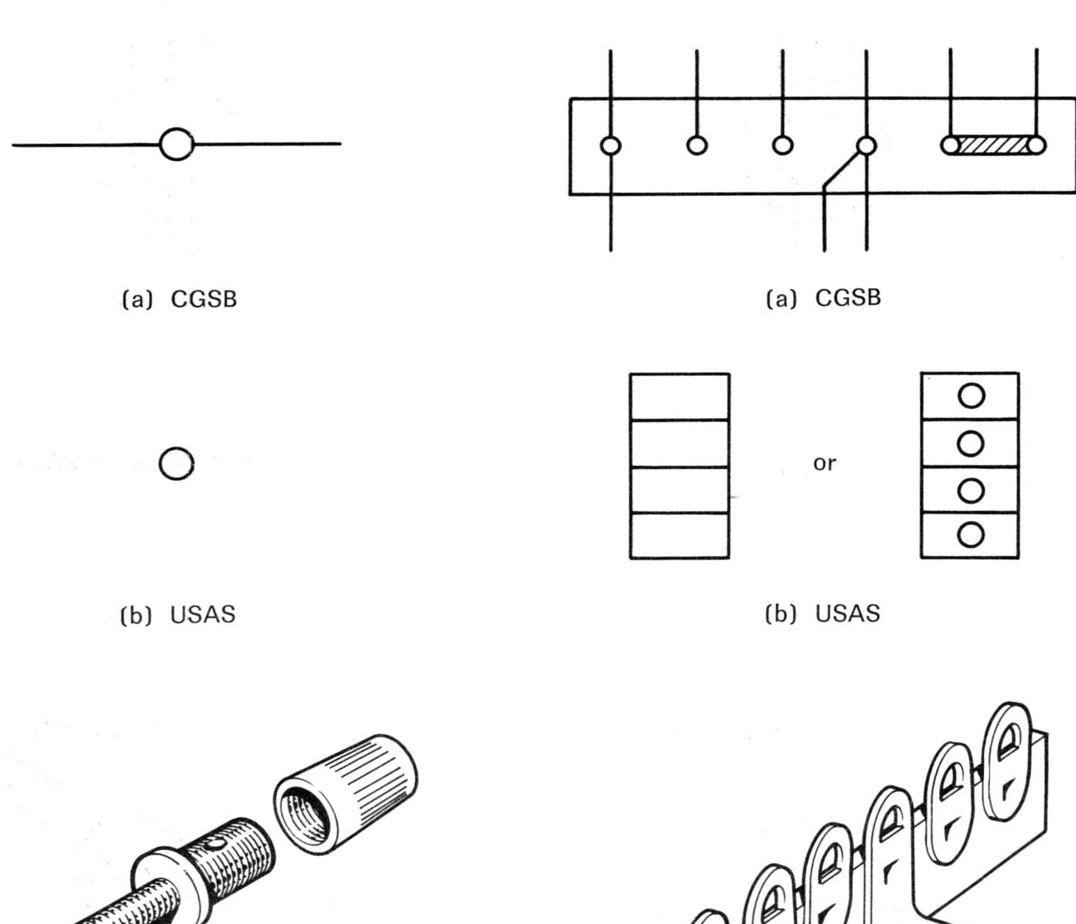

(a) CGSB

(b) USAS

(c) Pictorial

(a) CGSB

(b) USAS

(c) Pictorial

104. Terminal identification

If some form of identification is required at a terminal, such as a voltage value, the terminal symbol can be made larger in diameter to accommodate this information.

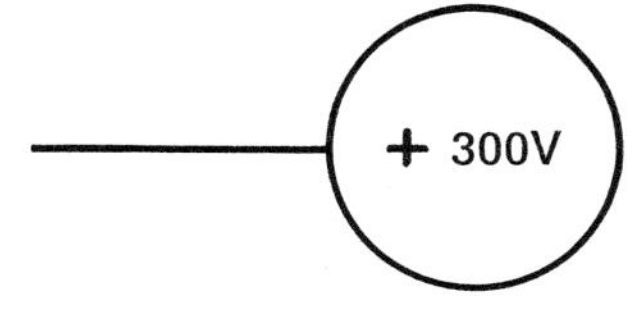

(a) CGSB

No equivalent symbol

(b) USAS

Not applicable

(c) Pictorial

THERMISTORS

105. Thermistor, no independent heater

This type of thermistor requires an external heat source to activate the manganese and nickel oxides within it. When this happens the thermistor, which is a type of bolometer, exhibits a tremendous decrease in resistance.

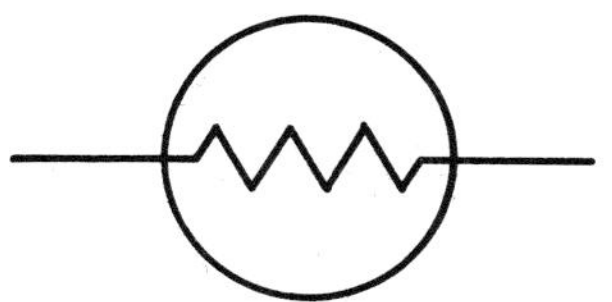

(a) CGSB

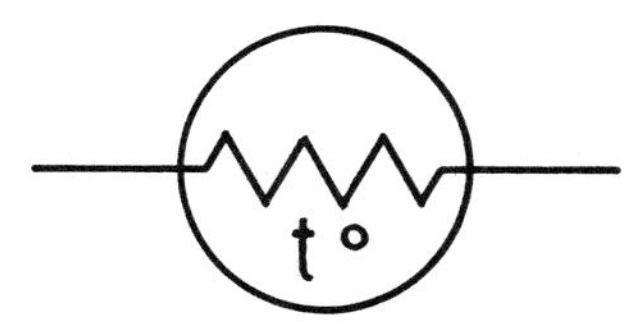

(b) USAS

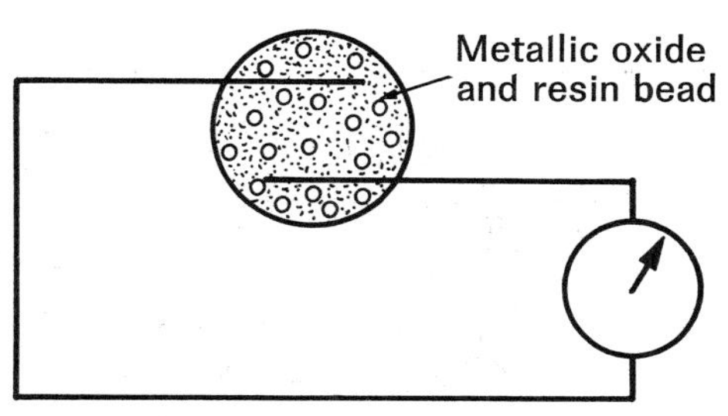

(c) Pictorial

106. Thermistor, independent integral heater
An integrally heated thermistor is similar to the one described for No. 105(a) but it has its own built-in heat source.

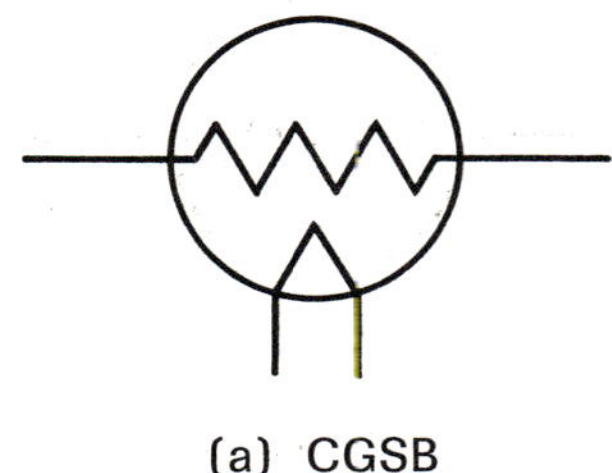

(a) CGSB

The same as the Canadian

(b) USAS

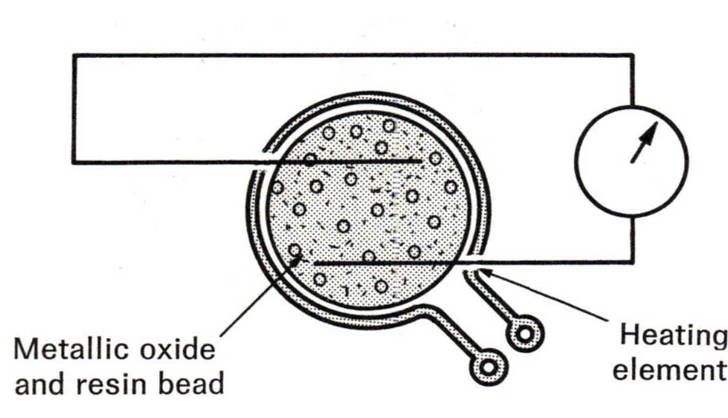

(c) Pictorial

THERMOCOUPLES

107. Temperature measuring thermocouple
This is a sensing device utilizing the principle of dissimilar metals. When an external heat source is applied to the free ends of this type of thermocouple, a voltage is developed proportional to the applied heat. In turn this voltage is correlated into the appropriate temperature units by suitable electrical circuitry.

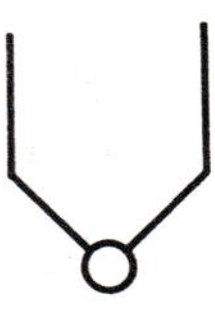

(a) CGSB

(b) USAS

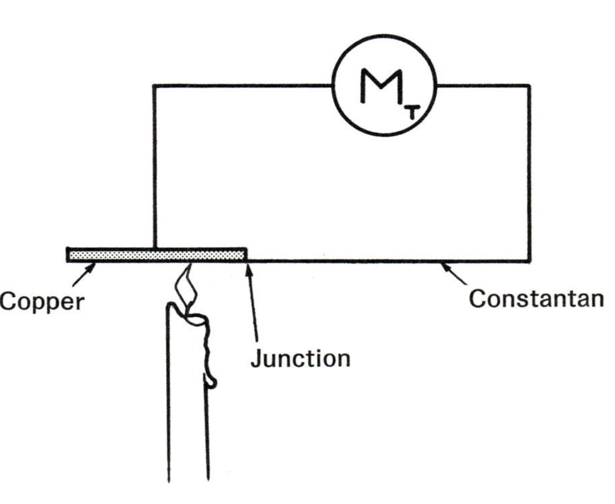

(c) Pictorial

108. Current measuring, integral heater, internally connected thermocouple

When it is necessary to measure the size of an electrical current remotely, a special thermocouple can be used. This type is referred to as a current measuring thermocouple, and it has its own internal heat source.

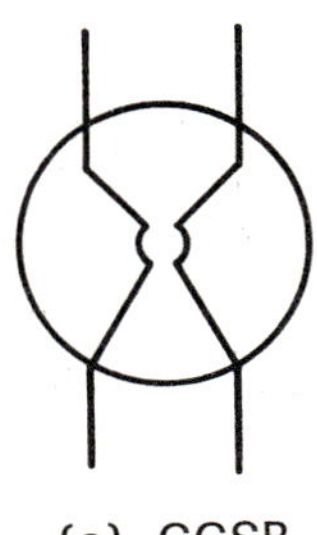

(a) CGSB

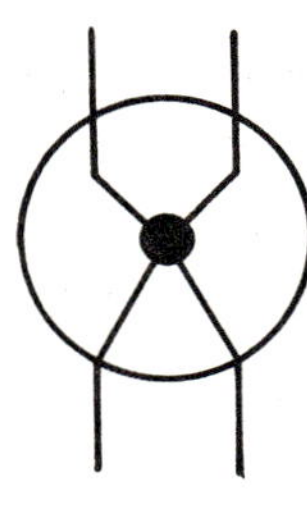

(b) USAS

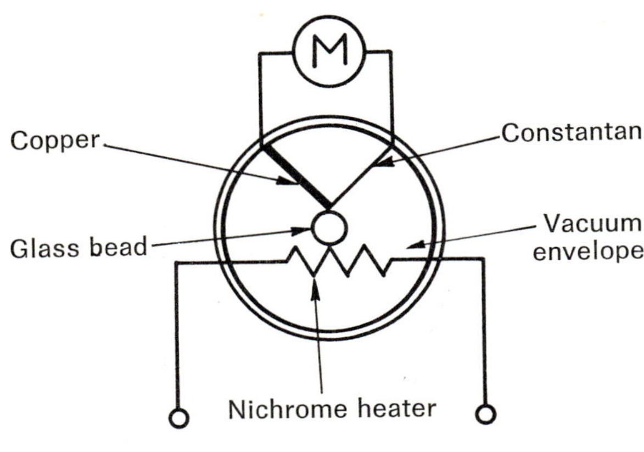

(c) Pictorial

TRANSFORMERS

109. Air-core fixed coupling transformer

A fixed coupling air-core transformer is a device consisting of two insulated coils wound on a form having a cylindrical space of air, known as an air core. The coils, known as primary and secondary windings, have their positions relative to each other permanently fixed.

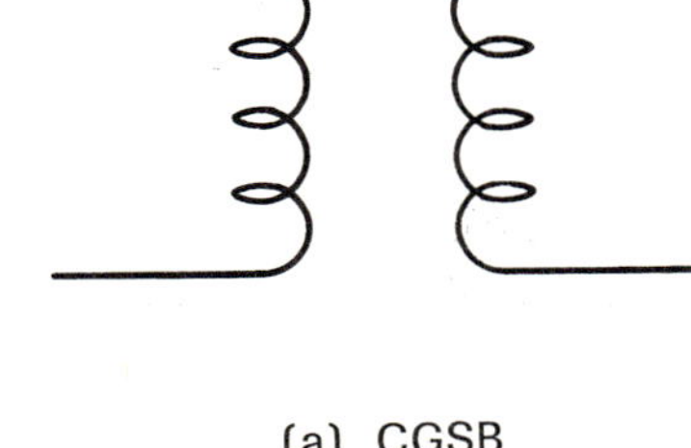

(a) CGSB

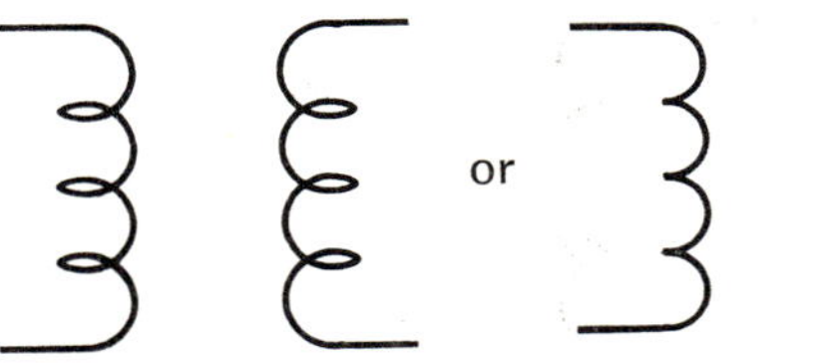

(b) USAS

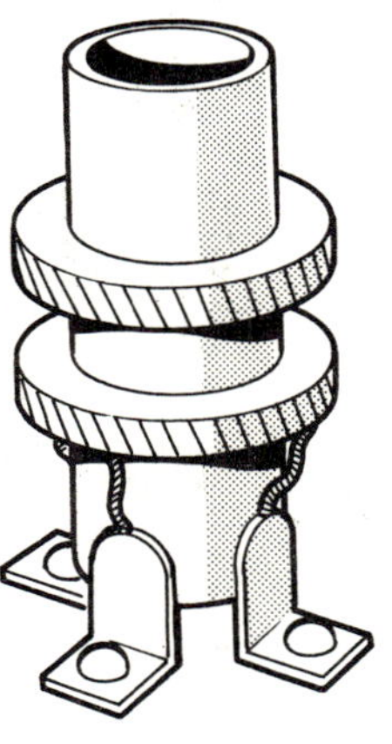

(c) Pictorial

110. Air-core variable coupling transformer
This is a type of transformer having an air core, and in addition having a coil arrangement whereby the distance or coupling between the primary and secondary windings can be varied at will.

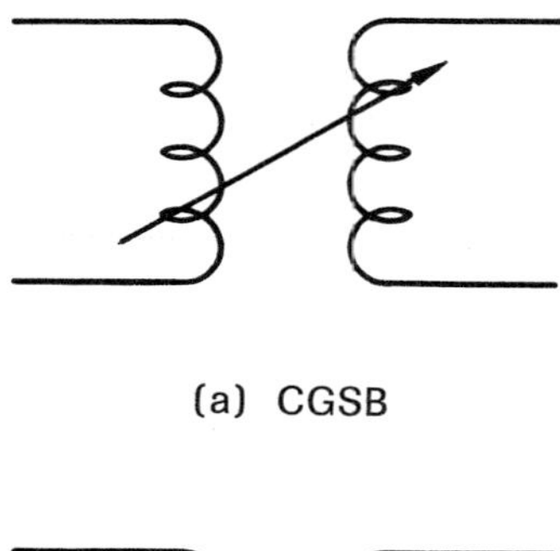

(a) CGSB

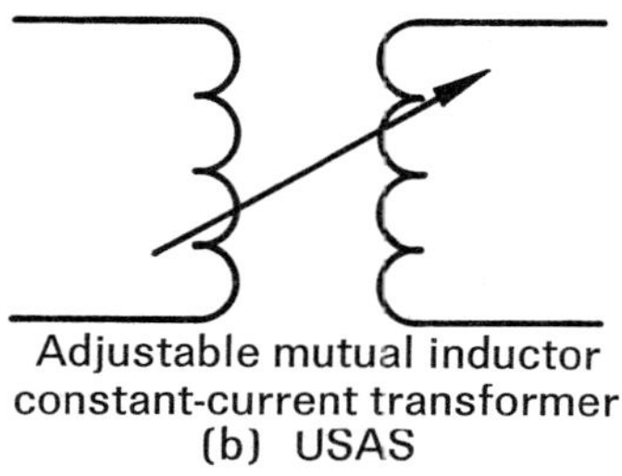

Adjustable mutual inductor
constant-current transformer
(b) USAS

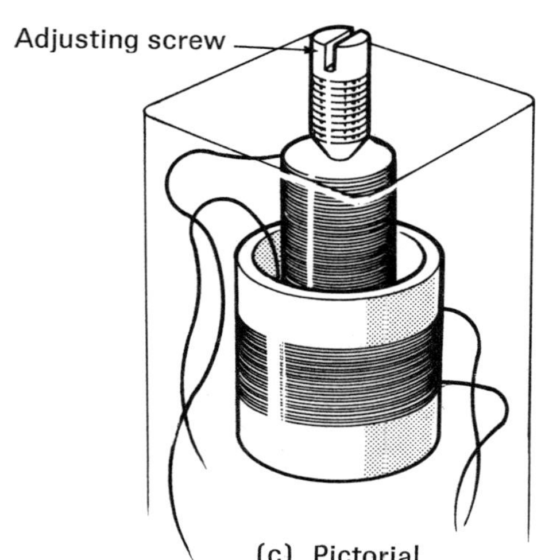

(c) Pictorial

111. Magnetic core, single-secondary, fixed coupling transformer
Another variation of the transformer is the type having a soft-iron core instead of air, a fixed coupling between the primary and secondary windings, and no centre tap. This type of transformer is illustrated below.

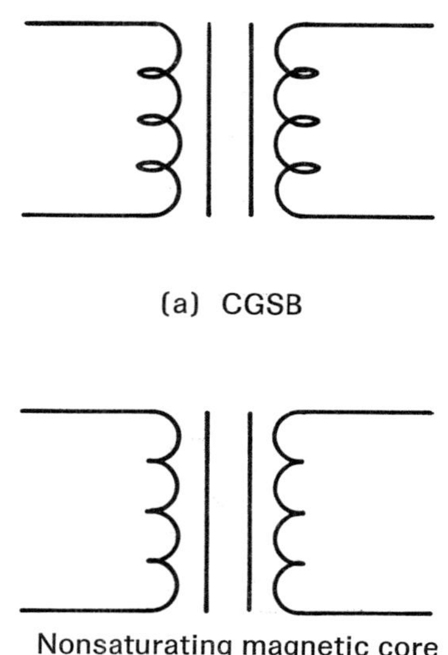

(a) CGSB

Nonsaturating magnetic core
(b) USAS

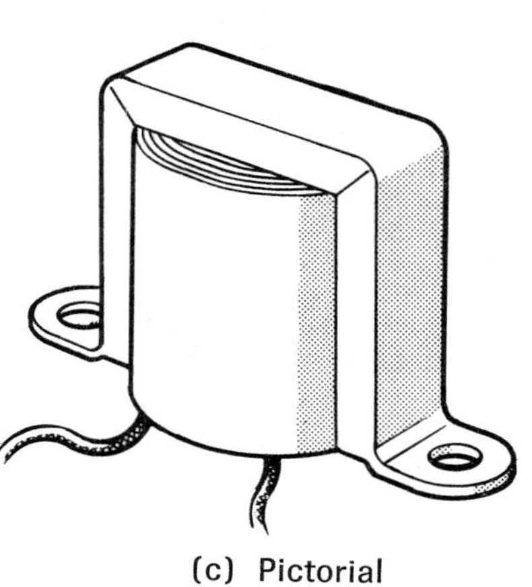

(c) Pictorial

112. Magnetic core, multiple-secondary with taps transformer

A further variation of the transformer is the type having a number of centre-tapped secondary windings, in addition to a soft-iron core through its coil form.

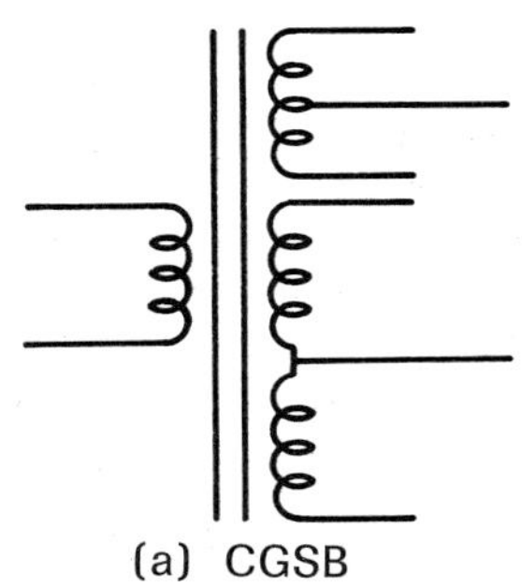

(a) CGSB

No equivalent symbol

(b) USAS

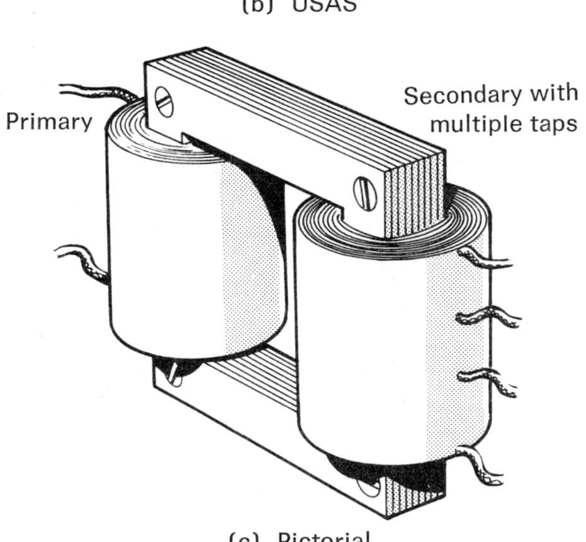

(c) Pictorial

113. Autotransformer

An autotransformer is a special type of transformer having only one winding and a sliding contact capable of being moved along this winding. As such it functions as a variable type of step-down transformer.

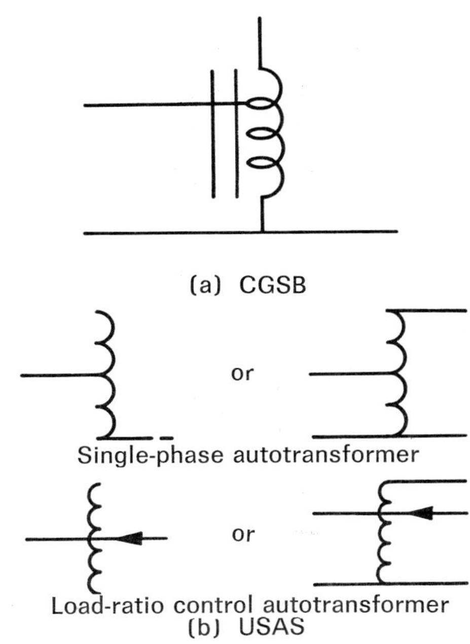

(a) CGSB

Single-phase autotransformer

Load-ratio control autotransformer

(b) USAS

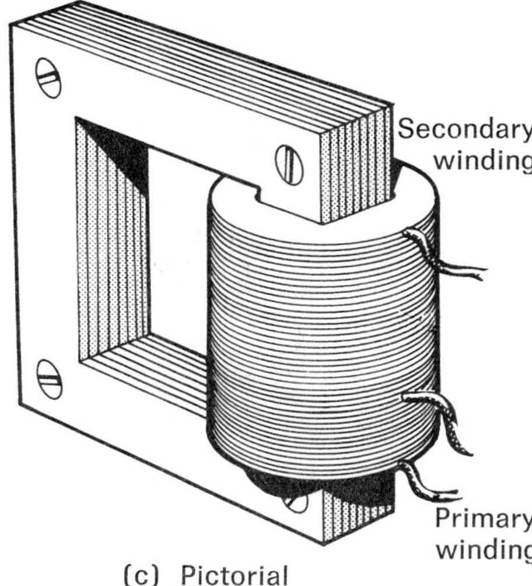

(c) Pictorial

114. Powdered iron or ferro-ceramic core transformer

Instead of a solid iron core or slug, powdered iron, or powdered iron formed with ceramic materials, is sometimes used in a transformer. The symbol for this type of transformer is shown below.

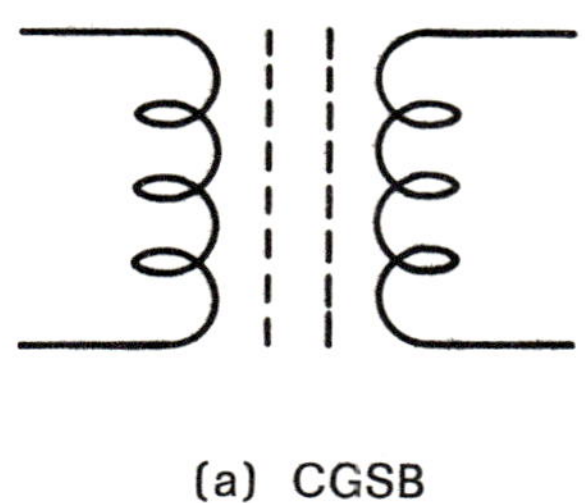

(a) CGSB

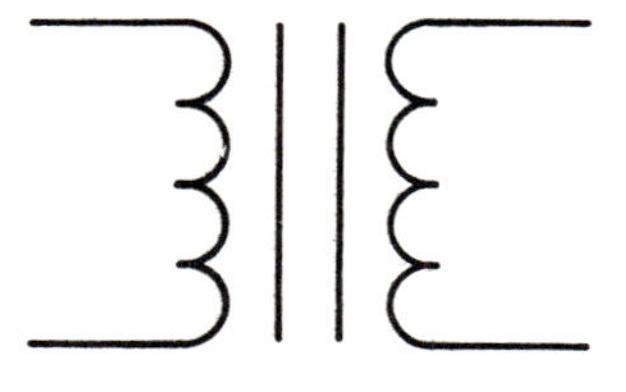

Nonsaturating magnetic core
(b) USAS

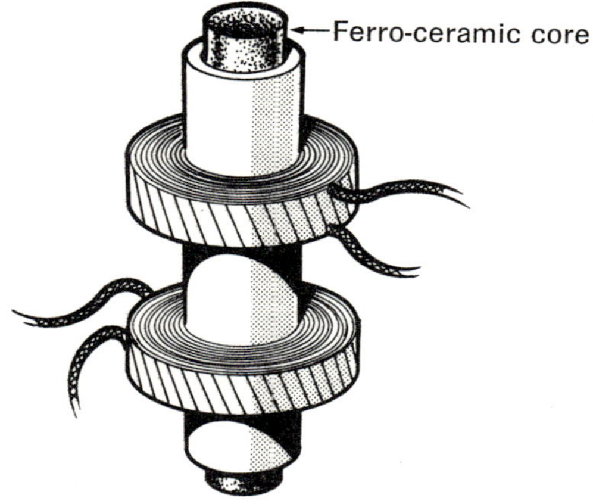

(c) Pictorial

115. Magnetic core, variable coupling transformer

This type of transformer has a soft-iron core and secondary windings constructed so that their coupling can be adjusted at will.

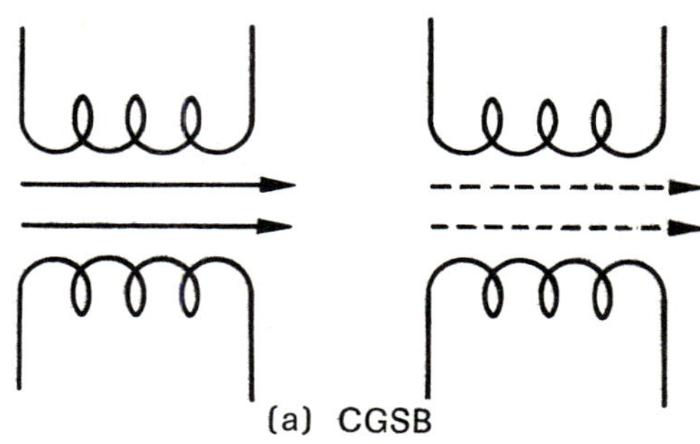

(a) CGSB

No equivalent symbol

(b) USAS

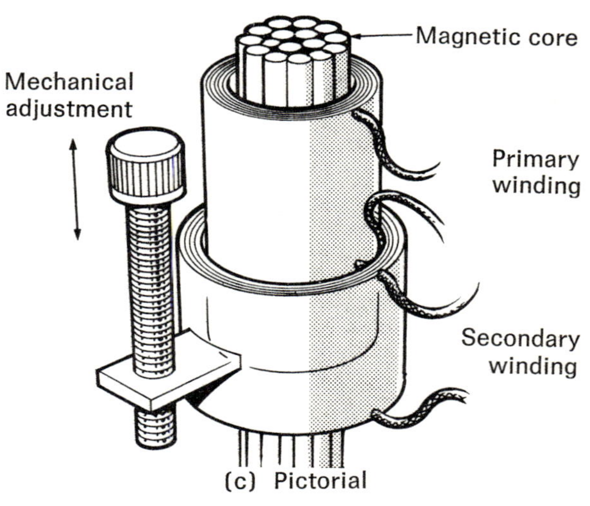

(c) Pictorial

TUBES, ELECTRONIC

EMITTERS

116. Directly-heated cathode or heater tube element

A heater or a directly-heated cathode element is one which is heated to incandescence by passing a suitable electrical current through the element.

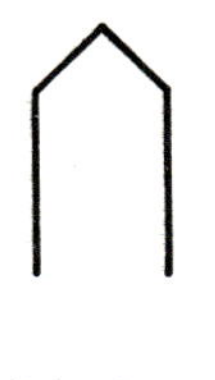

(a) CGSB

The same as the Canadian

(b) USAS

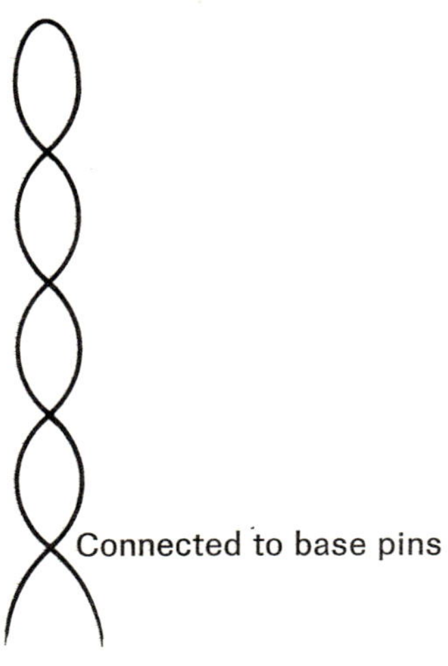

(c) Pictorial

117. Indirectly-heated cathode tube element

An indirectly-heated cathode is an element that is heated to incandescence by virtue of its close proximity to another directly-heated element, known as a filament.

(a) CGSB

(b) USAS

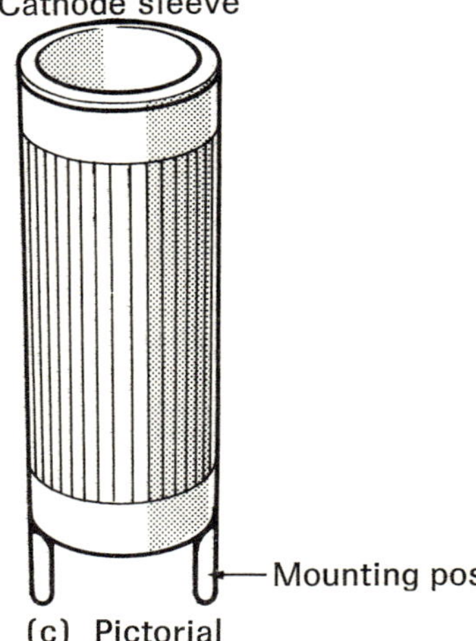

(c) Pictorial

118. Cold-cathode tube element
Glow tubes, such as are used in neon signs, utilize a single electrode known as a cold cathode. No. 118(a) shows the symbol for this type of cathode.

(a) CGSB

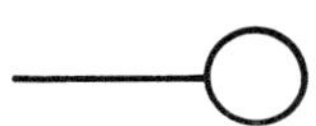

(b) USAS

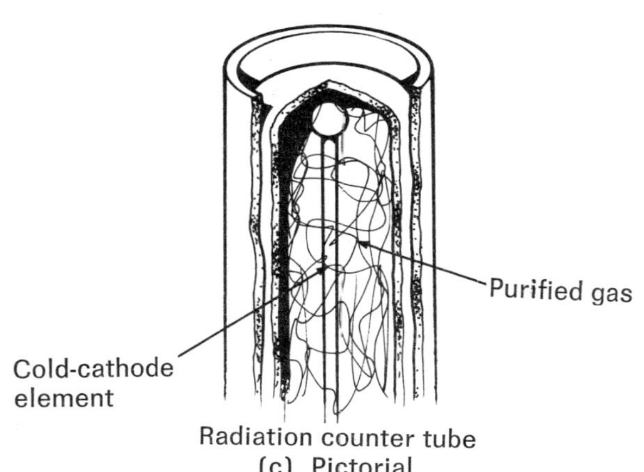

Radiation counter tube
(c) Pictorial

119. Photoelectric cathode tube element
A photoelectric cathode is the initial element in a photomultiplier tube upon which the light source impinges and is shown by the symbol in No. 119(a).

(a) CGSB

(b) USAS

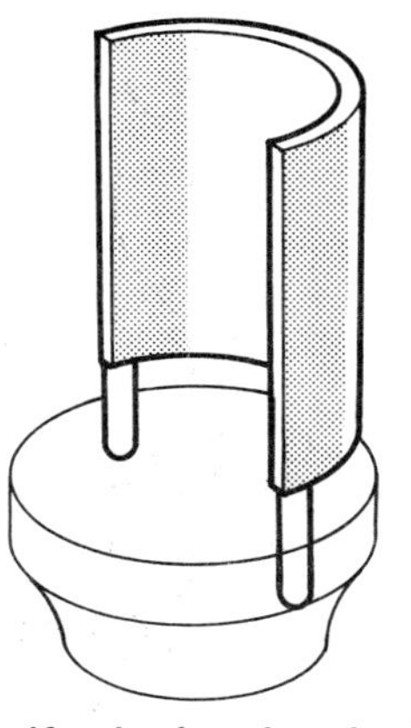

Half-cylindrical cathode
(c) Pictorial

120. Pool cathode tube element
A pool cathode utilizes a pool of mercury as an emitter, and it is used in ignitron tubes.

(a) CGSB

The same as the Canadian

(b) USAS

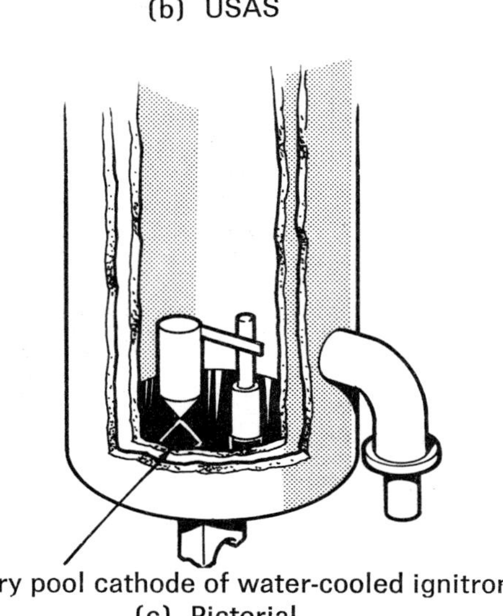

Mercury pool cathode of water-cooled ignitron
(c) Pictorial

121. Ionic-heated cathode tube element
In an ionic-heated cathode rectifier tube, the cathode is a coated element which becomes incandescent due to ionic bombardment from one of the anodes having a positive potential within the tube, rather than by heater or filament current from an external source. The type of tube containing this kind of cathode requires an inert gas at a reduced pressure.

(a) CGSB

The same as the Canadian

(b) USAS

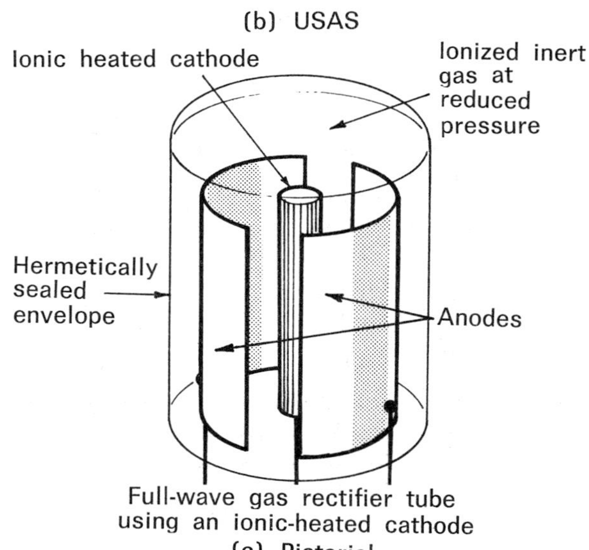

Full-wave gas rectifier tube using an ionic-heated cathode
(c) Pictorial

CONTROLLERS

122. Grid tube element

This is a tube element which is placed between the cathode and anode. By varying the electrical charge on the grid, control is maintained over the electron flow between the cathode and anode.

(a) CGSB

The same as the Canadian

(b) USAS

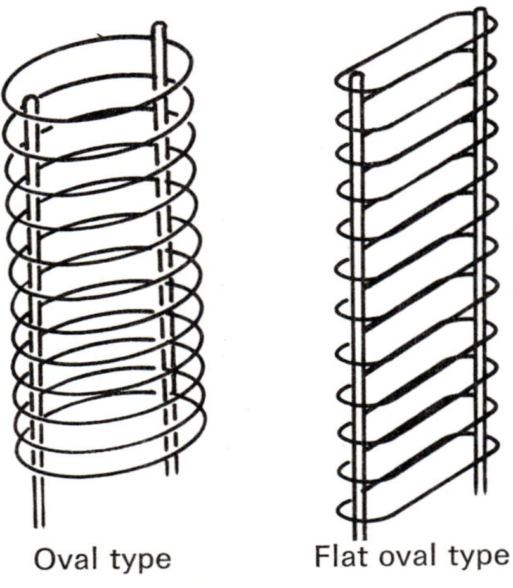

(c) Pictorial

123. Electrostatic deflector, reflector, or repeller tube element

These are flat, rectangular, metal elements within cathode-ray tubes, and they are positioned in sets of two, each set being at 90° to each other. Their purpose is to deflect the electron beam both horizontally and vertically as required.

(a) CGSB

The same as the Canadian

(b) USAS

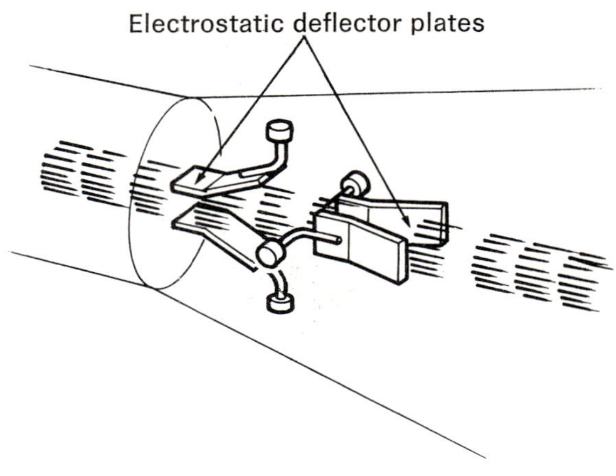

(c) Pictorial

124. Ignitor tube element
In arc tubes which use a mercury pool as a cathode, a special electrode known as an ignitor is used to start an arc to create the initial ionization of the liquid mercury.

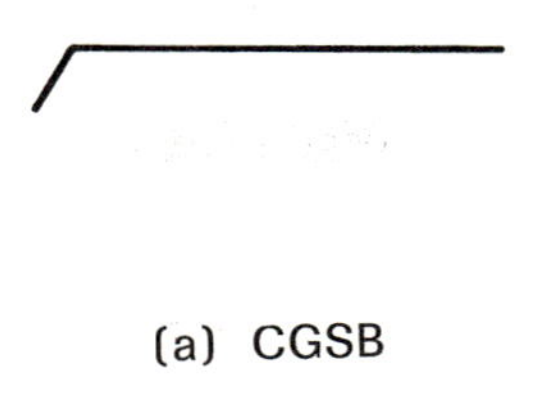

(a) CGSB

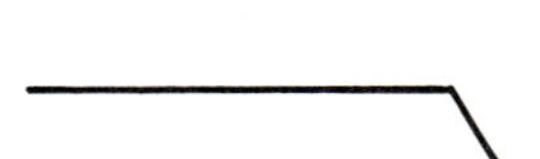

(b) USAS

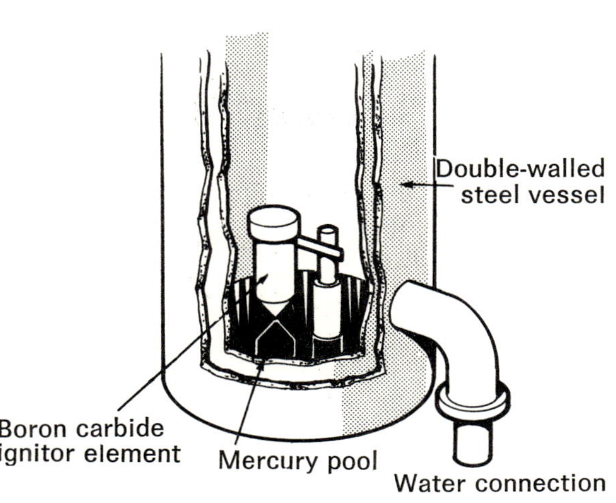

(c) Pictorial

125. Exciter tube element
This is a special element used in a pool cathode type of tube. It performs the function of an auxiliary anode in maintaining a continuous electrical discharge between itself and the cathode pool, even if the potential of the main anode becomes insufficient to maintain an arc.

(a) CGSB

The same as the Canadian

(b) USAS

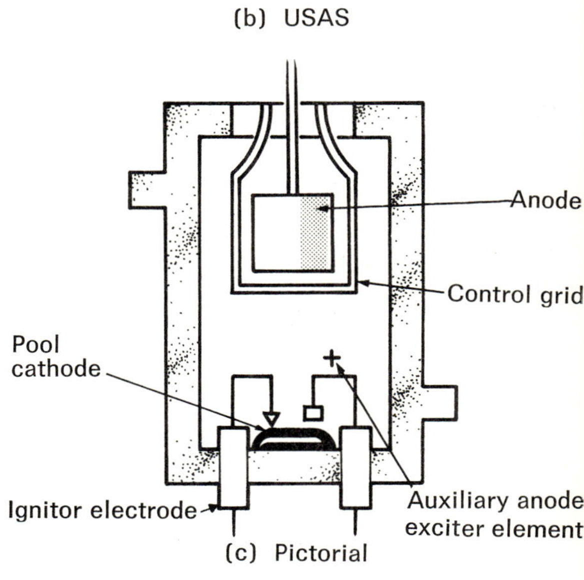

(c) Pictorial

126. Combined cold-cathode, trigger-electrode tube element

No. 126(c) shows the internal construction of a cold-cathode type trigger tube. The function of the trigger electrode is to initiate an electrical discharge between it and the cathode at a lower voltage than that required to do so between the cathode and the anode.

(a) CGSB

No equivalent symbol

(b) USAS

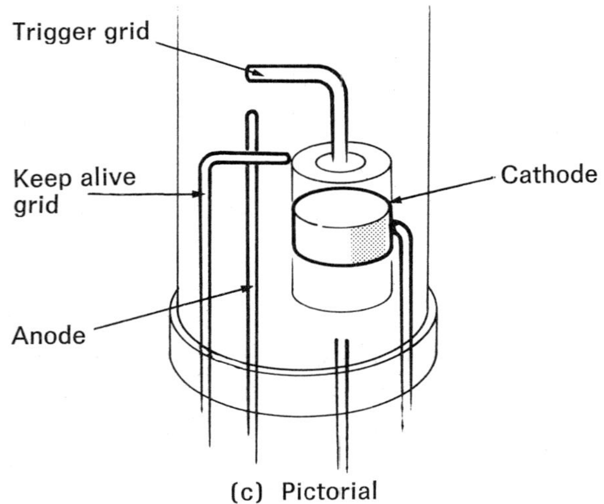

(c) Pictorial

COLLECTORS AND COLLECTORS/EMITTERS

127. Anode or plate tube element

It is that element within a vacuum tube which has a high positive potential with respect to the cathode. Electron flow is from the cathode to the anode or plate.

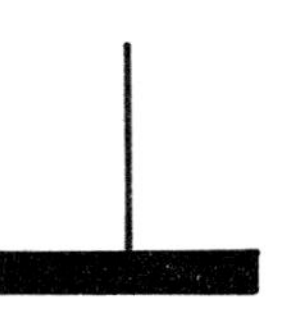

(a) CGSB

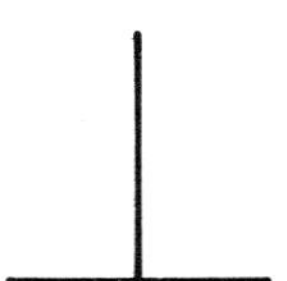

(b) USAS

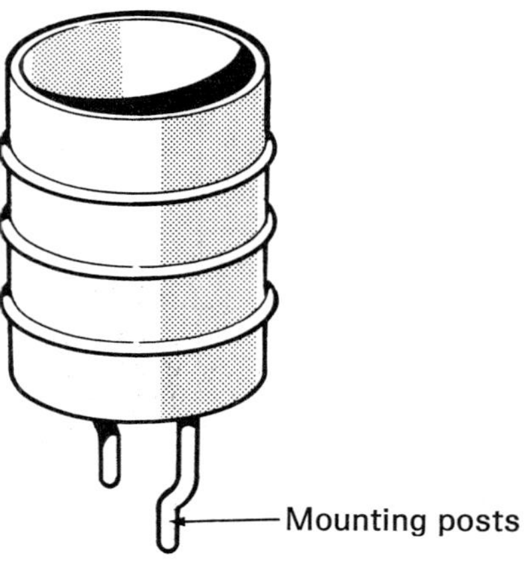

(c) Pictorial

128. Target, x-ray tube element

The target element in an x-ray tube acts as the final one in reflecting the x-rays being emitted from the focussing cup assembly. The symbol for this element is shown in No. 128.

129. Dynode with cold-cathode tube element

This device is a special element within a photomultiplier tube. Its purpose is to receive and to reflect electrons which are released from the photomultiplier cathode. The dynode has a low potential, and performs as an anode.

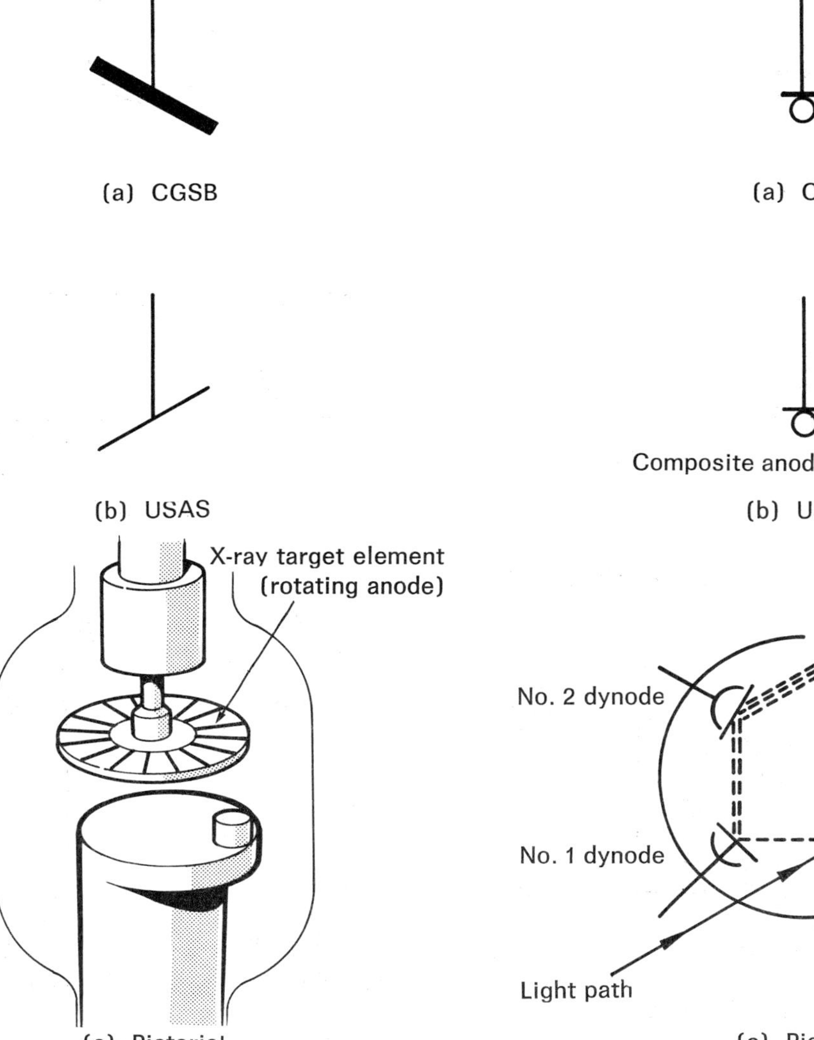

(a) CGSB

(b) USAS

(c) Pictorial

(a) CGSB

(b) USAS

(c) Pictorial

130. Dynode with photoelectric cathode-tube element

No. 130 (c) shows a pictorial drawing of an electron-ray tuning indicator tube in which the inverted cone-shaped target anode performs both as an anode and a photocathode. The attraction of electrons to this anode causes a further release of electrons which fluoresce.

ENVELOPES

131. Single-diode and single-triode tube envelope

A circle is used to illustrate the envelope or enclosure containing the elements for either a single-diode or single-triode tube.

(a) CGSB

Composite anode, photocathode

(b) USAS

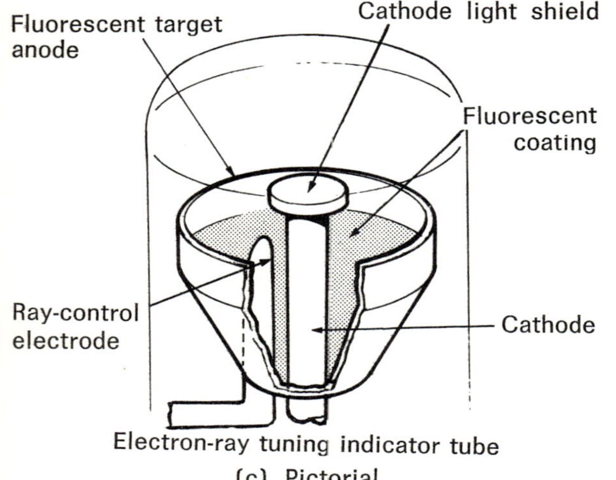

Electron-ray tuning indicator tube

(c) Pictorial

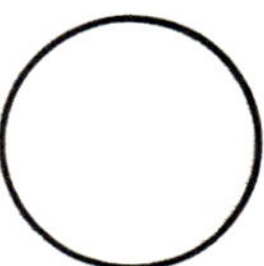

(a) CGSB

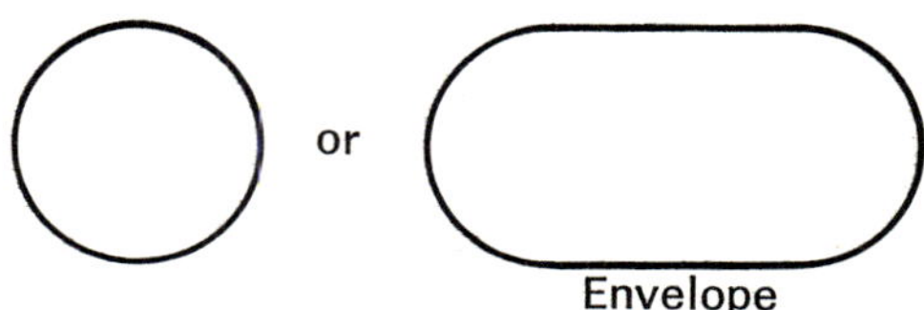

(b) USAS

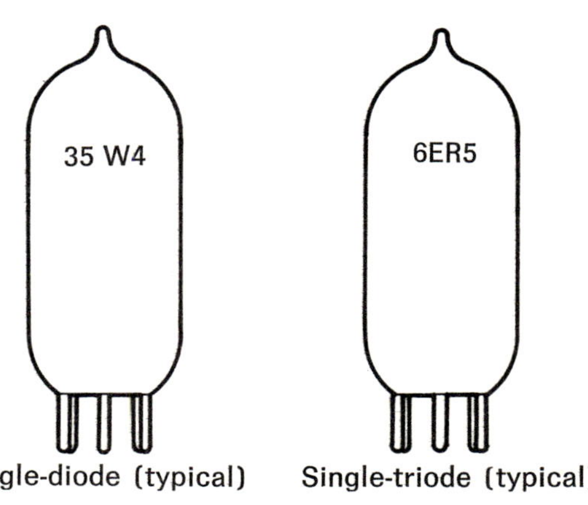

Single-diode (typical) Single-triode (typical)

(c) Pictorial

132. Double-diode and double-triode tube envelope When a tube is either a double-diode or a double-triode, the symbol form shown in No. 132 (a) is used.

133. Multi-grid and cathode-ray tube envelope Multi-grid tubes, that is, tubes other than single- or double-diodes or triodes, and cathode-ray tubes use the CGSB symbol form.

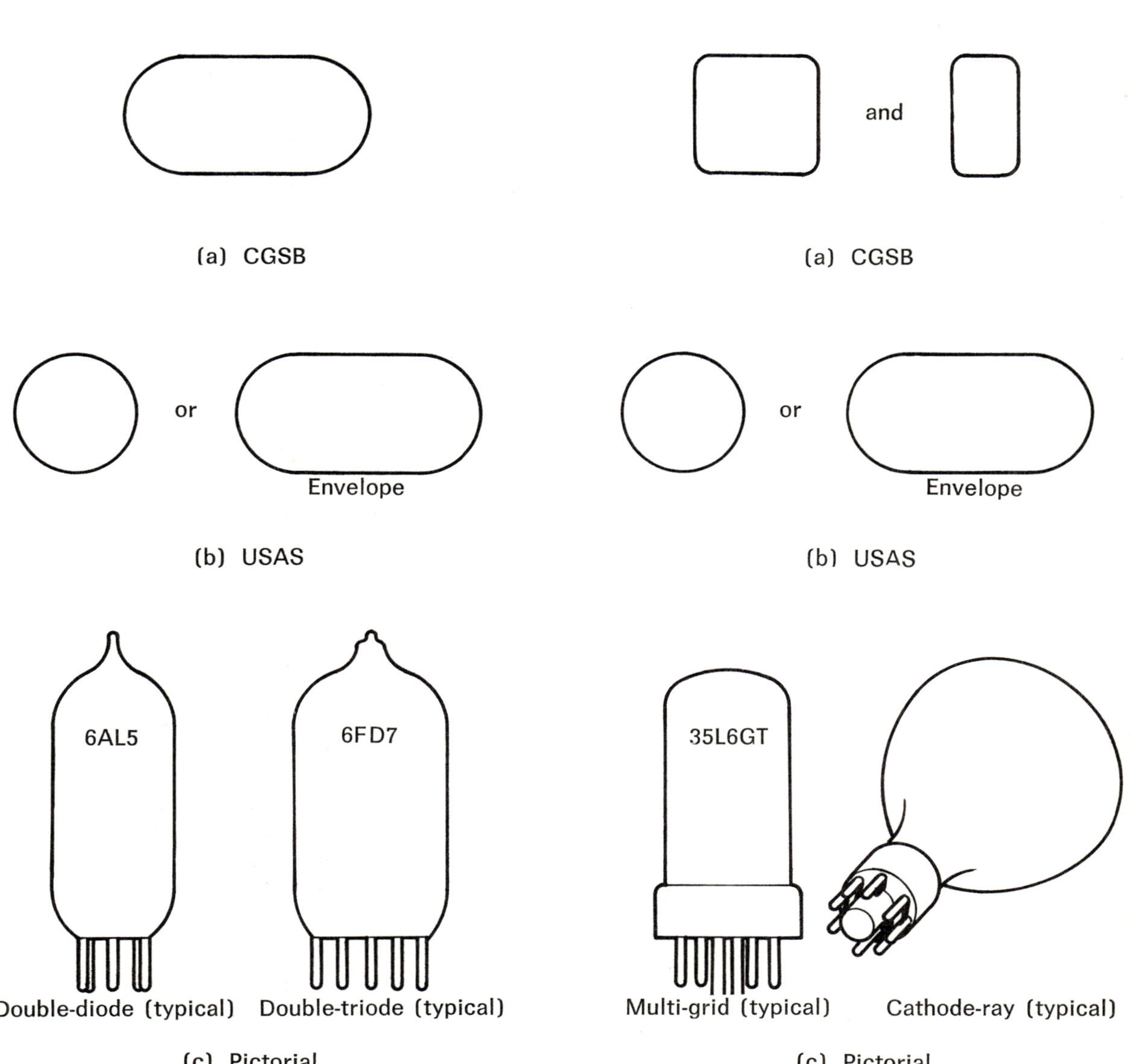

(a) CGSB

(b) USAS

Double-diode (typical) Double-triode (typical)

(c) Pictorial

(a) CGSB

(b) USAS

Multi-grid (typical) Cathode-ray (typical)

(c) Pictorial

134. Gas-filled tube indication
When a vacuum tube is filled with some type of gaseous material such as neon, the symbol shown in No. 134 (a) is placed within the basic tube envelope.

(a) CGSB

Gas-filled tube, indication

(b) USAS

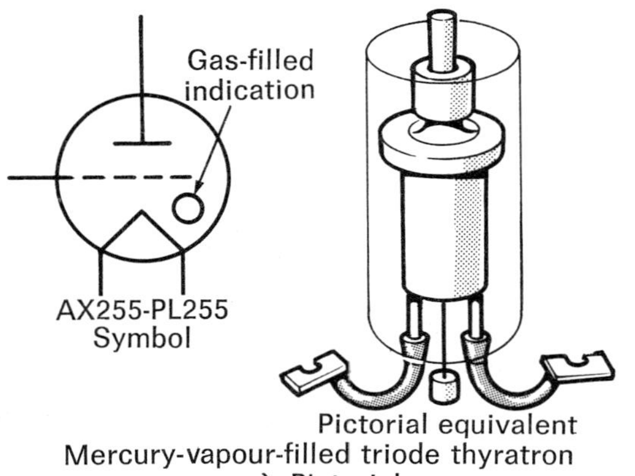

Pictorial equivalent
Mercury-vapour-filled triode thyratron
(c) Pictorial

135. Internal shield tube indication
Some special tubes are manufactured with internal electrostatic shields. If such is the case, a circle made of broken lines and grounded as shown in No. 135 (a) is added to the basic tube envelope.

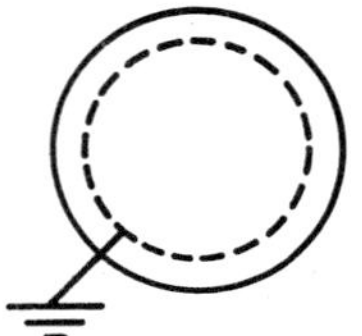

(a) CGSB

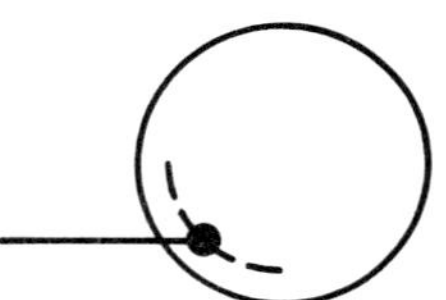

Internal shield tube, indication

(b) USAS

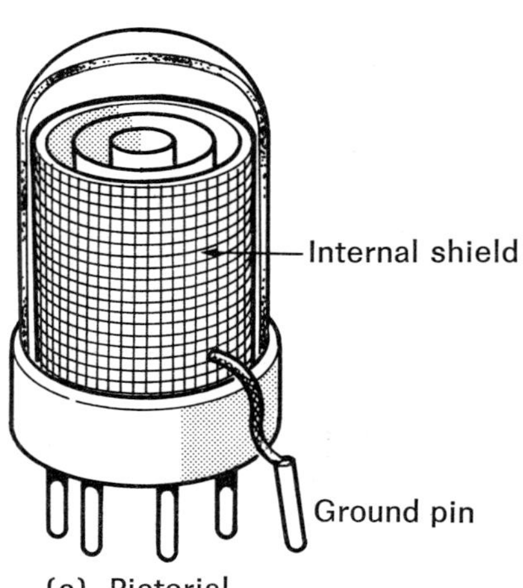

(c) Pictorial

TUBES

Element designation numbers

The numbers located about the periphery of the envelope symbol are to be shown as illustrated for each specific tube type. This procedure agrees with that of the Radio Manufacturers' Association (R.M.A.) graphic symbols for tubes.

136. Single-triode tube

A single-triode type tube is one having either a directly-heated cathode (filament) plus a grid and a plate, or, one having a filament plus an indirectly-heated cathode, grid, and plate.

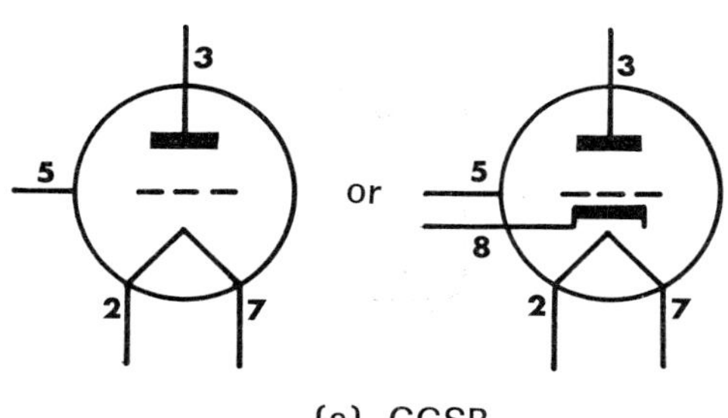

(a) CGSB

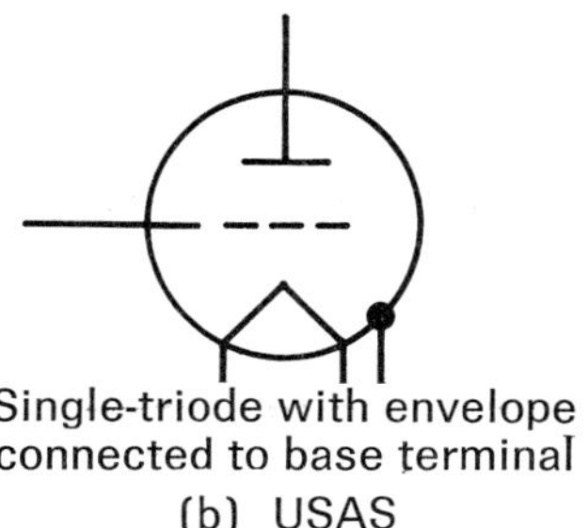

Single-triode with envelope connected to base terminal

(b) USAS

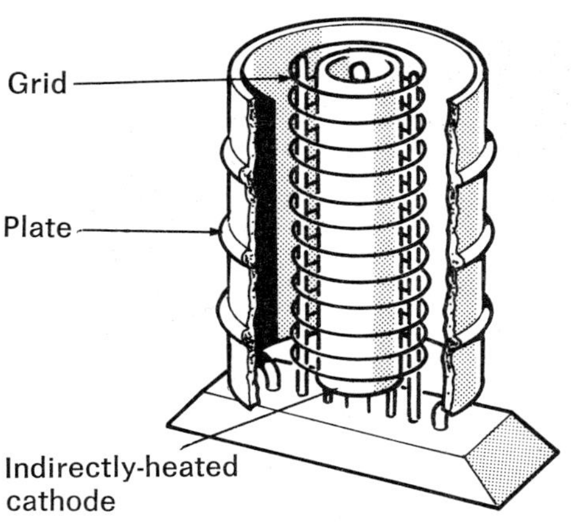

(c) Pictorial

137. Dual-triode tube

A dual-triode type tube, as the name implies, functions as two single triodes. There are three variations of this tube type. No. 137 (a), the top symbol, shows a dual-triode with a single filament, two grids, and two anodes. The middle symbol illustrates a dual-triode having a filament, one indirectly-heated cathode, two grids, and two anodes. The last variation as shown in the bottom symbol has a single filament, two indirectly-heated cathodes, two grids, and two anodes.

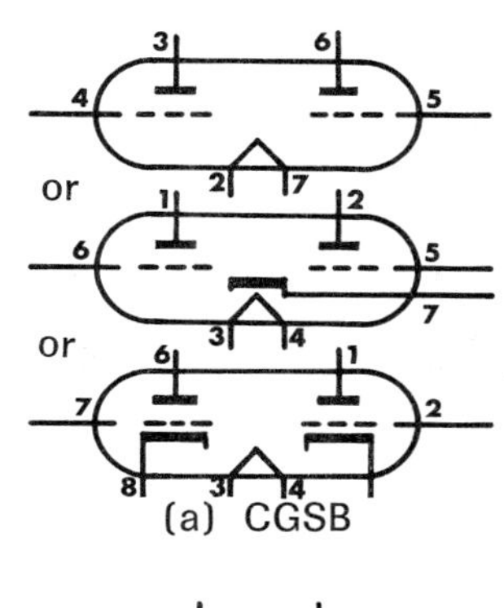

(a) CGSB

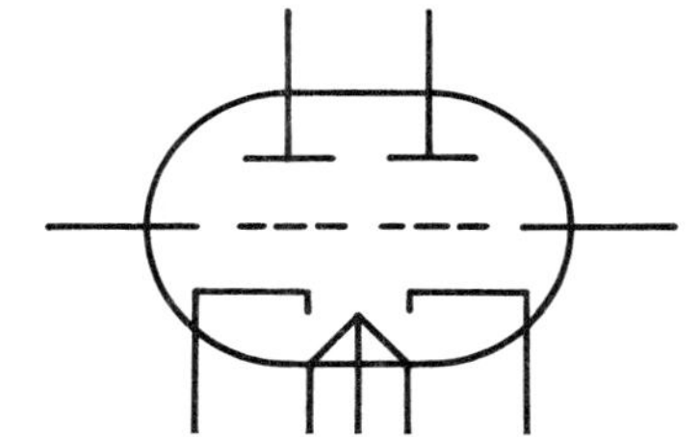

Equipotential-cathode twin triode tube

(b) USAS

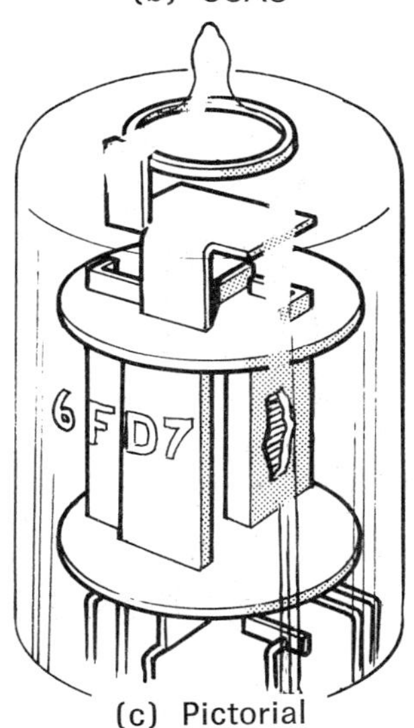

(c) Pictorial

138. Dual-diode with triode tube
This type of tube has an arrangement of elements within a single envelope that permits the operation of a triode and two diodes at the same time. It may have a filament, or a filament plus an indirectly-heated cathode. However, in both cases a total of three anodes, and one grid are required.

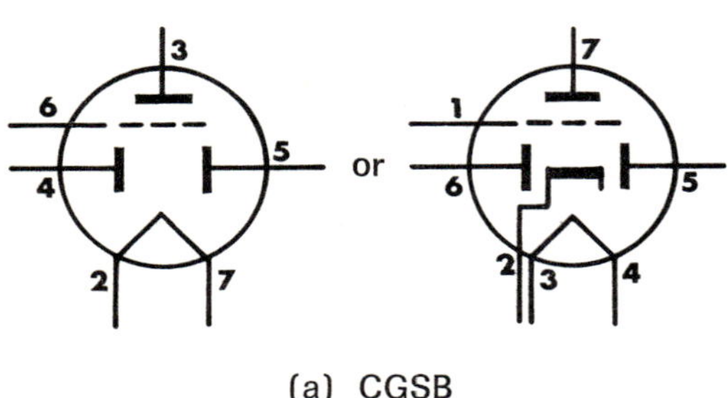

(a) CGSB

No equivalent symbol

(b) USAS

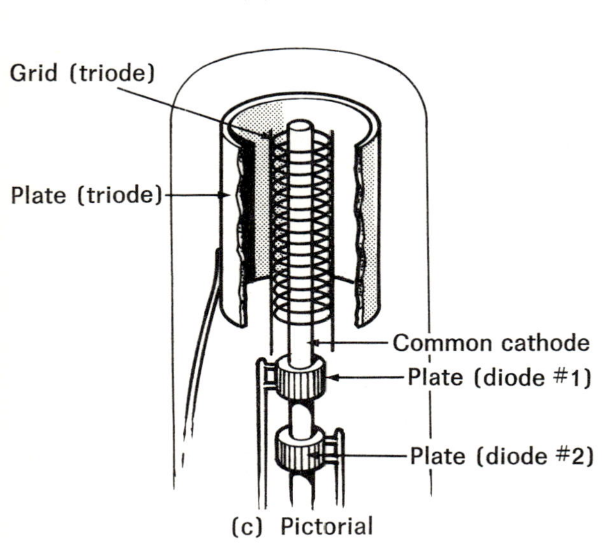

(c) Pictorial

139. Simple tetrode tube
A tetrode is a multi-element tube having one more element than an indirectly-heated cathode triode. This element is known as a screen grid. Its location in the tube symbol is between the control grid and the anode.

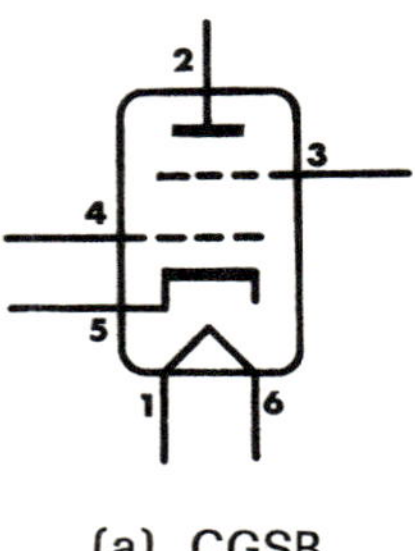

(a) CGSB

No equivalent symbol

(b) USAS

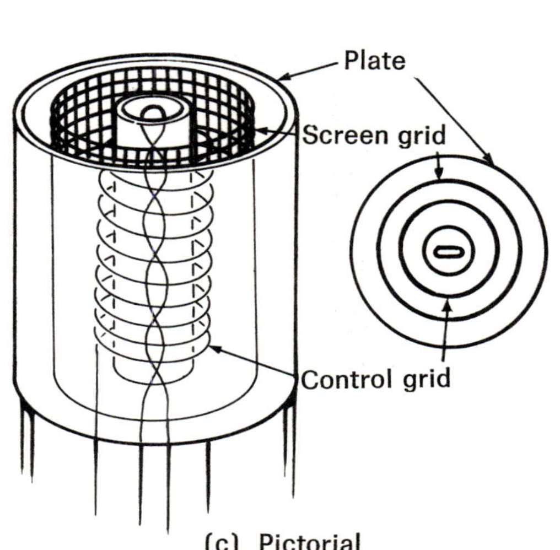

(c) Pictorial

140. Beam power tetrode tube
This is a special type of tetrode not having a third or suppressor grid, although its graphic symbol denotes a third grid. Upon inspection it will be noted that this third grid is 'tied' internally to the indirectly-heated cathode. In practice, this type of tube functions similarly to a pentode power amplifier.

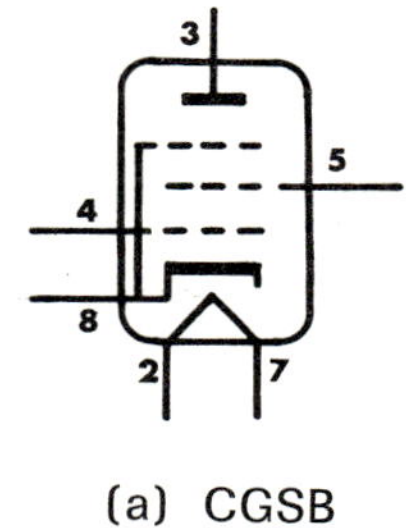

(a) CGSB

No equivalent symbol

(b) USAS

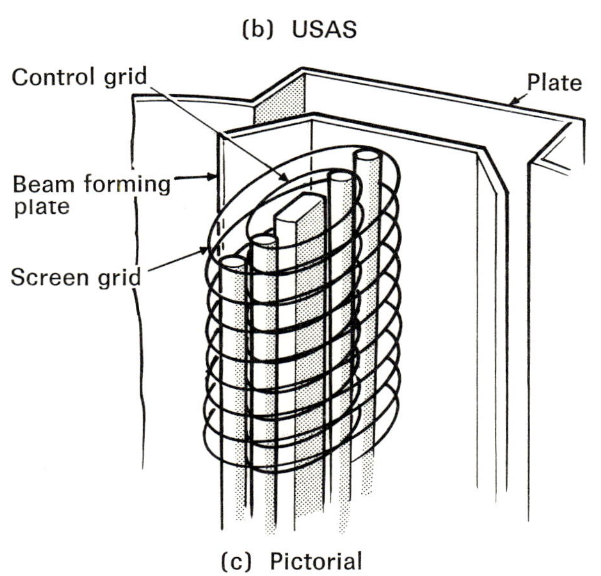

(c) Pictorial

141. Pentode tube
A pentode is a five-element tube used almost entirely as an r-f voltage amplifier. The suppressor grid which is located between the screen grid and the anode may be internally connected to the indirectly-heated cathode, or it may have an independent connection.

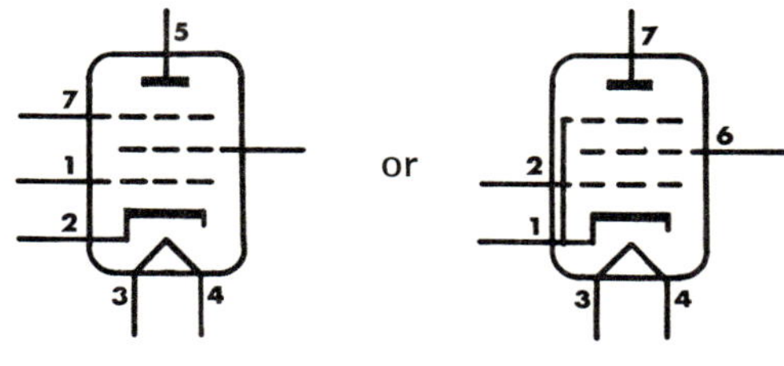

(a) CGSB

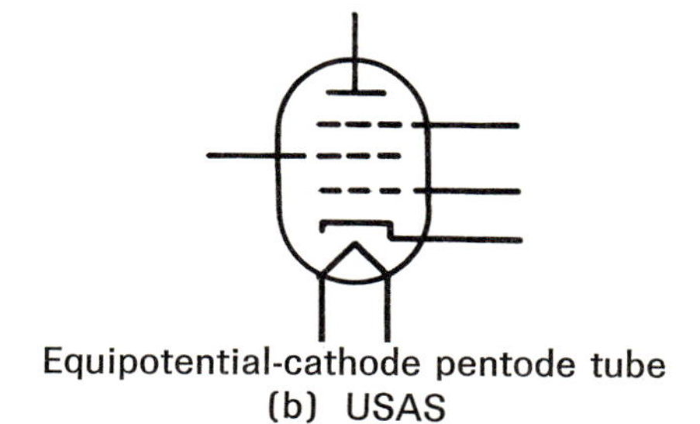

Equipotential-cathode pentode tube
(b) USAS

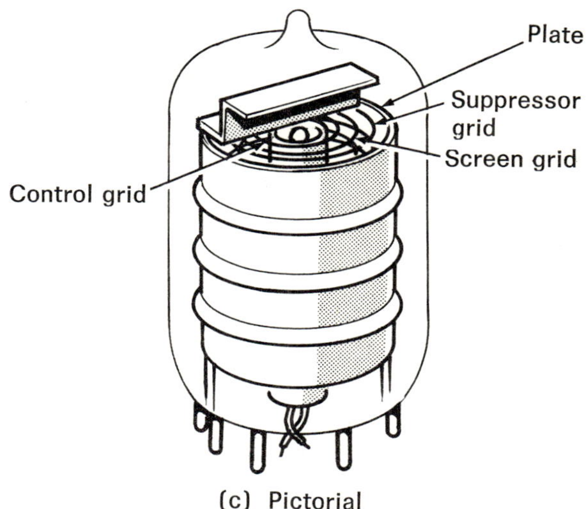

(c) Pictorial

142. Cold-cathode, gas-filled diode tube

This is a type of two element glow tube in which the envelope has been filled with a relatively dense gas, that is, gas which is at approximately 1/10,000 of normal atmospheric pressure. In addition, the cathode is a cold type and does not operate by thermionic emission.

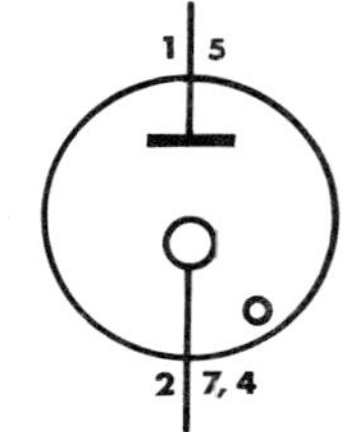

(a) CGSB

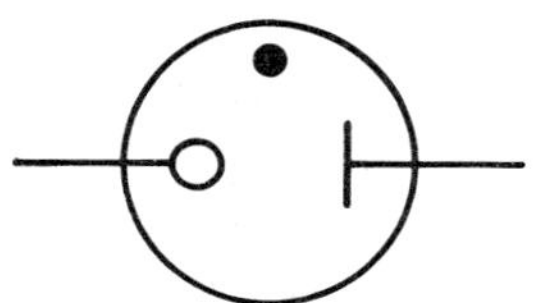

D.C. voltage regulator rectifier, cold cathode gas-filled tube
(b) USAS

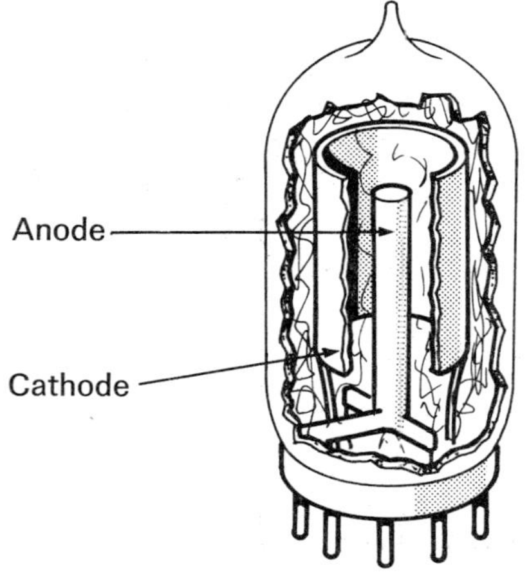

(c) Pictorial

143. Single-diode, heated-type tube

A diode is a two-element tube having no grid. It may have an indirectly-heated cathode for alternating-current operation, or simply a filament for direct-current operation.

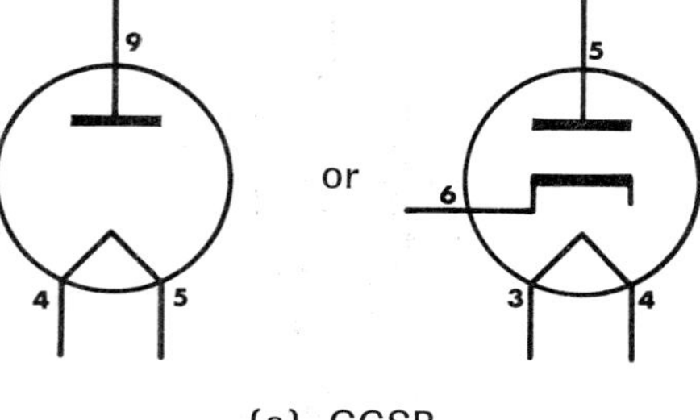

(a) CGSB

No equivalent symbol

(b) USAS

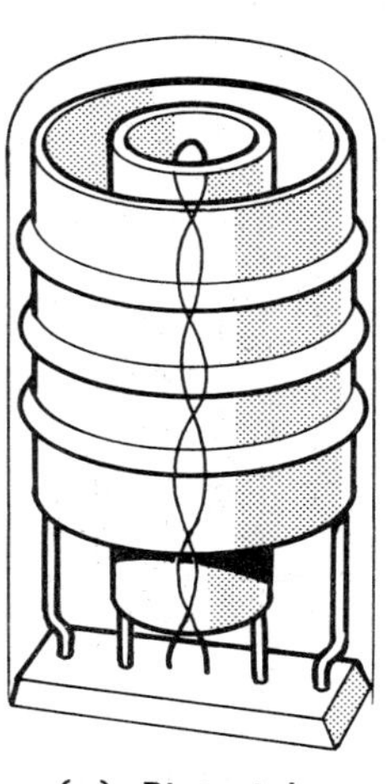

(c) Pictorial

144. Dual-diode tube

A dual-diode tube is a tube type having sufficient elements to perform as two single diodes would. It is manufactured in three basic types. For direct-current operation it is constructed with a single filament and two cathodes. For alternating-current operation it is manufactured with either a single indirectly-heated cathode, or two indirectly-heated cathodes, plus a single filament and two anodes.

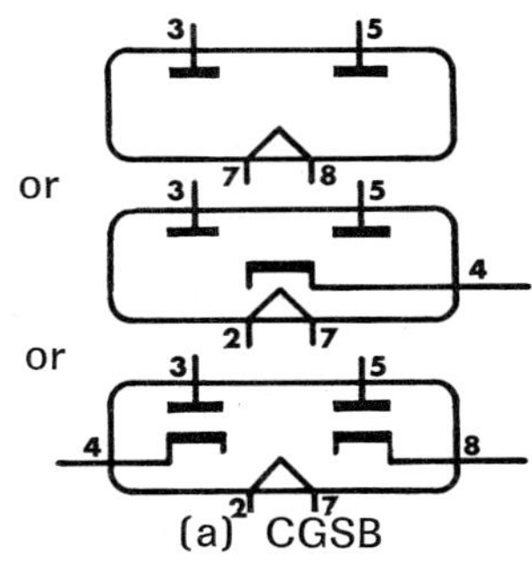

(a) CGSB

No equivalent symbol

(b) USAS

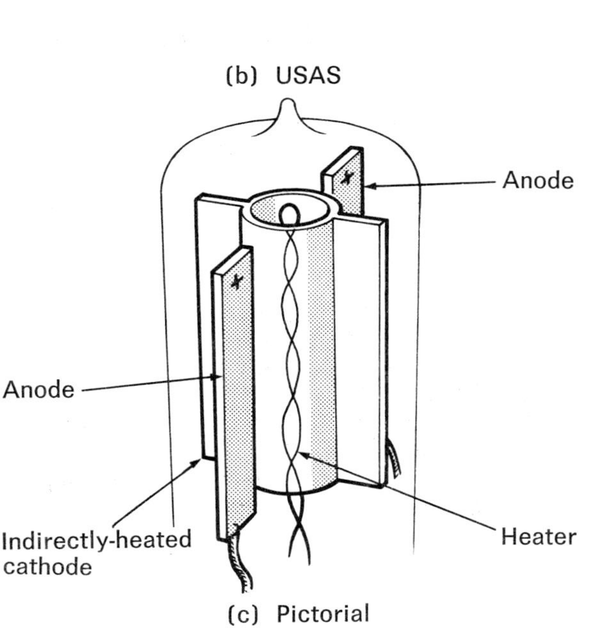

(c) Pictorial

145. Phototube

When a source of light strikes the photocathode of this two-element tube, electrons are displaced which stream off to the anode. The symbol for this simple phototube which develops only a very small current is shown in No. 145(a).

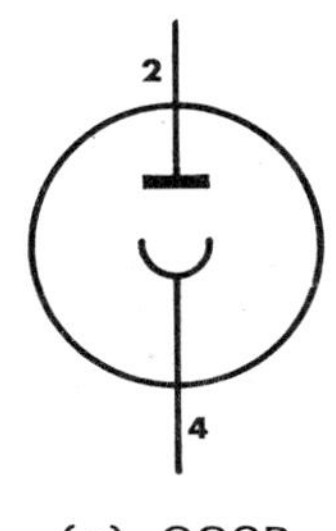

(a) CGSB

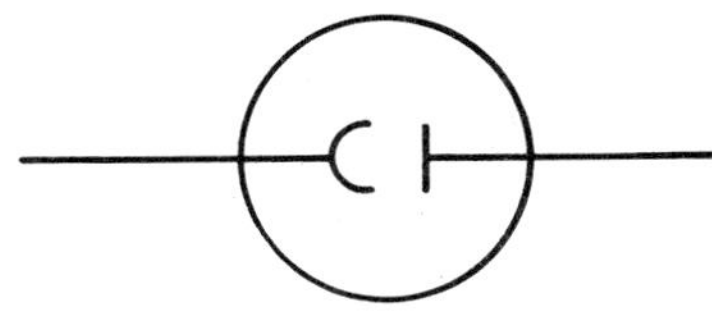

Single-unit vacuum type phototube

(b) USAS

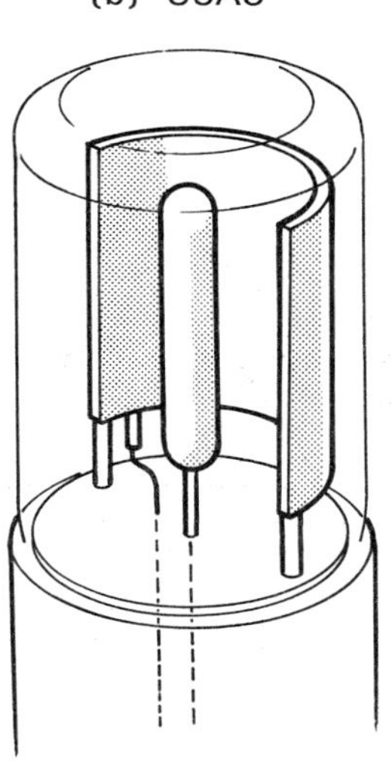

(c) Pictorial

146. Photomultiplier tube
This is a type of phototube having a number of dynodes which bring about a strong increase in the electron flow originally emanating from the photocathode by the principle of secondary emission.

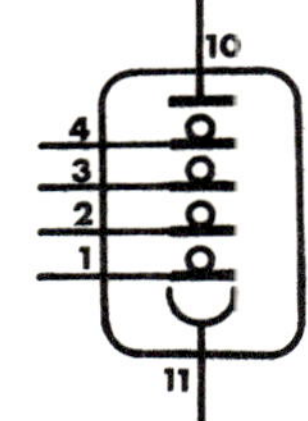

(a) CGSB

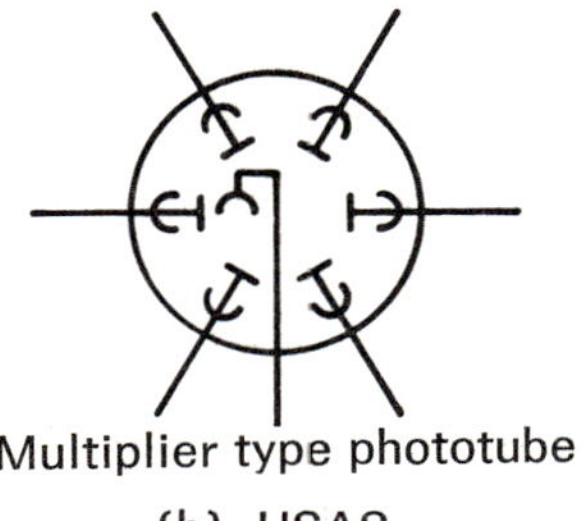

Multiplier type phototube
(b) USAS

Secondary plates
100V
300V
500V
700V
200V
400V
600V
800V
Cathode
Collector
Secondary plates

(c) Pictorial

147. Electrostatic deflected cathode-ray tube
An electrostatic deflected cathode-ray tube is one in which the cathode ray can be deflected both vertically and horizontally, by imposing an electrical charge on sets of deflector plates which are situated about the ray, and at 90° to each other.

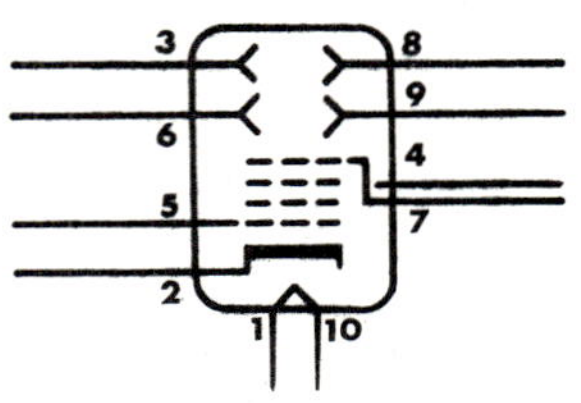

(a) CGSB

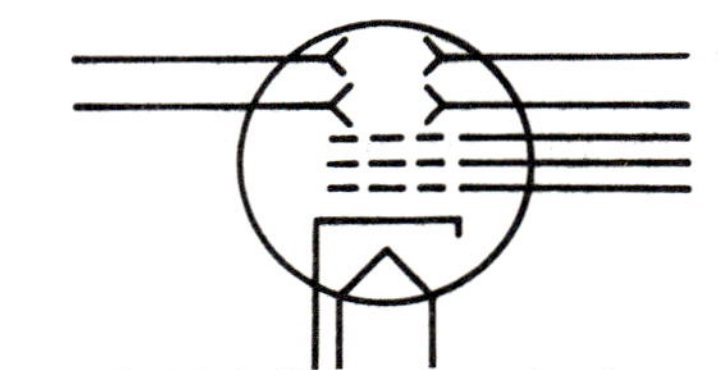

Electric-field deflection cathode-ray tube
(b) USAS

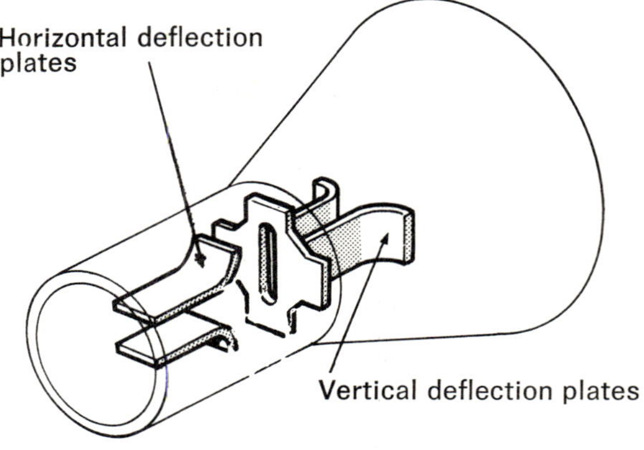

(c) Pictorial

148. Magnetic deflected cathode-ray tube
By using a controllable magnetic field about the cathode-ray beam, it can be bent or deflected according to will. The graphic symbol for the type of cathode-ray tube employing this method of beam deflection is shown in No. 148(a).

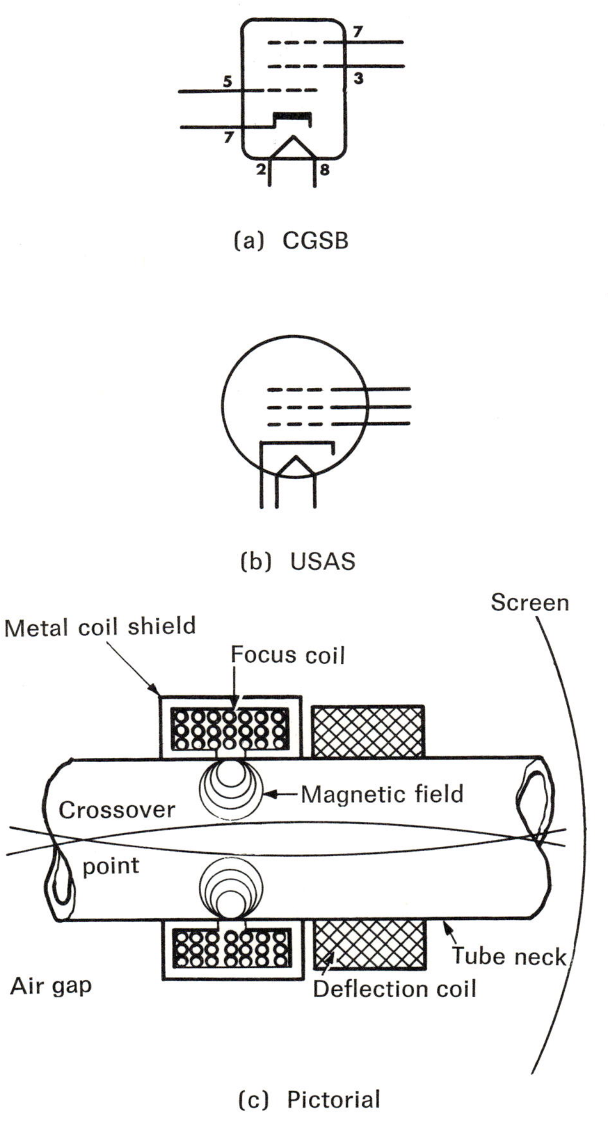

(a) CGSB

(b) USAS

(c) Pictorial

149. Mercury pool with ignitor and control grid tube
This type of tube is used in the control of industrial machinery where electrical current measured in hundreds of amperes is required. The cathode in this case is a pool of mercury. An ignitor needle penetrates slightly into the pool causing an arc to form thus ionizing the liquid mercury. The current then flows to the anode when voltage is applied to it, but this flow is made controllable by a grid. See No. 149(c).

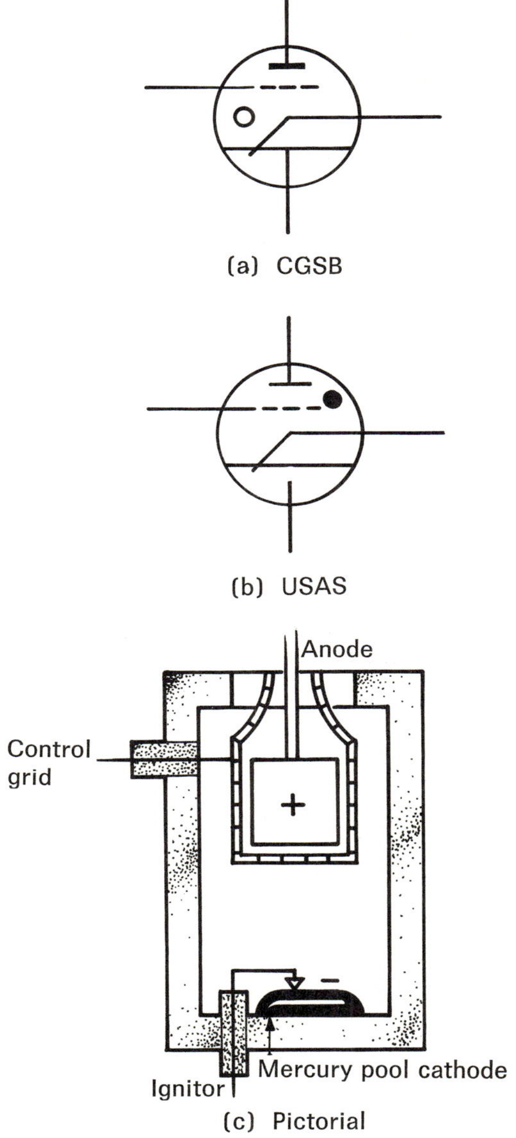

(a) CGSB

(b) USAS

(c) Pictorial

150. Trigger tube

When a type of tube is required in a counting or automatic controlling device circuit, a trigger tube is used. This type of device requires no cathode heating, and hence is always ready for instantaneous operation when the tube is triggered by electrical impulses produced by the phenomena to be controlled.

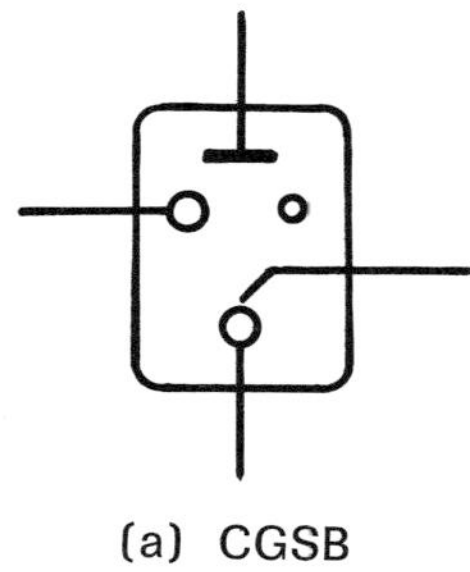

(a) CGSB

No equivalent symbol

(b) USAS

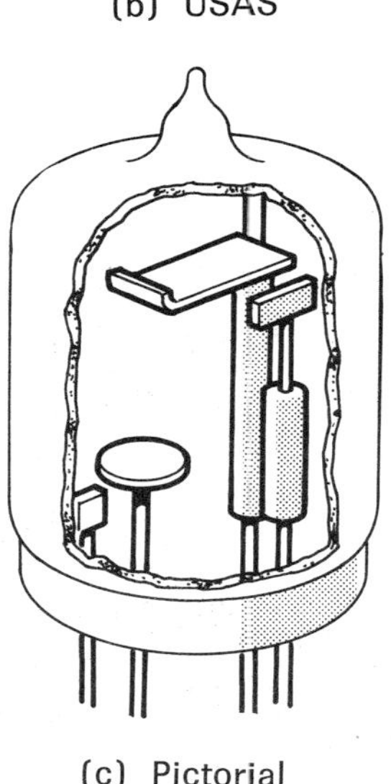

(c) Pictorial

151. Resonant type magnetron tube

A magnetron is an oscillator tube capable of operating at extremely high frequencies. The resonant type has a central core-shaped filament, which is surrounded by a number of cavities cut into an anode block. The purpose of the cavities is to serve as a resonant tuning circuit.

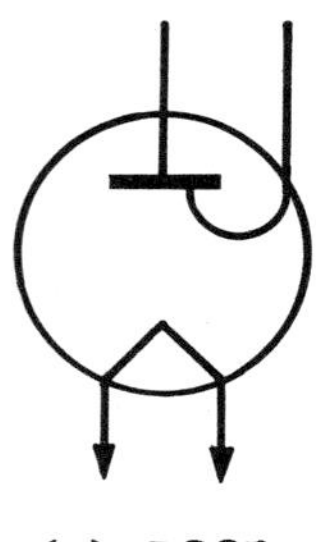

(a) CGSB

No equivalent symbol

(b) USAS

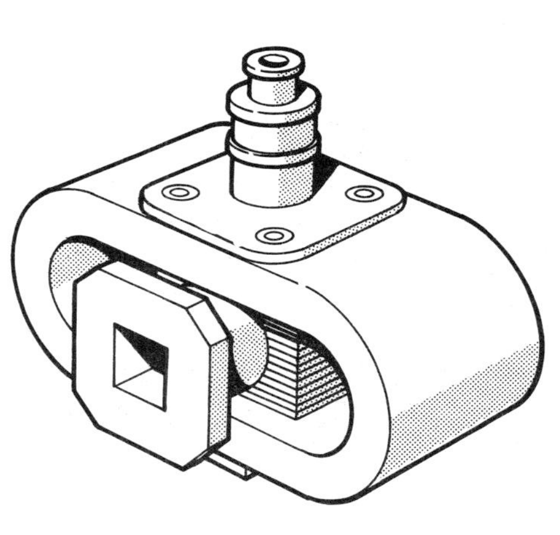

(c) Pictorial

WINDINGS

152. Relay, solenoid, or operating coil winding

A solenoid or operating coil is a specific number of turns of insulated wire wound about a soft-iron core to create an electromagnet. The direction of the turns must be consistently one way.

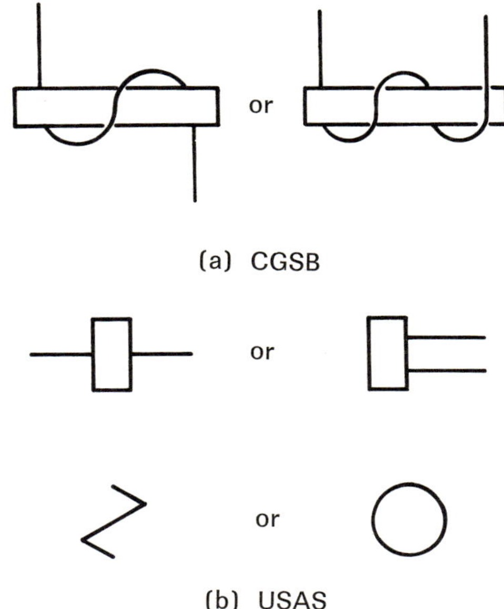

(a) CGSB

(b) USAS

(c) Pictorial

153. Inductor, transformer, or choke winding

The symbol for an inductor or for one winding of a transformer is shown in No. 153 (a). The addition of two parallel lines drawn adjacent to an inductor symbol as shown in No. 153 (a) indicates a choke. A choke is a device for restricting the flow of electrical current of a particular frequency, and finds application in rectifying circuits.

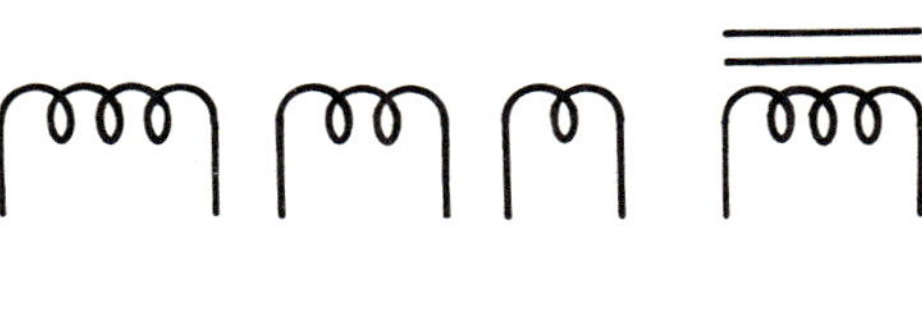

(a) CGSB

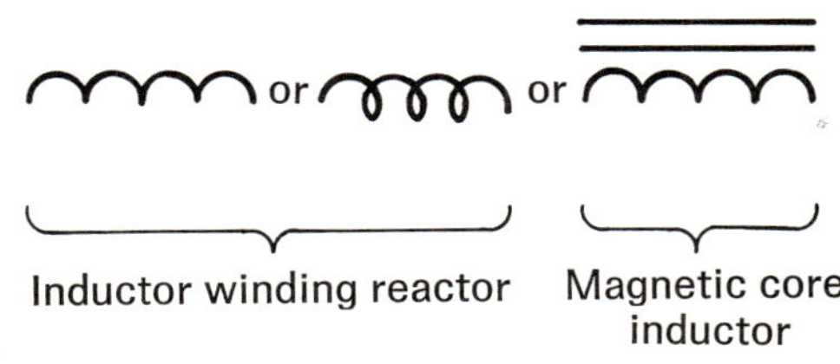

(b) USAS

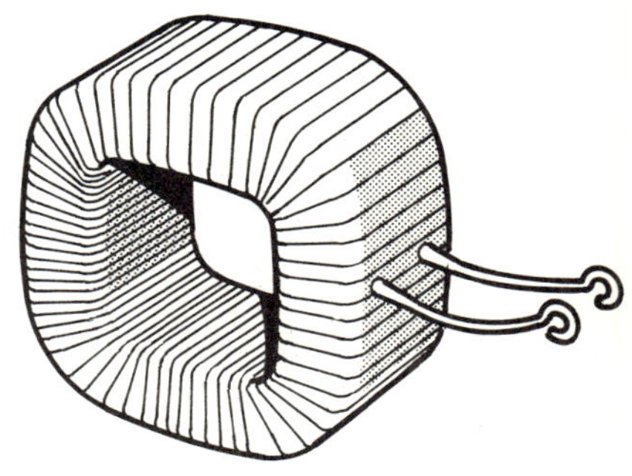

(c) Pictorial

154. ROTATING MACHINES

Devices such as motors, generators, or alternators are classified as rotating machines.

(a) CGSB

No equivalent symbol

(b) USAS

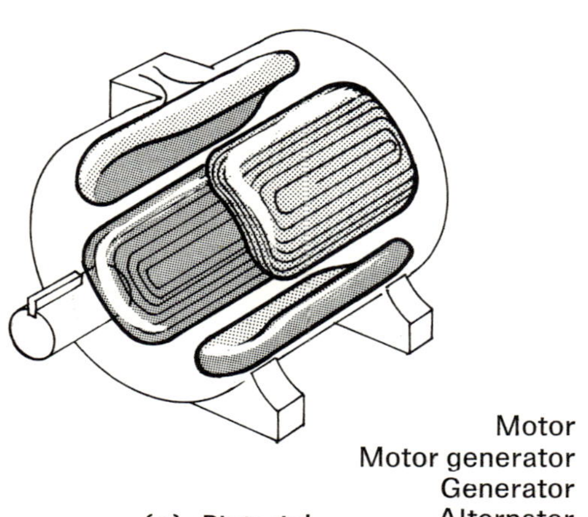

(c) Pictorial

155. BLOWOUT

This is an electromagnet constructed of flat-shaped metal in a coiled form which is used to deflect the arc between two contacts, thus blowing out or extinguishing it.

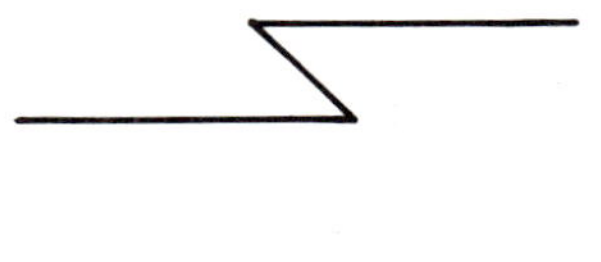

(a) CGSB

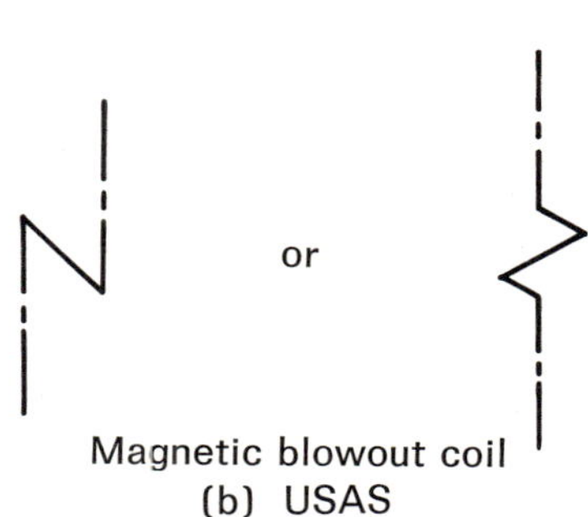

(b) USAS

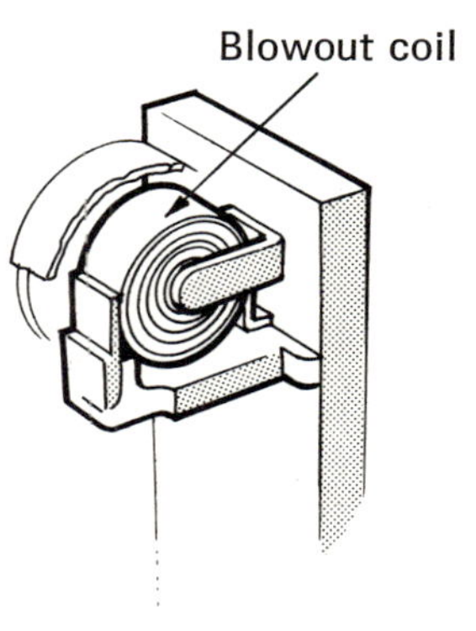

(c) Pictorial

WIRE

156. Crossovers

The method shown in No. 156(a) is to be exclusively used for a crossover, that is, where two conductors cross each other but are not electrically connected.

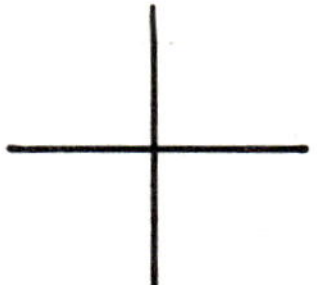

(a) CGSB

The same as the Canadian

(b) USAS

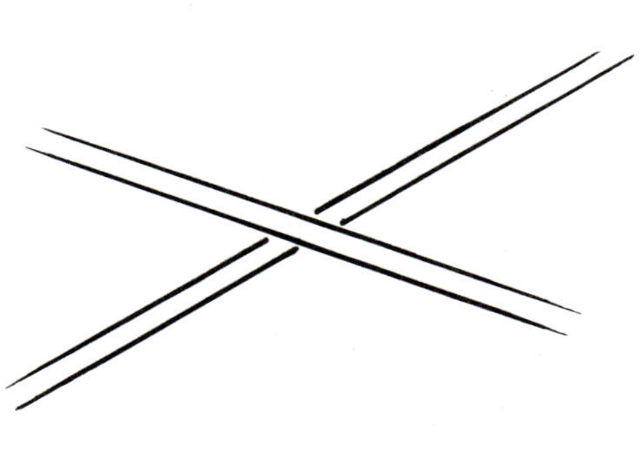

(c) Pictorial

157. Connections

The following methods are to be used exclusively to indicate an electrical connection between two conductors.

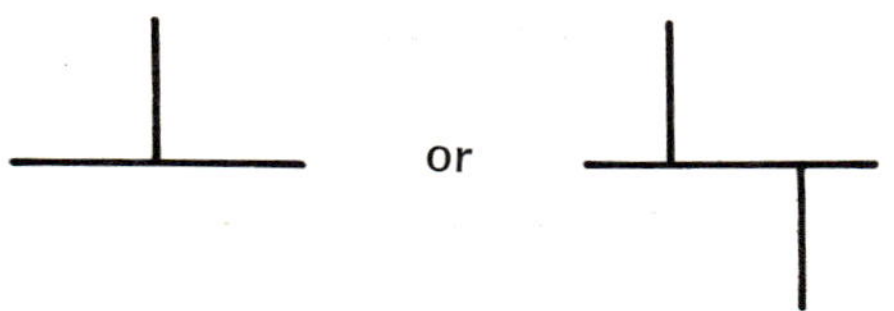

(a) CGSB

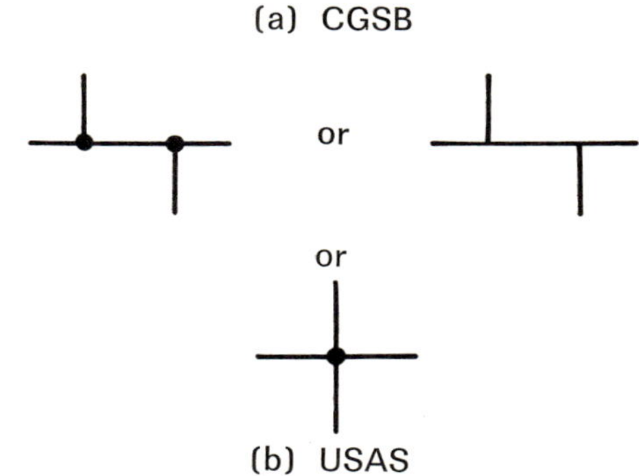

(b) USAS

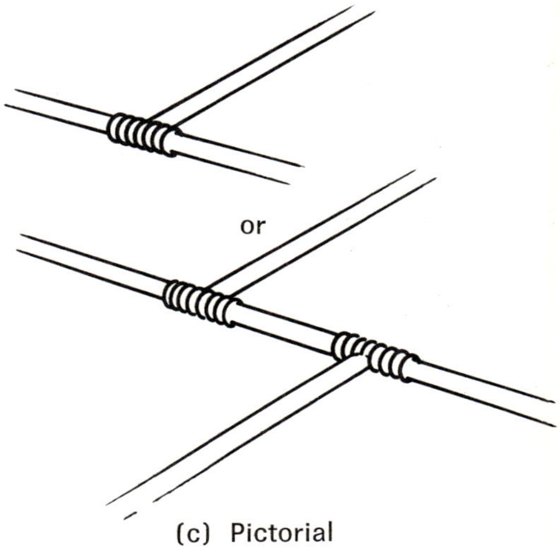

(c) Pictorial

158. Incoming conductor

No. 158(a) shows the correct symbol to illustrate a conductor or conductors coming from another electrical circuit or system.

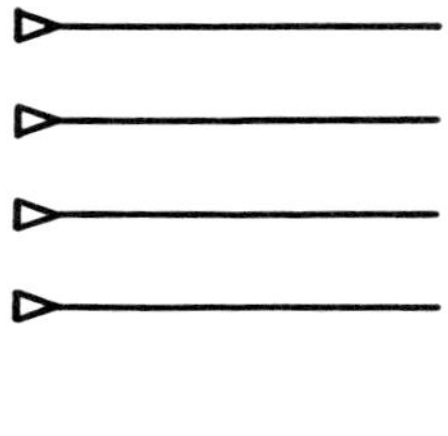

(a) CGSB

No equivalent symbol

(b) USAS

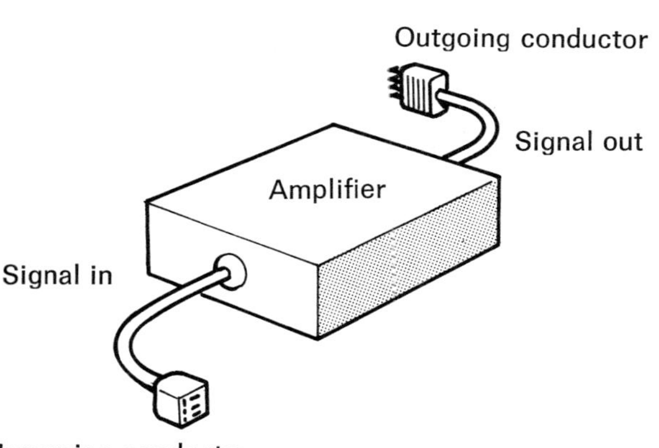

(c) Pictorial

159. Outgoing conductor

No. 159(a) illustrates the correct symbol to illustrate a conductor or conductors terminating at a circuit, but with the desired intention of indicating that they are outgoing to another circuit or system.

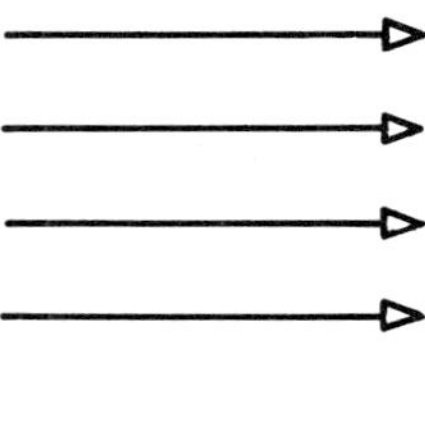

(a) CGSB

No equivalent symbol

(b) USAS

Outgoing conductor
Amplifier
Signal out
Signal in
Incoming conductor

(c) Pictorial

160. Transposed
When two insulated conductors are run parallel to each other, then crossed to opposite sides in an alternating manner, they are said to be transposed. The purpose of transposing conductors is to cancel out unwanted magnetic fields about them.

(a) CGSB

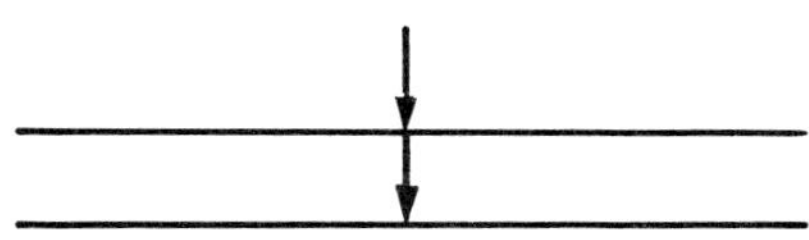

Pair (twisted unless otherwise specified)
(b) USAS

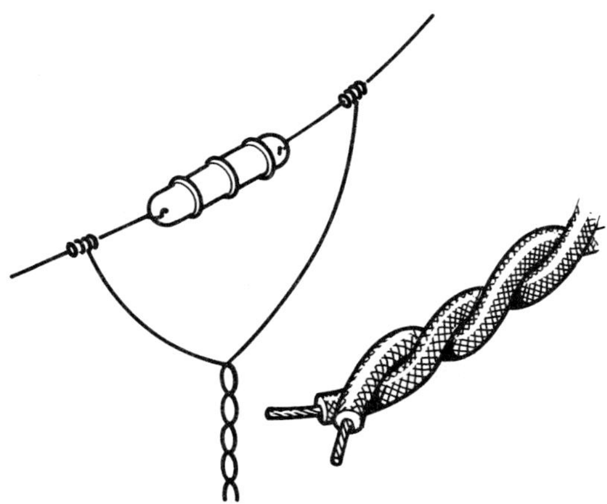

Antenna lead-in wires transposed
(c) Pictorial

161. Cabled or paralleled twin
When two conductors, insulated from each other, are enclosed in parallel fashion in a sheath, they are known as a cable. No. 161(a) illustrates the correct method of symbolically illustrating them.

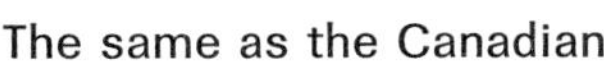

(a) CGSB

The same as the Canadian

(b) USAS

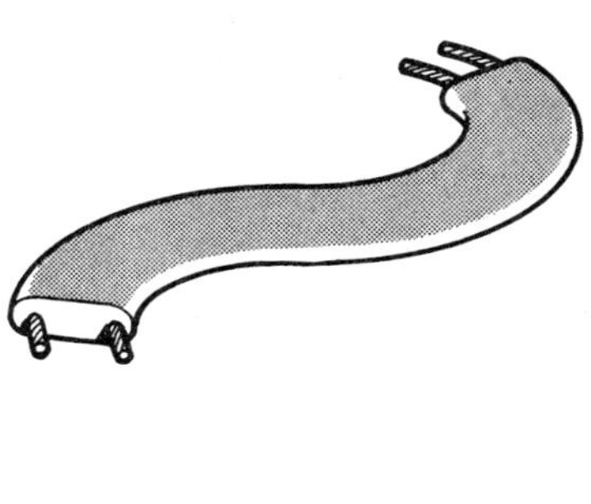

(c) Pictorial

162. Shielded twin

When two conductors, insulated from each other, are enclosed in parallel fashion in a suitably grounded metallic sheath, they are considered as being shielded. No. 162(a).

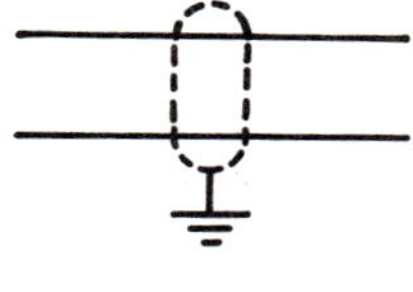

(a) CGSB

The same as the Canadian

(b) USAS

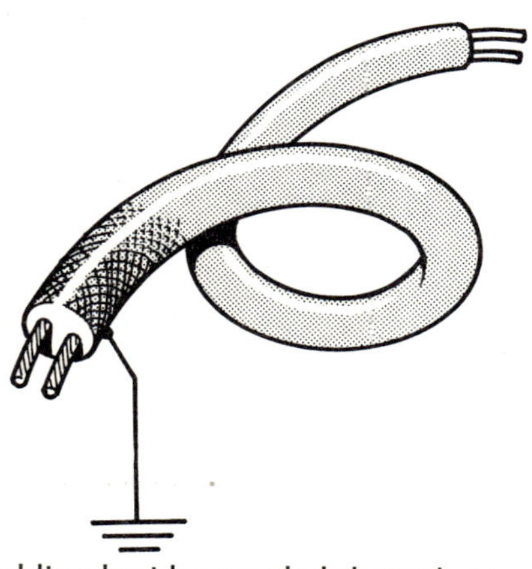

Shielding braid grounded through connectors
(c) Pictorial

163. Individual shield

Sometimes the individual insulated conductors within a cable have individual shields. The correct symbol to illustrate this condition is shown by No. 163(a).

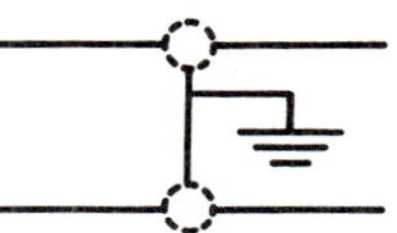

(a) CGSB

Shielded single conductor

(b) USAS

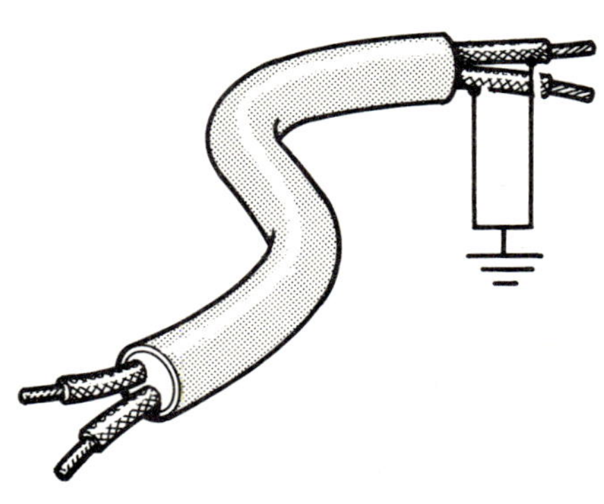

Shielding braids grounded through connectors
(c) Pictorial

164. Multiconductor cable
A multiconductor cable, that is a cable having more than two conductors, is shown by the example illustrated in No. 164(c). If additional conductors are required in the cable the size of the enclosing rounded oblong is increased.

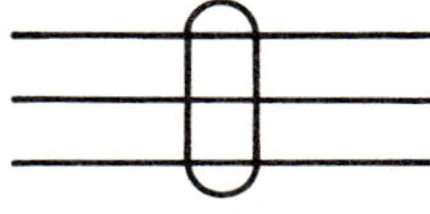

(a) CGSB

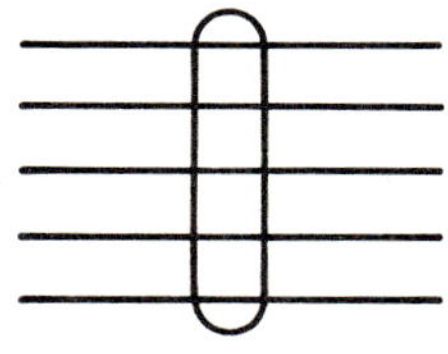

Assembled conductors
(b) USAS

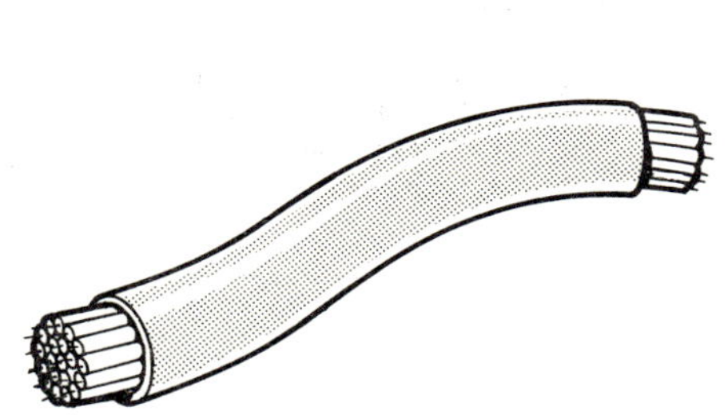

(c) Pictorial

165. Single-coaxial, radio frequency
In circuits carrying radio frequency energy, for example, from a transmitter to its antenna, a coaxial cable is used. This type of wire has a flexible inner conductor which is located centrally in a core of high-dielectric constant insulating material. Surrounding this material is a braided sheath which is in turn covered with a waterproof jacket.

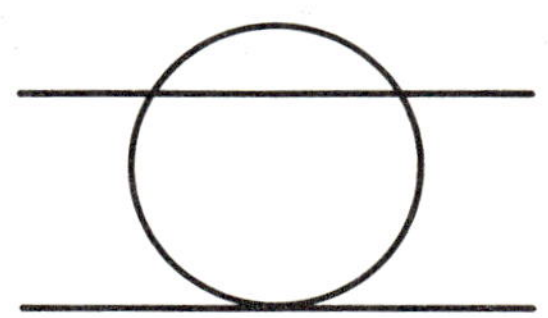

(a) CGSB

The same as the Canadian

(b) USAS

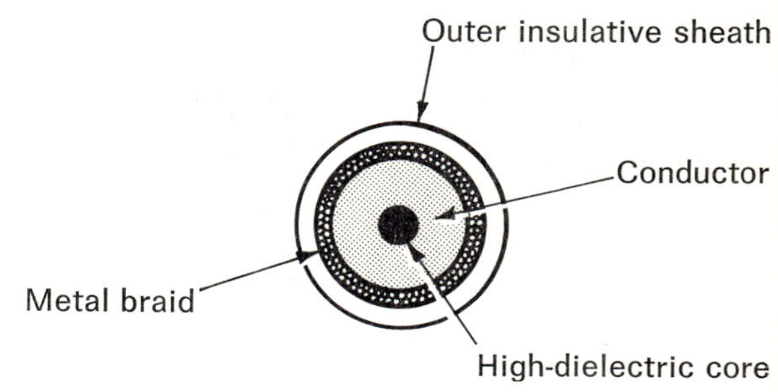

(c) Pictorial

166. Conduit and cabling

The correct form for drawing conduit or cabling is shown in No. 166(a). The actual direction in which the conductor joins the cable is indicated by the symbol.

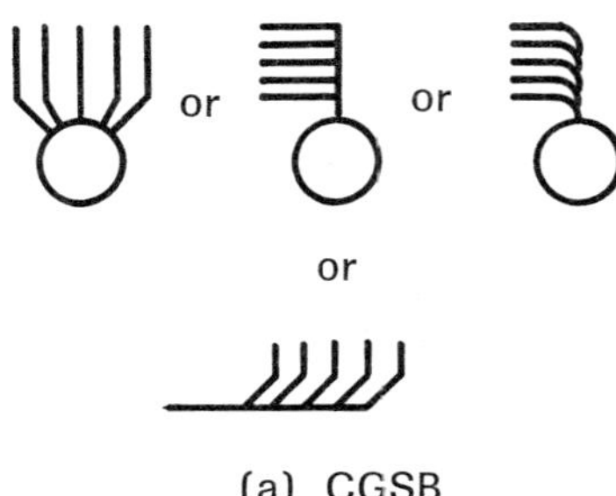

(a) CGSB

The same as the Canadian

(b) USAS

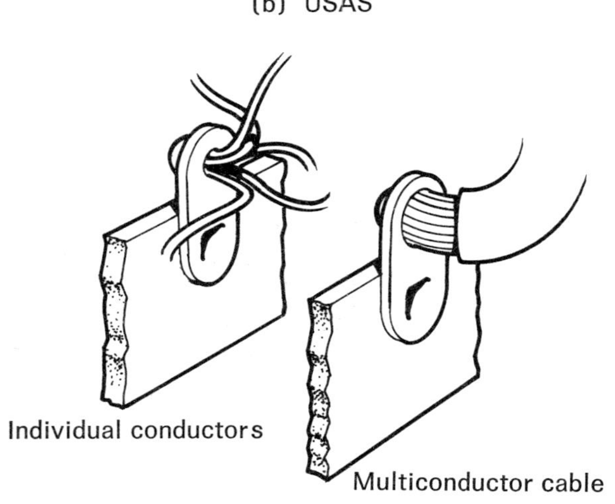

(c) Pictorial

SEMICONDUCTOR DEVICES

167. Semiconductor diode or semiconductor rectifier diode, new symbol

No. 167(a) illustrates two types of semiconductor diodes. The brokenout view shows a point contact type, while the view with the end exposed defines a rectifier type. The ability to carry electrical current in one direction is a function of either one.

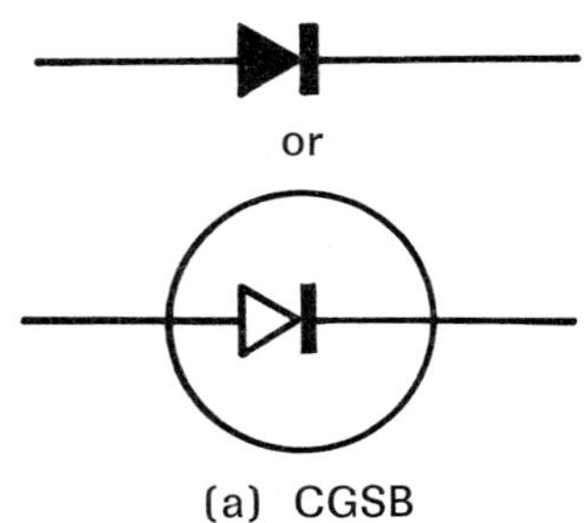

(a) CGSB

The same as the Canadian

(b) USAS

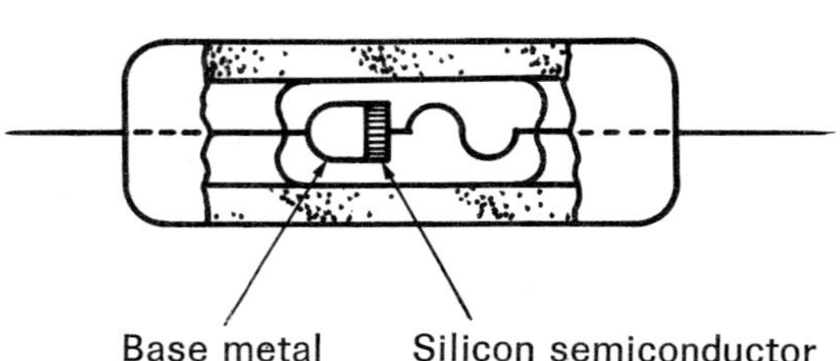

See also pictorial No. 85

(c) Pictorial

168. Capacitive diode, varactor
When a junction type diode is manufactured having a specific electrical capacity as a function of the reverse voltage that can be applied to it, it is known as a capacitive diode.

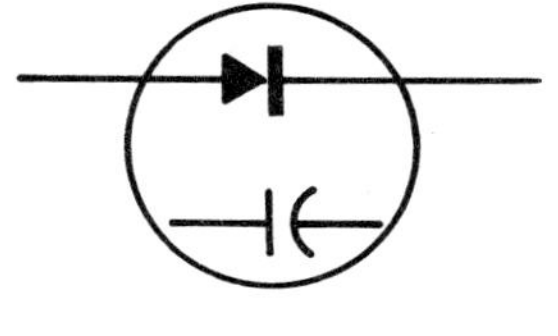

(a) CGSB

The same as the Canadian

(b) USAS

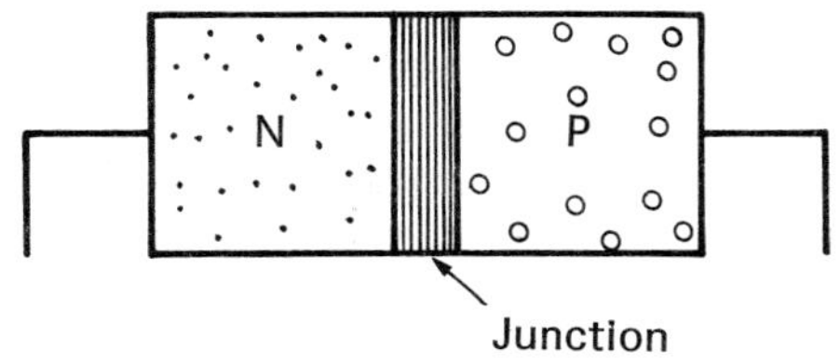

(c) Pictorial

169. Breakdown diode, unidirectional, new symbol
A junction type diode that can pass current in only one direction when the voltage applied to it reaches a specific level (breakdown voltage) is known as a unidirectional breakdown diode.

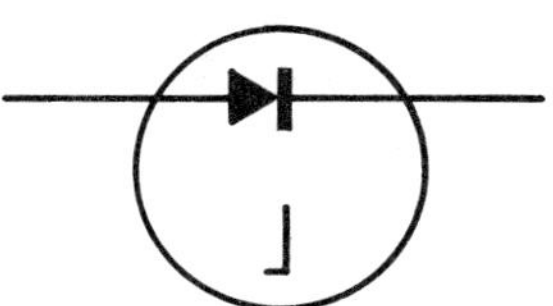

(a) CGSB

The same as the Canadian

(b) USAS

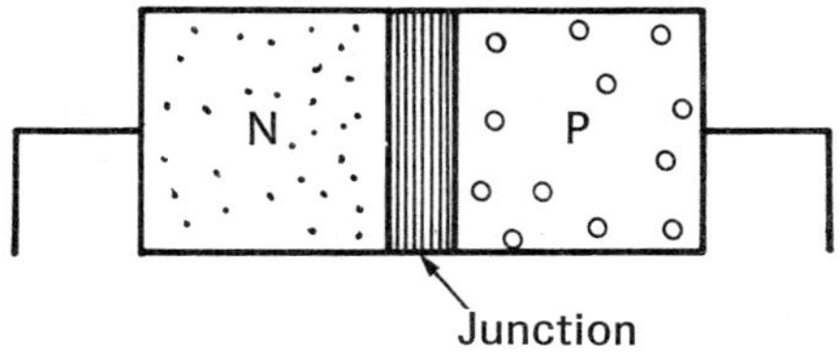

(c) Pictorial

170. Breakdown diode, bidirectional
A diode of the PNP type that can pass current in either direction only when the voltage applied to it reaches a specific level (breakdown voltage) is known as a bidirectional breakdown diode.

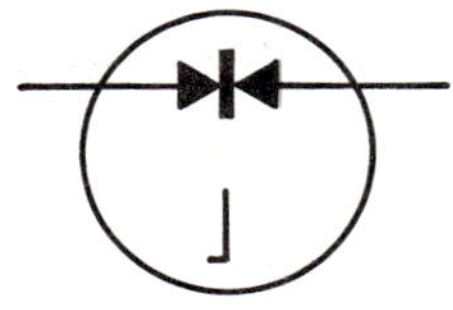

(a) CGSB

The same as the Canadian

(b) USAS

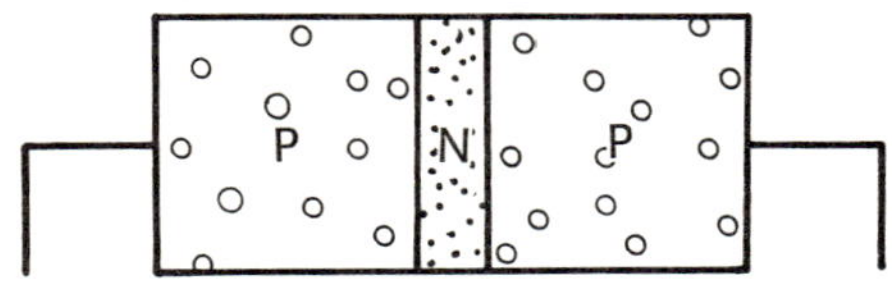

(c) Pictorial

171. Tunnel diode
No. 171(c) illustrates the voltage-current curve plotted for a tunnel diode. This special characteristic does not alter this type of diode's external appearance from that of an ordinary PN junction diode.

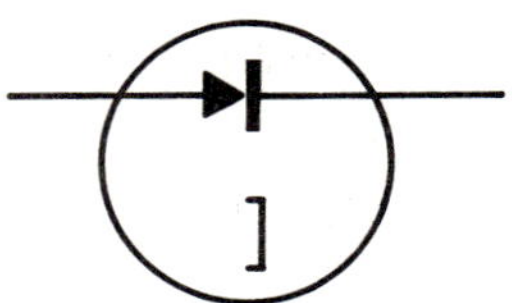

(a) CGSB

The same as the Canadian

(b) USAS

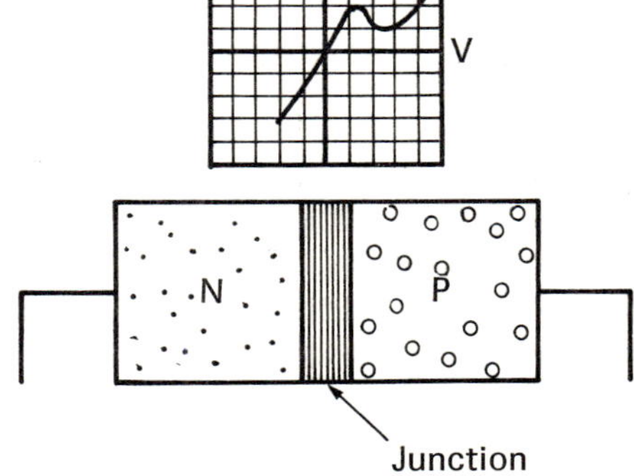

(c) Pictorial

172. Temperature dependent diode
No. 172(c) illustrates a specific type of junction diode, whose designed characteristics are a function of temperature.

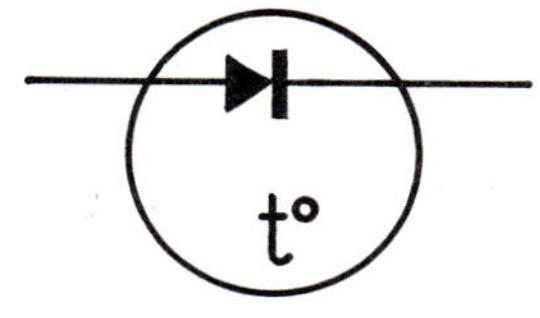

(a) CGSB

The same as the Canadian

(b) USAS

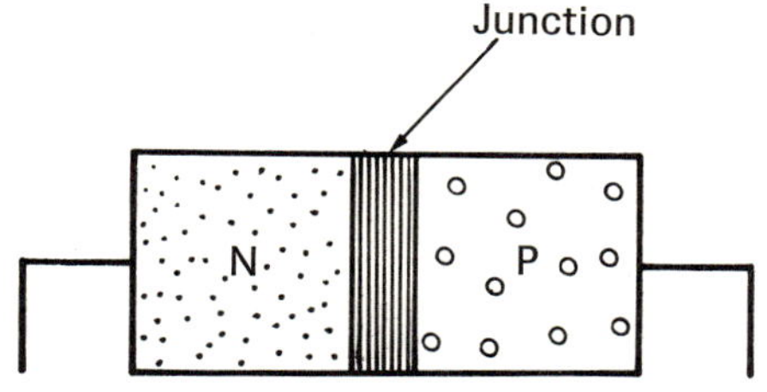

(c) Pictorial

173. Photodiode, new symbol
The graphic symbol illustrated by No. 173(a) refers to all types of photosensitive diodes that produce a small electrical current when struck by light rays. No. 173(c) illustrates only a point contact type.

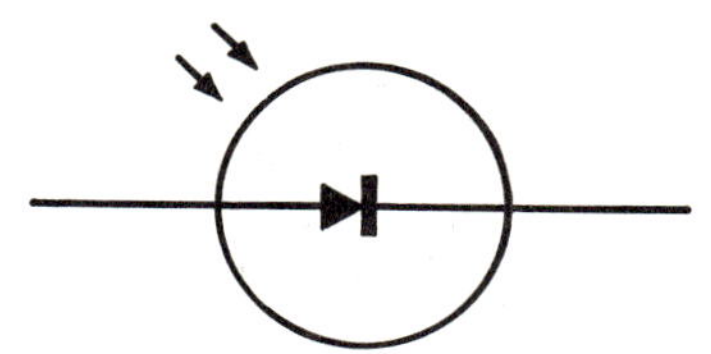

(a) CGSB

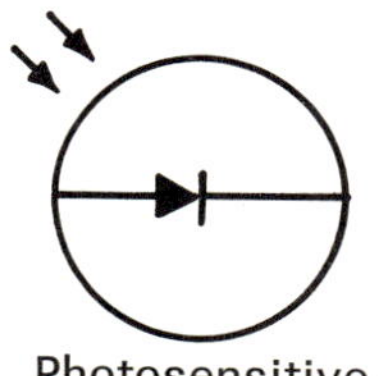

(b) USAS

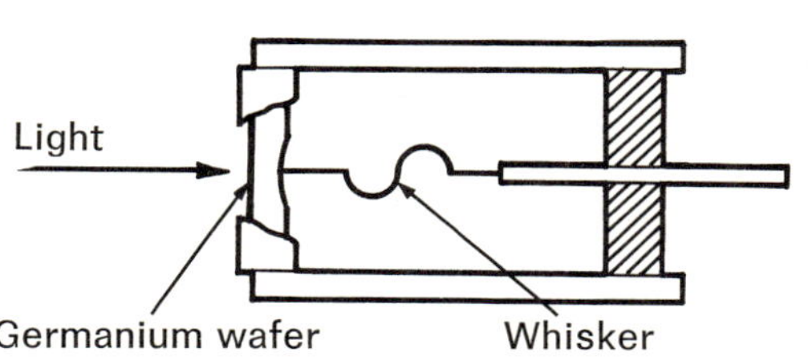

(c) Pictorial

174. PNP transistor
The transistor outlined by No. 174 (c) is a semiconductor device that is used frequently as an amplifier or oscillator and has three special connections as shown.

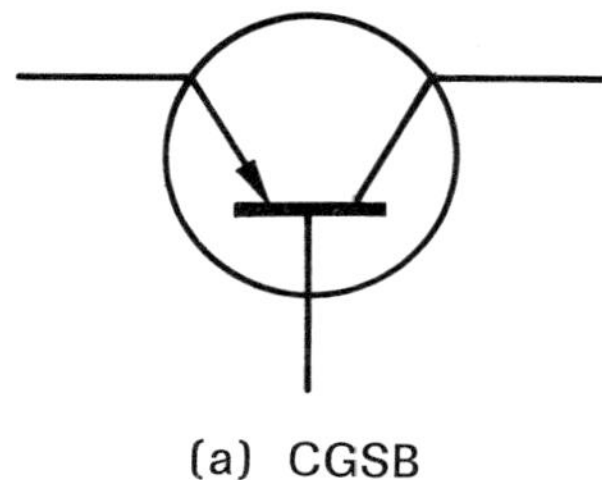

(a) CGSB

The same as the Canadian

(b) USAS

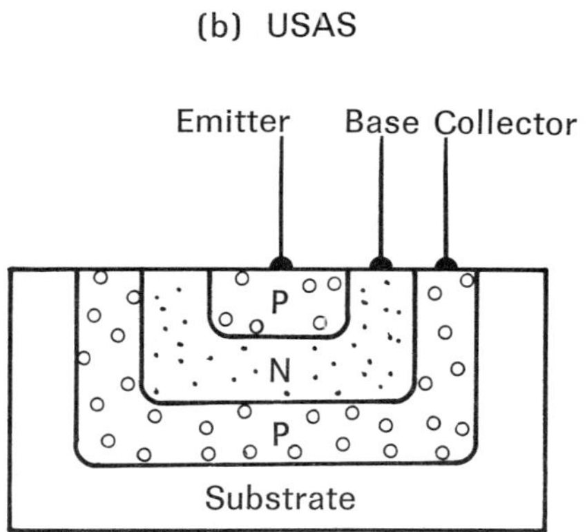

(c) Pictorial

175. Semiconductor diode PN-PN type switch
This semiconductor device is essentially two PN diodes layered on each other and having the characteristic of blocking current flowing in a reverse direction. Hence it performs as a switch.

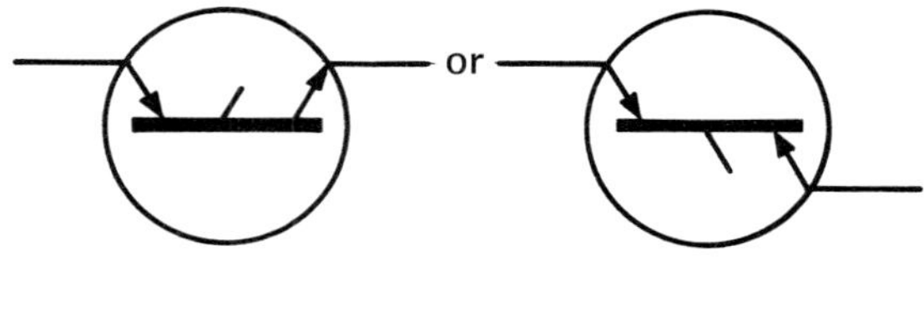

(a) CGSB

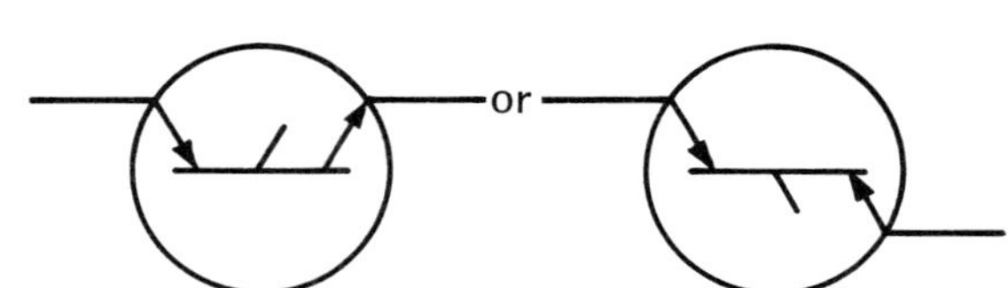

Thyristor, reverse-blocking diode

(b) USAS

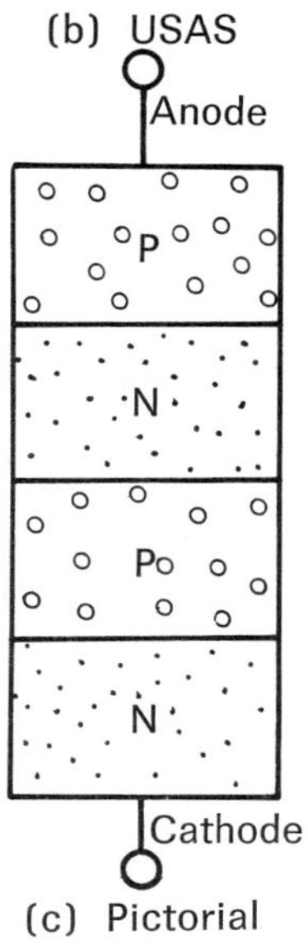

(c) Pictorial

176. PNP transistor, one electrode connected to envelope

No. 176(c) illustrates a standard PNP transistor with its collector lead electrically connected to a metallic envelope surrounding the transistor.

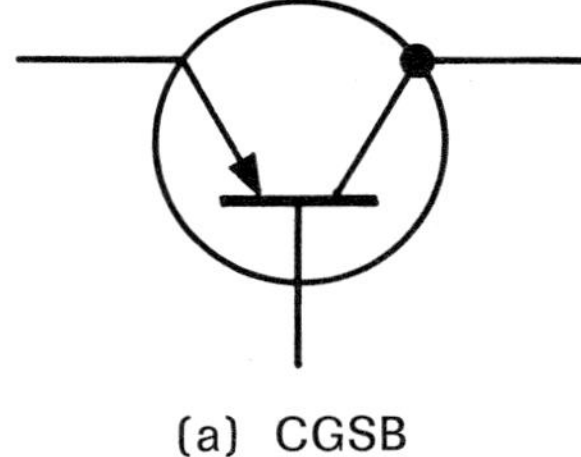

(a) CGSB

The same as the Canadian

(b) USAS

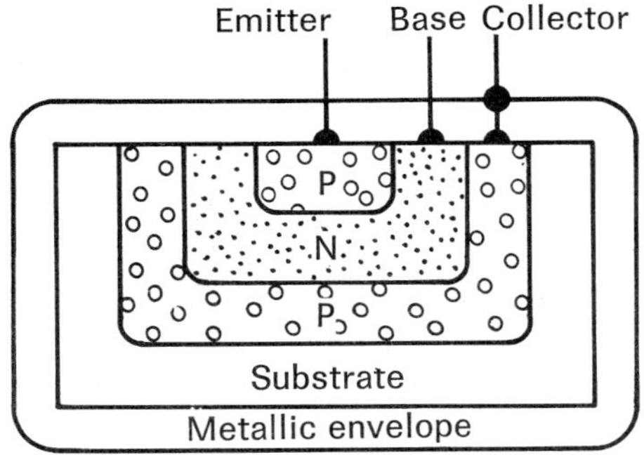

(c) Pictorial

177. NPN transistor

When the construction of a transistor is as shown in No. 177(c), it is known as an NPN type.

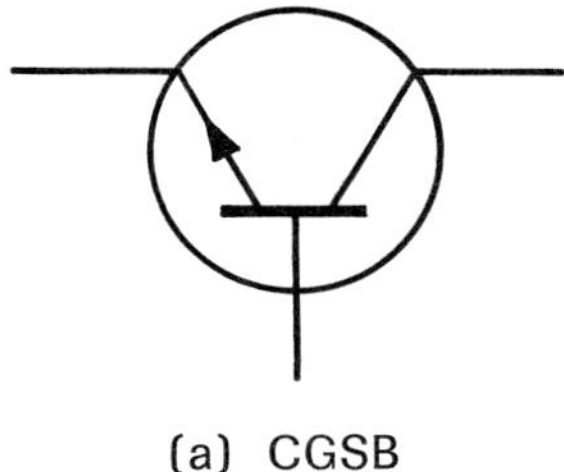

(a) CGSB

The same as the Canadian

(b) USAS

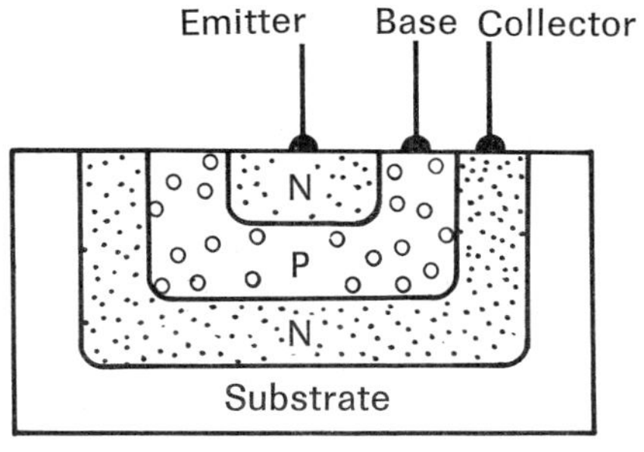

(c) Pictorial

178. Unijunction transistor N-type base

(a) CGSB

The same as the Canadian

(b) USAS

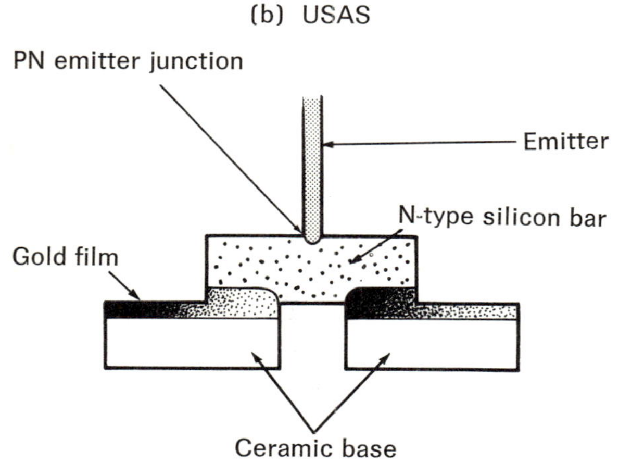

(c) Pictorial

179. Unijunction transistor P-type base
These types of transistors are used in relaxation oscillator circuits, and in trigger circuits of silicon controlled rectifiers.

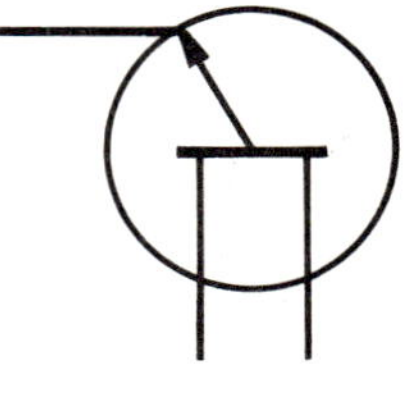

(a) CGSB

The same as the Canadian

(b) USAS

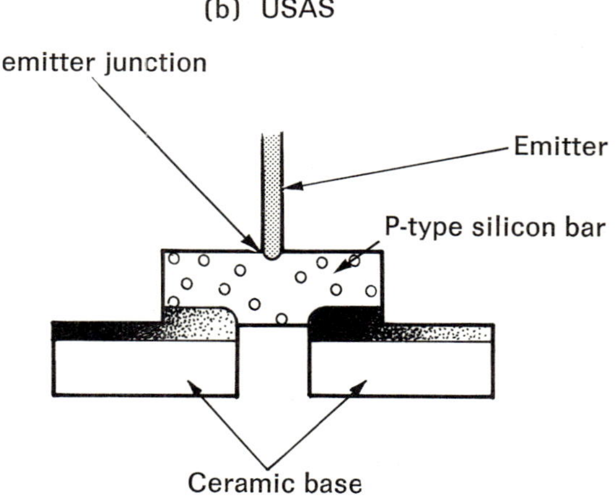

(c) Pictorial

180. Field-effect transistor N-type base

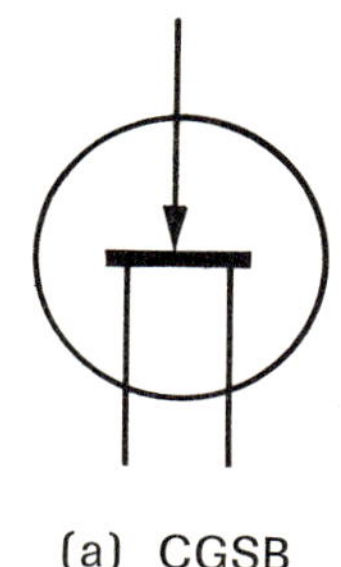

(a) CGSB

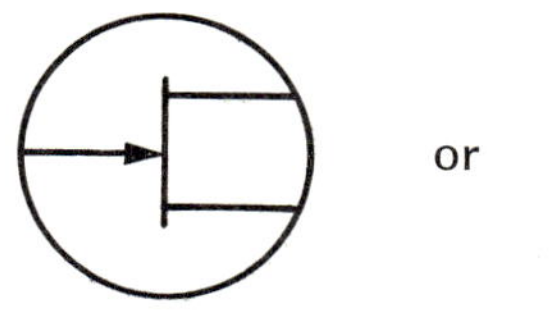

or

Field-effect transistor N-channel
(b) USAS

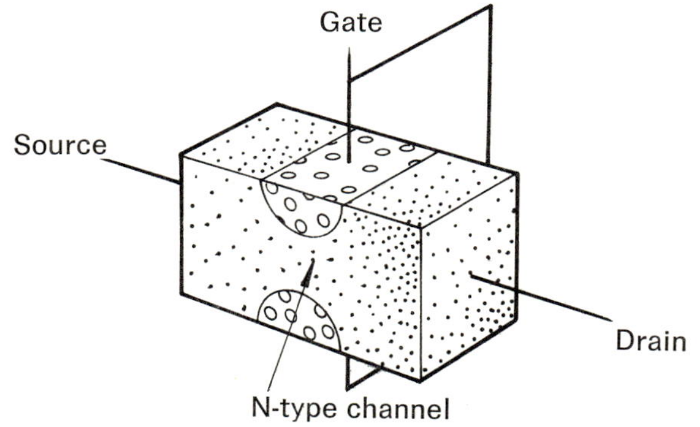

(c) Pictorial

181. Field-effect transistor P-type base
The creation of a neck or channel in the base material of these types of transistors causes a field-effect to commence when proper electrical voltages are applied.

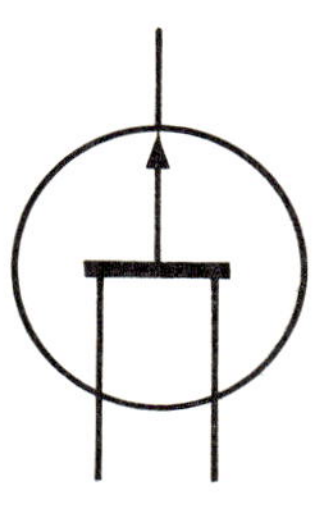

(a) CGSB

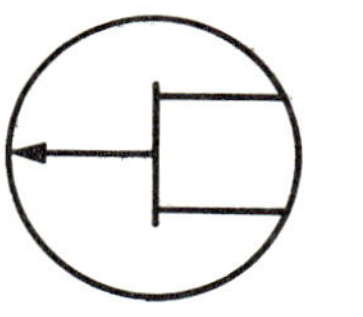

or

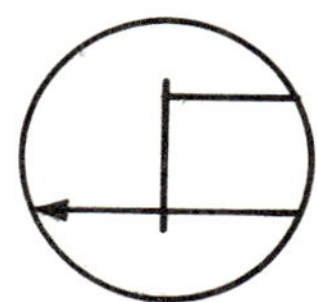

Field-effect transistor P-channel
(b) USAS

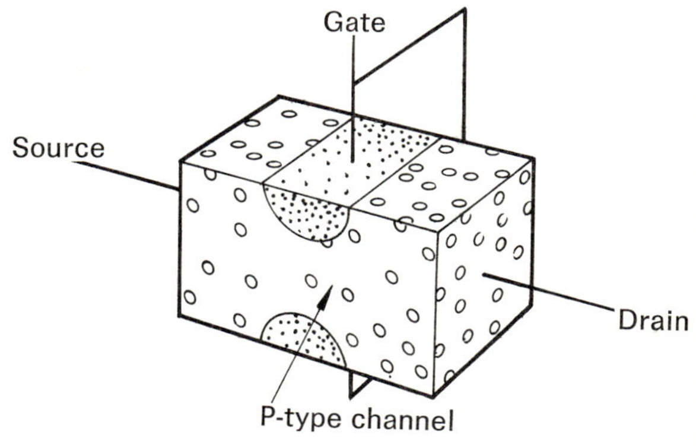

(c) Pictorial

182. Semiconductor triode NPNP type switch

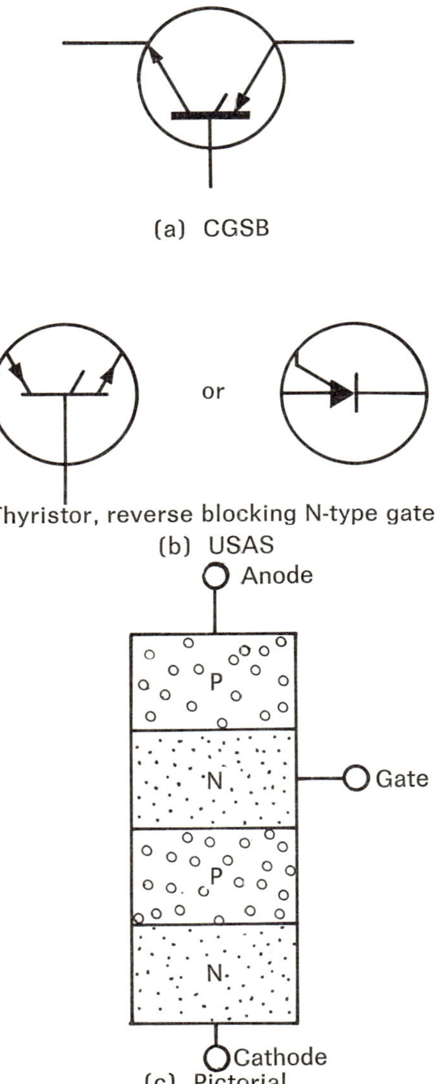

(a) CGSB

Thyristor, reverse blocking N-type gate
(b) USAS

(c) Pictorial

183. Semiconductor triode PNPN type switch
These transistors are constructed of two PN or NP layers. The tripping action brought about by increasing the collector voltage is used to operate switching and relay devices. Current multiplication is a characteristic of this type of transistor.

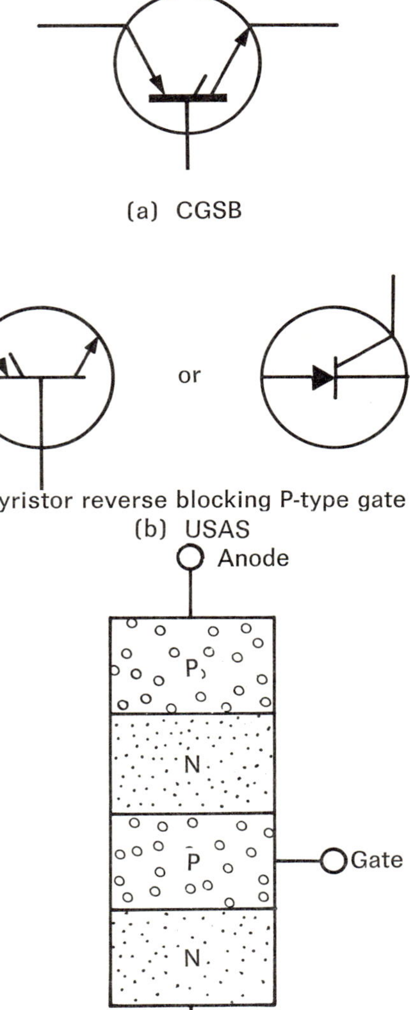

(a) CGSB

Thyristor reverse blocking P-type gate
(b) USAS

(c) Pictorial

184. NPN transistor with transverse biased base
This is a four-connection or junction tetrode type transistor having an extra base connection. The extra base connection is biased negative with respect to the normal base connection, and has a higher potential. The junction tetrode transistor finds application in high-frequency R.F. amplifiers, I.F. amplifiers, mixers, oscillators, and wide-band video amplifiers.

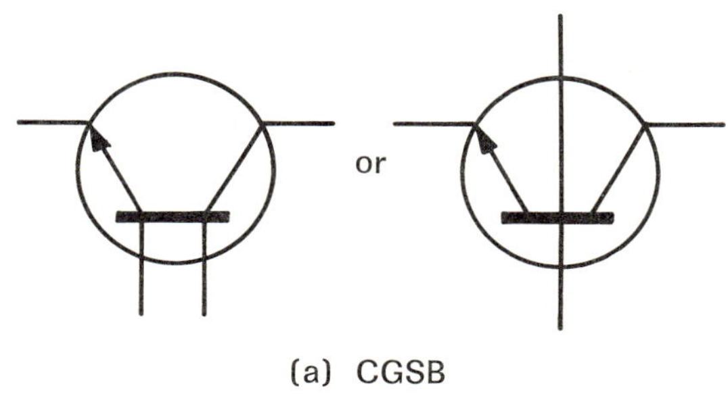

(a) CGSB

The same as the Canadian

(b) USAS

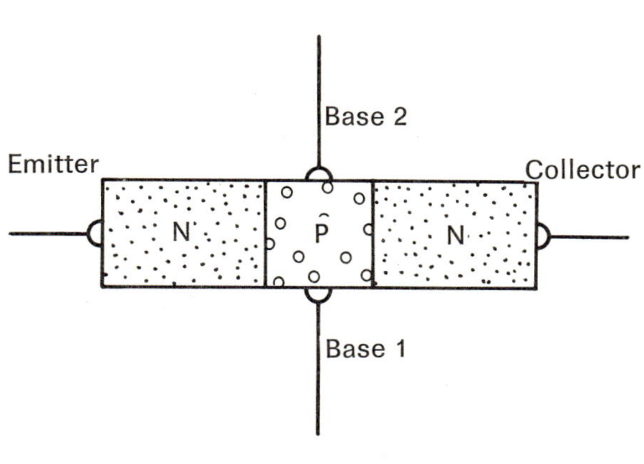

(c) Pictorial

185. PNIP transistor with ohmic connection to intrinsic region
The intrinsically pure semiconductor material in the 'I' connection of this tetrode transistor acts as an insulator and a gate. In microwave switching circuits this semiconductor device finds application.

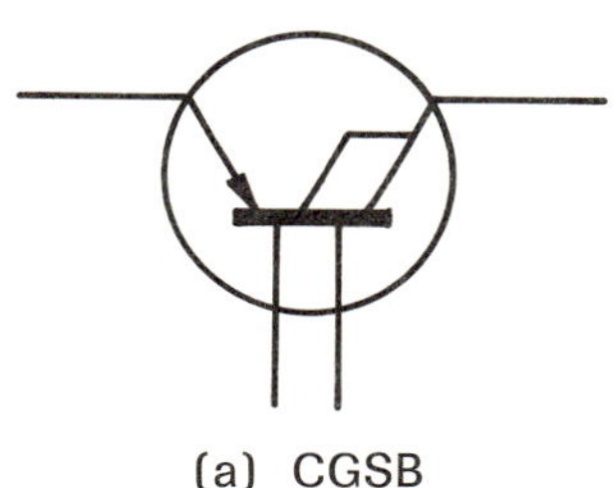

(a) CGSB

The same as the Canadian

(b) USAS

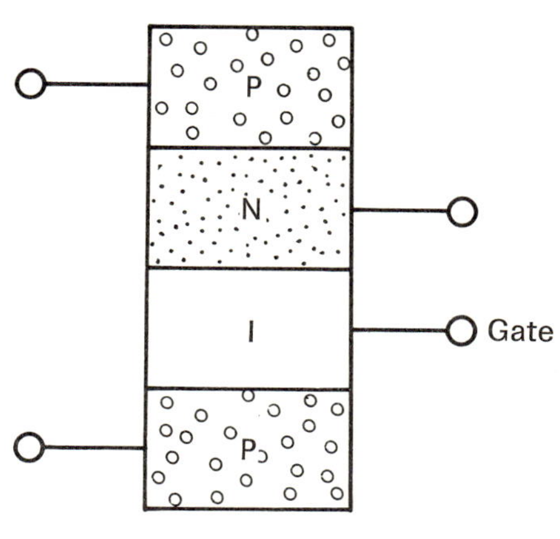

(c) Pictorial

186. NPIN transistor with ohmic connection to intrinsic region

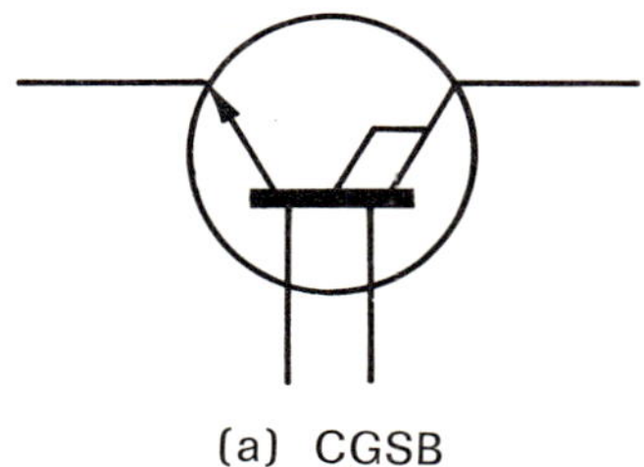

(a) CGSB

The same as the Canadian

(b) USAS

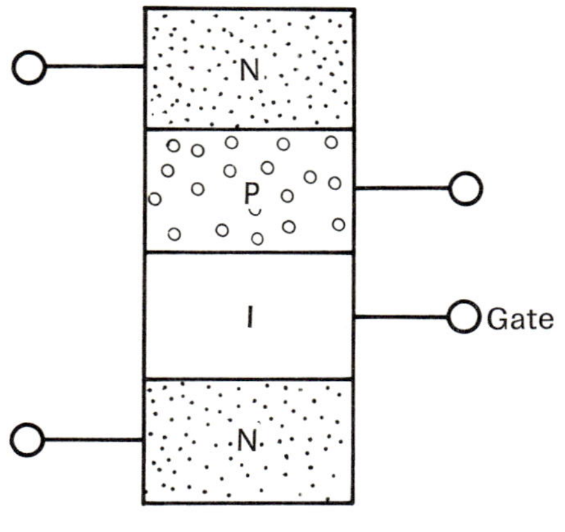

(c) Pictorial

187. PNIN transistor with ohmic connection to intrinsic region

The intrinsically pure semiconductor material in the 'I' connection of these tetrode transistors acts as an insulator and a gate. In microwave switching circuits these semiconductor devices find application.

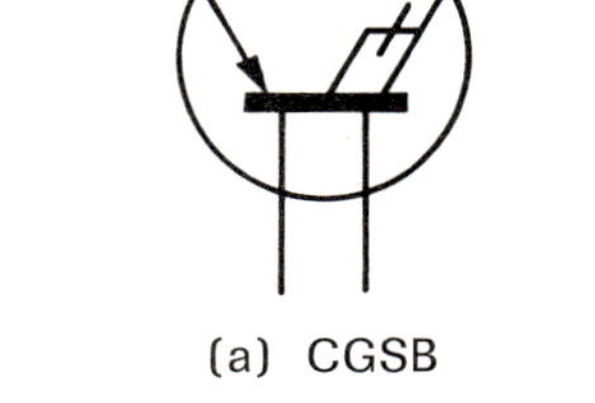

(a) CGSB

The same as the Canadian

(b) USAS

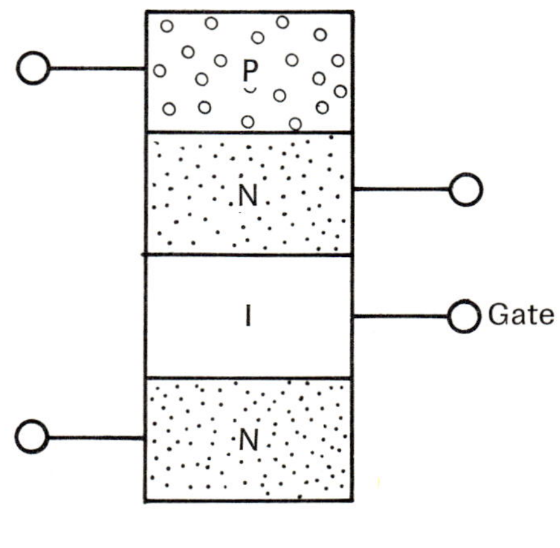

(c) Pictorial

188. NPIP transistor with ohmic connection to intrinsic region

The intrinsically pure semiconductor material in the 'I' connection of this tetrode transistor acts as an insulator and a gate. In microwave switching circuits this semiconductor device finds application.

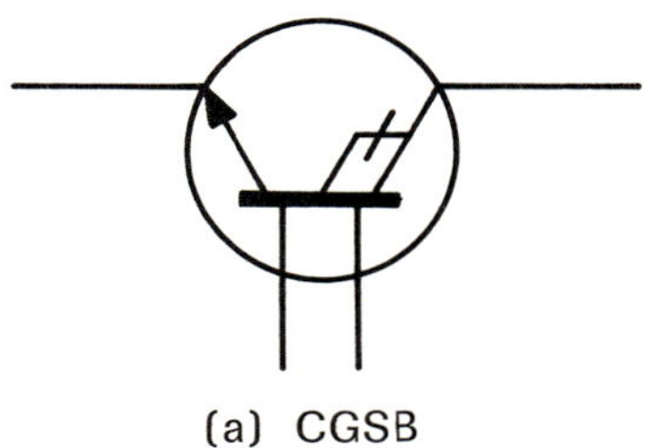

(a) CGSB

The same as the Canadian

(b) USAS

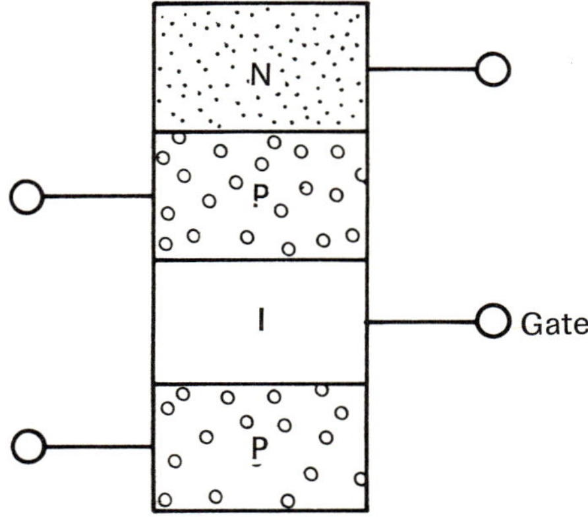

(c) Pictorial

KNOWLEDGE TESTERS

1. How many symbol sizes on a specific circuit drawing are permissible?
2. What is the relationship between electrical size and symbol size of a component?
3. Why is symbol ornateness not permitted?
4. When can a graphic symbol be separated?
5. How are conductor leads drawn to a symbol?
6. In what fashion is lettering placed about a symbol?
7. Outline the methods used in placing the numerical values about a symbol.
8. Where is the component name abbreviation located about a symbol?
9. What is meant by numerical sequencing of component symbols?
10. List the organizations standardizing electrical and electronic graphic symbols in Canada and the United States of America.
11. Name the actual standards handbooks for the two organizations asked for in Question 10.
12. What is the major difference between a bell and an annunciator?
13. How many elements does a dipole antenna have?
14. How many cells are shown to illustrate a battery graphically?
15. What is the purpose of a bus?
16. What is the function of a circuit breaker?
17. Give two important aspects of a good connector design.
18. If a N.O. power relay is in an energized state, what is the condition of its contacts?
19. What does the term 'ferro-ceramic' mean?
20. Does a chassis ground always imply that it is at a negative potential? Support your answer with a practical example.
21. Give two methods of making a good earth ground.
22. How is the polarity of a D.C. meter indicated?
23. What symbol is used to indicate an A.C. meter?
24. Differentiate in statement form between a motor and a generator.
25. How does a motor and a generator vary from a motor generator?
26. On the basis of the difference between the wire sizes used for the series and shunt-field motor coils, what can be deduced regarding the current flow?
27. What are the special peculiarities of a squirrel-cage induction motor?
28. Suggest why a capacitor-start motor is desirable for use in a compressor unit.
29. Why does a bimetal element bend under the influence of heat?
30. List some electrical appliances used in your home that use a bimetal element as a protection device.
31. What is the difference between a D.P.S.T. and a S.P.D.T. power relay?
32. In what device is an erasing head reproducer used?
33. How many electrical connections do the following variable resistors have?
    (a) Potentiometer
    (b) Rheostat
34. What does the term 'zener' indicate, as it applies to diodes?
35. In order for an electrostatic shield to be effective, what must be electrically connected to it?
36. Describe the difference between a push button and a toggle switch.
37. What is the basic difference between a moving coil, and a P.M. speaker?
38. What does the name 'synchro' suggest? How does it apply to the electrical device having that name?
39. Name a sensing device utilizing the action of dissimilar metals.
40. How many windings does an autotransformer have?
41. How does an indirectly-heated cathode operate?
42. What element is used in a pool cathode?
43. What is the function of the grid element in a vacuum tube?

44. What methods are used to control the positioning of the electron beam in C.R.T.'s?
45. In what type of tube is a dynode element used?
46. When a vacuum tube is filled with inert gas, what symbol is placed within the envelope to denote this state?
47. How many elements does a tetrode type tube have?
48. Describe in simple terms how a phototube performs.
49. Name some electrical devices that can be categorized as rotating machines.
50. What is the purpose of a blowout coil?
51. Why are two conductors sometimes transposed?
52. How is an individual conductor or a cable shielded?
53. What type of insulative material is used in a coaxial, radio frequency cable?
54. Give another name for a 'breakdown diode'.
55. A varactor appears externally as an ordinary junction diode. However, what is its special electrical characteristic?
56. How many connections does a PNP transistor have?
57. Name the lead that is electrically connected to the metallic case of some PNP transistors.
58. What special construction is used in a 'FET' to create the field effect?
59. How many 'base' connections does a NPN transverse biased base transistor have?
60. List two possible uses of a standard PNP transistor.

UNIT 5

Basic Direct-Current Circuits

IN SPITE OF THE WIDESPREAD USE OF ELECTRICAL DEVICES REQUIRING alternating current for their operation, a considerable number of devices must use direct current because of their design, such as an automobile. Also, in cases where a dependency on alternating current cannot be always assumed, such as in a burglar alarm system, direct current circuits find application. In addition, a strong likelihood exists that the beginning student may do experimentation using batteries as a source of direct current, and therefore a knowledge of the correct form of applicable direct current circuits will, no doubt, be of use.

Single-cell battery circuit

Circuits employing a single-cell battery are shown as illustrated in Fig. 5-1(a). Normally no polarity symbols, nor voltage values are displayed, but may be shown if required.

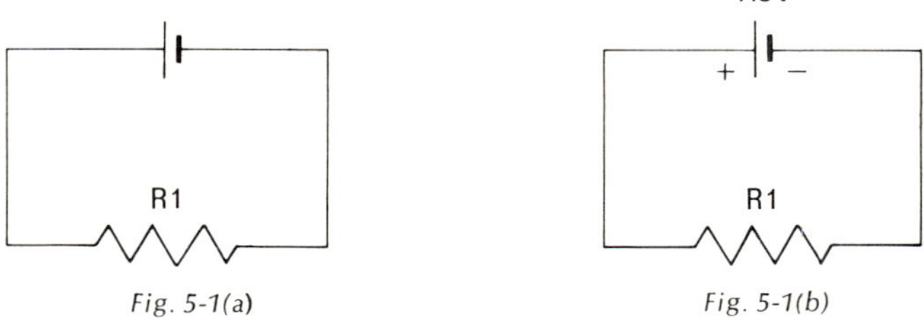

Fig. 5-1 (a) A single-cell circuit. (b) A single-cell circuit showing polarity and voltage

Multiple-cell battery circuit

When a battery must be illustrated to show a voltage greater than that of an individual cell, two cell symbols are shown in series. See Fig. 5-2. Regardless

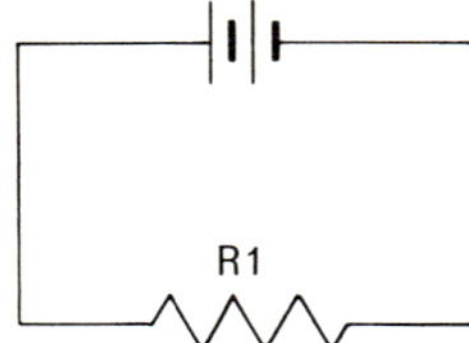

Fig. 5-2 A battery circuit

of the number of cells constituting the battery, it is still shown as in Fig. 5-2. Note that **no line** is shown connecting the two parts of the symbol together. Again, if it is necessary to illustrate polarity and to call up the voltage value, the same procedure is to be carried out as for a single-cell battery.

Single cells in a series circuit

Single cells can be wired in series to produce an increase in voltage. This is illustrated in Fig. 5-3. If such is the case, gaps are to be shown between each cell symbol. It is unlikely, however, that in actual practice six, 1.5 volt cells would be connected in series to produce a nine volt supply, unless a commercially manufactured nine volt battery was not available.

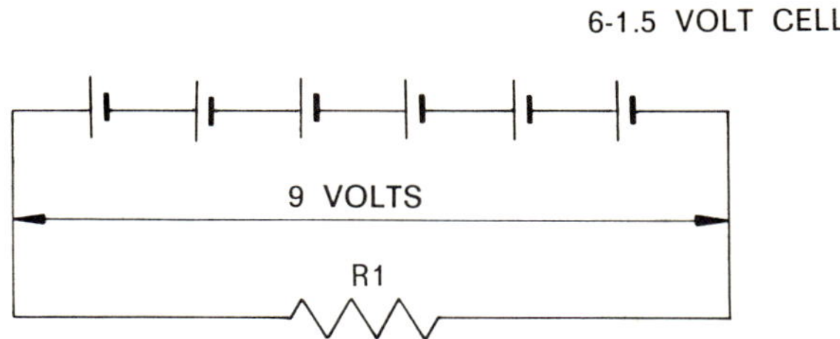

Fig. 5-3 Six single cells connected in a series circuit

On the assumption that a nine volt commercially manufactured battery was available, we could display the symbol for the battery as drawn in Fig. 5-4.

In actual practice, however, six individual cells would not be shown, but only two. This method would conform to the rule already pointed out, and illustrated in Fig. 5-2. Figure 5-5 shows the conventional method of illustrating a multiple-cell battery when a specific voltage is required to be shown.

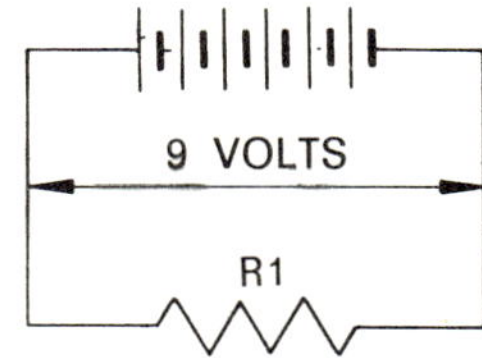

Fig. 5-4 A six-cell battery circuit

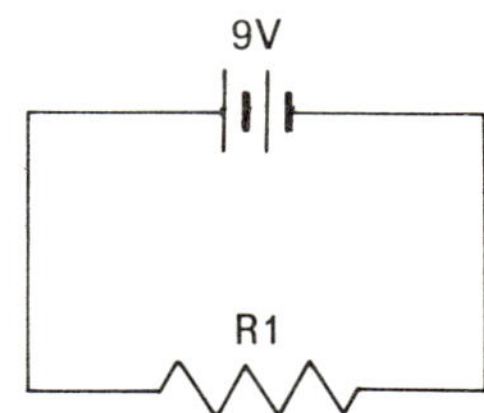

Fig. 5-5 The C.S.A. method of indicating a multi-cell battery

Batteries in a series-opposition circuit

On occasions batteries of different potential are wired in a **series-opposing** manner. Figure 5-6 illustrates the correct form of that type of circuit. In contrast to the series-addition circuit shown in Fig. 5-3, the student should be aware of the back-to-back relationship of the negative part of the battery symbols.

Cells or batteries in a parallel circuit

When it is necessary to display a number of cells or batteries wired in parallel fashion they are to be drawn as arranged in Fig. 5-7. Sufficient space must be left between the symbols in order that they can be stacked with their positive symbol element directly under each other.

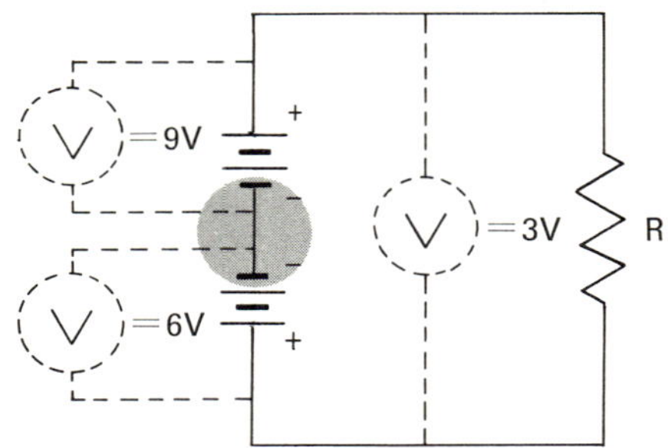

Fig. 5-6 A series-opposition battery circuit

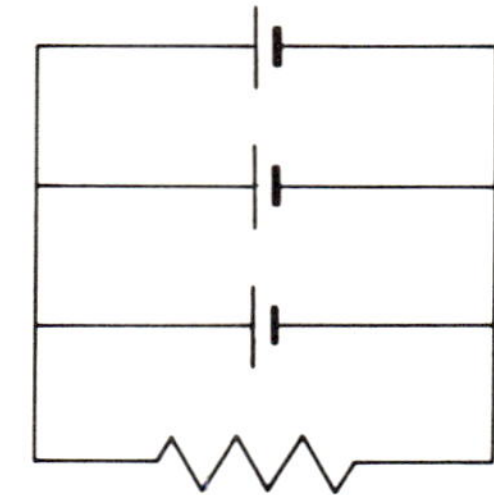

Fig. 5-7 Single cells in a parellel circuit

Cells or batteries in a series-parallel circuit

It may be a requirement to 'hook up' a number of cells or batteries in order to obtain increases in both voltage and current. The form for this situation is shown in Fig. 5-8.

By reversing the **orientation** of the cells in (B) of Fig. 5-8, an alternate method of displaying a series-parallel battery hook up can be made. See Fig. 5-9.

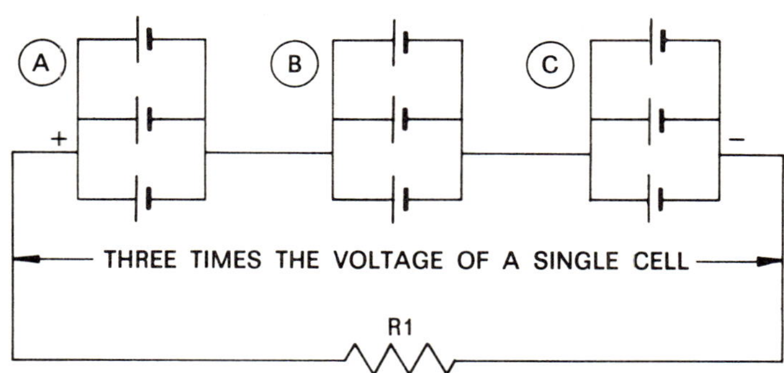

Fig. 5-8 Groups of single cells in a series-parallel circuit

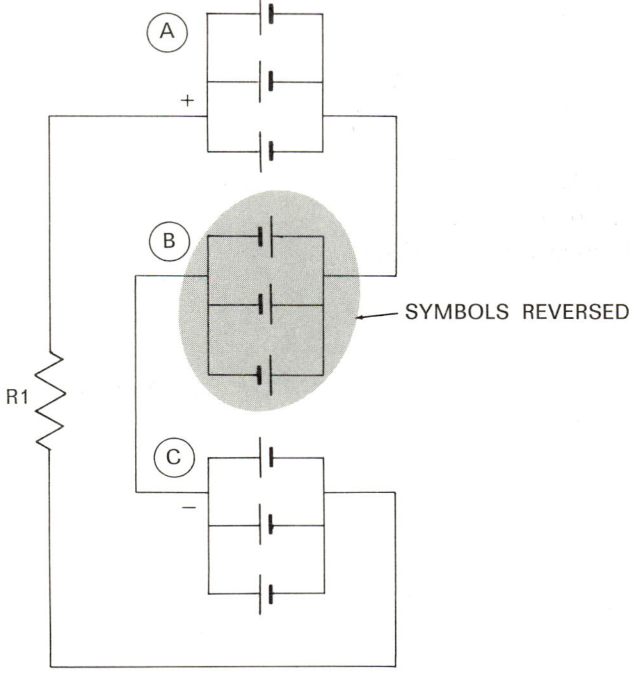

Fig. 5-9 An alternate method of connecting single cells in a series-parallel circuit

Resistances in a series circuit

In a simple series circuit involving two or more resistances, the placement of the resistances in either of the three branches of the circuit does not alter their electrical value. Figs. 5-10, 5-11, 5-12, and 5-13 show resistors R1 and R2

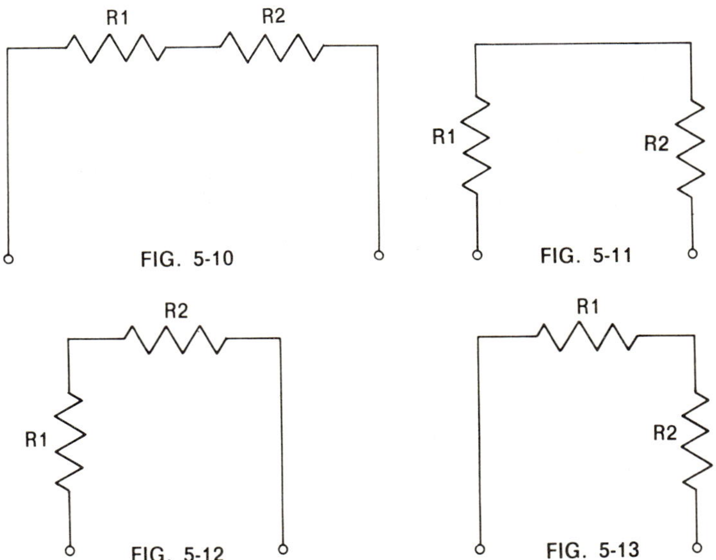

Figs. 5-10 to 5-13 Variations in the placement of resistors in a series circuit

drawn in different positions. It is, however, standard practice to balance the position of the resistors in the branches. Either of the four cases illustrated in Figs. 5-10, 5-11, 5-12, and 5-13 are acceptable. In the case of Fig. 5-10, ample space must be left between R1 and R2, otherwise misinterpretation of the values may result.

Situations like that shown in Fig. 5-14 must be avoided.

Similarly, the placement of a resistor about any corner must not be carried out. This extremely poor practice is illustrated in Fig. 5-15.

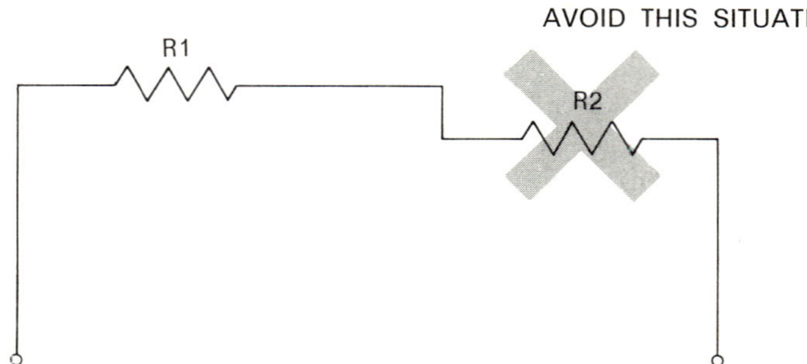

Fig. 5-14 The improper placement of the resistor

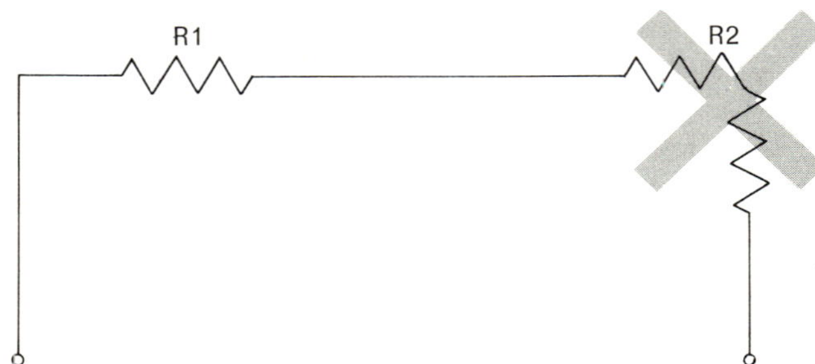

Fig. 5-15 The improper placement of the resistor about a corner

Resistances in a parallel circuit

In a parallel circuit involving two or more resistances, the resistors must be drawn centrally and of equal length, as shown in Fig. 5-16. Also, sufficient space must be provided between the resistors, otherwise misinterpretation of resistor numbers and values may result.

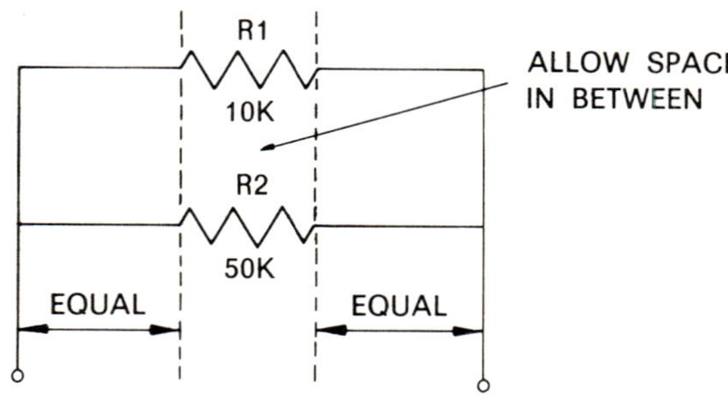

Fig. 5-16 The correct placement of resistors in parallel

Resistances in a series-parallel circuit

When drawing a series-parallel circuit, such as is shown in Fig. 5-17, it is preferable that the resistance R3 be placed centrally between R2 and the left-hand terminal for appearance reasons.

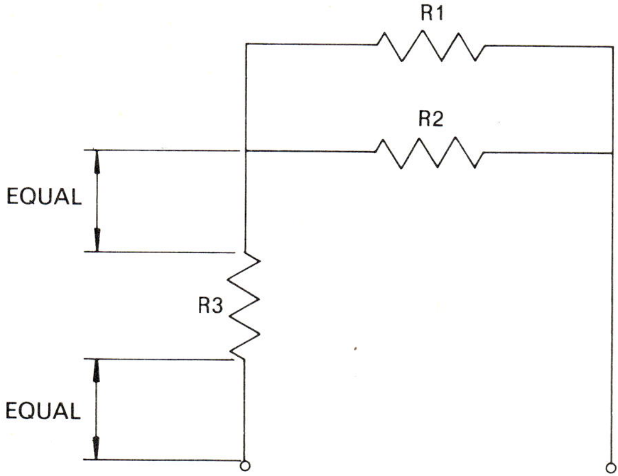

Fig. 5-17 Resistors in a series-parallel circuit

Grounded series circuit

Sometimes it is advantageous to use a common ground in a circuit. This is shown in Fig. 5-18. In this case, the negative side of the battery is grounded, as well as the right-hand terminal of resistor R2. The symbols indicating **ground** should be drawn at the same level for appearance and ease of reading.

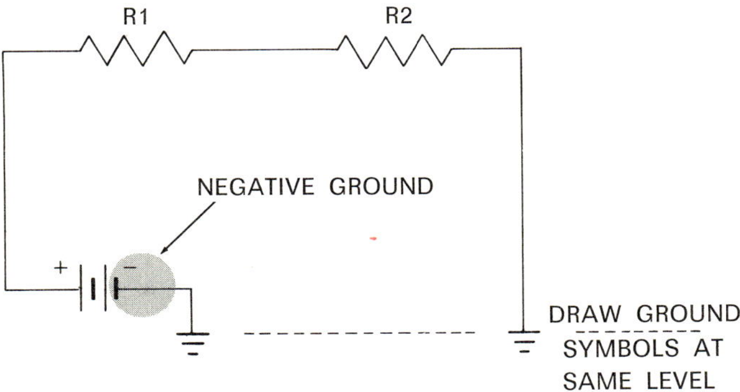

Fig. 5-18 A grounded negative series circuit

A positive ground is sometimes used in automobiles. This method is illustrated in Fig. 5-19.

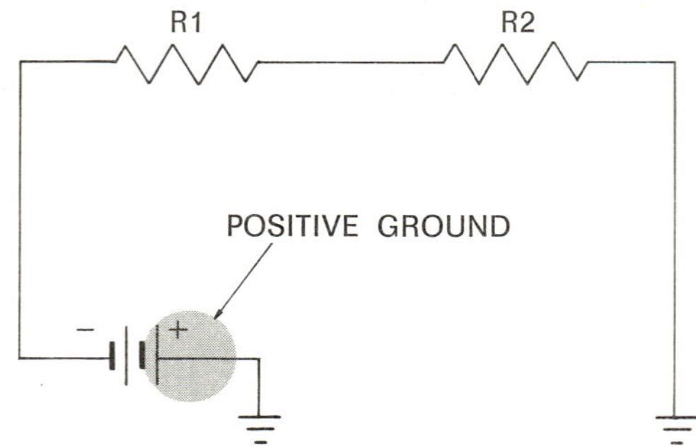

Fig. 5-19 A grounded positive series circuit

Single-pole, single-throw switched lamp circuit

A lamp circuit involving this type of switching device is shown in Fig. 5-20. Since this kind of switch is not usually a **normally-closed** switch, it is important to illustrate it in an 'off' position. It is possible to read a circuit wrongly if this procedure is not carried out. Fig. 5-20 shows the correct method of indicating a lamp controlled by a S.P.S.T. switch. Switch S1 could be placed in the opposing branch of the circuit without altering any interpretation of the circuit. This alternate position for switch S1 is shown in broken lines in Fig. 5-20.

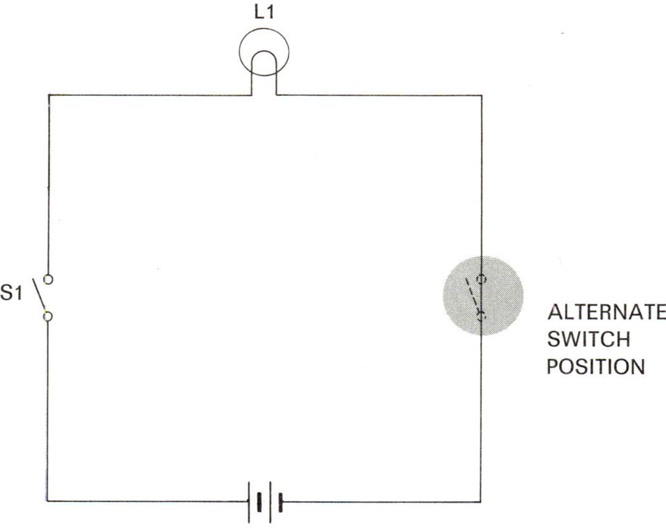

Fig. 5-20 A single-pole, single-throw switched lamp circuit

Single-pole, double-throw, switched lamp circuit

A S.P.D.T. switch can be effectively used to control two lamps so that one lamp is 'on' and one is 'off' in either switch position. Since this type of switch has no 'off' position it must be shown in either of its two closed positions. If the lamps L1 and L2 are arranged one above the other, the circuit shown in Fig. 5-21 is clearly followed. The connection at (A) should not be interpreted as a soldered connection since terminal screws and clips would be available at lamps L1, L2, and the battery.

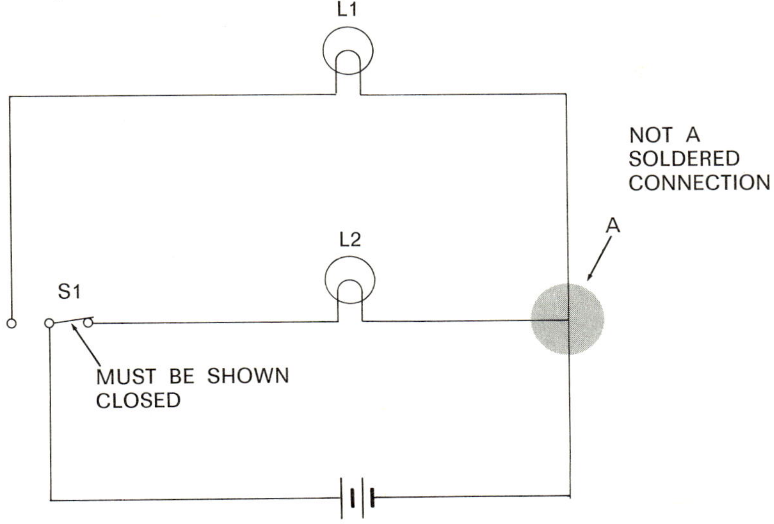

Fig. 5-21 A single-pole, double-throw switched lamp circuit

Double-pole, single-throw, switched lamp circuit

In Fig. 5-22, a single lamp L1 is shown connected to a battery through a D.P.S.T. switch. Since this type of switch has internally **ganged** switch elements, it must show a broken line between these ganged or linked parts to convey this

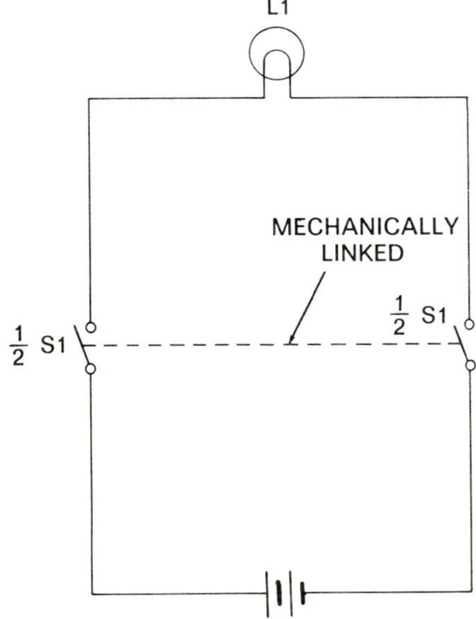

Fig. 5-22 A double-pole, single-throw switched lamp circuit

idea. In addition, the switch must be shown in its 'open' position for clearness of reading.

Double-pole, double-throw, switched lamp circuit

The proper method of placing symbols for two lamps controlled independently by a D.P.D.T. switch is illustrated in Fig. 5-23. By drawing lamp L2 in an inverted position as shown in Fig. 5-23 a number of crossovers are avoided. Notice that the battery is shown connected to the centre terminals of switch S1. This type of circuitry also incorporates the concept of **circuit isolation**, that is, in either of switch S1's positions, either lamp L1 or L2 is completely isolated from its source of power. In addition because this kind of switch is in an 'on' position at all times, unless it is a 'momentarily on' type of switch, it must be shown in a closed position. An alternate way of displaying two lamps controlled by a D.P.D.T. switch is illustrated in Fig. 5-24. While lamp L2 does not have to be inverted, it is necessary to bring the centre leads from S1 through the switch to the battery. There is little argument that the circuit presentation shown in Fig. 5-24 is not artistically better, but nevertheless, precaution must be taken in drawing the battery leads to avoid loss of clarity.

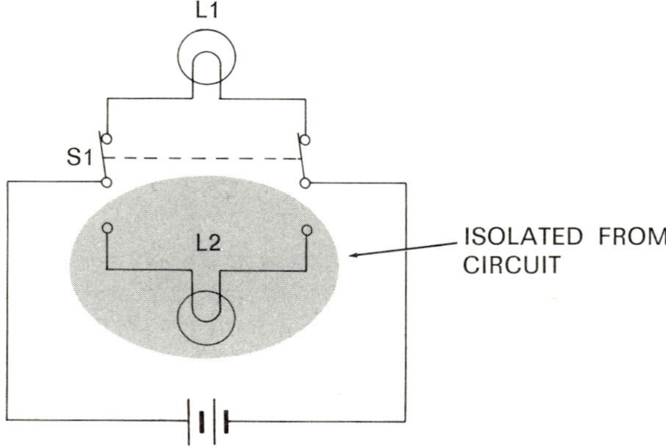

Fig. 5-23 A double-pole, double-throw switched lamp circuit

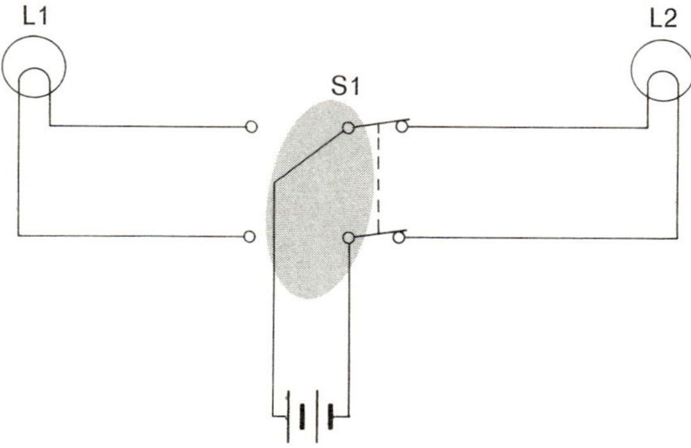

Fig. 5-24 An alternate method of showing a D.P.D.T. switched lamp circuit

MEASURING CIRCUITS

General

The treatment of measuring circuits involving the establishment of current and voltage values is not unique for direct-current circuits. The same circuitry holds good for alternating-current situations, but since the student of electricity becomes involved in battery powered circuits initially, it is desirable to introduce them at this point.

Ammeter circuit

Since an ammeter measures the flow of electrical current through a conductor, it is shown in series fashion as illustrated in Fig. 5-25. Fig. 5-26 shows an alternate method of drawing an ammeter in a circuit. Because the lamp and the ammeter are important parts of the circuit, their position at the top of the circuit is warranted. Variations of the ammeter symbol as shown in Figs. 5-27, and 5-28 are not to be used.

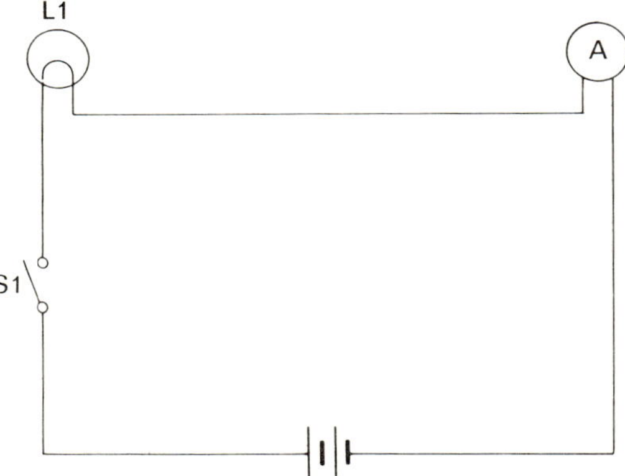

Fig. 5-25 An ammeter circuit

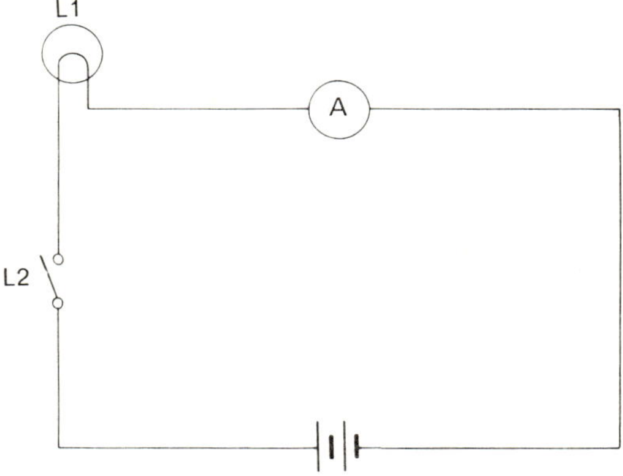

Fig. 5-26 An alternate method for an ammeter circuit

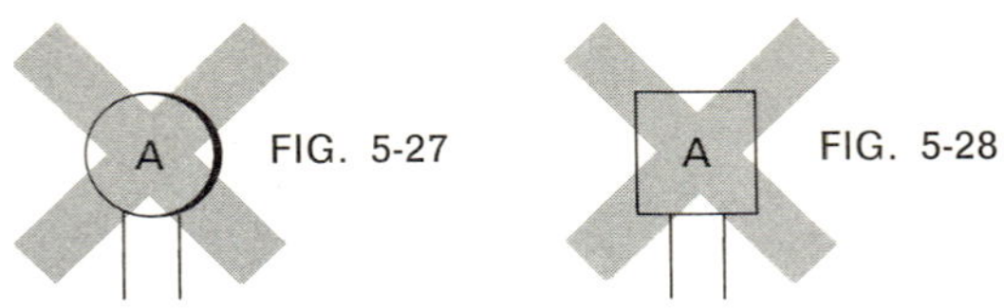

Figs. 5-27 and 5-28 Improper ammeter symbols

Voltmeter circuit

A voltmeter is an electrical measuring instrument designed to establish the pressure or E.M.F. across an electrical circuit. In order to graphically represent its function correctly it is drawn as shown in Figs. 5-29 and 5-30. Figure 5-29 displays both the lamp L1 and voltmeter in a horizontal fashion at the same level, but this introduces a crossover, and at the same time the parallel or **shunt** position of the voltmeter relative to the lamp L1 is obscured. Fig. 5-30 illustrates a vertical arrangement of the lamp and voltmeter. In this case, as shown in Fig. 5-30, the parallel connection of the voltmeter relative to the **load** (lamp L1) is readily seen. The practice of using artistic licence in varying the shape or style of the voltmeter symbol is not to be done, as was similarly pointed out in reference to the ammeter symbol.

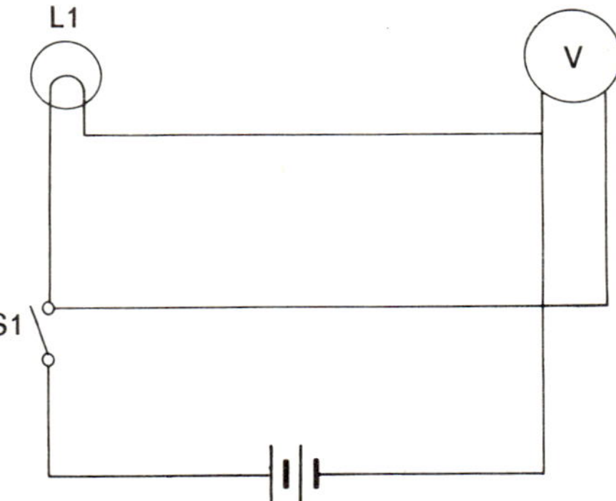

Fig. 5-29 A voltmeter circuit. The horizontal arrangement of L1 and V

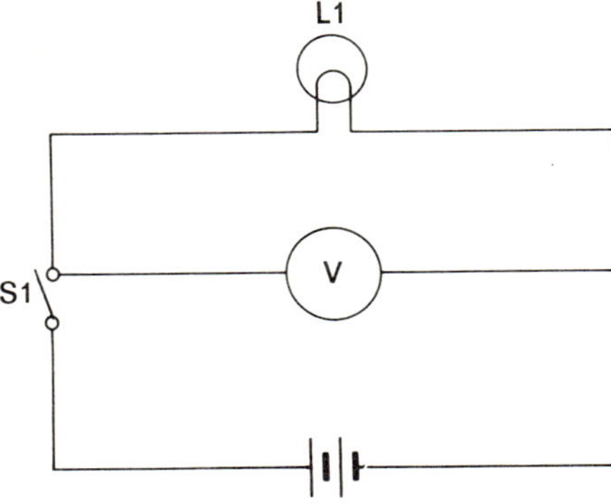

Fig. 5-30 A voltmeter circuit. The vertical arrangement of L1 and V

BATTERY POWERED ALARM AND WARNING CIRCUITS

Normally-open burglar alarm circuit

The circuit shown in Fig. 5-31 illustrates an alarm circuit powered by a battery. Switch S1 is the master switch in the circuit, and since the alarm is not activated until any of the other switches is closed, it is correctly shown in a 'closed' position. The layout of the circuit places the alarm in a prominent position at the top of the drawing. Switches S2, S3, and S4 are grouped in a vertical manner so that additional switches can be added simply. In regard to the leads connecting switches S2, S3, and S4 in parallel, the reader's interpretation should not be that the various connections are soldered.

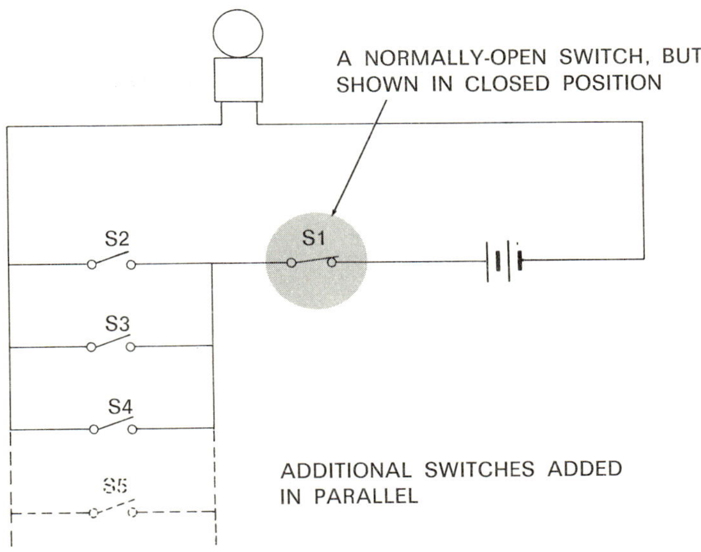

Fig. 5-31 A normally-open burglar alarm circuit

Burglar alarm circuit with a normally-open relay

Figure 5-32 illustrates a battery powered alarm circuit utilizing a normally-open relay. The contact part of the symbol is shown directly above the coil. There is no necessity of drawing a box about the contact and coil to indicate their integral nature. Also, a normally-open relay must be shown as illustrated in its **nonenergized** position. By following the layout as shown in Fig. 5-32 only one crossover is required. The layout of the four designated switches and battery are very similar to the circuit shown in Fig. 5-31. Once again, the alarm and in this case the relay are placed at the top of the circuit to indicate their importance.

Burglar alarm circuit with normally-closed relay

Fig. 5-33 shows an alarm circuit in which a relay stays energized as long as current flows through its coil. The opening of any one of the switches exclusive of S1 permits the relay contacts to close, and the alarm to ring. Because of this

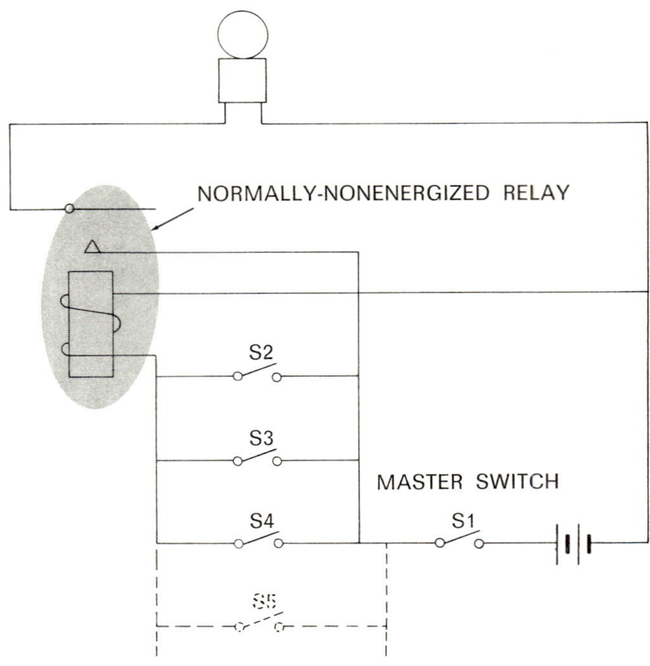

Fig. 5-32 A burglar alarm circuit with a normally-nonenergized relay

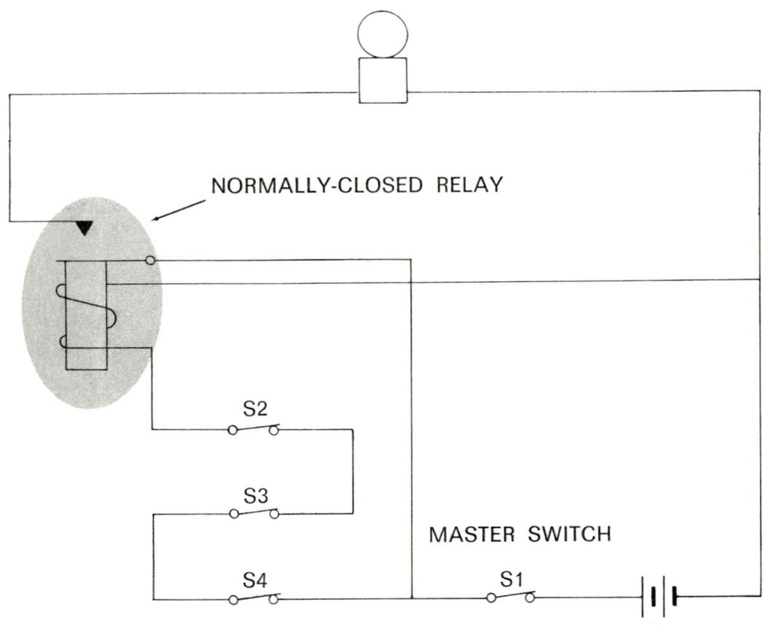

Fig. 5-33 A burglar alarm circuit with a normally-closed relay

action, the contacts must be shown in an inverted position in relation to that in Fig. 5-32. The positioning of the switches S2, S3, and S4 could be in a horizontal fashion, but this would spread out the circuit. All other component symbols have been placed in an identical manner to those of Fig. 5-32.

Automobile direction warning signal circuit

Fig. 5-34 shows a simplified diagram of an automobile direction warning signal circuit. The symmetrical placing of the symbols serves a number of purposes. First, crossovers are eliminated; second, appearance and ease of reading are improved; and third, the placement of the indicating lamps is made logical. Note that the flasher S3, switch S1, the fuse, and the battery are oriented vertically to eliminate potential crossovers. Since switches S1 and S2 are normally-open types, they must be shown as indicated.

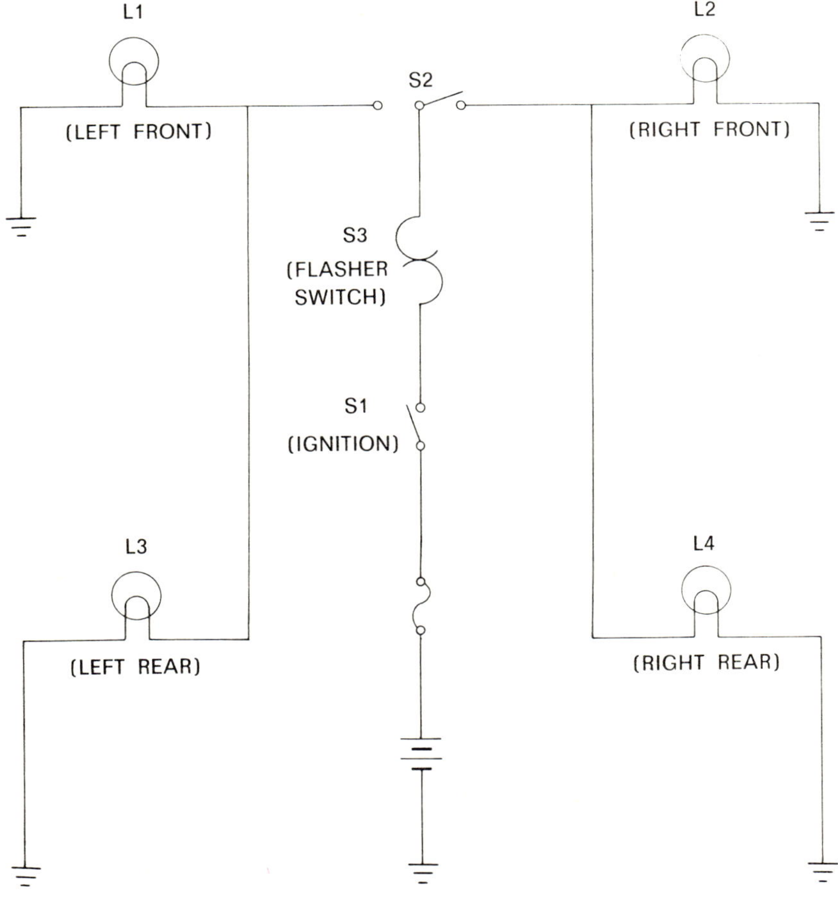

Fig. 5-34 An automobile direction warning signal circuit

DIRECT-CURRENT GENERATOR AND MOTOR CIRCUITS

Separately-excited field-coil generator circuit

When drawing the type of circuit shown in Fig. 5-35, it is important to locate the field coil directly behind the generator symbol. Also, the closed loops of the coil must be drawn towards the generator symbol. Unless the polarity is indicated at the field coil, the battery is shown as indicated. While on first glance it would appear that the field coil is not connected to the generator circuit, it must be understood that a physical connection does exist between these two components.

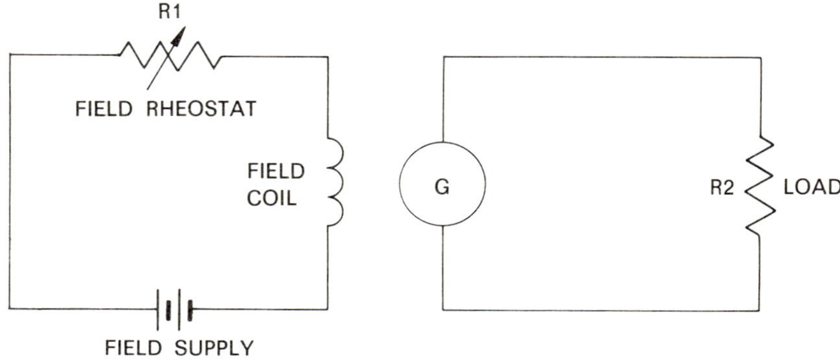

Fig. 5-35 A separately-excited field-coil generator circuit

Shunt-connected self-excited generator circuit

A shunt or parallel-connected self-excited generator circuit is shown in Fig. 5-36. The location and orientation of the field coil are the same as for the circuit illustrated in Fig. 5-35, but four loops are required instead of three. No battery symbol is required, and the coil, generator, and load R1 can all be located at the same level for ease of reading.

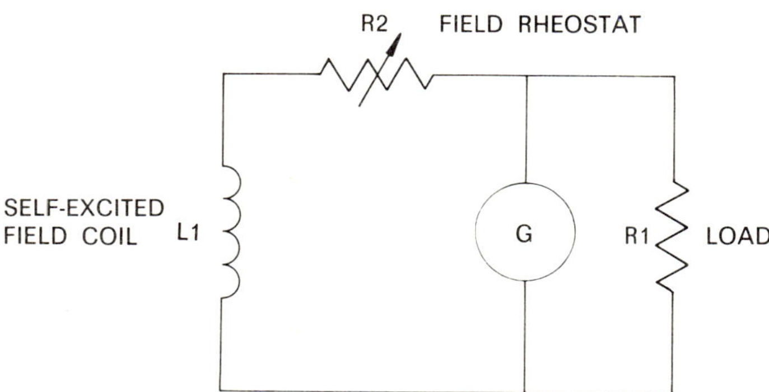

Fig. 5-36 A shunt-connected self-excited generator circuit

Compound-connected generator circuit

The circuit illustrated in Fig. 5-37 utilizes two separate field coils, one drawn in parallel to the generator, the other in series with it. The way in which the shunt-field coil is drawn is identical to that in Fig. 5-36. However, the series-field coil requires only three loops, and it is drawn with the closed side of its loops in an upwards position.

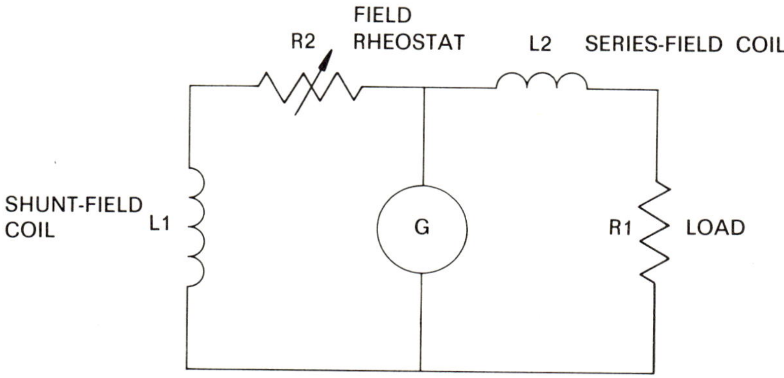

Fig. 5-37 A compound-connected generator circuit

Direct-current series motor circuit

Symbols for motors and their accompanying field coils are drawn in a similar manner to generator symbols. In Fig. 5-38 a series motor circuit is shown. Note that the motor symbol requires the letter 'B' to make it complete. Also, no polarity is shown at the source of supply, nor value of voltage given, since this would vary considerably, depending on the motor used.

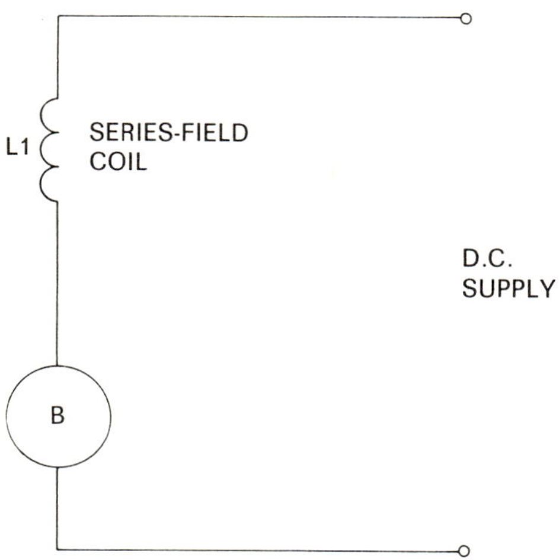

Fig. 5-38 A direct-current series motor circuit

Direct-current shunt motor circuit

The similarity of the layout of the shunt motor to the shunt generator is quite evident, as is apparent when Fig. 5-39 is studied. The treatment of the field coil is identical, but of course no field rheostat is required. In such simplified drawings as Fig. 5-39, the absence of polarity symbols at the D.C. supply means little. However, in actual practice, motor rotation would be affected and therefore means of identification would be required at both the coil and the supply terminals.

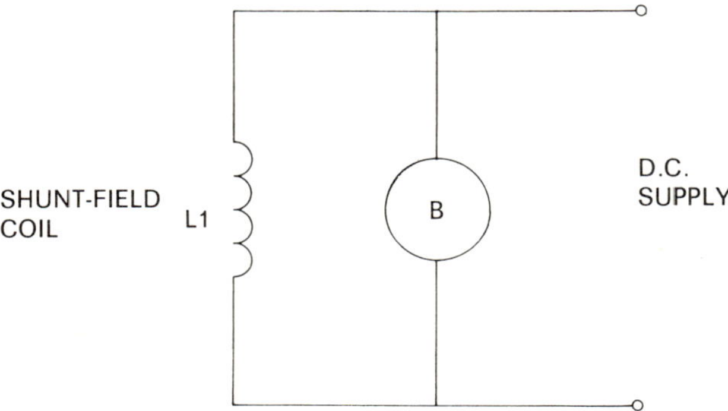

Fig. 5-39 A direct-current shunt motor circuit

Direct-current compound motor circuit

The layout of symbols for the compound motor shown in Fig. 5-40 is very similar to that of the compound generator. As would be expected, the absence of the shunt-field rheostat and a replacement of the load with a direct-current supply is evident.

It should be made clear that the D.C. motor circuit diagrams shown in Figs. 5-38, 5-39, and 5-40 are drawn without any form of overload protection, and motor starting devices. In Unit 8, detailed explanations and accompanying diagrams will explain the correct drawing procedures for them.

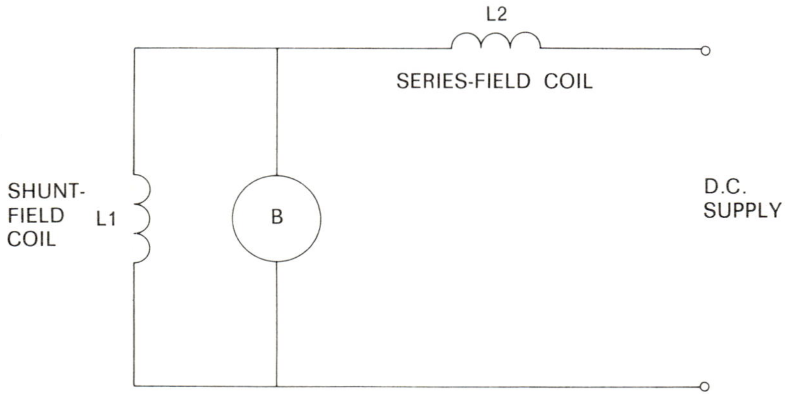

Fig. 5-40 A direct-current compound motor circuit

KNOWLEDGE TESTERS

1. How many cells are drawn to indicate all multi-cell batteries?
2. What part of a cell symbol indicates the positive pole?
3. Under what circumstances would cell symbols have interconnecting leads drawn between them?
4. If two multi-cell batteries are wired in 'series opposition', what would be the relationship of the battery poles?
5. What electrical advantages are gained by wiring a number of batteries in series parallel?
6. List at least three incorrect ways of placing resistances.
7. Compare the terms, 'negative ground', and 'positive ground'.
8. Is a switch controlling a lamp in a single-family residence, a 'normally-open', or a 'normally-closed' type?
9. In what position is the moveable arm of a S.P.D.T. switch shown?
10. What does the broken line joining the moving contacts of a D.P.S.T. switch indicate?
11. What is the main difference between a regular D.P.D.T. switch, and a 'momentarily-on' D.P.D.T. switch?
12. Why is a D.P.D.T. switch particularly useful in controlling two independent electrical circuits?
13. How is an ammeter drawn in a circuit in relationship to a voltmeter?
14. What would be the advantage of using battery power for an alarm circuit?
15. In Fig. 5-31 a master switch is used. Describe its purpose.
16. Why is switch S1 in Fig. 5-31 shown closed, although it is a 'normally-open' switch?
17. If a relay is a 'normally-open' type, is it energized or nonenergized?
18. Where is the correct drawing position of a relay coil relative to its contacts?
19. What type of relay would be used with the burglar alarm systems employing windows taped with thin metallic foil and wired into the circuit?
20. In Fig. 5-34 the location of the fuse is next to the battery. What is the purpose of doing this?
21. In Fig. 5-34, is switch S3 'normally closed', or 'normally open'?
22. What logical arrangement of the symbols is made in Fig. 5-34?
23. Is a separate field supply used in a self-excited shunt-connected D.C. generator?
24. What criteria establishes the supply voltage for a D.C. motor?
25. If the polarity of the supply for a D.C. shunt motor is given, and the shunt-field coil terminals are designated, what problem would result if the draftsmen altered them by mistake?

PROJECTS

1. Draw a circuit employing single-cell 1.5 volt batteries that will provide the necessary voltage to operate a 9 volt transistor receiver, and at the same time will give a current life of three times that of a single cell.
2. Redraw the following schematic **lumping** resistors wherever possible, but without altering circuit values.
3. Study the following pictorial drawing carefully then redraw as a schematic wiring diagram using proper graphical symbols.
4. Make a schematic wiring diagram for a bicycle having a permanent magnet field generator (no field coil), a front headlamp having an 'off', 'long-range lamp', 'short-range lamp' positions, and a tail lamp. The system is to be a single-conductor negative ground type.
5. Make a wiring diagram for the lighting system required for the night operation

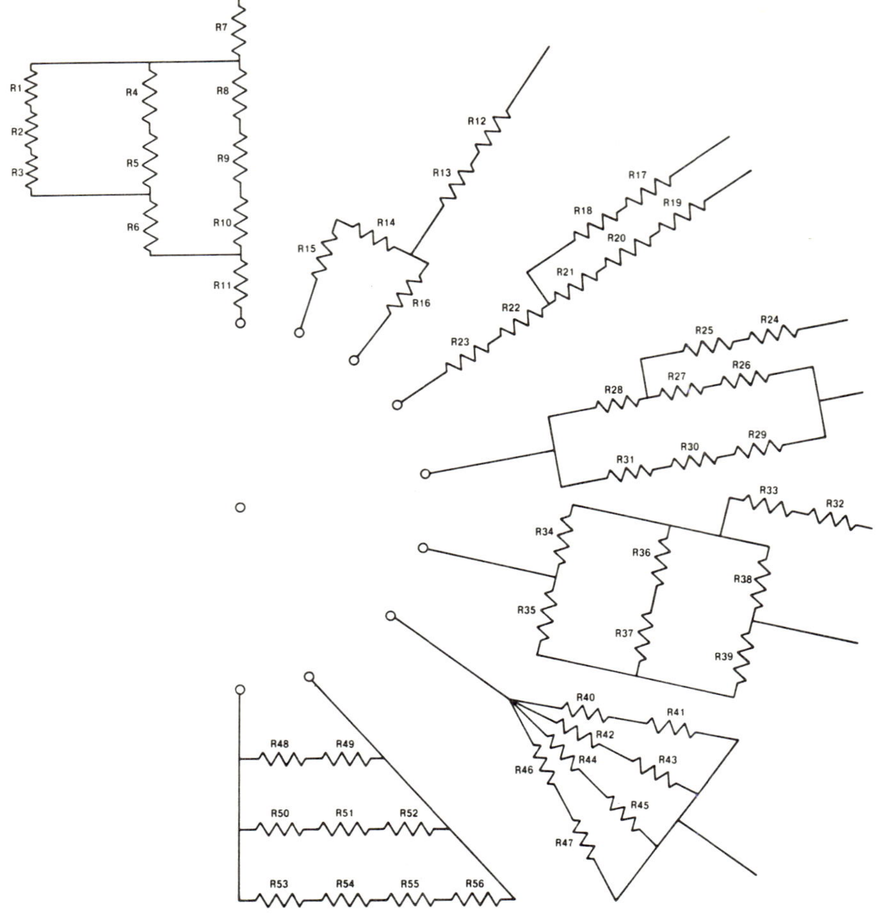

Fig. 5-41 Diagram for Project 2

of a small motorboat. The system is to be powered by a 12 volt storage battery, and each lamp is to be individually switched and fused. Port (red), starboard (green), and stern (white) lamps are required. In addition the complete system is to be fused and controlled by a master switch.

6. Design and draw a simple circuit showing three battery powered lamps controlled independently by a double-pole, triple-throw switch.

7. Redraw the following circuit showing the correct placement of an ammeter to measure the current flow through lamps L2 and L1 and the voltage across resistors R1 and R2.

8. Make a schematic wiring diagram of a high-temperature warning light and alarm system for a large food freezing storage. The temperature sensing device is normally open, and closes at 40°F. It can also be bypassed by a S.P.S.T. switch for testing purposes. In order to provide reliability the system is battery operated. The light has a flasher built into its system to attract attention, and the complete system is controlled by a master switch.

9. Trace the motor starting circuitry in your family's car. Include only the starting motor, starting motor relay, battery, and ignition switch. Redraw your sketch using proper drafting techniques.

10. Investigate the electrical wiring on your, or your friend's, motorbike. Trace the wiring in its entirety. Make a sketch on the job, and redraw using correct symbols and techniques.

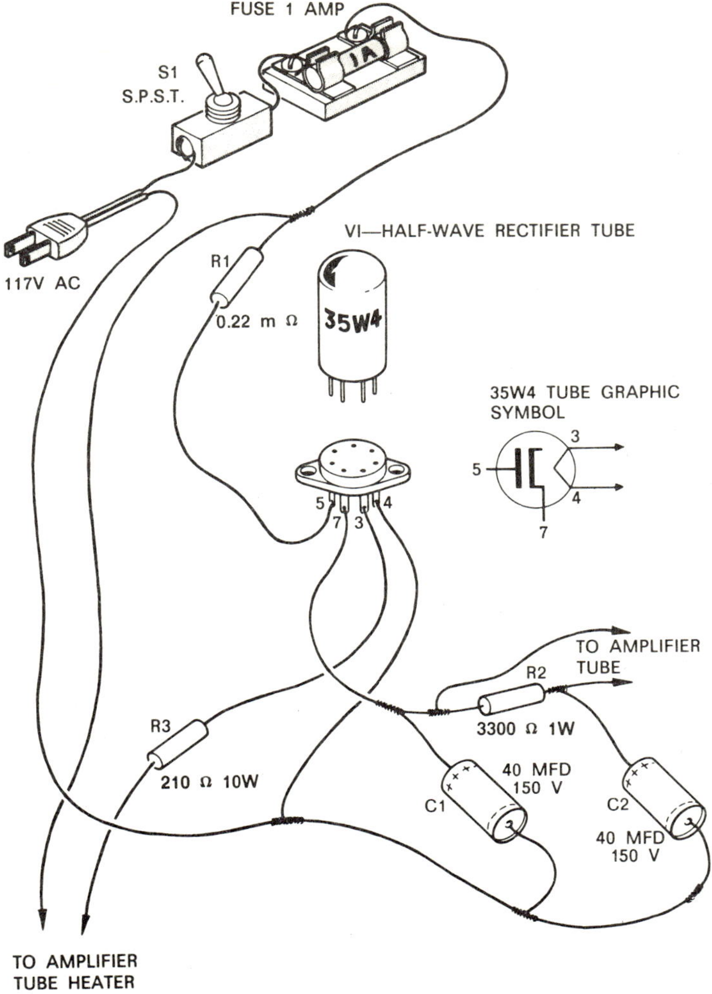

Fig. 5-42 Diagram for Project 3

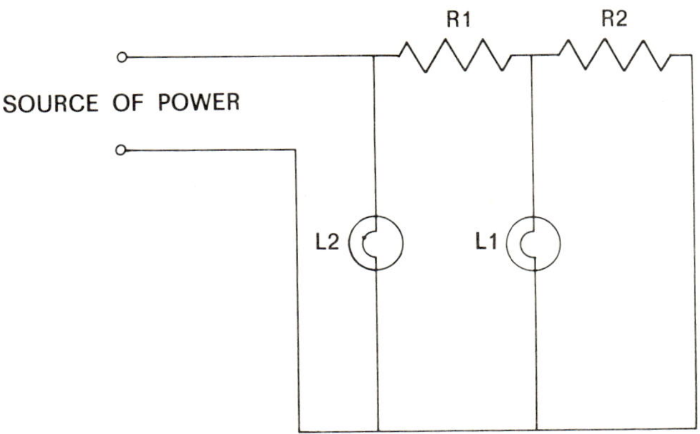

Fig. 5-43 Diagram for Project 7

UNIT 6

Basic Alternating-Current Circuits

THE FOLLOWING CIRCUIT DIAGRAMS HAVE BEEN INCLUDED IN THIS UNIT in order to aid the beginning student of applied electricity as well as of electrical drafting. The diagrams illustrated have been chosen to include a wide range of **basic** alternating-current circuits. As a consequence, the complexity of the circuits illustrated has been purposely limited.

Pure resistive circuit

Fig. 6-1 illustrates an elementary circuit containing a single resistance R1 connected in series with a D.P.S.T. switch. In this case, R1 would appear in either of the horizontal branches of the circuit, but its location as shown is considered good practice. The switch S1 must be shown in an 'open' position. In order to indicate that the circuit is an alternating-current one, suitable identification must be included at the source of power.

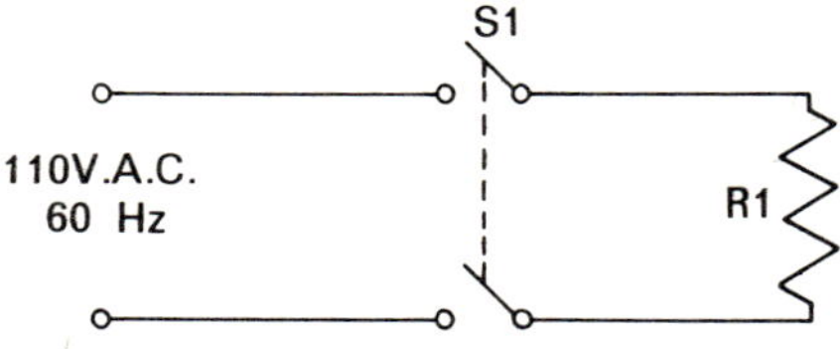

Fig. 6-1 A pure resistive circuit

Pure inductive circuit using an air-core or iron-core reactor

Fig. 6-2 illustrates an inductance connected in series with a D.P.S.T. switch. The reactor must be drawn with short vertical elements at the ends of the coil, before the horizontal branches are constructed. In addition, the coil is normally

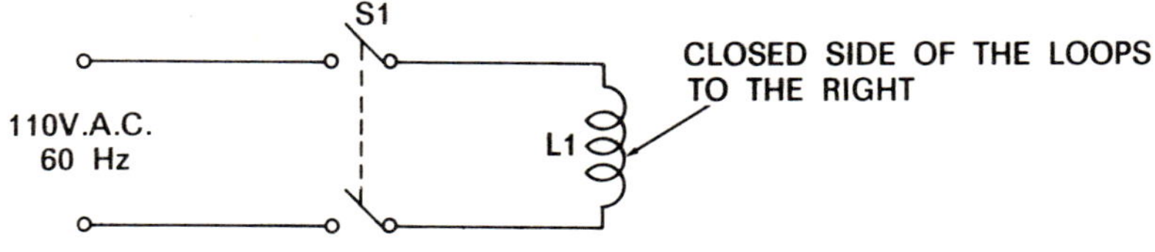

Fig. 6-2 A pure inductive circuit using an air-core reactor

drawn with the closed side of its loops facing upwards, and/or to the right. Placement of the reactor, and details regarding the switch are identical to those concerning the switch and resistance in Fig. 6-1. Figure 6-3 shows the only change required if the reactor is an iron-core type.

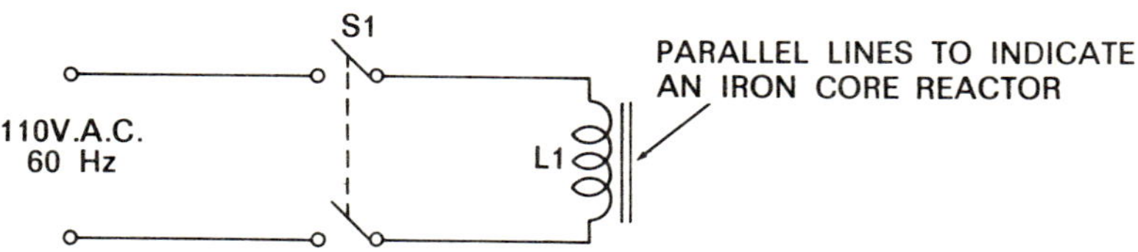

Fig. 6-3 A pure inductive circuit using an iron-core reactor

Pure capacitive circuit

The pure capacitive circuit displayed in Fig. 6-4 illustrates three possible locations for the capacitor C1. The position of C1 shown in solid lines is preferable. However, C1 could be correctly placed in either branch. The important point to remember in the placement of a capacitor is that when it is shown in a vertical branch, the straight element of the symbol is on top. When the capacitor is located in a horizontal circuit branch, the straight element of the symbol is normally placed to the left.

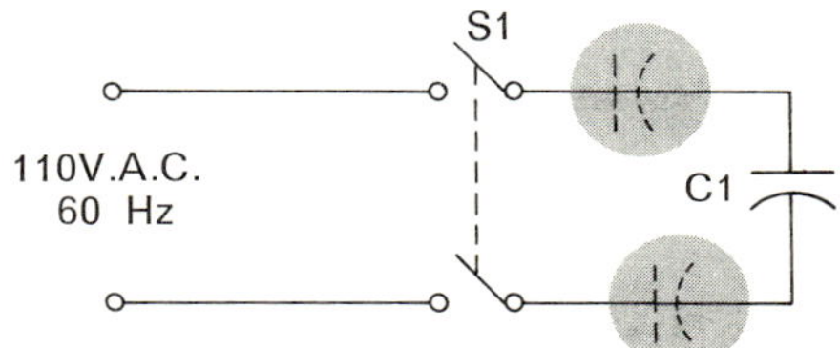

Fig. 6-4 A pure capacitive circuit

Resistive, inductive, and capacitive loads in a series circuit

The diagram illustrated in Fig. 6-5 shows a series circuit containing a resistance, an inductance, and a capacitance, in that order. The drafting problems concerning this type of circuit are negligible. However, sufficient space must be left between the symbols for the sake of clearness.

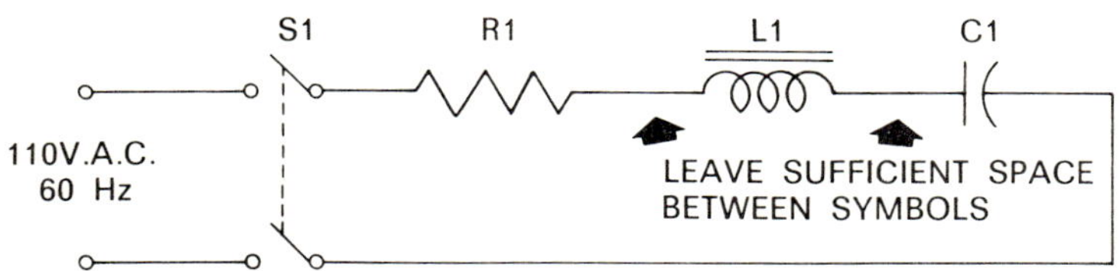

Fig. 6-5 Resistive, inductive, and capacitive loads in a series circuit

Resistive, inductive, and capacitive loads in a parallel circuit

In drawing the type of schematic shown in Fig. 6-6, it is desirable to place R1, L1, and C1 equidistant apart for appearance reasons. Each symbol must be placed centrally in its vertical branch circuit. Also, the orientation of the symbol

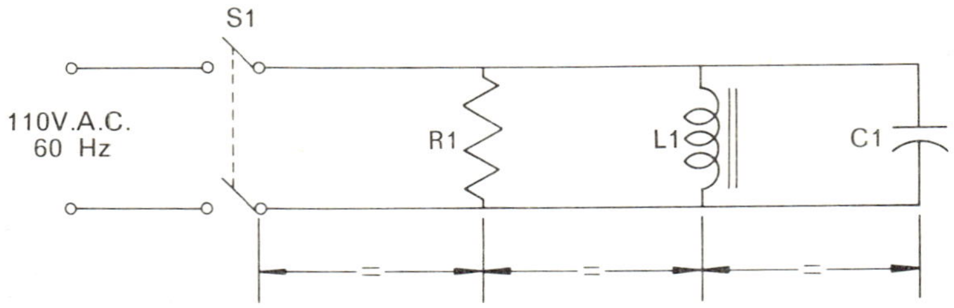

Fig. 6-6 Resistive, inductive, and capacitive loads in a parallel circuit

for reactor L1 should be made, such that the closed side of its coil loops is to the right. It is also good practice to carry out the procedure or orienting all reactors in the same direction in a given schematic, keeping in mind the difference between horizontally and vertically positioned symbols.

Resistive, inductive, and capacitive loads in a series-parallel circuit

The circuit displayed in Fig. 6-7 involves a combination of parallel and series circuitry. In the parallel branch, the symbols must be drawn centrally with ample vertical separation to permit the labelling of each component. The symbols shown in the series branch must have ample horizontal separation for similar reasons. If they are spaced equidistant apart the appearance of the circuit is improved.

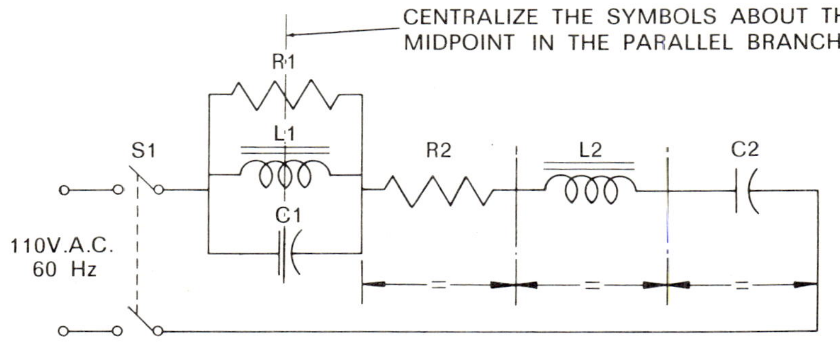

Fig. 6-7 Resistive, inductive, and capacitive loads in a series-parallel circuit

LIGHTING CIRCUITS

Series lamps circuit using a step-down transformer

In Fig. 6-8, two 10 volt lamps are shown connected in series, and controlled by a S.P.S.T. switch. The switch must be shown in its 'normally-open' position. Also, there is not a necessity to extend lines downward from the lamp symbols when joining them in series, as is required when connecting two buzzer symbols in series. See Fig. 6-14. The transformer symbol T1 is ideally located in the central position of the lower branch circuit to indicate its relative importance.

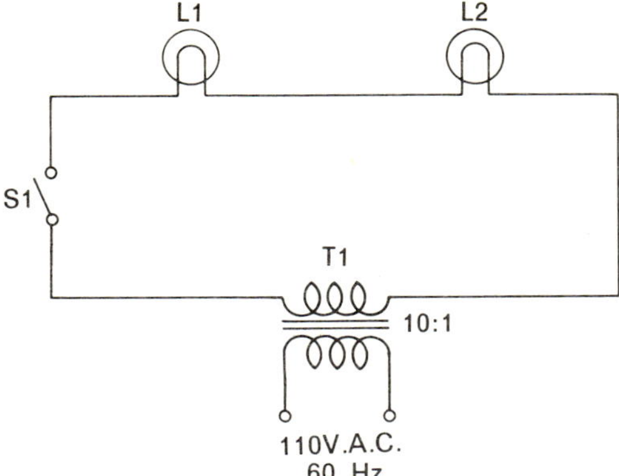

Fig. 6-8 A series lamp circuit using a step-down transformer

Lamp circuit controlled by a remote control relay

In Fig. 6-9 a circuit is illustrated whereby low-voltage switches are used in a lighting circuit for safety reasons. The relay contact appears to be in a separate circuit, but it is actually a physical part of relay R1. The coil part of the relay is placed directly below the contact, although the coil could be moved eleswhere for convenience, provided a suitable explanation is given. Switch S2 is merely indicative that lamp L1 can be controlled from another location. Note the angular method of drawing the leads from switch S1 to switch S2. The angle selected should permit the vertical portion of the leads to be entirely above the horizontal ones. Also, the same angle is normally used at both switch terminals. The type of switch used in this circuit has a neutral centre position, and hence must be shown as it is illustrated. In either closed position, one-half of the coil is energized, which in turn opens or closes the contact in the lamp circuit.

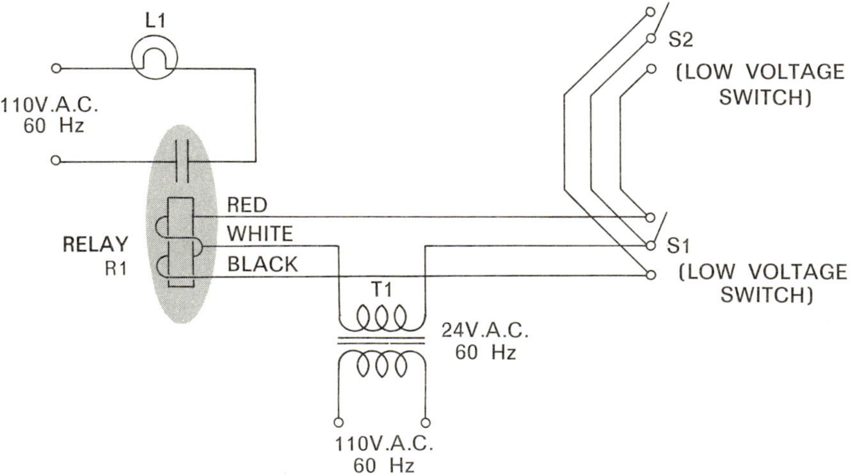

Fig. 6-9 A lamp circuit controlled by a remote control relay

Lamp circuit controlled by two three-way switches

When a lamp is to be controlled from two switching positions, special switches are used. In the electrical trade these are known as 'three-way switches'. Essentially, they act as D.T.S.P. types. Since these have no 'open' position, they must be shown in either of their 'closed' ones. Figure 6-10 illustrates such a circuit, and is shown with the lamp in its energized state. Artistic license has been used here to emphasize this point, but this practice is purely for that reason and under any other circumstances is not to be used. Colour coding of the individual conductors is shown. In the trade, the red and black conductors are referred to as the traveller leads.

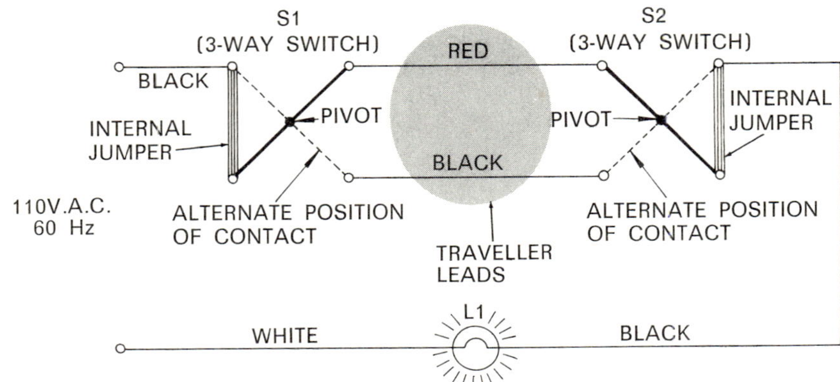

Fig. 6-10 A lamp circuit controlled by two three-way switches

Lamp circuit controlled by two three-way and one four-way switch

When a requirement exists to control a lamp from more than two locations, a third switch is inserted into the circuit. See symbol S2 in Fig. 6-11. In the electrical trade this type of switch is known as a 'four-way switch', because each switch leg or arm has two paths or ways. A switch of this sort has no 'open' position, hence it must be shown in either of its closed positions. The graphical presentation shown in Fig. 6-11 illustrates a simple, and clear cut drafting method of displaying the circuit, in spite of the variety of other ways shown in other sources.

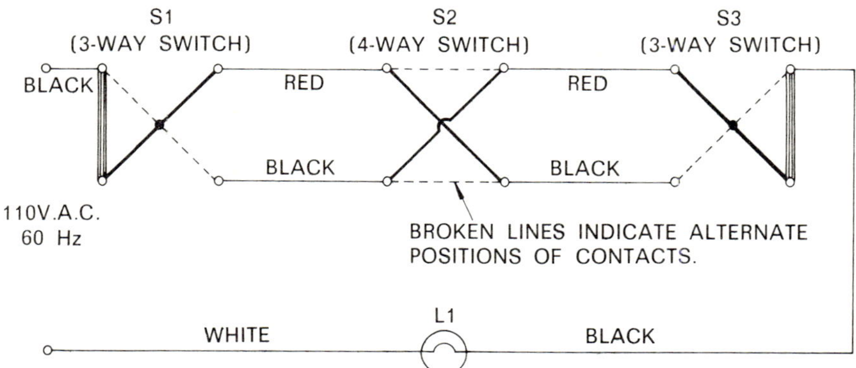

Fig. 6-11 A lamp circuit controlled by two three-way switches and one four-way switch

CONVENIENCE AND CEILING OUTLET CIRCUITS

Single-convenience outlet and one switched lamp outlet circuit

The circuit diagram illustrated in Fig. 6-12 should be considered as a hybrid. This is so, because while standard symbols are used for the lamp and switch, the others are not standard. In addition, the electrician who is involved in doing the actual wiring would not have the need of such a detailed schematic. The purpose then of this schematic is primarily to assist the student of electricity.

Note should be taken of the fact that S1, O1, and L1 are physically contained in their respective metallic boxes or enclosures, but are drawn to one side as is illustrated in Fig. 6-12. Colour coding of the individual conductors is shown, and also the grounding of each enclosure should be noted. The author has intentionally used a rectangular shape for all boxes or enclosures, even if this is contrary to their actual one. The purpose for doing this is dictated by the fact that confusion would result between lamp symbols and these boxes. In addition, since the wiring between the various boxes is in the nature of a cable, an oblong shaped symbol is drawn about these wires to indicate this condition.

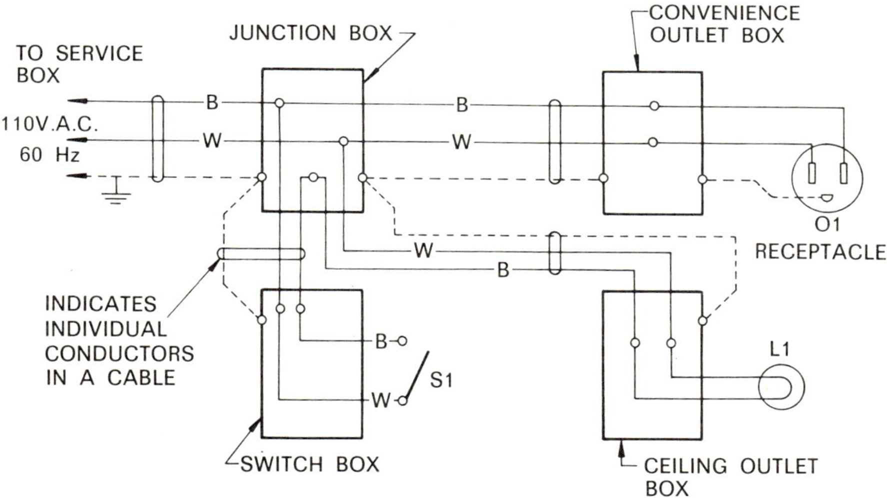

Fig. 6-12 A single-convenience outlet and one switched lamp outlet circuit

Single-switched convenience outlet circuit

In similar fashion to Fig. 6-12, the schematic illustrated in Fig. 6-13 has been included to show the circuitry involved when a convenience outlet is required to be switched from a remote position. No additional concepts are involved in this cricuit. However, the student should be aware that the conductors in the cables running from the switch and the convenience outlet boxes must first enter the junction box before being rerouted to their ultimate destinations.

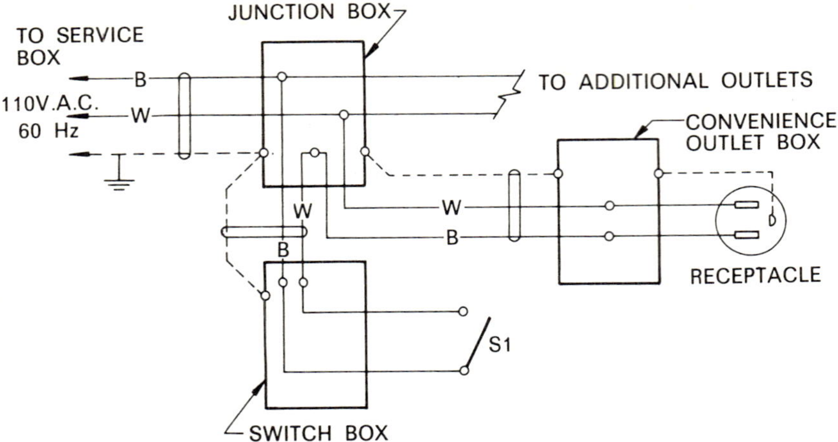

Fig. 6-13 A single-switched convenience outlet circuit

ALARM CIRCUITS UTILIZING STEP-DOWN TRANSFORMERS

Series buzzers alarm circuit

The schematic illustrated in Fig. 6-14 shows the correct form for drawing two buzzers wired in series and controlled by a single push button. Since approximately 10 volts is required to operate the buzzer solenoids, a step-down transformer is used. It is of no electrical significance if the push button is placed in the opposing leg of the circuit. If the components are symmetrically placed as shown in Fig. 6-14, the use of crossovers is not required.

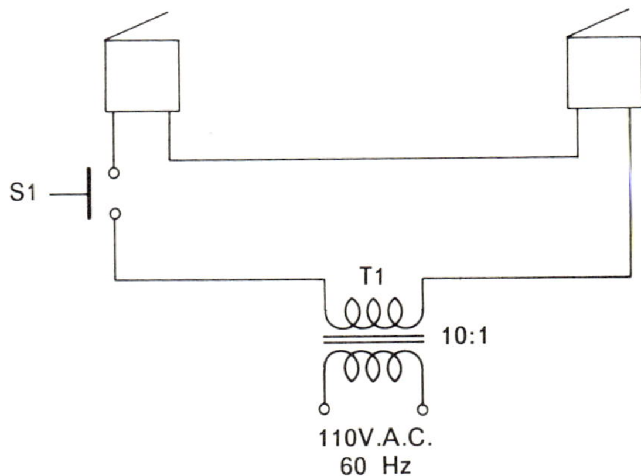

Fig. 6-14 A series buzzers alarm circuit

Single-buzzer alarm circuit controlled by two push buttons in series

Sometimes it is desirable to place two push buttons in series in an alarm circuit as shown in Fig. 6-15. In military applications, S1 is used as an **arming** switch, while S2 is the activating switch. In industrial machine applications, frequently two individually hand-operated series switches are used for safety reasons. By having the switches located in safe positions, both hands of the machine operator are isolated from any dangerous area.

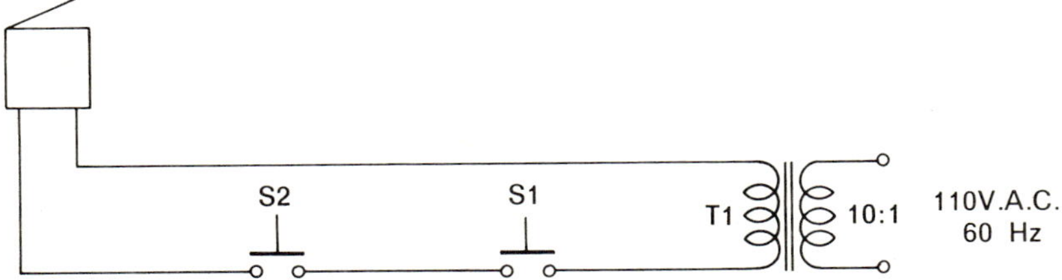

Fig. 6-15 A single-buzzer alarm circuit controlled by two push buttons in series

Single-buzzer alarm circuit controlled by two push buttons in parallel

When one buzzer is to be controlled from two positions, the schematic shown in Fig. 6-16 can be used. Sometimes it is desirable to reverse the orientation of a push button in order to gain space, and improve the readability of the circuit. This procedure is shown in Fig. 6-16. In addition all crossovers are eliminated.

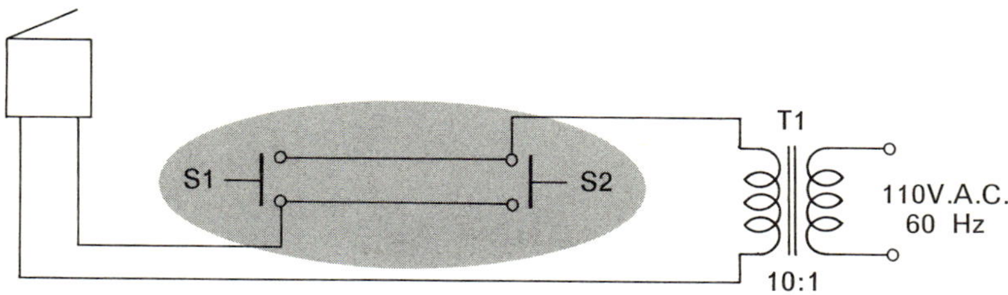

Fig. 6-16 A single-buzzer alarm circuit controlled by two push buttons in parallel

Front and rear door signal circuit

Figure 6-17 illustrates a circuit in which a crossover must be used.

Also, the connection directly below the bell must not be interpreted as a soldering point, since screw connections normally exist at both the transformer and the bell.

Apartment door lock and buzzer alarm circuit

In Fig. 6-18 it is necessary to use at least one crossover. The important components are placed at the top of the schematic, with the push buttons located directly below, in order that vertical leads can be made. The position of the connection shown encircled with a broken line can be made in actual practice at either the door lock or the buzzer. As in Fig. 6-17, it would be possible to eliminate the crossover from the secondary side of transformer T1, but this would entail the drawing of a lead around the buzzer. By doing so, the clarity of the buzzer symbol would be obscured.

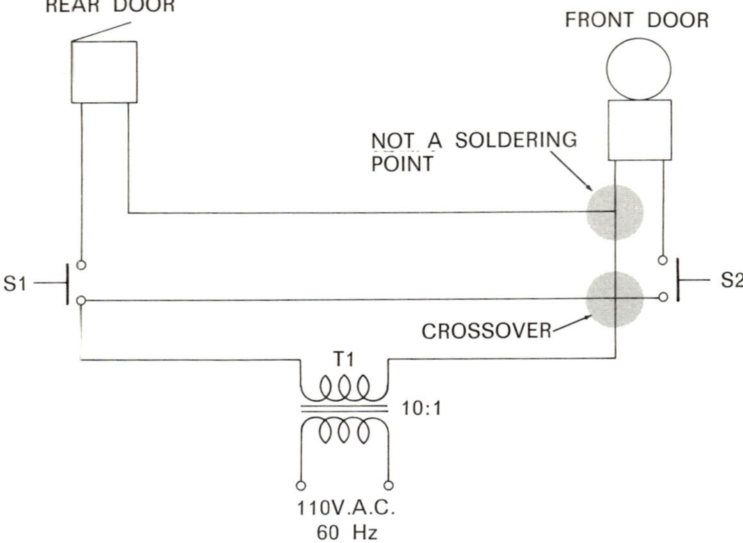

Fig. 6-17 A front and rear door signal circuit

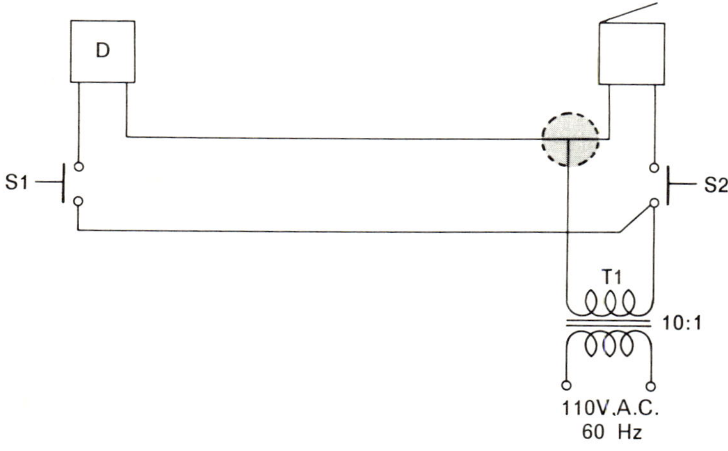

Fig. 6-18 An apartment door lock and buzzer alarm circuit

Four-point annunciator circuit

In drawing a simple annunciator circuit as is illustrated in Fig. 6-19, the annunciator symbol is shown above the push buttons. Directly below that symbol and in a horizontal fashion, the push button symbols are located. It is also a good drafting procedure to arrange them in numerical sequence.

In actual practice jumpers are connected to the bottom terminals of the four push buttons, but the method illustrated in Fig. 6-19 makes the interpretation of the schematic simpler. Since this piece of electrical equipment is largely obsolete or obsolescent, the likelihood of a drawing requirement for this circuit is probably very remote.

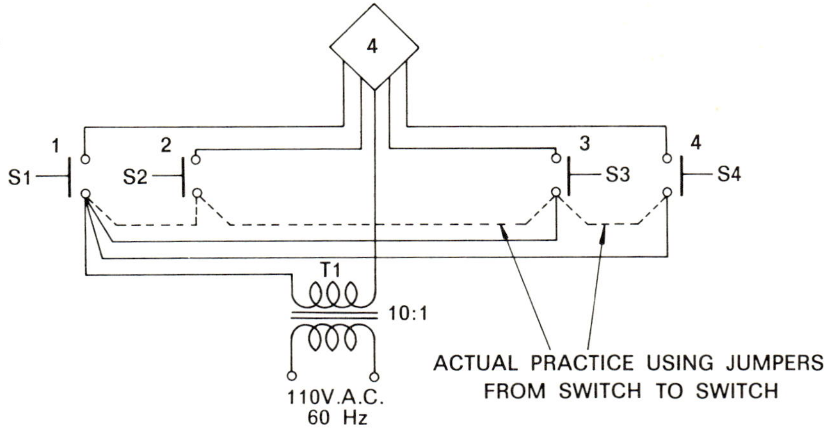

Fig. 6-19 A four-point annunciator circuit

KNOWLEDGE TESTERS

1. Under what circumstances is a D.P.S.T. switch shown in a 'closed' position?
2. Describe the correct orientation of either an air-core or iron-core reactor symbol in a circuit.
3. Capacitor symbols are placed in a particular way or method in a schematic diagram. Briefly describe this way or method.
4. Outline the correct method for symbol placement when constructing a parallel circuit consisting of a capacitor, a resistance, and a reactor.
5. What is the chief purpose in providing adequate separation between symbols that are stacked in parallel form?
6. What variation in form is used to denote a relay that has normally-closed contacts, rather than normally-open ones?
7. Why is a neutral centre position necessary for the switches used in Fig. 6-9?
8. Where is a relay coil normally placed in relation to its contacts?
9. What type of relay coil is used in Fig. 6-9?
10. Under what classification of switches should a 'three-way' switch be included?
11. How many, and what kind of switches will be required to design a circuit which will permit a lamp to be controlled from four positions?
12. Why are the diagrams shown in Figs. 6-12 and 6-13 considered hybrids?
13. In Fig. 6-13, why are the ground lines always returned to the junction box J1?
14. What does the closed loop surrounding the individual conductors in Fig. 6-12 imply?
15. Explain two purposes for using the type of circuitry illustrated in Fig. 6-15.
16. In Figs. 6-17 and 6-18 joined connections are shown. What precautions must be taken in the interpretation of these connections?
17. What is the difference between an obsolete and an obsolescent circuit?
18. Describe the actual method of wiring the common terminals of the switches in the circuit shown in Fig. 6-19.

PROJECTS

1. An electronics technician finds that he is handicapped in having only 25 ohm and 1000 ohm resistors. Design and draw a circuit that he might have used in which four 4K ohm resistances are in in parallel, and these in turn are preceded by a 0.075K ohm resistance, and followed by a 3.1K ohm resistance.
2. Draw a network of pure capacitances as follows: three groups of three capacitors wired in series parallel, followed by three capacitors wired in series.
3. Make a circuit showing a resistance, an inductance, and a capacitance connected in series. In addition, show a capacitance shunted about the resistance, an inductance paralled with the inductance, and a resistance in parallel with the capacitance.
4. Make a properly labelled circuit diagram showing two 10 volt alternating-current lamps connected in series. The switching arrangement of the circuit must be such that the lamps can be controlled independently as well as together. The source of power (S.O.P.) for the circuit is 110 volt, 60 Hz. alternating current.
5. Redraw the pictorial diagram shown in Fig. 6-20 as a schematic diagram using standard symbols.
6. Make a wiring diagram showing how three 110 volt lamps connected in parallel can be controlled from three separate locations.
7. A client desires the following electrical features for a playroom recently added to his home; two ceiling outlets independently switched, three wall mounted duplex receptacles, and a fourth wall mounted duplex receptacle controlled by a switch. Construct a wiring diagram using Fig. 6-12 as a guide.
8. An eight-unit motel has an annunciator circuit to permit each guest to call for room service. In conjunction with the annunciator, a single buzzer acts as an audible alarm. Make a suitable wiring diagram containing these features.
9. Draw a circuit diagram for a single buzzer capable of being operated from five separate locations.
10. A six-unit garden home has front and rear entrances. Design and draw a circuit in which each unit has a 10 volt bell alarm at the front entrance, and a 10 volt buzzer alarm at the rear entrance. The S.O.P. is 110 volts alternating current.

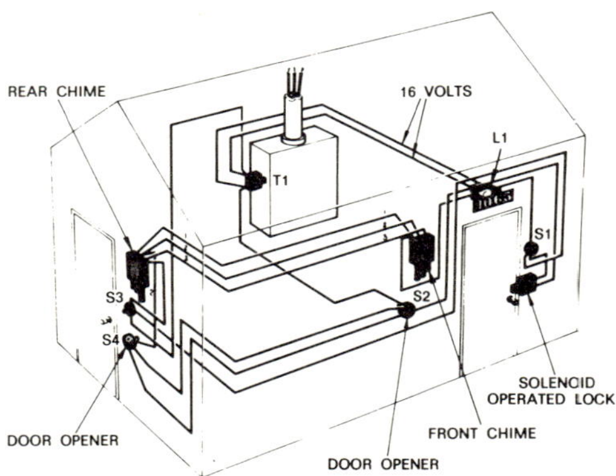

Fig. 6-20 Drawing for Project NO. 5

UNIT 7

Building Wiring Diagrams

THE ELECTRICAL DRAFTSMAN, WHEN INVOLVED IN THE DRAWING OR design of building wiring diagrams, must be conversant with the various electrical codes of his community, in order to carry out his work effectively. These codes consist of those set down by federal government agencies as well as ones formulated by local municipal authorities.

Purpose of the code

The necessity of having an electrical code is, of course, determined by the need of having a safe electrical installation. This aspect of safety includes two areas. The first area of safety involves that of possible fires brought about by circuit overloading; the second, the loss of human lives brought about by electrocution.

An electrical code is a comprehensive set of regulations published in booklet form prescribing the **mandatory** way in which safe electrical construction practices are to be carried out. To insure that electrical construction practices are carried out according to the letter of the code, all installations require inspection by an authorized inspector.

Canadian Electrical Code

In Canada, the national code has been set by the Canadian Standards Association, and is entitled the Canadian Electrical Code, CSA Standard C22.1-1969.

American Electrical Code

In the United States of America, the 'National Electrical Code' published by the American Standards Association lays down the minimum design criteria for safe electrical installations and practices.

Figure 7-1(a) shows the title page from the Canadian Electrical Code. Note that this page indicates the number of the edition. Draftsmen using this publication must always be aware of this fact, as changes are made in the various editions published. Figure 7-1(b) is a sample page from the same publication, and relates to the grounding of D.C. systems. Note that each section has an important preamble which covers the 'scope' and the 'object', before details regarding systems and circuits are given.

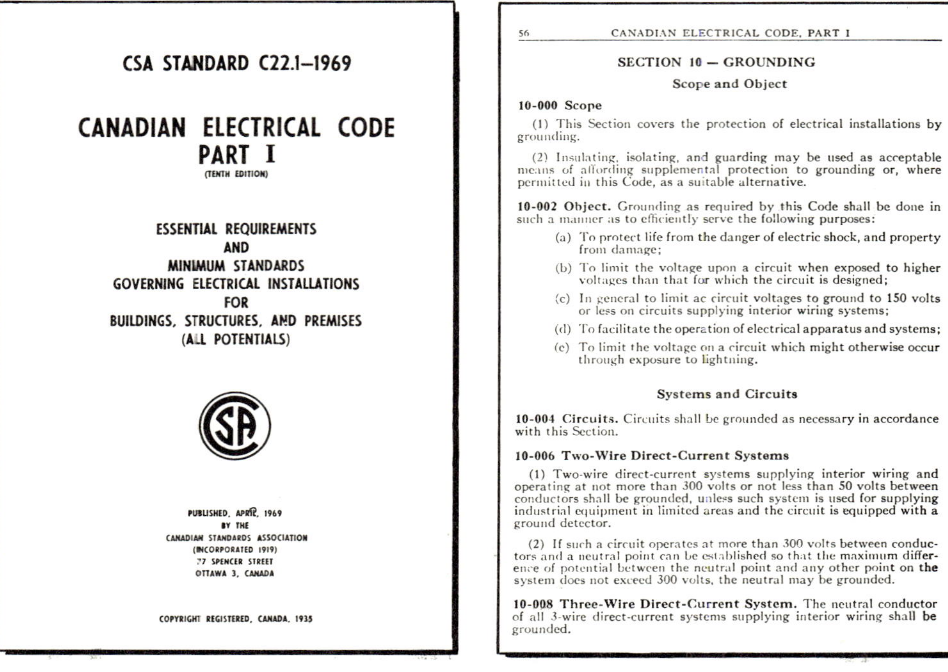

CSA STANDARD C22.1–1969

CANADIAN ELECTRICAL CODE
PART I
(TENTH EDITION)

ESSENTIAL REQUIREMENTS
AND
MINIMUM STANDARDS
GOVERNING ELECTRICAL INSTALLATIONS
FOR
BUILDINGS, STRUCTURES, AND PREMISES
(ALL POTENTIALS)

PUBLISHED, APRIL, 1969
BY THE
CANADIAN STANDARDS ASSOCIATION
(INCORPORATED 1919)
77 SPENCER STREET
OTTAWA 3, CANADA

COPYRIGHT REGISTERED, CANADA, 1935

56 CANADIAN ELECTRICAL CODE, PART I

SECTION 10 — GROUNDING

Scope and Object

10-000 Scope

(1) This Section covers the protection of electrical installations by grounding.

(2) Insulating, isolating, and guarding may be used as acceptable means of affording supplemental protection to grounding or, where permitted in this Code, as a suitable alternative.

10-002 Object. Grounding as required by this Code shall be done in such a manner as to efficiently serve the following purposes:

(a) To protect life from the danger of electric shock, and property from damage;

(b) To limit the voltage upon a circuit when exposed to higher voltages than that for which the circuit is designed;

(c) In general to limit ac circuit voltages to ground to 150 volts or less on circuits supplying interior wiring systems;

(d) To facilitate the operation of electrical apparatus and systems;

(e) To limit the voltage on a circuit which might otherwise occur through exposure to lightning.

Systems and Circuits

10-004 Circuits. Circuits shall be grounded as necessary in accordance with this Section.

10-006 Two-Wire Direct-Current Systems

(1) Two-wire direct-current systems supplying interior wiring and operating at not more than 300 volts or not less than 50 volts between conductors shall be grounded, unless such system is used for supplying industrial equipment in limited areas and the circuit is equipped with a ground detector.

(2) If such a circuit operates at more than 300 volts between conductors and a neutral point can be established so that the maximum difference of potential between the neutral point and any other point on the system does not exceed 300 volts, the neutral may be grounded.

10-008 Three-Wire Direct-Current System. The neutral conductor of all 3-wire direct-current systems supplying interior wiring shall be grounded.

Fig. 7-1 Extracts from the Canadian Electrical Code. The title page. A sample page. (Courtesy Canadian Standards Association, Ottawa, Ontario)

INTERPRETATION OF A BUILDING WIRING DIAGRAM

A building wiring diagram is essentially a cabling diagram, although it is seldom considered in those terms. The reason it may be considered as such, is that the wiring used contains at least two insulated conductors, plus a noninsulated ground wire. Accordingly then, one cannot read a drawing of this sort without this realization. In order to obtain a better mental picture of the foregoing, it is necessary to show the relationship between a 'true' wiring diagram and an installation or simplified wiring diagram.

True wiring diagram

Figure 7-2 shows a true wiring diagram of two ceiling outlets connected in parallel. In Unit 6 it was pointed out that the type of drawing shown in Fig. 7-2 must be considered a hybrid since it shows conventional symbols in conjunction with nonconventional ones. Unfortunately this type of drawing would

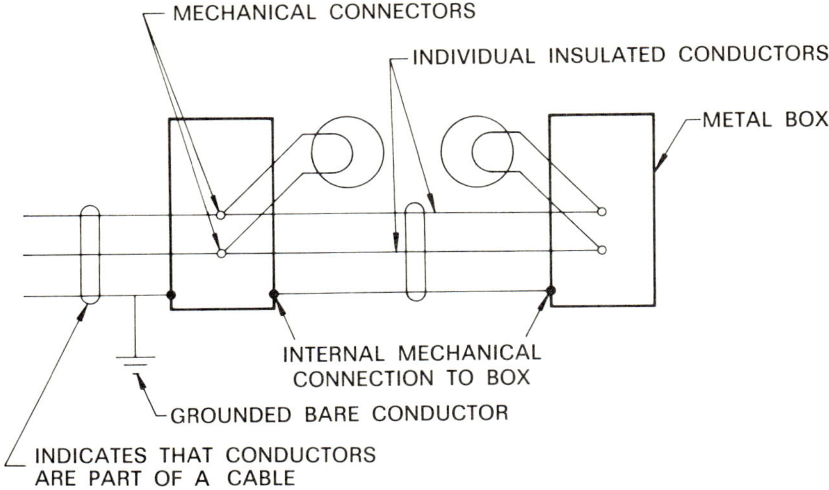

Fig. 7-2 A true wiring diagram of two lamps connected in parallel

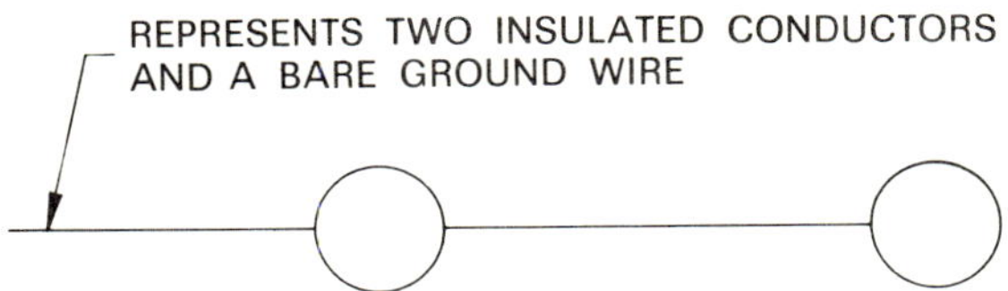

Fig. 7-3 A simplified or installation drawing of two lamps connected in parallel

require too much drafting time to convey a relatively small amount of information. Secondly, it would require far too much space. To eliminate this problem a simplified wiring diagram is used.

Installation diagram

The drawing referred to in Fig. 7-3 illustrates a simplified or installation diagram of the same circuit shown in Fig. 7-2.

It is obvious from the simplicity of diagram Fig. 7-3, that considerable drafting time and space will be conserved by using this recognized method.

In general, the same procedure is carried out consistently for all building wiring diagrams; but of course, a knowledge of all symbols used, electrical codes involved, and a basic understanding of architectural floor plans will be a prior requirement before drafting competency will be achieved.

SINGLE AND MULTI-FLOOR RESIDENTIAL BUILDINGS

Plan scale used

The architectural scale used for single and multi-floor residential buildings, but not including large apartment or commercial buildings, is usually $1/4'' = 1'-0''$.

Position of symbols on plan

The location of the symbols on either basement, ground floor, and subsequent floor plans is only approximate. The electrical contractor, by virtue of his practical experience and adherence to codes, establishes the actual position of the service box, outlets, and switches. He will not scale off the position of the symbol relative to the room in which it is located.

Single-floor or bungalow residential building wiring diagram

Figure 2-1 illustrates a typical architectural plan of a single-family residence which, it has been previously pointed out, incorporates the required electrical services. Note that the only symbols used to illustrate a cable are the broken lines joining switches to either ceiling light fixture or wall duplex outlets. Ceiling light fixture outlets are shown frequently in the centres of rooms, in particular, bedrooms. In some cases three-way switches are used, such as in the kitchen, where a ceiling light can be controlled from either the vicinity of the rear entrance door, or the bifold door to the dining room. Also, note the back-to-back arrangement of wall duplex outlets between Bedroom #2 and Bedroom #3.

The basement plan of a single-family residence would show the symbols indicating the service box, hot-water tank (if electrically heated), the furnace (if electrically ignited), wall and ceiling outlets. In addition, it might include additional outlets for a playroom or family room. However, nothing of a very special or unique nature would be involved.

Multi-floor residential building wiring diagram

Beyond the fact that additional floor plans are required for a two-storey, single-family residence, a duplex apartment (two individual units stacked vertically), or a triplex apartment, (three individual units stacked vertically) nothing of new importance is essentially involved from an electrical standpoint. In a similar fashion to the bungalow layout, the electrical symbols are superimposed on the plan views. The basement plan for the apartment buildings might show individual service boxes if the metering of the electrical current was on an individual basis. A second point worth mentioning is the almost identical semilarity of the location of electrical symbols on each individual apartment floor plan, no doubt an entirely logical approach for reasons of economy.

Small commercial building wiring diagram

A small commercial building containing both offices and stores, in terms of electrical design, may present the consultant and his draftsmen with some unique problems.

For the offices, proper fluorescent lighting and adequate wall and/or floor receptacles for business machines are extremely important. Also, facilities for a considerable number of telephones are necessary, and in this area the public telephone utility company is available to offer extensive technical assistance.

Whether or not complete air conditioning is to be incorporated initially in the design of the building, or if only air conditioning window units are to be added later, must be considered carefully.

Also, many business offices are rapidly becoming engaged in the use of computers for billing and other purposes and this probability must be considered. Fortunately, the current generation of computers using solid-state devices, and microcircuitry techniques have lessened the requirement greatly for large amounts of electrical current, than did first generation computers containing hundreds of vacuum tubes. Nevertheless, the facilities for such a device in terms of circuit availability and protection must ultimately appear properly delineated on a drawing.

If the small commercial building is of the type having stores on the ground floor, the electrical engineering staff must consider the electrical situation accordingly. In many cases the stores are operated on a rental basis, hence the layout of receptacles and ceiling lights must provide for the maximum usefullness and flexibility. Figure 2-13 shows the layout for one kind of commercial building where offices are located at the front. Note that the lighting fixtures are called up with a letter and number designation. The letter indicates the type of luminaire, and is shown in the schedule on Fig. 7-4.

LUMINAIRE SCHEDULE

KEY	MAKE AND CAT. NO.	LAMPS			MOUNTING		DESCRIPTION AND/OR LOCATION
		TYPE	NO.	WATTS	LOCATION	HEIGHT	
A	C & M #6048M WILSON #C148	RS-CW	1	40	CEILING	SUSPEND AS NOTED	SUPPLY TANDEM UNITS WHERE POSSIBLE FIXTURES C/W REFLECTOR
B	WILSON #6131-M12 C & M #G7644-4	RS-CW	4	40	CEILING	LAY-IN	OFFICES
C	C & M #M1-248 WILSON #8900	RS-CW	2	40	CEILING	SUSPEND AT 11'0"	STORAGE AREA
D	WILSON #3402 C & M #CA2624-4	RS-CW	2	40	CEILING	SURFACE	
F	WILSON #2104	RS-WWX	1	40	WALL	SURFACE	WASHROOM
G	ELECTROLIER B-20-6	I.F.	1	100	WALL	SURFACE	VESTIBULE
H	ELECTROLIER, PHAROS #B-30	A-25	1	200	WALL	SURFACE	OUTDOOR — MT'D 90" ABOVE FIN. FLOOR
J	ELECTROLIER, VAPOURLUX #KFP-76-24B	RS-CW	2	40	CEILING	SURFACE	
L	WILSON #2606	INC.-PAR 38	1	150	CEILING	RECESSED	VESTIBULE — DOWNLIGHTS
M	C.G.E. #59778G2	QUARTZ IODINE	1	1500			SIGN FLOOD LIGHTING
X	ELECTROLIER #X-2-S						EXIT LIGHT C/W 6 VDC 3RD LIGHT

Fig. 7-4 The luminaire schedule for the commercial building in Fig. 2-13

Farm building wiring diagram

Modern dairy farms require a multitude of electrically powered machines. For example, a typical dairy loafing barn of modern design requires a silo unloader, portable hay elevator, hay drying fan, stock tank water heater, augur bunk feeder, feed grinder, grain elevator, and of course electrical lights. Because of such an array of equipment, it is necessary to have a master switch box of large capacity. In practice, a 200 ampere master switch box is frequently used and this is located on the transformer pole at a central location in the farmyard. Figure 7-5 shows a typical pictorial layout of a well designed electrically powered farm installation. In addition, Fig. 7-6 illustrates a part of the arrangement of electrical components for a modern tie stall dairy barn. In designing such a layout, specific information regarding the operations within the barn must be considered. This will involve liaison with the professional farmer, the farm machinery manufacturer or dealer, and possibly an agricultural representative from a government agency. The illustration shown by Fig. 7-7 is a specialized one, in that it superimposes the wiring over the two orthographic views of the farm building. This is necessary in order to provide those persons installing the lightning arresting equipment with sufficient information.

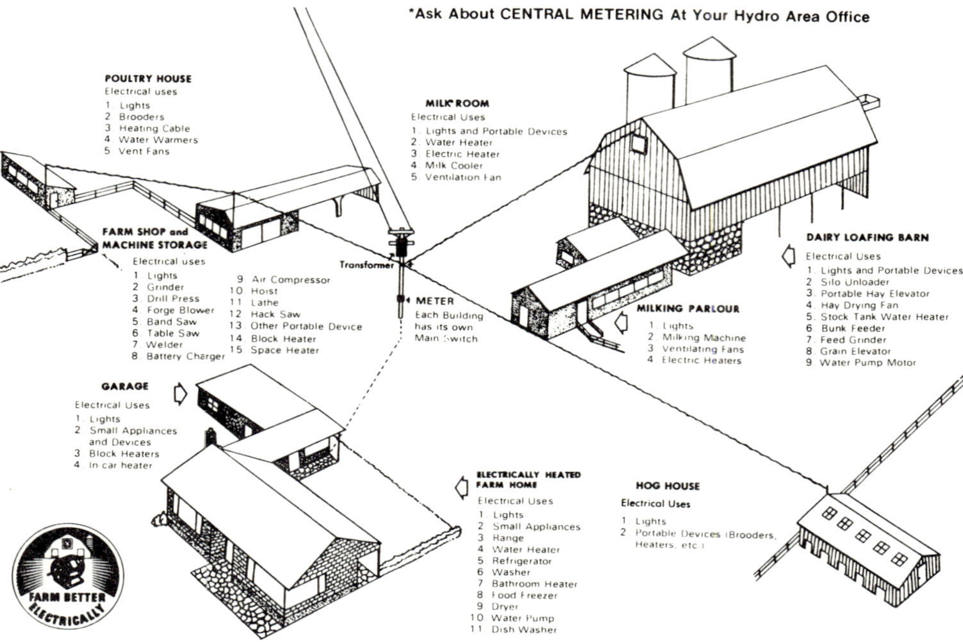

Fig. 7-5 A well-designed, electrically powered farm installation. (Courtesy Ontario Hydro)

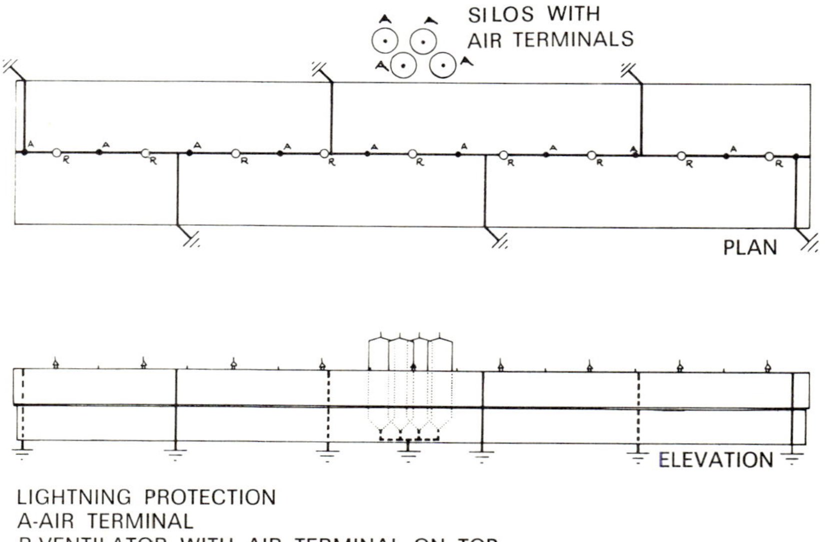

Fig. 7-7 Plan and elevation views of a large modern farm building showing the location of the lightning rods, or arrestors, and grounds at the Greenbelt Farm, Research Branch, Canada Department of Agriculture

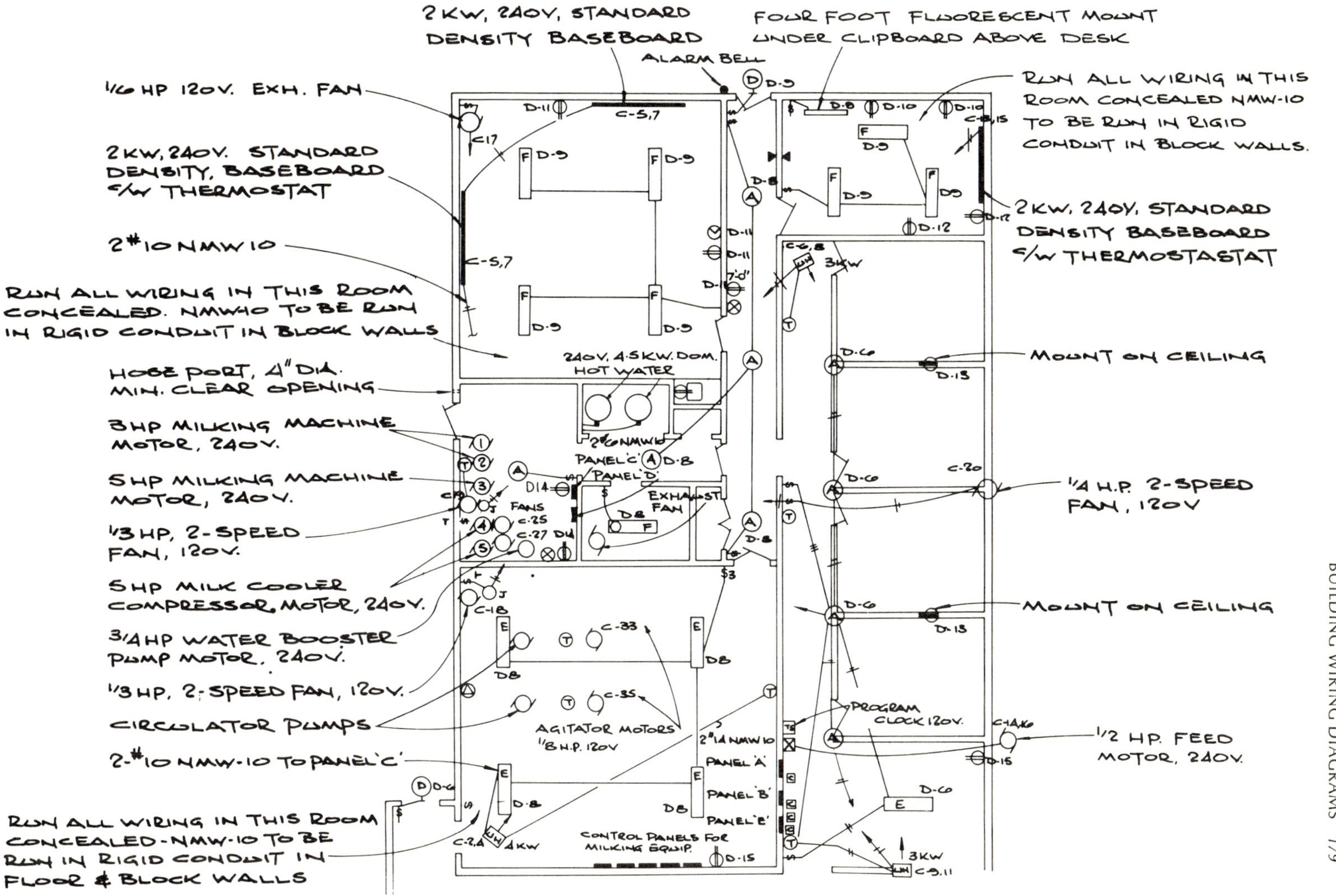

Fig. 7-6 Part of an electrical diagram for a modern tie stall milking barn in the Greenbelt Farm, Research Branch, Canada Department of Agriculture

Fig. 7-6 and 7-7 (Courtesy Brais, Frigon, Hanley, Brett & Minty, Consulting Engineers/ Ingénieurs Conseil; Murray and Murray — Architects and Town Planners, Ottawa, Ontario)

School building wiring diagram

The electrical design and drawings for schools are often undertaken by a consulting firm of professional electrical engineers. When the requirements of a particular school are known, action is taken to establish the minimum size of protective devices capable of handling the total electrical load. Since schools vary considerably in size and in function, their electrical requirements also, of course, vary. For example, an academic school would not require the amount of electrical outlets, electrical motors, and other electrical devices as would a technical school. Also, in a technical or vocational school where woodworking and machine shops exist, as for example, underfloor ducting for electrical cables are frequently a requirement. In addition, each work station having an electrically operated piece of machinery requires a circuit breaker type of switch which will trip out on power failure. This, of course, acts as a safety device, since the machine or machines running at the instant of power failure would not be energized until the circuit breaker switch was retripped. In addition, such a classroom design requires one or two emergency switch stations which when triggered cut off power to all machines. These emergency switches are carefully located in strategic positions.

NOTE. Luminaire designations are interpreted as follows:

EXAMPLE: 11 — B-L1-10
11 — Number of luminaires.
B — Type of luminaire as per schedule (not shown).
L1 — Luminaire circuit connected to lighting panel No. L1.
10 — Circuit No. 10 on the lighting panel.

In addition, the electrical design includes classroom lighting, chalkboard and showcase lighting, specialized auditorium lighting, interclassroom communications, and E.T.V. cabling, but to mention a few.

For a large educational institution a number of individual drawings are required. Fig. 7-8 shows part of the layout for the lighting of classrooms in an elementary school. Note the method of designating the various luminaires in a particular circuit. The illustration listed as Fig. 7-9 is typical of a well-planned secondary school machine shop. This system uses 575 volts, hence locked disconnect switches are mandatory at each work station. Other unique features that are worth noting are the use of underfloor ducts and conduits. Note also the strategic location of the two emergency stop stations.

Hospital building wiring diagram

The electrical requirements for a large modern hospital cover such basic items as room and hall lighting and convenience outlets. However, specialized lighting for operating room theatres, explosion proof switches and lamps for areas where volatile gases are in use, and x-ray equipment necessitate special design criteria. Intercommunications, paging systems, television monitoring systems, elevators, food preparation equipment, and sterilizing equipment add to the problems of the consultant and his drafting staff.

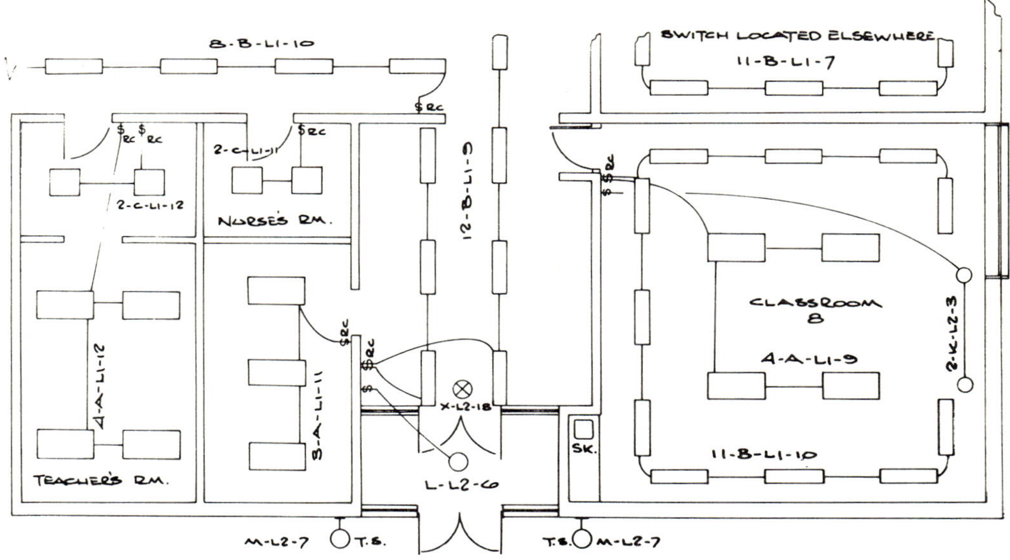

Fig. 7-8 *Part of the lighting system for Gouldbourn Public School, Gouldbourn Township, Ontario. (Courtesy Balharrie, Helmer and Associates, Ottawa, Ontario)*

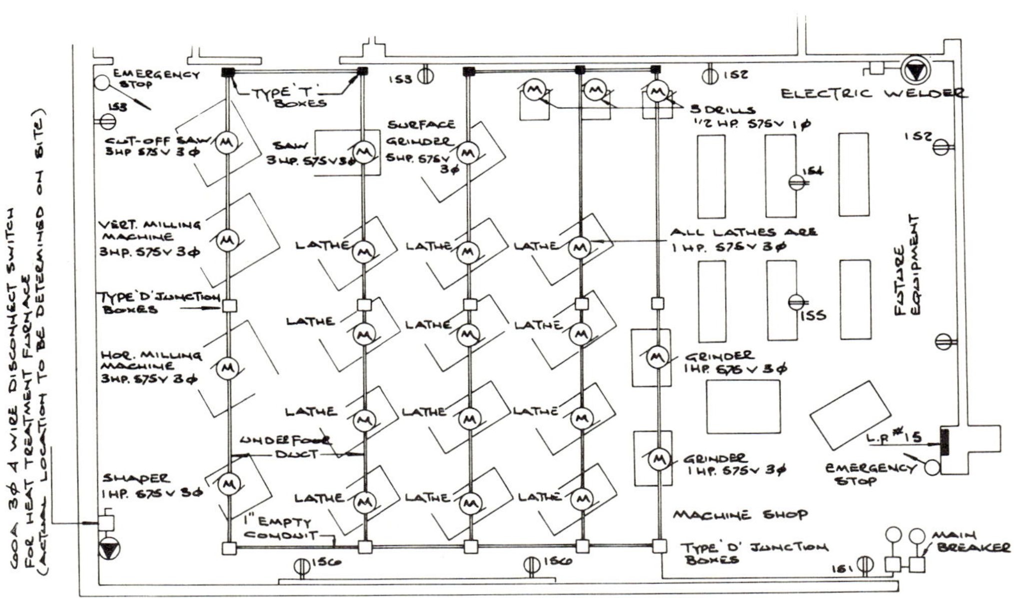

Fig. 7-9 *An electrical diagram showing the layout of conduit, underfloor ducts, outlets, and the position of the motors for a machine shop in the Sir John A. Macdonald High School, Ottawa, Ontario. (Courtesy Lithwick, Lambert, Sim and Johnston, Architects, Ottawa, Ontario)*

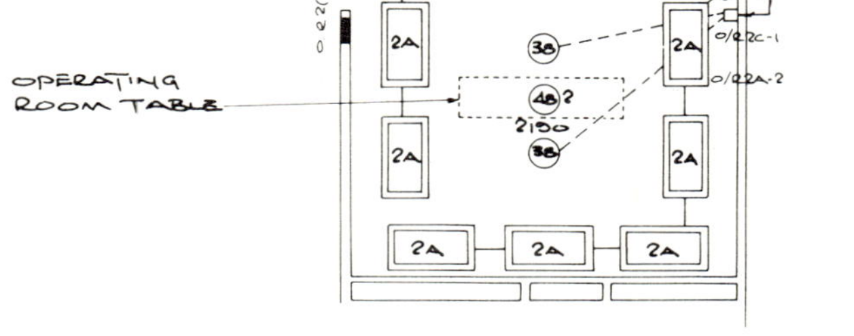

Fig. 7-10 *An electrical schematic of two adjoining operating rooms showing variac controlled lights, and emergency fluorescent lights in the Riverside Hospital, Ottawa, Ontario*

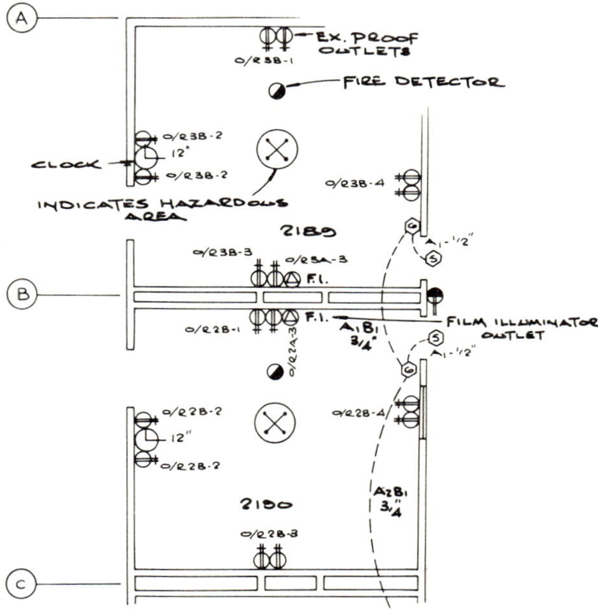

Fig. 7-11 *An electrical schematic of two adjoining operating rooms showing explosion proof outlets, fire detectors, film illuminator outlets, and clocks in the Riverside Hospital, Ottawa, Ontario*

Figs. 7-10 and 7-11 (Courtesy Balharrie, Helmer and Associates, Ottawa, Ontario)

The drawings for such an installation would indeed be extensive in number, since each major system per floor of the hospital requires individual sheets.

Extensive collaboration between the manufacturer and/or supplier, medical specialists, and the design team of the electrical consultant are demanded for a hospital installation. The illustrations listed as Figs. 7-10, and 7-11 cover some of the areas mentioned above; but for reasons of space, show only two adjoining operating rooms. Since oxygen is available in these rooms, special attention is paid to the use of explosion proof outlets and fire detectors. Note also that the incandescent lights on either side of the operating table can be adjusted in intensity by the use of variacs. See Fig. 7-10.

Church building wiring diagram

Contemporary church design is frequently so unique that some problems in concealing electrical conduit may arise. Lighting design is often difficult to solve because of vaulted ceilings, and of course, the consultant must keep in mind the overall aesthetics of the church interior. Also, in harmony with the atmosphere within the church, light dimming circuits are normally a requirement.

Again, the size of the church and its accompaying wings and annexes dictate the complexity of the electrical circuits. In addition, it is frequently common for church halls, and christian education wings to have quite an extensive kitchen, and dishwashing and drying facilities. Beyond these items, circuits for ceiling lights, wall outlets, and those necessary for the heating system are also a requirement.

Figs. 7-12 and 7-13 detail some of the areas involved in the electrical design of a house of worship. In Fig. 7-12, a dimming system is shown. Note that a

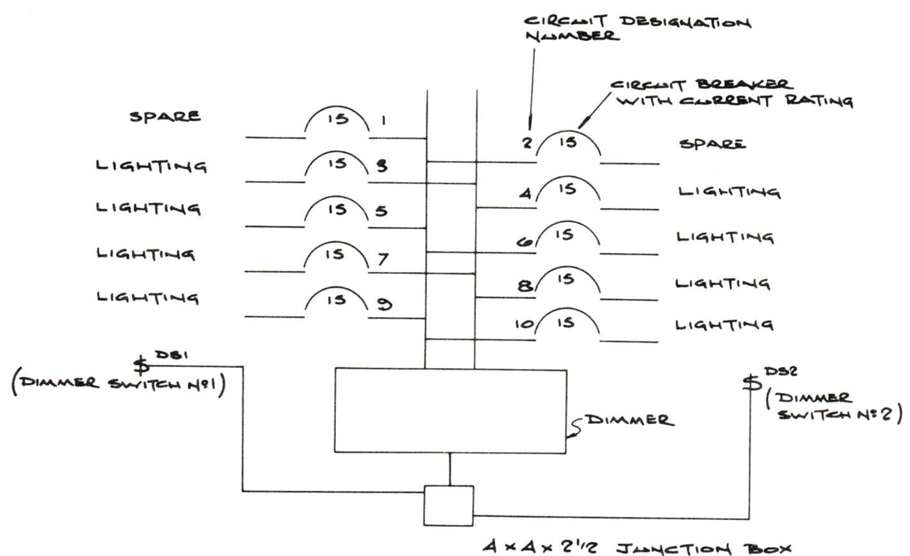

Fig. 7-12 An electrical schematic showing lighting circuits to a church nave with a dimmer and dimmer switches in Saint John's Anglican Church, Nepean Township, Ontario

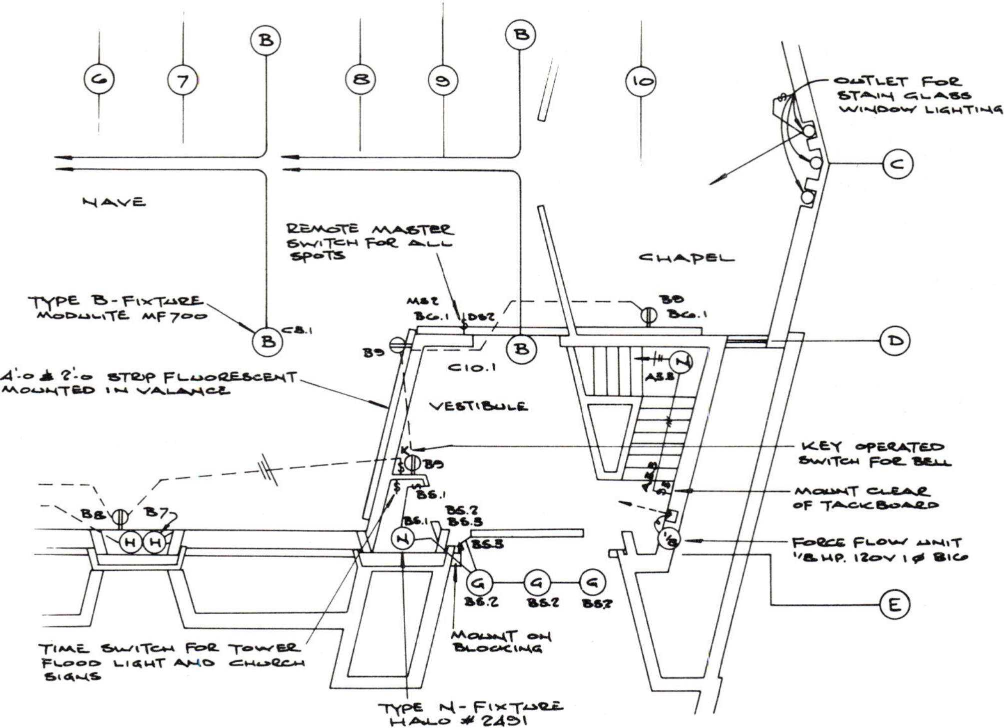

Fig. 7-13 Part of a church wiring diagram in Saint John's Anglican Church, Nepean Township, Ontario

wise designer has included two spare circuits for future use. This is a good practice. Figure 7-13 illustrates a portion of a church floor plan, upon which is superimposed the electrical symbols and circuitry. In this case two kinds of special switches are used. The time switch provides a form of electrical automation; the key switch, electrical security. Note that fluorescent strip lights are located in valances in the church nave. The valances serve to diffuse the light from these strips towards the floor, thus preventing a harsh glare.

Riser diagram

The electrical riser diagram is a special type of schematic which shows the various constituent components of an electrical system with their interconnecting buses, raceways, and cables. The special feature of a riser diagram that makes it unique is that it is drawn as an elevational view. In addition to the placement of the component symbols in a vertical sense, they are also located in their approximate horizontal position. However, since this type of drawing is a schematic, a drawing scale is not used. Buildings that are multi-floor in design frequently include horizontal datum or reference lines to indicate the various floor levels, and are so designated. Figure 7-14 shows a simple, electrical riser diagram. This particular one is for a microphone system, and shows both loudspeaker and microphone outlets, as well as other necessary components.

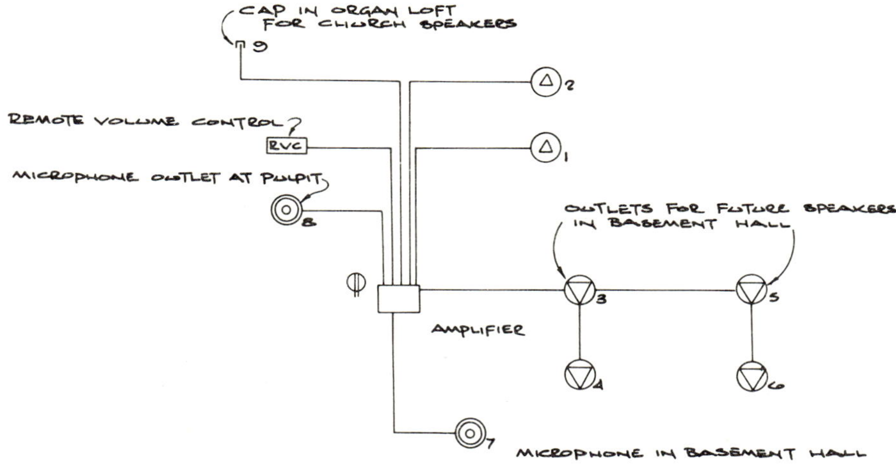

Fig. 7-14 A simple riser diagram for a church microphone system in Saint John's Anglican Church, Nepean Township, Ontario

Figs. 7-12 to 14 (Courtesy Craig and Kohler, Architects; J. L. Richards and Associates, Consulting Engineers, Ottawa, Ontario)

An interesting aspect of this schematic is the note regarding the installation of a fish wire in the conduit. The purpose of this wire is because some of the speakers are not to be installed until some later date. When this time comes, it will be necessary to use the fish wire to pull leads through the conduit to the speakers.

Transformer vault diagram

When a building of large size requiring extensive electrical services is designed, a transformer vault or room becomes a mandatory feature for convenience and safety reasons. Certain regulations govern the construction of the room, and, of course, admittance is restricted to only authorized personnel. The purpose of the room is to hold the step-down transformers and their ancillary equipment, that are necessary to reduce the high primary voltage being fed to the building to safer and hence useable voltages.

The drawings necessary to explain the layout of equipment are drawn to scale, and are orthographic in nature. Both plan and elevation views are frequently required. Details of concrete mounting pads for securing heavy equipment, as well as any other structural elements, are shown. The drawing listed as Fig. 2-3 portrays a typical transformer vault diagram in the plan view. In addition, a single-line schematic diagram is employed to show the electrical wiring. See Fig. 7-15.

Single-line schematic diagram

Figure 7-16 shows a single-line schematic diagram for the main electrical service entrance of a commercial building. Information regarding conductor and conduit sizes, panel types, and distribution from the main panel are readily

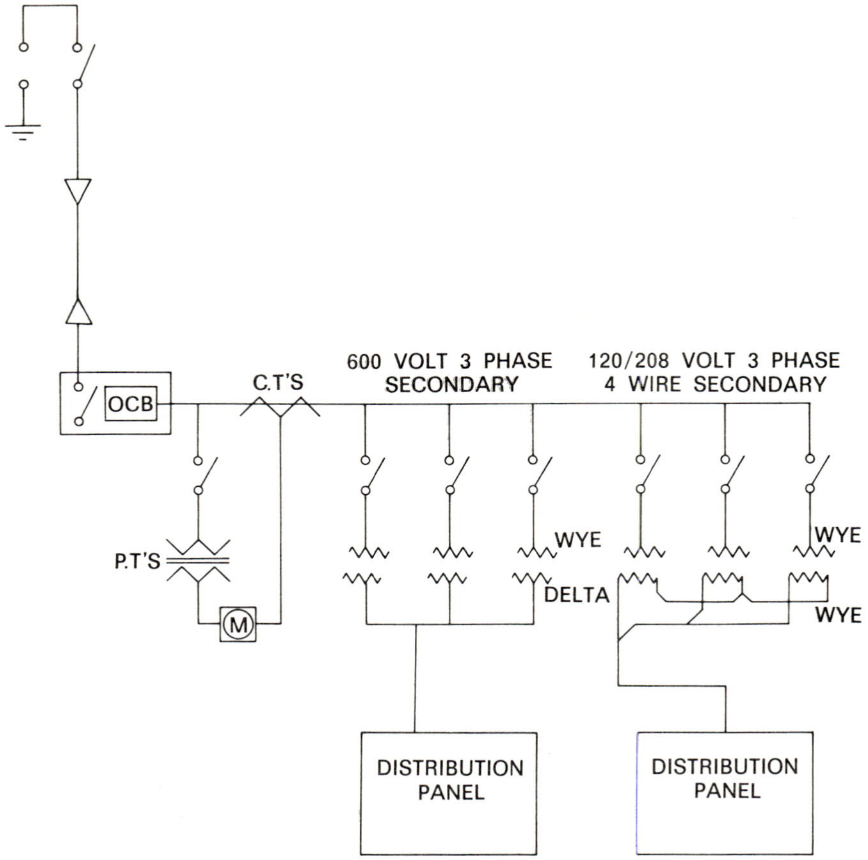

600 VOLT 3 PHASE SECONDARY
120/208 VOLT 3 PHASE 4 WIRE SECONDARY
OCB
C.T'S
P.T'S
M
WYE
DELTA
WYE
WYE
DISTRIBUTION PANEL
DISTRIBUTION PANEL

Fig. 7-15 A single-line schematic diagram for transformer vault equipment and connecting circuitry in the Sir John A. Macdonald High School, Ottawa, Ontario. (Courtesy Lithwick, Lambert, Sim and Johnston, Architects, Ottawa, Ontario)

obtainable from this kind of drawing. Also this type of drawing is advantageous, in that it is easy to trace the path of the incoming electrical current as it is fused, metered, and finally distributed.

MANUFACTURERS' INSTALLATION DIAGRAMS

Manufacturers of electrical heating equipment frequently include small illustrated brochures or pamphlets covering the specifications, mechanical installation, and wiring procedure of the components they manufacture. In addition, schematic wiring diagrams illustrating the various phases of the circuit operation may be included.

Since these pamphlets are used external to the manufacturing plant throughout the country and perhaps in other countries, the quality of the drawings must be of high order. Also, notes and any instructions included with the drawings must be explicit.

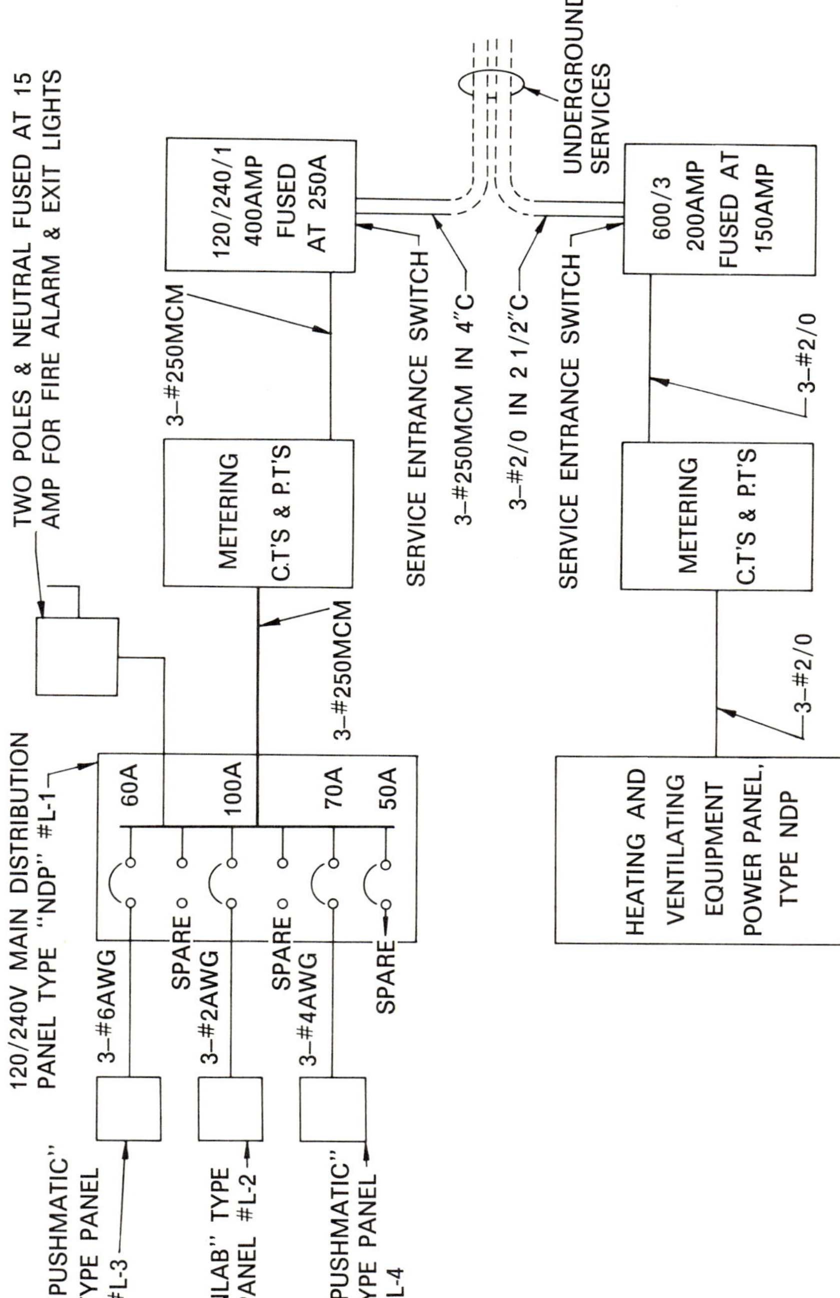

Fig. 7-16 A typical single-line electrical diagram of a main service entrance of the Grand and Toy Building, Ottawa, Ontario. (Courtesy Balharrie, Helmer and Associates, Architects and Engineers, Ottawa, Ontario)

Besides these schematic wiring diagrams, pictorial drawings and photo-drawings are often used. Fig. 7-17 shows the installation drawing of an electrical control unit for a furnace. Fig. 7-18 illustrates a wiring diagram used to show the necessary connections between the room thermostat and its relay, and connections from the line to the fan and stepdown transformer. These are made at the terminal panel of the furnace control unit. All wiring for these connections must agree with the applicable codes and ordinances. Two conduit knockouts are provided in the back of the case to facilitate wiring where all wiring will be inside the furnace casing. Three additional knockouts are provided in the bottom of the case for exterior wiring. Convenient terminal panels are provided to simplify making connections. Approved barriers are supplied on low-voltage models, and available on the LA401A terminal panel.

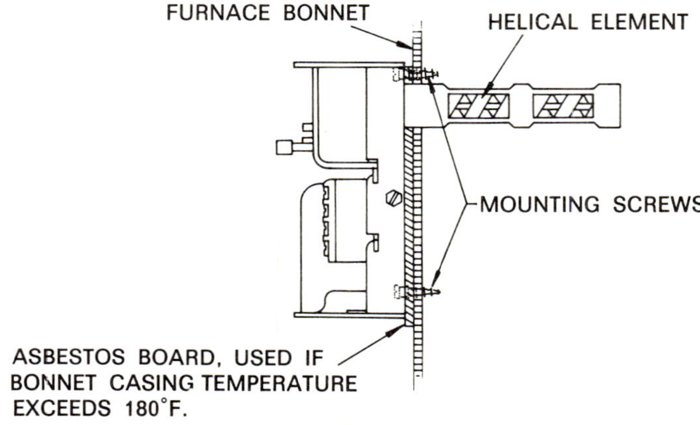

Fig. 7-17 A manufacturer's installation diagram showing the correct method of mounting the furnace control unit directly on the surface of the furnace, or plenum chamber

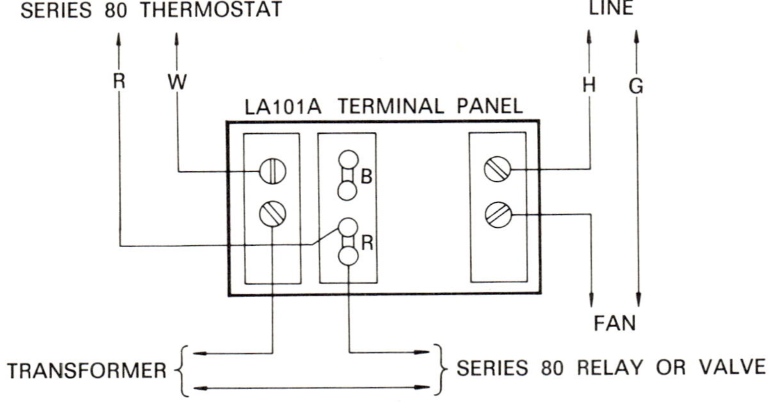

Fig. 7-18 A wiring diagram showing the connections to a specific type of furnace thermostat, and the accompanying terminal panel of a furnace control unit

The diagram, Fig. 7-19, is quite similar to Fig. 7-18. However, it varies in that it has inset drawings showing the position of the contacts in the furnace control unit. Contact 'A' breaks at limit indicator (Fig. 7-20) setting — makes 25°F. below. Contact 'B' makes at fan 'on' indicator (Fig. 7-20) setting — breaks at fan 'off' indicator (Fig. 7-20) setting. Also, a different thermostat is used which requires a three-conductor circuit. Finally, Fig. 7-20 typifies the kind of pictorial drawings used. This one is a combination type of furnace control unit. That is to say, it controls both the combustion and fan units within the furnace. It is shown with its front cover removed. The helical element shown in the upper right of the drawing is frequently located in the furnace smoke pipe. It operates on a bimetal principle.

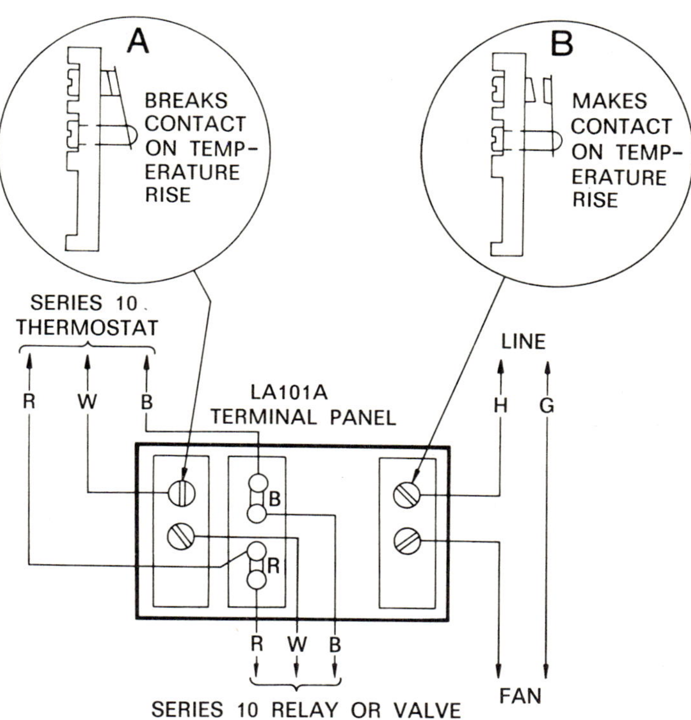

Fig. 7-19 A circuit diagram and accompanying note used to explain the operations of the temperature-controlled contacts in a furnace control unit

UNDERFLOOR DUCTING OR RACEWAYS

Large buildings utilizing poured concrete floors normally use some type of underfloor ducting or raceway. This is particularly true of installations where a large number of electrically powered machines are used.

The ducting or raceway serves to carry the insulated conductors so that they are below the level of the finished floor. This method provides for a

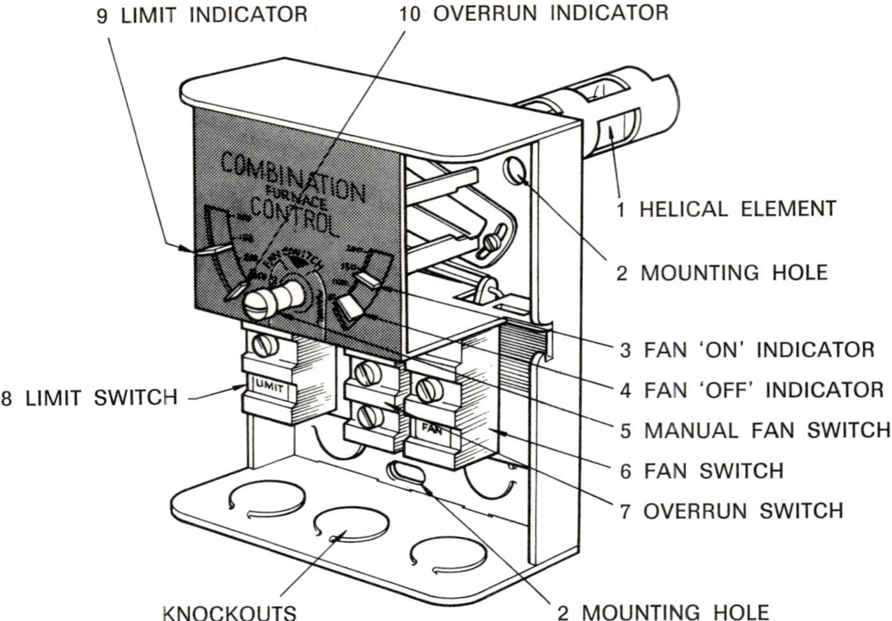

Fig. 7-20 A pictorial diagram of a furnace control unit shown with its cover removed. This particular unit is known as a combination one, since it controls both the operation of the combustion unit within the furnace and the furnace fan

Figs. 7-17 to 7-20 (Courtesy Honeywell Controls Limited)

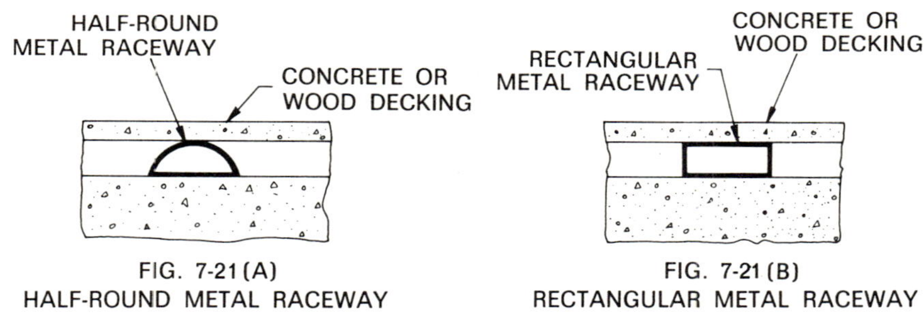

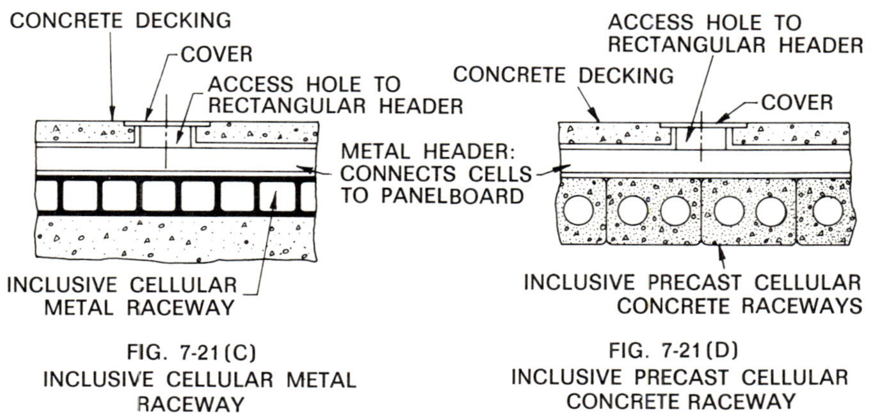

Fig. 7-21 Underfloor ducting or raceways. Not to scale, and simplified

mechanically sound, fireproof arrangement. It also permits the installation of additional electrical conductors at a future date. In addition, the elimination of hanging cables or loose cables on the floor provides for a neat and hazardproof installation.

A number of different types of ducts or raceways are used.

Single-channel raceways fabricated from heavy-gauge sheet steel, and having either a half-round or rectangular shape can be obtained. See Figs. 7-21(a) and 7-21(b).

Multi-channel or cellular steel raceways are also available. This type of raceway permits the installation of conductors in a side-by-side, parallel fashion, but separated by the baffles of the raceway. Fig. 7-21(c) illustrates this type.

Fig. 7-21(d) shows a concrete type of inclusive raceway. In this case a number, frequently two, of parallel, cylindrical holes are precast in a rectangular-shaped concrete slab. These parallel holes are used to convey the electrical conductors.

Access holes, with covers flush with the finished floor, are preset at specific locations in the ducting or raceway. Also, the distance between the parallel runs is accurately maintained during installation in order that entry to the raceways can be carried out without extreme difficulty at a future date.

The raceways are levelled during installation, and subsequently anchored to the floor. Ends of the runs are frequently marked by a metal plug set in the floor. Dead ends are closed. In some systems using the steel, rectangular-shaped raceway, junction boxes are installed at strategic locations. Also, patented inserts are located along the centre of the raceways at a fixed pitch or distance. These are installed flush with the finished floor level, and sealed. At a later date they may be opened and conductors brought through them. Fig. 7-22 shows such a system.

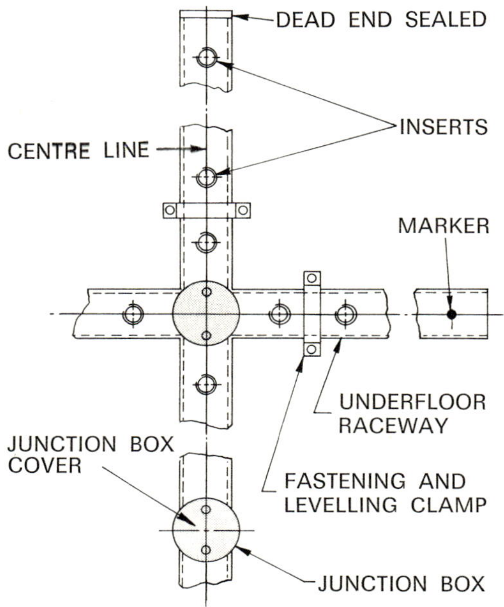

Fig. 7-22 A plan view of a metal raceway installation included during the initial floor construction

Regulations governing the size of the different types of raceways, the permissible spacing distance between parallel runs, the kind and thickness of material covering the raceway, and the maximum size of the conductors being used in a particular raceway are to be found in the electrical code handbooks. These regulations are not geographically constant. Hence, cognizance of this fact must be taken.

KNOWLEDGE TESTERS

1. What organizations or agencies formulate electrical codes?
2. Why is an electrical code necessary?
3. Give the names of the Canadian and the American codes.
4. Describe the major differences between true and installation wiring diagrams.
5. Why are installation wiring diagrams used in preference to true ones?
6. What common architectural scale is normally used for single-family residential building drawings?
7. What precaution must be observed with regard to the electrical symbols on an architectural floor plan?
8. Define the term, 'duplex outlet'.
9. Why are three-way switches used?
10. What type of line is used to delineate a cable on a single-family residential building plan?
11. What is meant by a duplex apartment?
12. List the various electrical loads that could be involved in the design of a commercial building.
13. What characteristics should be built into the electrical design of a commercial building, if it is to be a rental type?
14. List a number of the labour saving devices that are electrically operated on a modern farm.
15. What capacity is the master switch box on a farm? Where is it usually located?
16. What advantages would underfloor ducting give?
17. Why are automatic trip type circuit breakers required on power operated machines, especially in a school situation?
18. Where are emergency switches located?
19. Why are explosion proof switches used in a hospital operating room?
20. Why is the electrical design for a modern church sometimes difficult to resolve?
21. What is meant by the term 'aesthetics'?
22. Although a riser diagram is a schematic and uses symbols, in what special way is it drawn?
23. In what two spatial senses does a riser diagram conform?
24. Name the two types of drawings that are used to convey the graphic information for a transformer vault design.
25. Specific wall outlets in Fig. 2-13(a) are designated 42" and 48". Suggest what these dimensions mean.
26. In Fig. 7-10, the incandescent lamps on either side of the operating room table (shown by an oblong broken line) are variac controlled. Why is this so?
27. How many fluorescent fixtures are controlled by each switch in Fig. 7-10?
28. Why are explosion proof outlets used in the operating rooms as indicated in Fig. 7-11?
29. From how many locations can the circuits shown in Fig. 7-12 be controlled? What is the function of Circuits #1 and #2?

30. What is the purpose of mounting the fluorescent strips under the valances in Fig. 7-13?
31. Why is the bell circuit in the church, Fig. 7-13, key operated?
32. Why is a fish wire included in the conduit in Fig. 7-14? See the note below the schematic.
33. What size are the conductors feeding panel L-2 from the main distribution panel? See Fig. 7-16.
34. What size of conduit is connected to the service entrance switch box from the underground services? See Fig. 7-16.
35. Why must manufacturers' installation diagrams be especially well drawn?
36. What basic types of drawings are used for manufacturers' installation diagrams?
37. What is a combination type furnace control unit?
38. By what principle does the helical element frequently located in the furnace smoke pipe operate? Explain the nature of this principle.
39. Why are underfloor ducts or raceways used in large building construction?
40. What is an inclusive raceway?
41. Describe the construction of a precast, cellular, concrete raceway.
42. What is the purpose of a raceway header duct?

PROJECTS

1. Redraw the floor plan of the single-floor residential building shown in Fig. 7-23, to a scale of 1/4″ = 1′0″. Superimpose on this drawing the symbols for lighting outlets, wall outlets, switches, and other necessary electrical components, to make your drawing a logical and well designed one.
2. Using trade magazines and brochures as your reference material, make a luminaire schedule for the light fixtures required for the building in Question 1.
3. Make a true wiring diagram of the simplified one shown in Fig. 7-24.
4. Using the blank outline of the automotive dealer's showroom and office space shown in Fig. 7-25, subdivide this area as you see fit to include four interviewing offices, a sales manager's office, an area for a receptionist-switchboard

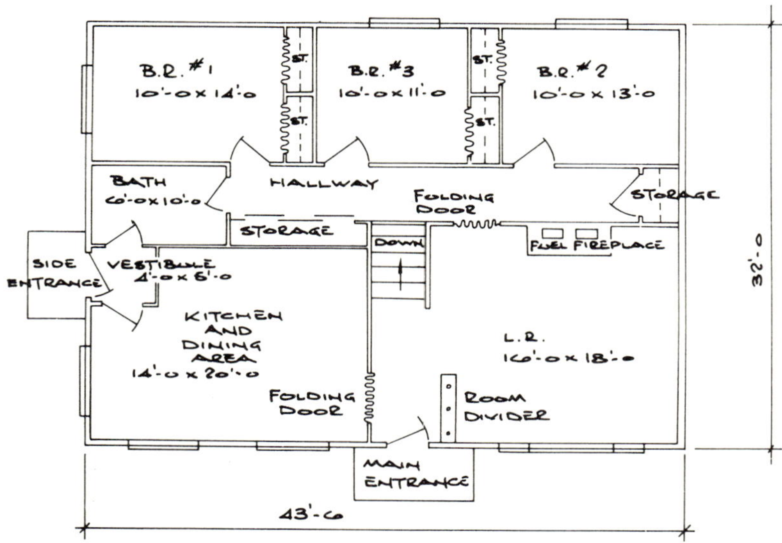

Fig. 7-23 The drawing for Project No. 1

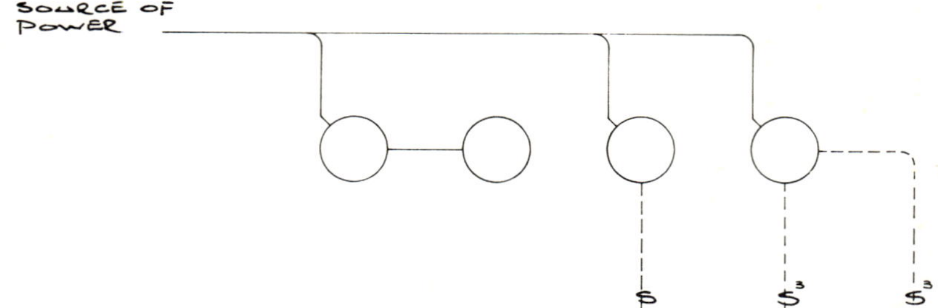

Fig. 7-24 The drawing for Project No. 3

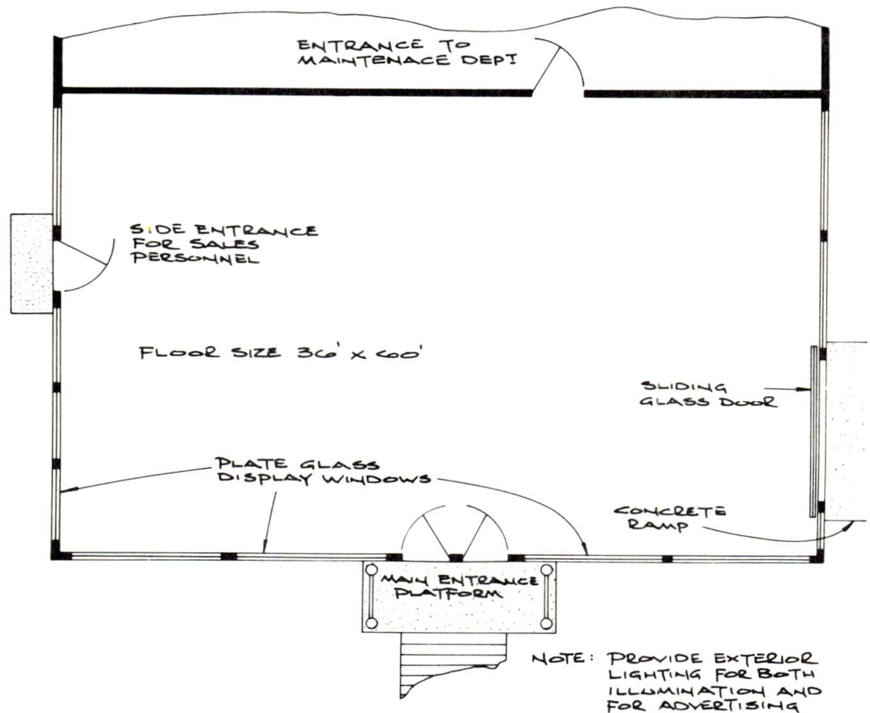

Fig. 7-25 The drawing for Project No. 4

operator, and a suitable display area for at least three American-size automobiles. After this subdivision has been completed, design and draw a suitable electrical system using standard symbols.

5. Make a riser diagram for the electrical system of a custom-designed country home given the following information:

> The system is 120/230 volt, single phase. Through a weather head fitting mounted on a roof mast, the cable is brought to a meter box, located on the outside wall at eye level, and then to a main 200 ampere switch box located on the basement wall at the west end of the building.
>
> From this switch box the cable travels to a 225 ampere splitter box mounted to the right of the main switch box. Directly above the splitter box are smaller individual switch boxes. From left to right, they have the following capacities: 60 amperes, 60 amperes, and 200 amperes.

The first switch box is connected to a panel (panel 's') in a small service building located approximately 90 feet north of the home. The second switch box is connected to a panel (panel 'b') in a barn located 500 feet north of the home. The third switch box feeds electrical current to the house distribution panel, (panel 'h'). It is located on the same basement wall to the right of the third switch box.

A 2500 watt 120/230 volt, single phase, emergency generator also feeds into panel 'h'. It is located on the basement floor to the right of panel 'h'.

One of the bathrooms on the first floor at the east side of the house is located such that it demands an exhaust fan controlled by a timer switch.

Also, an additional attic fan at the extreme east side of the building, and mounted in the gable end is required. It is also controlled by a timer switch.

For the sake of convenience the following cable information is shown in a schedule form.

CIRCUIT	NO. OF CONDUCTORS	SIZE	DESCRIPTION
Meter to main switch box	3	3/0	Corflex with PVC Jacket
Switchbox to panel 's'	3	#6	"
Switchbox to panel 'b'	3	#1	"
Panel 'h' to each fan	2	#14	—

6. In consultation with the teacher of applied electricity at your school, and using Fig. 7-16 as a guide, draw a single-line electrical diagram for the main service entrance of the residential building listed in Question #1 of this Project Section.

7. Obtain and study some trade literature on underfloor ducting systems. Also, investigate the type of system that is in use in the machine shop or other technical classrooms of your school. Following this research, make both dimensioned plan views and necessary elevations to correctly delineate its form. Some installations may be ceiling-hung, tray type. If such is the case, utilize only the trade literature.

8. Redraw the view of the transformer vault shown in Fig. 2-3 to a scale of $1/2''$ $= 1'\ 0''$. The original engineering drawing was to a scale of $1/4'' = 1'\ 0''$.

9. Figure 7-5 shows a pictorial illustration of an electrically powered farm installation, in which a poultry house is located. As a research project consult your local provincial, state, or federal agricultural department for help in designing the electrical layout for such a building. Consider this poultry house to be an integral unit of a mixed farming operation, and not a specialized unit of a farm strictly devoted to the raising of poultry.

10. Using the illustration, Fig. 7-26, of a winterized summer home, make a complete set of electrical drawings. These are to include a plan view, a single-line main service entrance diagram, a luminaire schedule, and schedules for other main components. This building is to be electrically heated. Outlets for an electrically powered hot water tank, range, and refrigerator are to be included. Ample service outlets are also to be provided.

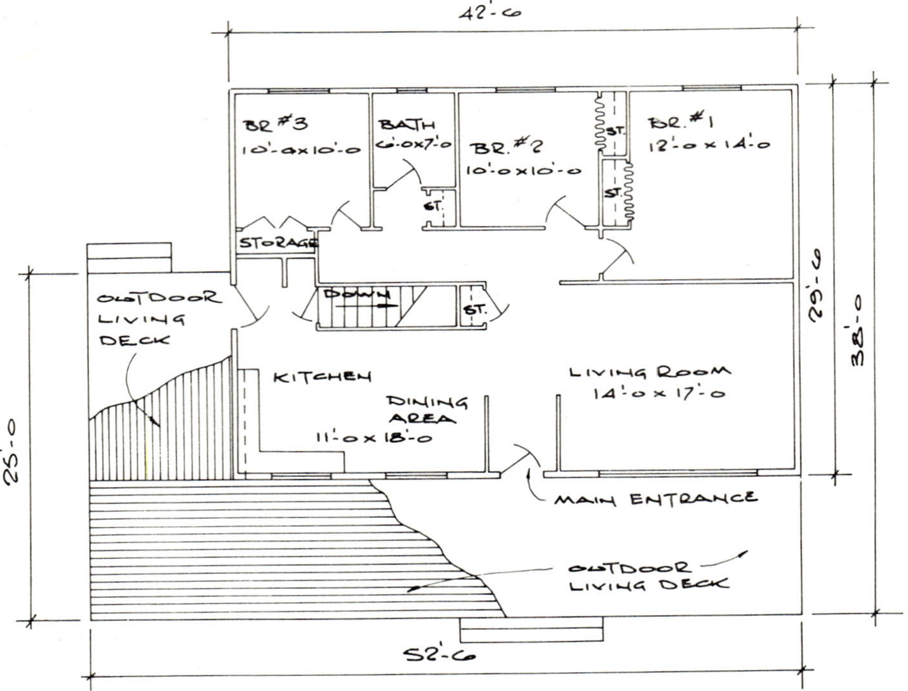

Fig. 7-26 The drawing for Project No. 10

UNIT 8

Industrial Wiring Diagrams

INDUSTRIAL WIRING DIAGRAMS ARE CONCERNED CHIEFLY WITH RELAY and breaker circuits, motor control circuits, timing, sequencing, and feeding circuits. In the past these circuits were almost exclusively electrical, that is to say, did not involve electronic components. Today, the field of electronics has pervaded what was once an exclusive electrical domain. Because of this, some reference will be made in this unit to systems involving electronic components.

It is necessary for the electrical draftsman to appreciate the wealth of data and pertinent schematics shown in the manufacturers' brochures and pamphlets of industrial wiring equipment. This is so, because each installation will call for electrical equipment manufactured by a specified industrial firm. As such, it is necessary to interpret the schematic shown for a particular type of control and adapt it to the overall design. Fortunately, the manufacturers of electrical equipment adhere quite closely to the graphic symbols laid down by the CGSB or A.S.A. On some occasions, interpolation of symbols may be required because of the failure of some manufacturers to adhere to symbol standards.

BASIC CIRCUIT BREAKER DIAGRAMS

Series-trip circuit

The diagram illustrated in Fig. 8-1 shows the correct layout when the circuit breaker coil and contacts, as well as the line and load are in series with each other.

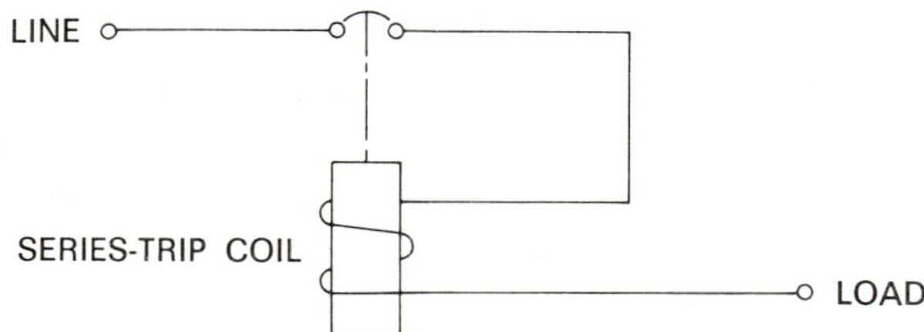

Fig. 8-1 A series-trip circuit breaker diagram

Relay-trip circuit

In Fig. 8-2, a circuit diagram is shown whereby the relay-trip coil can be actuated remotely, and in addition is electrically isolated from the line and load terminals. Because of this, a voltage of different value than that being switched through the contacts can be used.

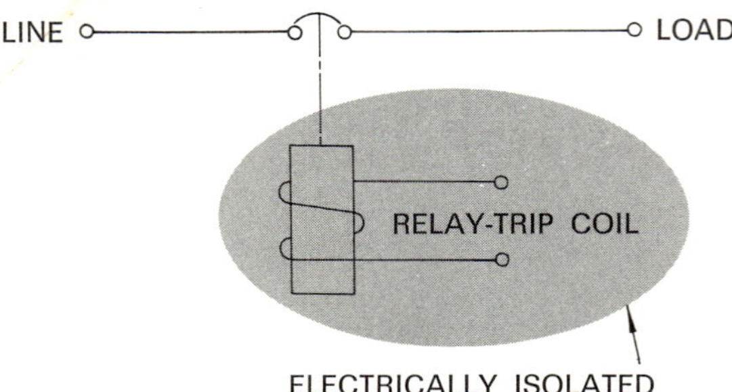

Fig. 8-2 A relay-trip circuit breaker diagram

Shunt-trip circuit

The shunt-trip circuit shown in Fig. 8-3 is wired so that while remote tripping is possible, the shunt-trip coil utilizes the line voltage, rather than a separate source.

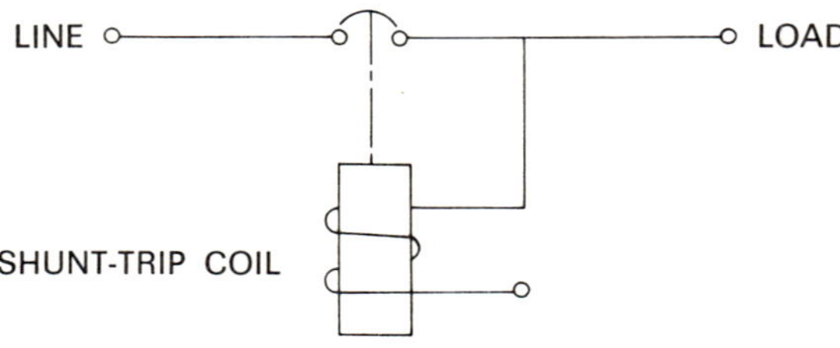

Fig. 8-3 A shunt-trip circuit breaker diagram

Circuit breaker construction

Some understanding of circuit breaker component parts and their function is necessary before proceeding into their application in specific controlling circuits. While each manufacturer designs and fabricates circuit breakers on an individual basis, hence providing for uniqueness in design, all circuit breakers have some more or less common features. The pictorial drawing shown in Fig. 8-4, illustrates a modern circuit breaker.

The circuit breaker shown in Fig. 8-4 is manually set by a handle (2) which has only two operational positions, 'on' and 'off'. When the circuit breaker is triggered by a fault, a flick of the handle resets the latch mechanism and service is restored. The latch mechanism (3) is designed for trip-free operation; and even if the handle is held in the 'on' position when the circuit breaker is in an overload condition, its contacts will not close. The contacts (4) are especially designed to remain clean by virtue of their sliding action while under pressure. To eliminate **oxidation** and **pitting** of the contacts caused by **arcing**, a series of

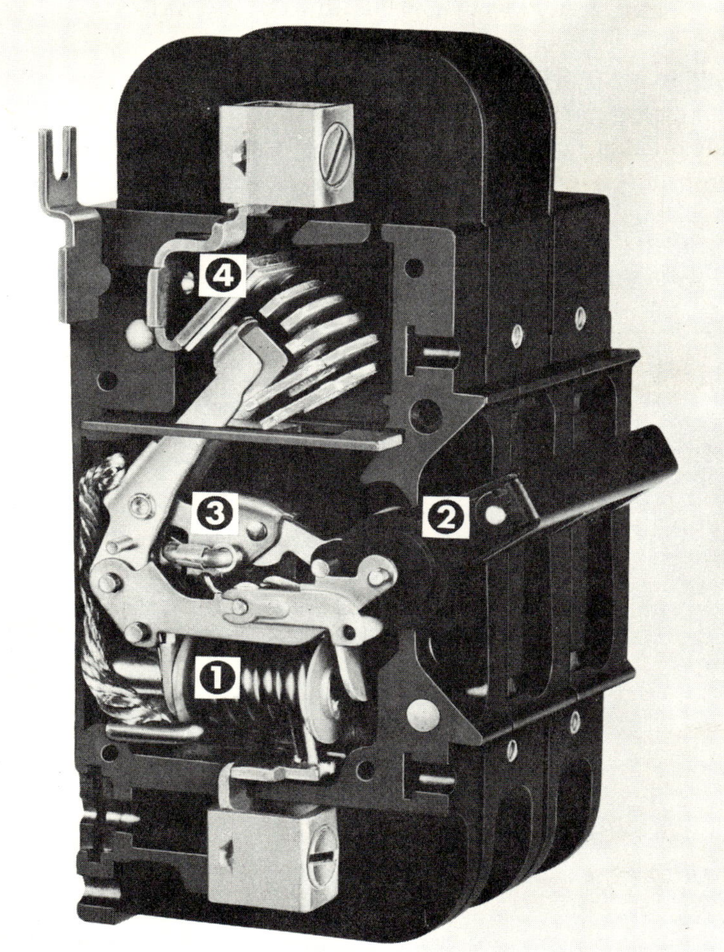

Fig. 8-4 A modern multi-pole circuit breaker. (Courtesy The Heinemann Electric Co., Trenton, New Jersey)

grid plates formed from U-shaped magnets quickly draw out, fragmentize, and extinguish the arc. In addition, the arcing is prevented from taking place on the contacts themselves, but does so on the larger surfaces of the grid plates. The actual tripping of the circuit breaker contacts is brought about by an overload in the load-sensing coil (1). This coil is **nonthermal** in operation, that is to say, its operational conditions are not affected by **ambient temperature**, and therefore nuisance tripping cannot occur.

An additional feature worthy of mention is the **time** delay aspect built into the coil mechanism. In the centre of the coil is a nonmagnetic, **hermetically sealed** tube, containing a spring-loaded iron core capable of moving in a **silicone** liquid. The strength of the magnetic field created by the overload current moves the iron core towards the freely pivoting armature. However, it is restricted in its travel by the silicone fluid. This restricting or dampening effect creates a controlled time delay, that is inversely proportional to the strength of the overload.

APPLICATION OF CIRCUIT BREAKERS

Application of a series-trip, circuit breaker

When the primary side of a transformer is connected in series with a series-trip circuit breaker, and across the lines of the source of power, overcurrent sensing and subsequent circuit interruption can take place. The example shown in Fig. 8-5 illustrates the form that the circuitry must take for this electrical protection. The delineation of the line voltage circuitry is intentionally drawn heavy to capture the reader's attention.

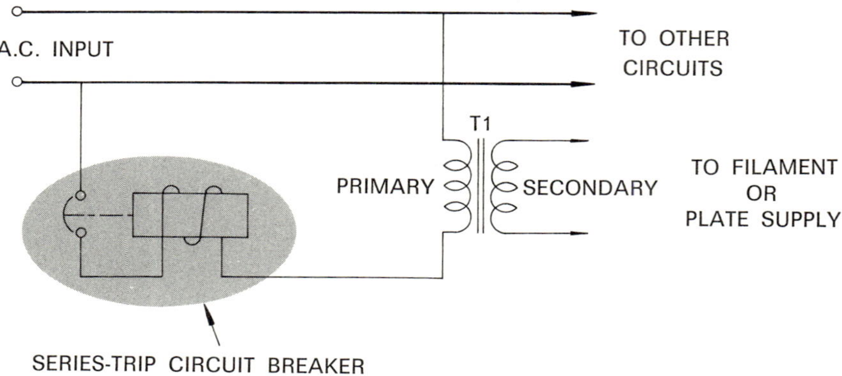

Fig. 8-5 The application of a series-trip circuit breaker

Application of a relay-trip, circuit breaker

The form of the schematic utilizing a circuit breaker in a relay-trip circuit is shown in Fig. 8-6. Note that the coil for the relay-trip circuit breaker is disassociated from the primary side of the transformer T1. In its location, an overload current flowing from the **centre-tap** connection of the **plate or anode** winding of transformer T1 to ground, will cause the contacts to open, thus interrupting the circuit. In addition, the student should realize that the D.C. output voltage circuitry, (A), could be drawn otherwise, but the method shown is quite typical. Another point that must be mentioned is that the circuit breaker enclosure (shown by a broken line) contains two separate units, and each one is actuated by an individual switch lever.

Application of a shunt-trip, circuit breaker

The diagram shown in Fig. 8-7 displays a circuit which permits control of current flowing to the supply transformer from a remote place. In addition a special feature has been added. A **safety interlock switch** which might be actuated by temperature, pressure, or air flow **variants** has been included in series with the circuit breaker coil.

Application of a series-trip/shunt-trip and auxiliary switch circuit breaker

In some industrial applications where sequential operations are performed on a product under manufacture, it is mandatory that a control system be used

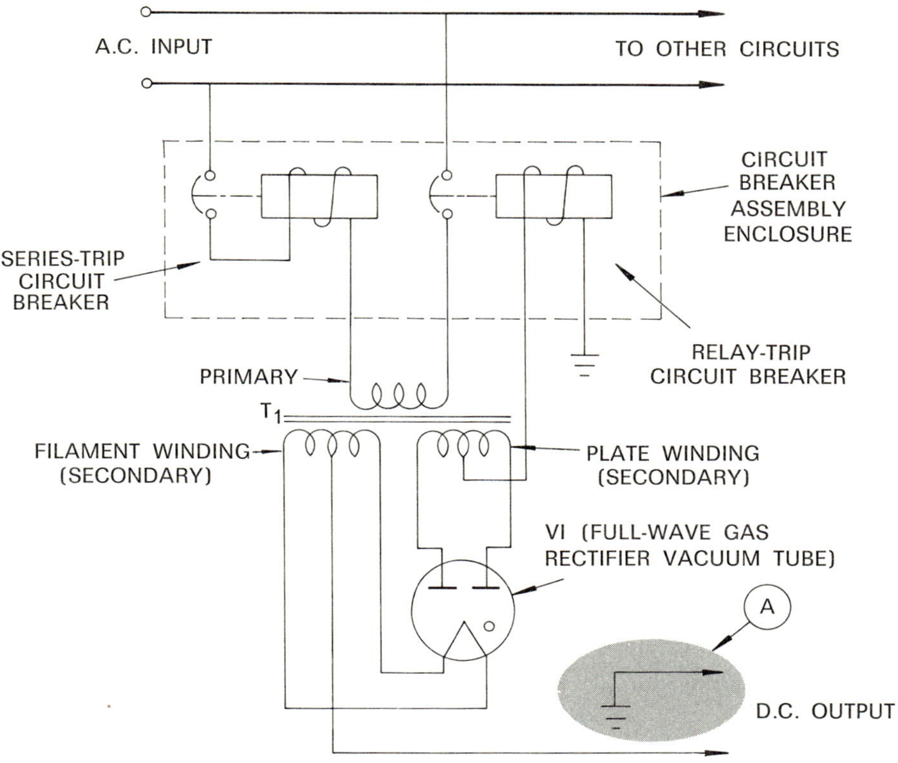

Fig. 8-6 The application of a relay-trip circuit breaker

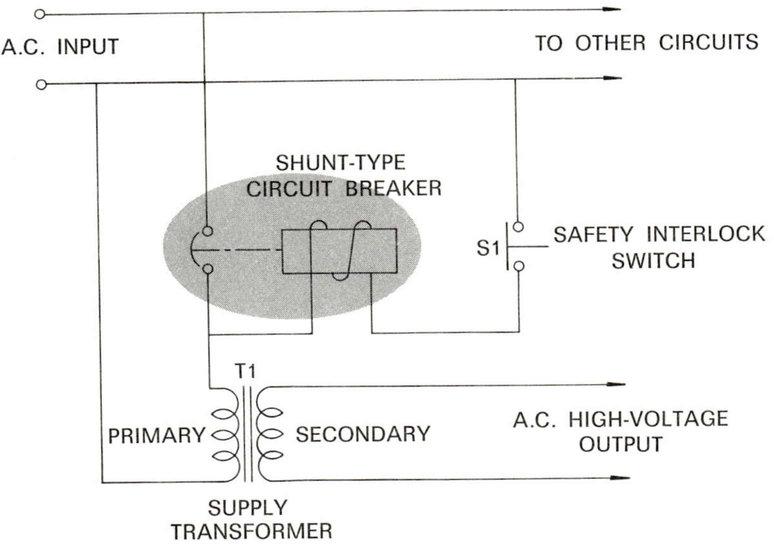

Fig. 8-7 The application of a shunt-trip circuit breaker

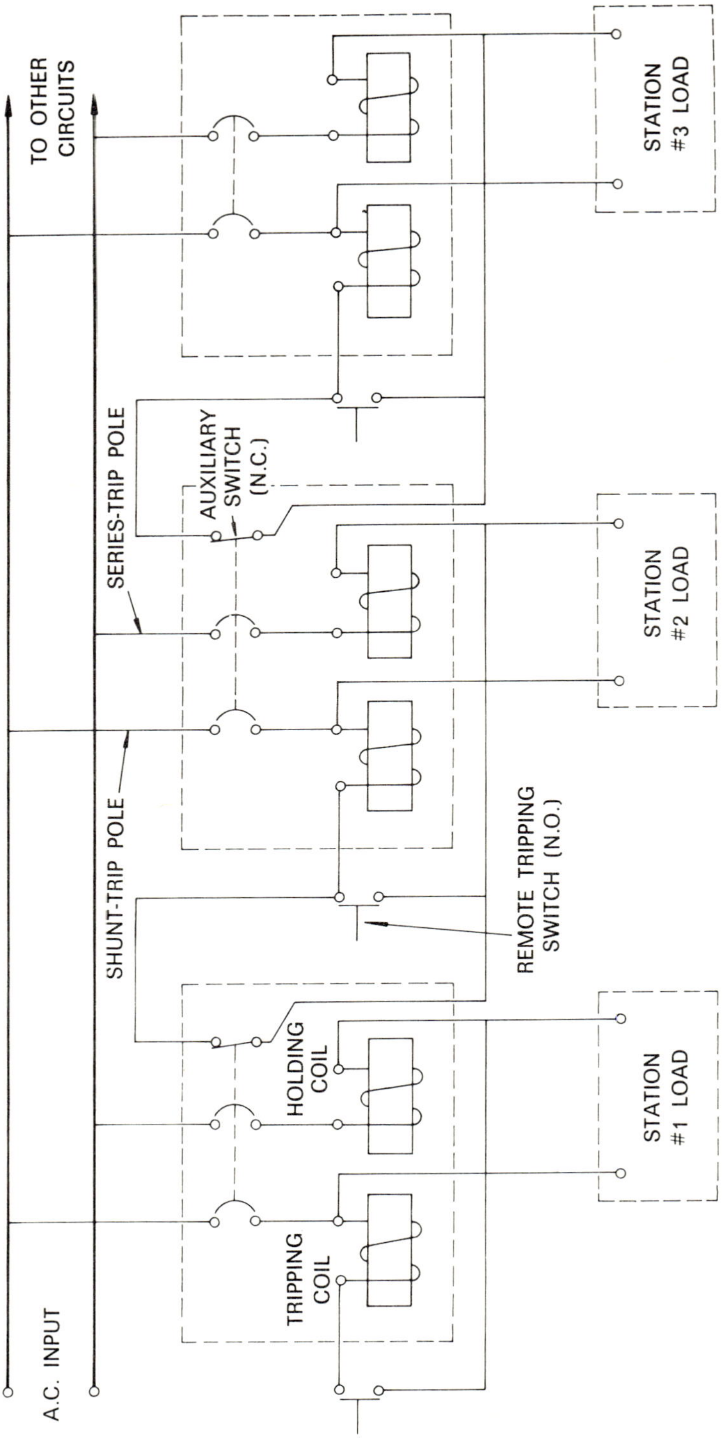

Fig. 8-8 The application of a series-trip/shunt-trip and auxiliary switch circuit breaker

that only permits the various stations in the manufacturing process to be turned 'on' in a 1-2-3, etc. sequence, and turned 'off' in a 3-2-1 sequence. A circuit involving a three-step or station process is illustrated in Fig. 8-8. It is apparent that the two-pole circuit breakers with their accompanying auxiliary switch have been triplicated, except that the third stage has no requirement for an auxiliary switch.

In this schematic, no details are given regarding the type of load, hence the leads are terminated in the outline blocks representing the loads. These conceivably might be electrical motors. Note that by placing the line voltage leads at the top of circuit, the circuit breakers in the centre, and the loads at the bottom, a pleasing, easily read, and logical schematic results.

PUSH-BUTTON STATIONS

Fig. 8-9(a) *Push-button stations. (Left: Courtesy Cutler-Hammer Canada Limited. Right: Courtesy Allen-Bradley Canada Limited)*

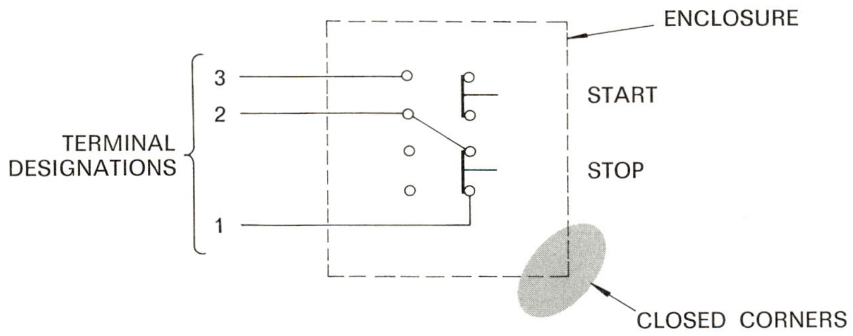

(b) A start-stop push-button station

Push-button station with start-stop functions

The simplest unit in a motor control circuit is the push-button station. See Fig. 8-9(a). This unit is essentially two individual spring-loaded, double-pole, single-throw switches mounted in a neatly designed steel enclosure. The outline for the enclosure is drawn with a broken line, and the corners are formed with closed L's. The internal wiring is shown in Fig. 8-9(b).

Push-button station with forward and reverse functions

Frequently it is desirable to have a motor installation in which the direction of the motor rotation can be changed at will. The push-button station associated with such a circuit when illustrated schematically would appear as in Fig. 8-10. Note that the terminal designation numbers are not sequential. This is a prerogative of the manufacturer. In addition, the weight of the enclosure lines must be drawn lighter than the graphic symbols.

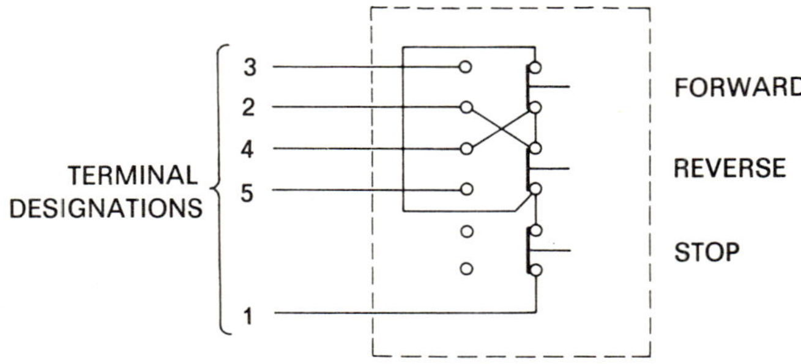

Fig. 8-10 A forward-reverse-stop push-button station

MOTOR STARTERS

Single-phase starter

In conjunction with the push-button station, a device known as a starter is used to control the stop-start action of a motor. If the motor is a **single-phase** type the circuit diagram shown in Fig. 8-11 is applicable. The components illustrated by the symbols in the circuit are **engineered** into a suitable enclosure, exclusive, of course, of the motor. Besides the fact that the circuits through (2) to (3), and L1 and L2 are protected by normally-open breaker contacts, the student should observe the symbol marked 'OL', which is an overload protection element. The heat generated by an overload current passing through it opens the N.C. reset contact, thus protecting the motor circuit. It should also be noted that the primary circuit from line to load is drawn with a much thicker line than the remainder of the circuit.

Three-phase, nonreversing starter

When a **polyphase**, nonreversing motor is used, the schematic illustrated by Fig. 8-12 is required. In this particular case an additional block is shown on

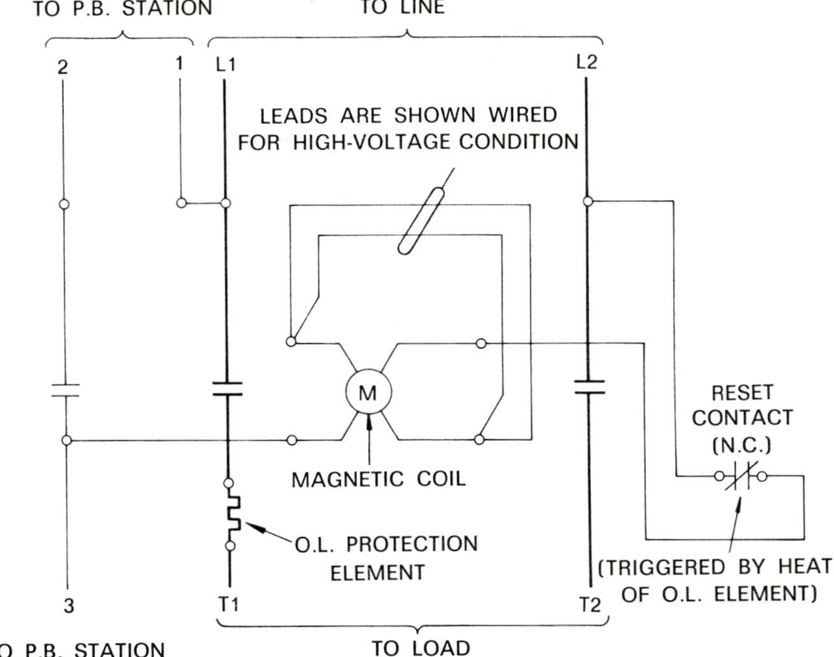

Fig. 8-11 A single-phase motor starter circuit

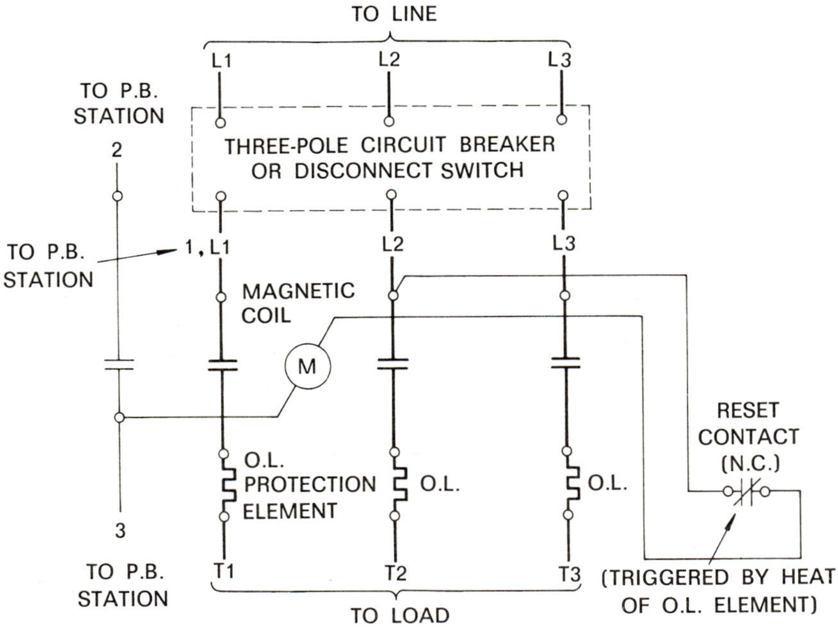

Fig. 8-12 A three-phase, nonreversing motor starter circuit

the schematic which provides for a fused or unfused **disconnect switch** on each branch of the line circuit, or in place of a disconnect switch, a circuit breaker. Note that overload protection is provided on each main branch of line to load including the centre branch. It is normal practice to locate the magnetic coil symbol as shown, although it does tend to confuse the interpretation of the schematic. Also, the primary part of the circuit is delineated by a thick line in contrast to the other circuitry.

Three-phase, reversing starter

The schematic shown in Fig. 8-13 would demand the use of the push-button station shown in Fig. 8-10. It should be noted that as the number of functions increases, accordingly the circuit complexity does also. Once again the primary circuitry is drawn in bold lines. Also, some symmetry is maintained in the placement of contacts and overload coils. In addition, it is common practice to place the line terminals at the top of the schematic, and the load terminals at the bottom. This was also the case in the simpler schematics shown in Figs. 8-11, and 8-12. The letter symbols (R) and (F) are to indicate the reverse and forward aspects of the starter.

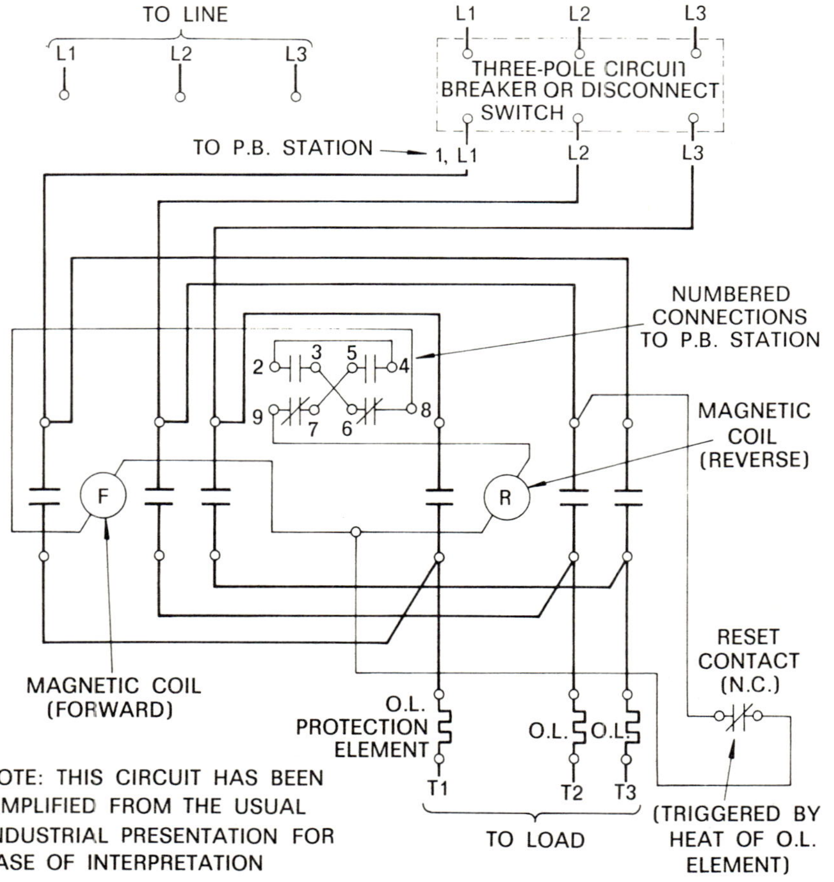

Fig. 8-13 A three-phase, reversing motor starter circuit

PROTECTION CIRCUITS INVOLVING ELECTRICAL AND ELECTRONIC COMPONENTS

Transistor overheating protective circuit

A simple but practical circuit using a relay-trip breaker to protect the transistors in a small compact desk-side computer is shown in Fig. 8-14. The **thermistor** which acts as a temperature sensing device is located in the transistor ventilating duct of the computer. The breaker which has N.C. contacts is triggered when a lowering of the thermistor resistance due to overheating permits an actuating voltage to flow through the circuit breaker coil. This in turn breaks the circuit to the computer from its power source.

In a circuit of this sort it should be noted how the energization of the relay coil in one circuit affects the function of another circuit not connected by any electrical means. Also, when it is necessary to leave off a part of a circuit, arrows are drawn at the terminations of the conductors.

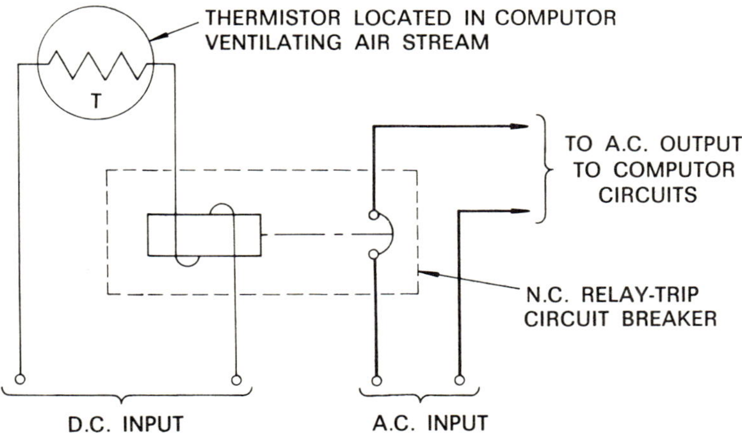

Fig. 8-14 A computer overheating protection circuit

Transducer high-pressure sensing circuit

A circuit incorporating a relay-trip circuit breaker and a pressure sensing **transducer** is illustrated in Fig. 8-15. The purpose of the circuit is to act as a back-up in case the mechanical relief valve in the high-pressure tank fails. Should this happen, a low-voltage signal is transmitted to the circuit breaker coil in the transducer circuit. This in turn causes the circuit breaker contacts to open, thereby interrupting the flow of electrical current to the tank heating element, and in turn preventing a further build up of tank pressure.

Schematic drawings, such as illustrated in Fig. 8-15, which incorporate mechanical components should be drawn so that these parts are fully distinguishable from the electrical graphic symbols. The purpose of the heavy-weight lines for the heating element is for reasons of emphasis and identity.

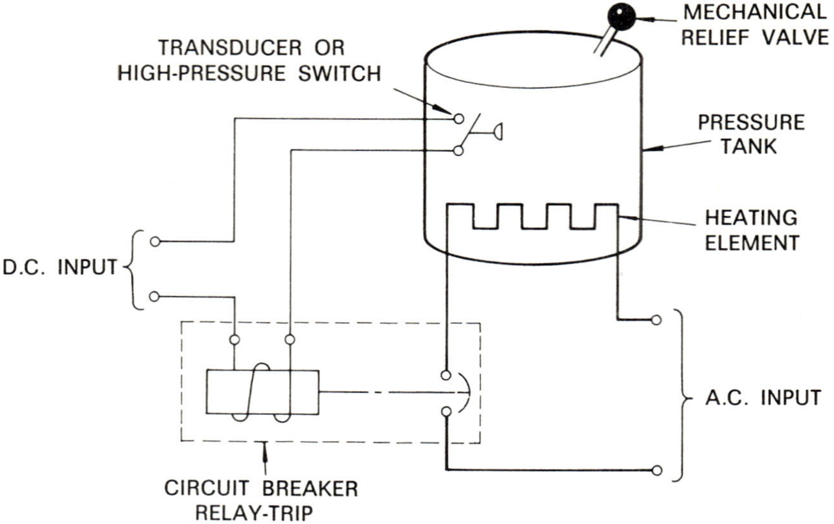

Fig. 8-15 A transducer, high-pressure sensing circuit

INDUSTRIAL TIMING CIRCUITS

Circuits involving timing devices incorporating either cam-operated switches, vacuum tubes, or solid-state devices are used in a variety of industrial applications. Welding machines, vending machines, dryers, microwave ovens, and aerospace equipment are but a few that could be mentioned.

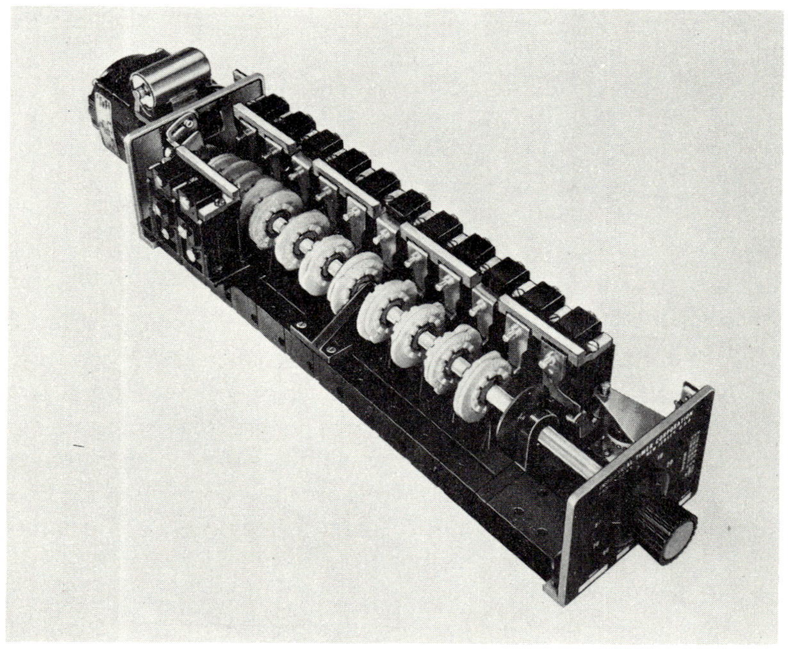

Fig. 8-16 A typical cam-operated timer or step programmer. (Courtesy Industrial Timer Corporation)

Cam-operated timer or step programmer

MECHANICAL DETAILS

Figure 8-16 shows a typical mechanical timer. Essentially this device consists of a series of switches operated by an identical number of cams. The cams are attached to a motor driven shaft, and are capable of being oriented about the shaft to give a different timing position for the operation of each switch. The reliability of the entire assembly is of high order. In particular, the life rating of each individual precision switch permits at least 1,000,000 operations.

CIRCUIT DETAILS

The circuitry illustrated in Fig. 8-17 shows the electrical connections required for the automatic sequential operation of a number of different loads. The vertically stacked orientation of the load switches should not be construed as their position of operation, that is, a 1-2-3 sequence. Note that in a schematic

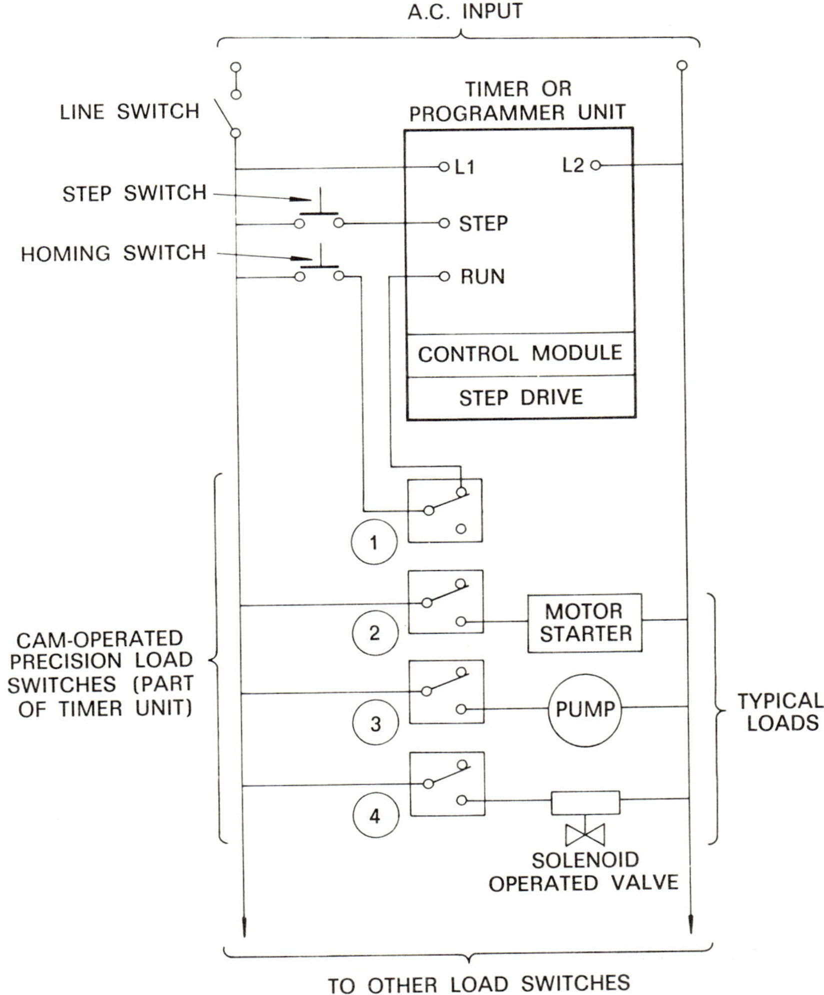

Fig. 8-17 A timer or step programmer

of this type, no internal wiring of the timing unit is required to be drawn, since all internal wiring of the timing unit is carried out at the place of manufacture. Also, it should be pointed out that when the first load switch is used in series with the 'run' switch, the timer will automatically come to halt when this load switch is tripped by its cam. Finally, this type of circuit drawing is primarily intended for use in a manufacturer's brochure, and for this reason detailed connections to the loads have been purposely neglected.

Sequence chart

When designing and drawing circuits involving timers or programmers a sequence chart is drawn to establish when various loads must be brought into play. A format for such a chart is shown in Fig. 8-18. In this case Load 1 is cycled four times at four-step intervals, and Load 2 is cycled four times at three-step intervals, the load being held for a two-step period. By studying the remainder of the chart shown in Fig. 8-18 it is self-evident regarding the sequencing of Loads 3, 4, 5, and 6. The production of such a chart would of necessity precede the actual design of the circuit diagram.

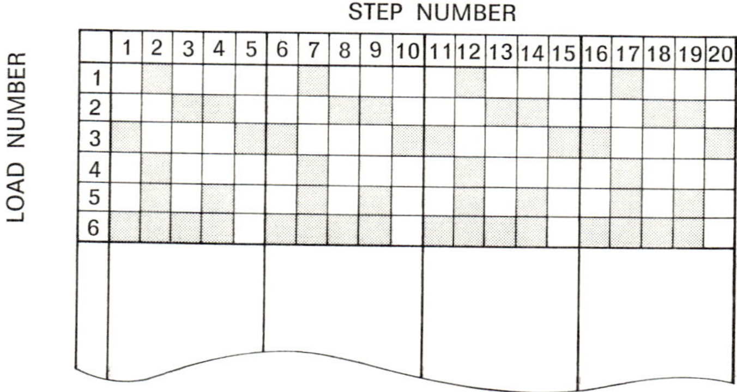

Fig. 8-18 A timer or step programmer sequence chart

Thyratron controlled, welding timing circuit

In industrial welding, frequently a timing system is necessary. By utilizing a special gas-filled vacuum tube, known as a **thyratron**, and an appropriately designed circuit, a quick-acting electronic switch action can be achieved. The schematic shown in Fig. 8-19 represents a typical circuit of this nature. Note the apparent isolation from the main circuit of the contacts in the 'weld control' circuit. These contacts are activated by the energization of the contact relay. The broken lines connecting either set of contacts to its respective relay are presented for the student's ease of interpretation. Under normal circumstances identity would be made by affixing an abbreviation to both contact and relay, which would have a common connotation, for example, CR contacts and CR relay.

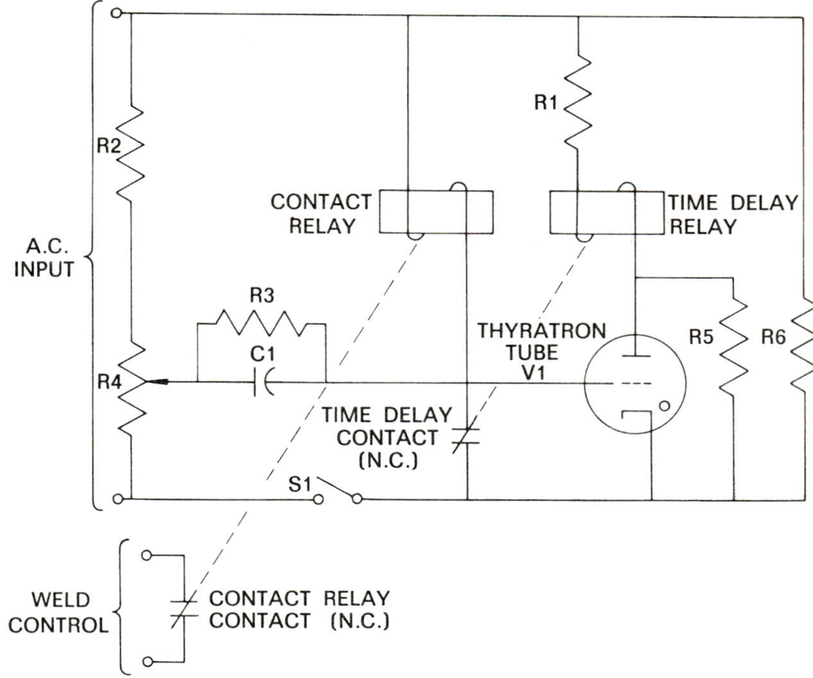

Fig. 8-19 A thyratron controlled welding timing circuit

Solid-state timer

A solid-state timer operates on purely electronic principles and is devoid of relays and other moving parts. Other advantages are its high resistance to vibration or shock, and its ability to operate properly under conditions of dust and humidity. The reason for this is that the electronic components are **encapsulated** in a material known as **epoxy**. Figure 8-20 shows such an electronic timing device. Different models or types of such devices are manufactured to

Fig. 8-20 A solid-state electronic timer. (Courtesy Regent Controls, Inc., Stamford, Connecticut)

perform specific switching functions. The NEMA (National Electrical Manufacturers Association) solid-state schematic outlined in Fig. 8-21 relates to a timer with a normally-open switch, having a time delay on energization of its closing circuit, and resets to open instantly when this circuit is de-energized. Other important features of this device are its timing ranges which vary in increments from 0.02 seconds to 300 seconds. In addition the manufacturer offers a guaranteed claim of 500,000,000 operations.

The wiring diagram shown in Fig. 8-22 illustrates the simplicity with which a solid-state timer is wired into a circuit.

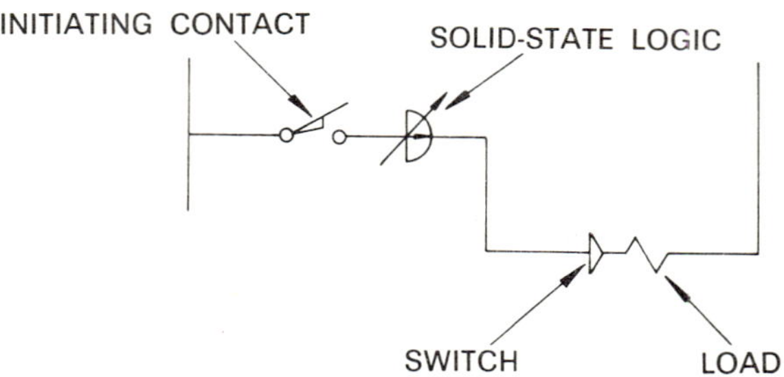

Fig. 8-21 A solid-state timing circuit schematic

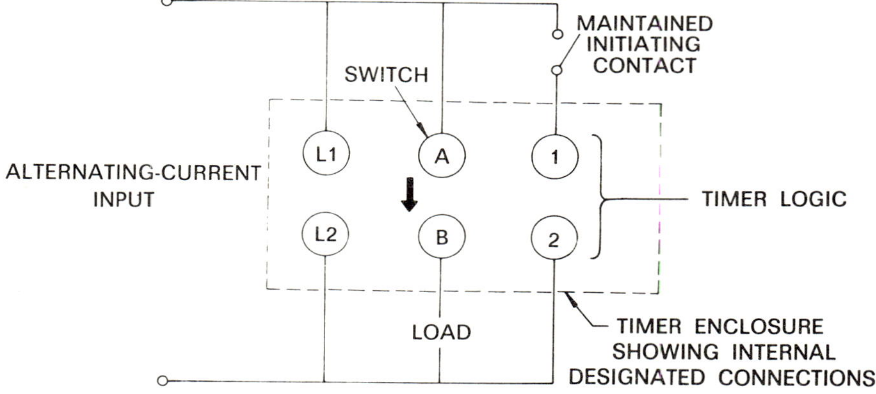

Fig. 8-22 The method of wiring a solid-state timer in a circuit

Machine tool, feed-control circuit

The diagram illustrated in Fig. 8-23 is shown to provide an example of the type of layout required for the electronic feed-control system of an industrial machine tool. A certain number of crossovers are unavoidable in the layout of this schematic, but its electrical interpretation is nevertheless not difficult. The variable resistance controls are ideally shown one above the other, while the gas diode vacuum tube, because of its importance in the circuit, is located in such a position as to attract the reader's attention. With regard to the solid-state diodes, care must be taken not to orient them except as they are shown, otherwise their rectifying action will be in the wrong direction.

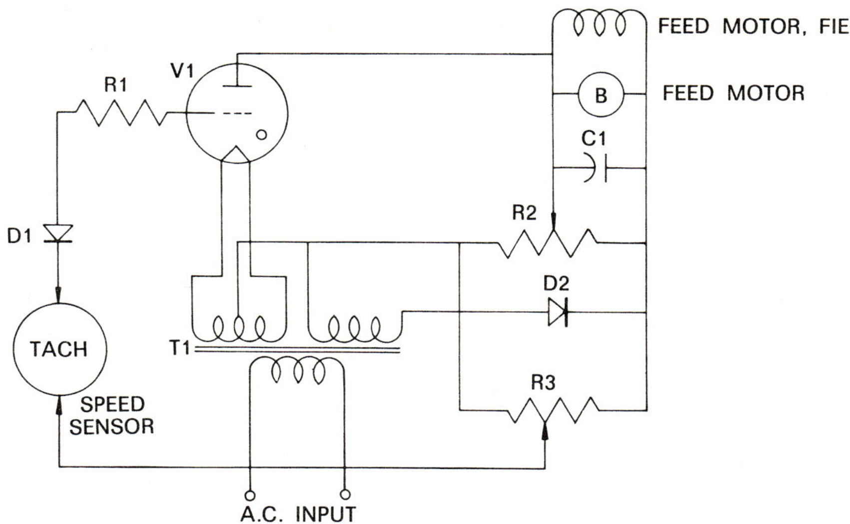

Fig. 8-23 A machine tool feed control circuit

KNOWLEDGE TESTERS

1. What function or purpose do manufacturers' brochures on industrial wiring equipment serve?
2. If the coil in a series-trip circuit breaker circuit burns out, what condition results? See Fig. 8-1.
3. What does the broken line between the armature of the coil and the contacts of a relay-trip circuit breaker indicate?
4. List two advantages in using a relay-trip circuit breaker.
5. What source of electrical current is normally used to energize the coil of a shunt-trip circuit breaker?
6. Describe the safety aspect of a circuit breaker when it is in an overload condition.
7. How is arcing eliminated at circuit breaker contacts?
8. What type of material would be used for the handle and the moulded body or housing of a circuit breaker?
9. If arcing persists at a circuit breaker contact, what condition will result?
10. Why is the load-sensing coil of a circuit breaker nonthermal in operation?
11. If a circuit breaker in a closed condition is located in an area of high **ambient** temperature, what result will take place?
12. What purpose does the silicone fluid serve in a circuit breaker coil, time delay mechanism?
13. In Fig. 8-6, what type of secondary winding in the transformer T1 is used in the plate circuit?
14. What is the significance of the small open circle within the envelope symbol of vacuum tube V1 in Fig. 8-6?
15. Explain the use of the safety interlock switch in the circuit illustrated by Fig. 8-7.
16. In Fig. 8-8, explain how the remote tripping switch is instrumental in preventing the electrical current reaching the load.
17. By studying the circuit in Fig. 8-8, explain why the N.C. auxiliary switches are required.

18. How many poles are required in a push-button station controlling a motor having both forward and reverse functions?
19. In a single-phase motor starter circuit how is the primary circuit delineated?
20. How does the overload element protect the motor circuit in Fig. 8-11?
21. By studying the circuit in Fig. 8-11, what conclusion is reached regarding the nature of the contacts between line and load, and between 2 and 3?
22. What major circuitry changes are involved in drawing a three-phase motor circuit in relationship to a single-phase circuit?
23. What additional protection is incorporated in the three-phase motor circuit shown in Fig. 8-12?
24. Under normal operating conditions how does a thermistor prevent the energizing of the circuit breaker coil? See Fig. 8-14.
25. In Fig. 8-15, explain how the heating element is electrically de-energized if the tank pressure exceeds its maximum operating limits.
26. How are the precision switches in a mechanical timer or programmer activated?
27. How can the timing of each individual precision switch in a programmer be altered?
28. What is the purpose of a timer sequence chart?
29. If the program shown in Fig. 8-18 takes place during the full rotation of the timer cam shaft, how many lobes are there, and how many degrees apart are the lobes located for load #2?
30. In Fig. 8-19, what type of resistor is R4?
31. By studying Fig. 8-19, what conclusion is reached regarding the electrical size of resistor R5?
32. List three advantages of a solid-state electronic timer.
33. What does the abbreviation 'NEMA' mean?
34. What is the purpose of the diode in the grid circuit of vacuum tube V1? See Fig. 8-23.
35. Why are two separate magnetic coils required for the motor starter circuit illustrated by Fig. 8-13?

PROJECTS

1. Make a wiring diagram of a circuit to control a single-phase, 1/4 h.p., 110 V.A.C. motor. Both a stop-start, push-button station, and a magnetic starter with O.L. protection are required.
2. Draw a circuit in which a 12 V.A.C. relay-trip, circuit breaker, when activated by the closing of a remote S.P.S.T. switch, closes the D.P. contacts of the circuit breaker. In doing so it permits a 110 V.A.C. supply to flow to an electrical load. In drawing the circuit, assume that the only electrical supply available is 110 V.A.C.
3. Make a wiring diagram of a circuit to control a three-phase, 1 h.p., 208 V.A.C. motor. Both a stop-start, push-button station, and a magnetic starter with O.L. protection on the two outer branches of the supply leads to the motor are required. In addition, a fusible disconnect T.P.S.T. switch on the line source is required.
4. A short-wave radio transmitter requires 700 volts D.C. for the operation of certain of its tubes. Since a definite safety hazard from electrocution exists, design and draw a circuit which will offer the necessary protection when the access door to the enclosure surrounding this equipment is opened. Assume that the primary source of power is 110 V.A.C.
5. Make a wiring diagram of a circuit to control a reversible, three-phase, 3/4 h.p.,

220 V.A.C. motor. Overload protection on each branch, as well as a fusible disconnect switch in the line circuit is required. The wiring diagram is also to include a suitably wired push-button station.

6. A special oven used in a chemistry laboratory requires an automatic means to control its temperature, and which is independent of the source of power for the oven elements. The temperature controlling device must be other than that depending on the action of a bimetallic strip. Design, and draw such a circuit.

7. A cam-operated programmer is required to control five identical, light industrial manufacturing machines. Each machine carries out eight different, sequentially timed operations or steps. As each step or operation is performed an electrical load according to the following information is activated:

Load 1 commences at step 1 for a 1 step period.

Load 2 commences 1 step after step 1 for a 2 step period.

Load 3 commences at step 2 for a 3 step period.

Load 4 repeats load 2.

Load 5 commences at step 5 for a 5 step period.

Load 6 is energized during step 8 period only.

Load 7 repeats load 2.

Load 8 is energized except for step 8.

With this information, design and draw a sequence chart for the programmer.

8. An electrically operated advertising sign is required by a television servicing company. The sign is to spell out the words, 'TV SERVICES'. The letters in the sign are to be illuminated in sequence as follows:

(a) Each letter, starting with 'T' is to be illuminated for a two-second period, nonilluminated for a two-second period, and illuminated for a further two-second period.

(b) The complete sign is then to be illuminated for five seconds at a time, followed by one-second periods of nonillumination. This action will require one minute of time, and is to take place after the illumination of the last individual letter in the sign.

Given the above information do the following:

(i) Make a suitable sequence chart for the system.

(ii) Make a wiring diagram showing how such a system could be devised.

(iii) Indicate the nature of the switch cams required by means of illustrations, and specify the number of them.

9. In consultation with the woodworking instructor of your school, draw a block diagram showing the electrical components required to operate the major fixed pieces of power machinery in the woodworking shop. These are to include planers, jointers, saws, and lathes. Particular emphasis is to be placed on starters, disconnect switches, and the main emergency circuit breaker. The diagram must also include the distribution circuit breaker panel in the room.

10. Make a wiring diagram of a full-wave rectifier circuit having overload protection in its primary circuit, an interlock switch in its high-voltage D.C. circuit and a device which will sense abnormally high ambient temperatures in the enclosure surrounding this equipment, and in turn will open the circuit from the source of alternating current.

UNIT 9

Electronic Wiring Diagrams

Definition

An electronic wiring diagram is a type of drawing associated with either vacuum-tube components or semiconductor devices. It may in its entirety be concerned with industrial control equipment, entertainment equipment, measuring and detecting equipment, or communications equipment.

Circuitry types

In general, the major types of electronic circuitry fall into four classes, and as such offer particular problems to those people involved in their design, which include the draftsman.

The first class is of the type where individual components are soldered to separate, free standing, insulated conductors. We shall refer to this class as **conventionally wired** circuits. This method might now be considered out-of-date, but is again establishing itself as a suitable mode of wiring for high-quality home entertainment equipment.

The second class is concerned with printed circuits. In this case the pattern for the connecting circuitry is formed by a special silk screen process on a **phenolic laminate matrix** board. Then the individual electronic components are soldered in their correct places on this board, by dipping the noncomponent side of the board to a proper level in a hot, liquid-solder bath. Miniaturized components and semiconductor devices such as diodes and transistors are used in conjunction with printed circuit boards almost exclusively.

The third class is referred to as integrated circuitry or IC and involves advanced semiconductor devices. These devices are very minute in size, and are designed as complete and functional circuits. Essentially they consist of a germanium, or silicon, chip of extremely small size (measuring typically sixty by eighty thousandths of an inch) upon which are formed circuit combinations of transistors, diodes, capacitors, and resistors by processes known as vacuum deposition or cathode sputtering. Some present day applications for microminiaturized integrated circuits lie in the computer and aerospace equipment fields. Here their extreme reliability, compactness of size, and small weight are of prime importance.

The fourth method is known as LSI, or **Large Scale Integrated** circuitry. This is a process of producing a variety of different functional circuits on a single slice of silicon measuring an inch or so in diameter. These individual but different circuits are not sliced into chips, such as is done under normal integrated circuit processes, but remain as a full, or subsystem, already interconnected on the slice by the deposition of aluminum conductor pathways. With the present state of the art it is possible to build an electronic system having as many as 1000 transistors with their associated diodes and resistors on a single one inch diameter slice of silicon.

In order to produce an LSI circuit on a silicon slice or wafer, drawings of the type used for normal IC chip design would be required, but of more complexity, since a much larger number of electronic components would be built up. There are so many and varied electronic systems that it is impossible to include sufficient representative examples for all of them. Hence, only a select number will be illustrated and referred to within the scope of this unit.

CONVENTIONALLY WIRED CIRCUITS

Tube-type circuit

The circuit diagram shown in Fig. 9-1 illustrates a typical tube-type **superheterodyne** receiver circuit. Superheterodyne receivers having vacuum tubes will in time become obsolete. The inclusion of such a circuit in this unit is for the purpose of introducing the student of drafting to the electronic field in a comprehensive manner. In laying out such a circuit, the symbols for the tube envelopes are placed on the same horizontal level, but with sufficient space to permit circuit details about them. Also, it is standard practice to place the I.F. transformers between tubes V1 and V2, and V2 and V3 as shown. The method, by which the filaments of all the vacuum tubes except the rectifier are shown, is included in the interest of clarity, since these filaments are actually wired in series. A not uncommon procedure is shown regarding the mechanical linkage of the two variable capacitors. While in different circuits, their two sections of rotor plates are on a common shaft, and this, of course, is indicated by the broken or dashed line. One of many of the problems faced by the draftsman in making a finished electronic schematic from a very rough sketch is that he must exercise caution when reorienting leads for the sake of clarity. As was initially pointed out, a respectable grasp of electrical fundamentals is entirely necessary in order to prevent the misplacement of circuit leads about components, thereby altering the correct functioning of a circuit.

Transistor type circuit

The drawing layout of a transistor receiver is similar to that of a tube type. The various transistors are placed along a horizontal axis with suitable spacing for the air-core transformers between them. See Fig. 9-2. In similar fashion to the tube-type superheterodyne receiver, the progressive stages are drawn from left to right. Note the iron-core tuning coil at the extreme left. Note also the speaker or earphone jack at the extreme right. Also, broken lines are used for

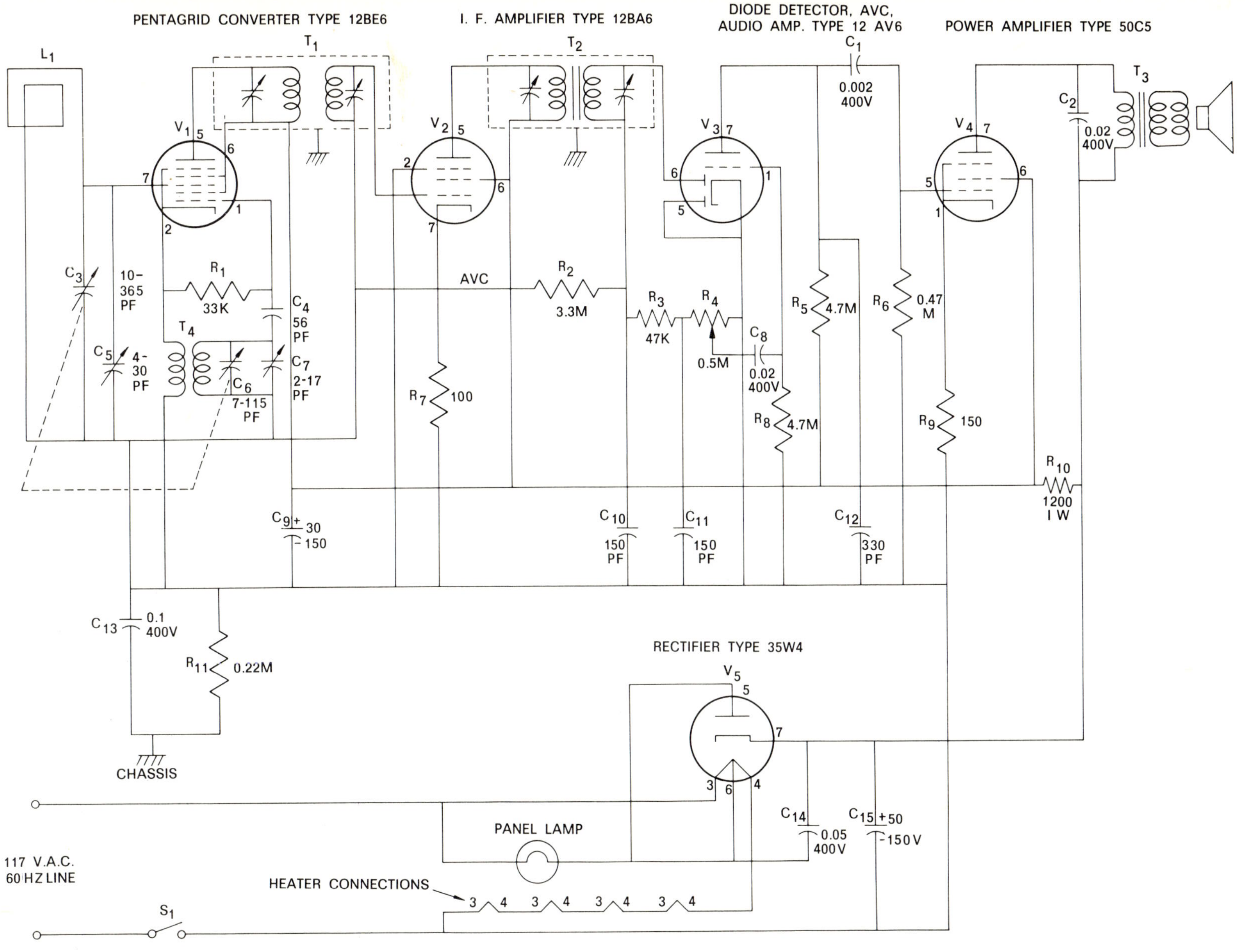

Fig. 9-1 A circuit diagram of a typical tube-type superheterodyne radio receiver. Adapted from the RCA Receiving Tube Manual. (Courtesy R.C.A. Corporation)

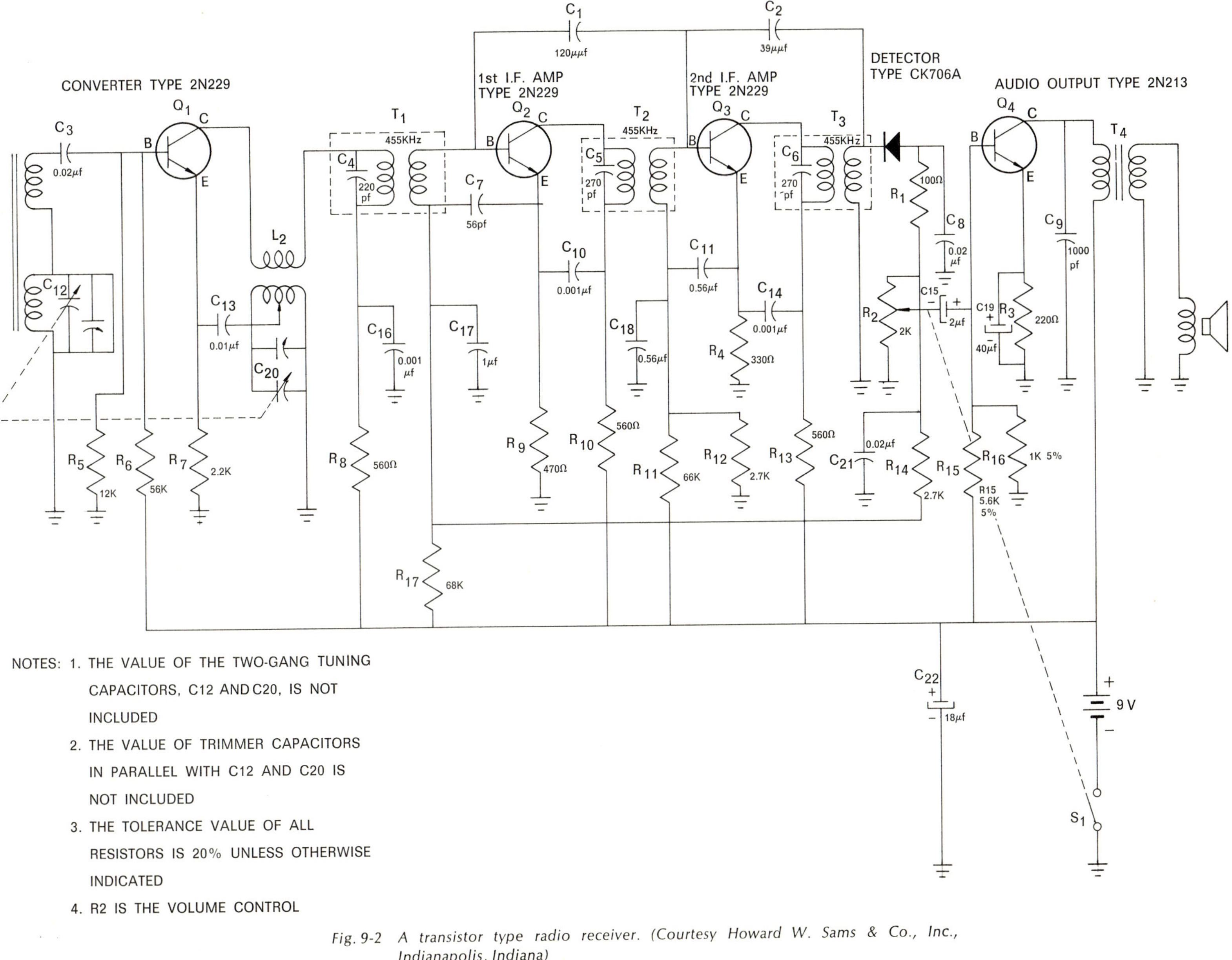

Fig. 9-2 *A transistor type radio receiver. (Courtesy Howard W. Sams & Co., Inc., Indianapolis, Indiana)*

two purposes: shielding about the 455 KHz I.F. transformers, and mechanical linkage between the two stages of the tuning capacitor.

Some other features must also be pointed out, that have a significance to the student of electronic drafting. These are as follows:

1. The notation that switch, S1, is on the volume control, (Resistor R1).
2. That the system is a chassis grounded one.

Generally speaking, it is no more difficult to draw a 4-transistor, plus diode detector receiver, than it is to draw a comparable receiver using five vacuum tubes.

PRINTED CIRCUITS

Printed circuits using miniaturized components

While many electronic components have undergone size shrinkage simply because of the use of new materials and improved or new manufacturing processes, other electronic components have been reduced in size or **miniaturized**

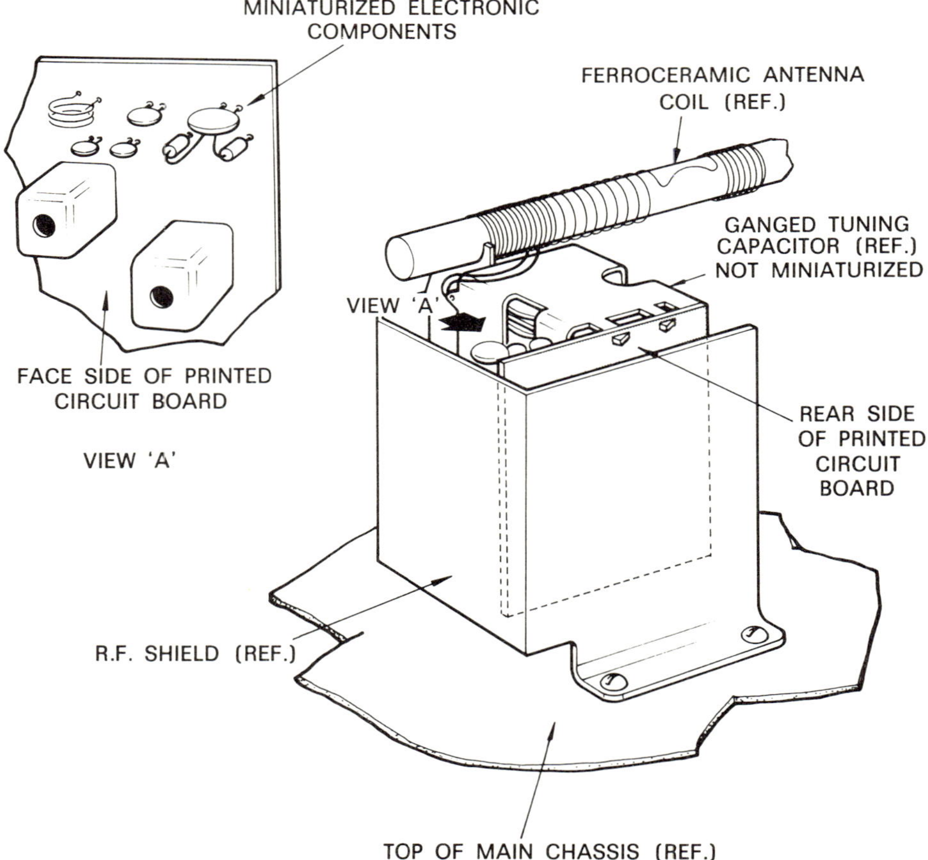

Fig. 9-3 Part of a wired radio receiver chassis utilizing a printed circuit board and miniaturized components

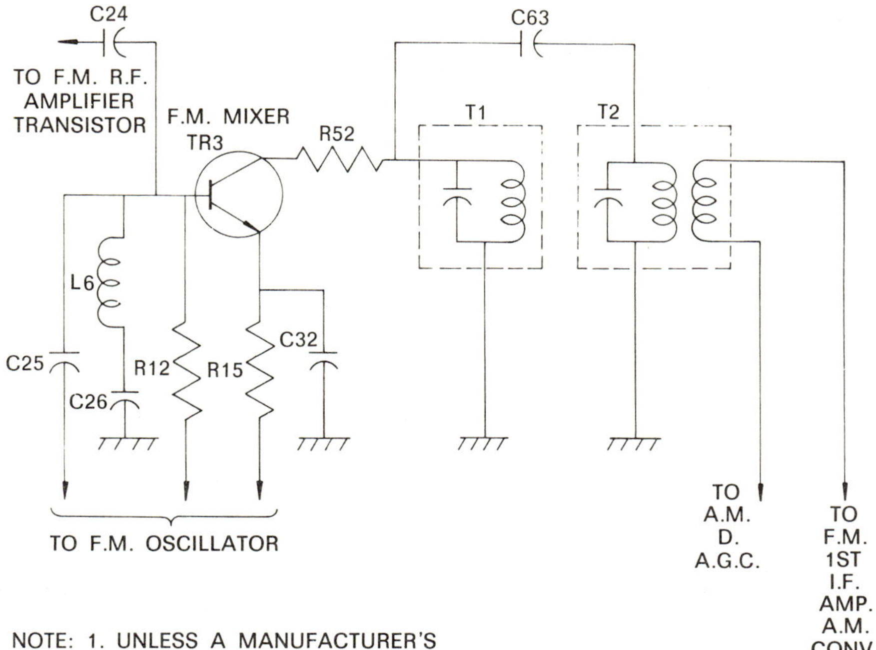

NOTE: 1. UNLESS A MANUFACTURER'S DESIGNATION IS INCLUDED IN THE SCHEMATIC, NO ATTENTION IS DRAWN TO A COMPONENT'S MINIATURIZATION OR LACK OF IT

2. A SCHEMATIC DOES NOT NORMALLY INDICATE WHETHER A CIRCUIT UTILIZES A PRINTED CIRCUIT BOARD OR NOT

3. COMPONENT VALUES HAVE BEEN OMITTED INTENTIONALLY FOR CLARITY.

Fig. 9-4 A partial circuit diagram of a radio receiver utilizing a printed circuit board and miniaturized components

because a genuine engineering attempt has been made to reduce their proportions without reducing their effectiveness. In fact, in many cases electrical characteristics have been improved. The inclusion of miniaturized components in printed circuit board applicatons is a convenient arrangement. Because of the reduced size of the components, the maximum size of the board is automatically made smaller. In addition, the weight of the assembled board that is complete with components is reduced. This factor will have a particular significance if the board is used in portable electronic equipment. The drawings illustrated in Figs. 9-3 and 9-4 depict both the pictorial representation and the basic schematic of such a circuit board.

Printed circuit board construction

Prior to designing a printed circuit board, engineers and technicians are involved in the original or **prototype** design and construction of the complete electronic unit, or package. In this stage, engineering 'bugs' are worked out, circuit designs altered, and no doubt some components substituted for others. Wiring at this stage of operation is normally done by hand, probably using a **vector board** (an insulating board with holes punched in it) as the medium for mounting components. See Fig. 9-5.

When the design has been **fixed**, the engineering department produces an up-to-date schematic wiring diagram. The schematic illustrated by Fig. 2-2(a) in Unit 2 suitably defines this type of drawing.

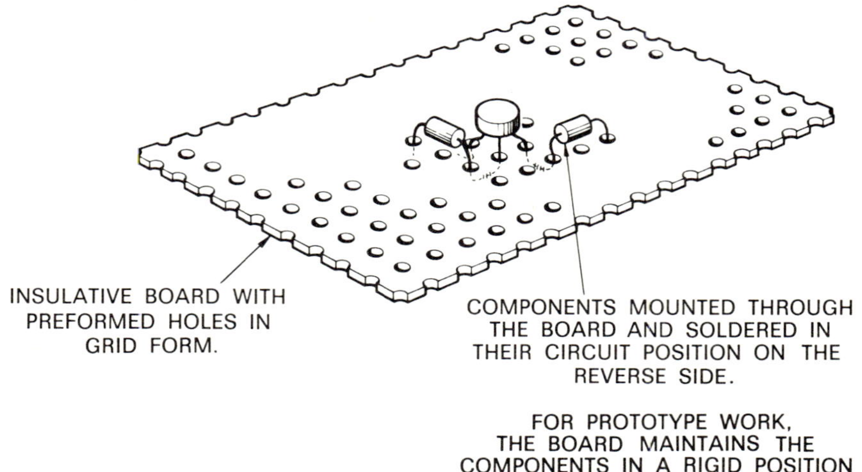

Fig. 9-5 The use of a vector board

Certain information must be given to the draftsman at this stage, and this includes the actual trim size of the circuit board. Final drawings concerning a particular circuit board design include both the overall trim sizes shown to three places of decimals, inclusive with **bidirectional tolerances**. See Fig. 9-6. Accurate locational dimensions for register circuit targets must also be included on the final drawings. These are also illustrated in Fig. 9-6. Also, if the holes for component leads in the board or substrate are to be drilled using a **jig** or by means of drilling machines having **numerical control**, discrepancies between the printed circuit master and the substrate with regard to hole location often result. Thus is due to the scale inaccuracy brought about by the photoreduction.

To overcome this problem, the finished drawings should also include accurate coordinate dimensions to locate the holes in the board. See Fig. 9-6.

From this schematic, the draftsman establishes a preliminary sketch showing the layout of components with their related wiring. With respect to the components, their position conforms roughly to their final location on the actual board.

Once again a knowledge of fundamental electronics is desirable, if not indispensable, since certain components may produce interacting electrical fields with other closely positioned ones. As such, the positioning of the components is not always simply a fundamental problem in spatial layout. At this point, a large scale or master layout drawing of a ratio of perhaps four to one

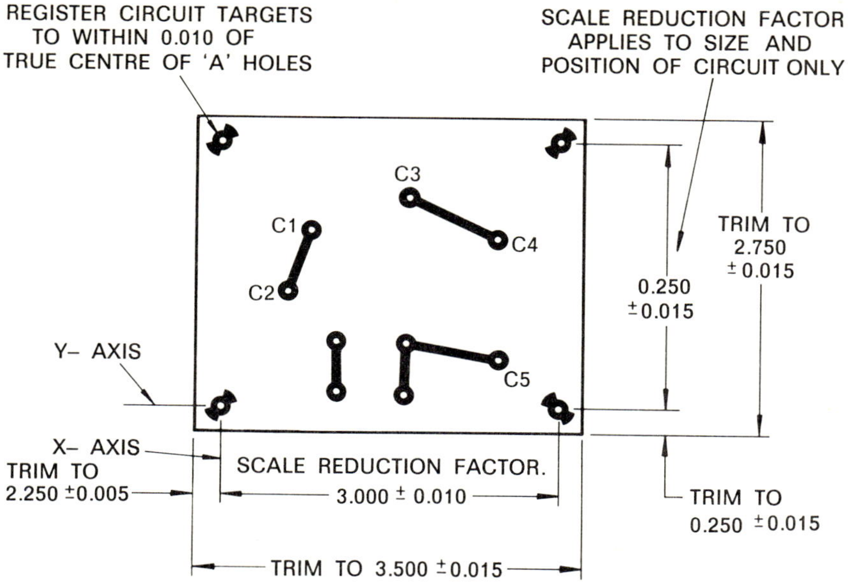

REGISTER CIRCUIT TARGETS TO WITHIN 0.010 OF TRUE CENTRE OF 'A' HOLES

SCALE REDUCTION FACTOR APPLIES TO SIZE AND POSITION OF CIRCUIT ONLY

C3

C1

C4

C2

TRIM TO 2.750 ±0.015

0.250 ±0.015

Y- AXIS

C5

X- AXIS

TRIM TO 2.250 ±0.005

SCALE REDUCTION FACTOR.

3.000 ± 0.010

TRIM TO 0.250 ±0.015

TRIM TO 3.500 ±0.015

HOLES SIZES

SYMBOL	DRILL SIZE	TOLERANCE
A	0.125	±0.003
B	0.060	±0.003
C	0.082	±0.005

HOLE POSITIONS

SYMBOL	X-COORD.	Y-COORD.	TRUE CENTRE TOLERANCE
A1	0.000	2.250	0.008
A2	3.000	2.250	
A3	3.000	0.000	
A4	0.000	0.000	
B1	0.880	0.800	0.015
B2	0.880	0.350	
B3	1.700	0.800	
B4	1.700	0.350	
C1	0.550	1.650	0.010
C2	0.350	1.000	
C3	1.800	1.950	
C4	2.400	1.600	

Fig. 9-6 Tolerance trim dimensions for a printed circuit board, and decimal notation of hole sizes, also indicated as tolerance dimensions

or larger is made which represents the actual conductors in the schematic, but of course does not include the electronic components. Figure 9-7 illustrates the layout of the circuit previously shown in Fig. 2-5(b) in Unit 2. From this stage, the large scale drawing is photoreduced to a much smaller size. The effect of doing so is to sharpen the line work to a high degree. Figure 9-8 represents the actual size of the photoreduced image. Since all parts of the original large-scale drawing diminish in size during the photoreduction process, it is imperative that the initial layout be of a precise nature. This is especially true where conductors are running close to each other.

On many occasions printed circuit boards must contain circuitry on both sides. If this is the case, then the draftsman is required to analyze the initial

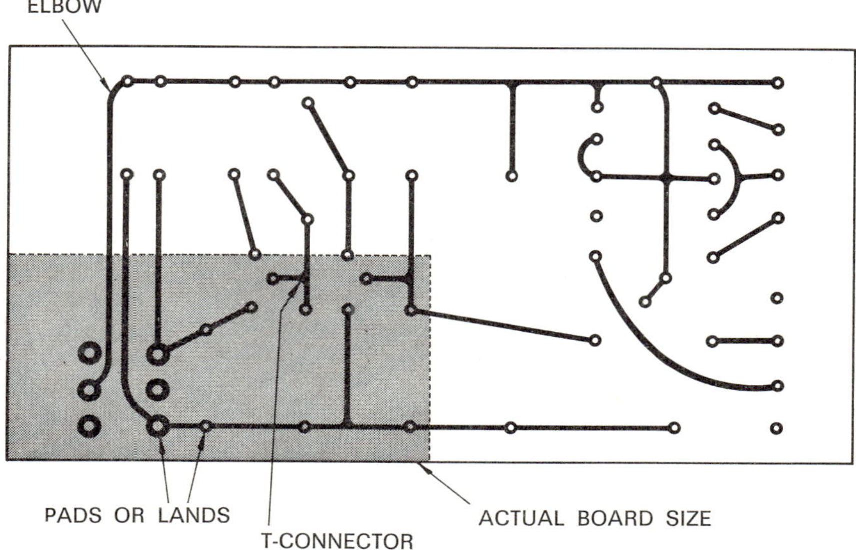

Fig. 9-7 A large scale drawing of conductor paths for a printed circuit board construction. For the sake of convenience, the drawing has been made only twice full size. In actual practice, scales of 4:1 or greater are used

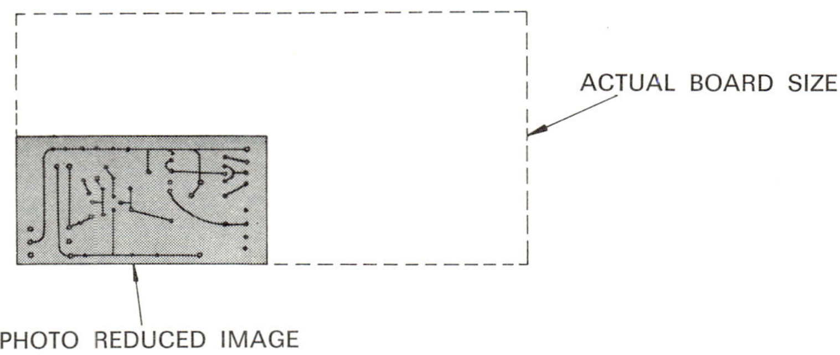

Fig. 9-8 A photoreduced image of the conductor paths of a printed circuit board. The scale shown is in correct relationship to Fig. 9-7, but the actual board is twice as large

schematic in order that he can design necessary connections that are to be made through the board from one side to the other.

When the master photoreduced circuits have been made, they are utilized in a silk screen process to transfer the circuit path image to the board. Other image transferring processes used are **offset printing** and **photoengraving**. The matrix or board is made of phenolic laminate or **methyl acrylic** plastic, which is covered on either one side or both with a metallic foil made from copper, aluminum, or brass. The result of any of these processes described is to imprint an acid-resistant image on the board that is the exact shape of the photoreduced master. Upon completion of this step, the circuit board is immersed in an **etchant** which dissolves away those blank areas not acid-resistant, leaving the circuit paths intact.

The advantages of printed circuit boards are as follows:

1. The elimination of hand wiring is brought about, with its subsequent savings in cost.
2. The process lends itself ideally to mass-production techniques.
3. Compactness and weight reduction are brought about, with the accompanying elimination of possible mistakes in the wiring due to human error.

Finally the various electronic components are soldered in their designed places on either side of the board to complete the circuit. The illustration listed as Fig. 2-5(a) in Unit 2 shows a completed board.

TECHNIQUES IN MAKING PRINTED CIRCUIT BOARD DRAWINGS

Conductor path spacing

Since the conductor paths are not insulated in the same fashion as a normal insulated wire, it is necessary to maintain fixed distances between the conductors. These distances vary according to the size of the electrical potential. For example, at a potential of from 0 to 150 volts, the suggested spacing is 1/32 of an inch. From 150 to 300 volts, 1/16 of an inch is required. While these values are indicated, individual firms establish their own conductor path spacing standards.

Conductor path width

The width of the conductor path is usually held at 1/16 of an inch. Obviously if a circuit is carrying a very-high current this value is increased.

Conductor path thickness

Two thicknesses of conductor path are frequently used. These are 0.00135″ and 0.027″. For a fixed current value, the width of the path can be reduced in half if the thicker one is used. As an example, if the path is carrying 2.6 amperes of electricity, a 1/32″ wide path is used for the smaller thickness, then a 1/64″ path is suitable when using the larger.

Conductor connections

Connections must be of a specified size, and are referred to as a **pad** or **land.** Sometimes they are also called **doughnuts** because of their shape. Also, as a path approaches a pad it must be gradually increased in size. Illustration Fig. 9-9 shows an enlarged view of a pad.

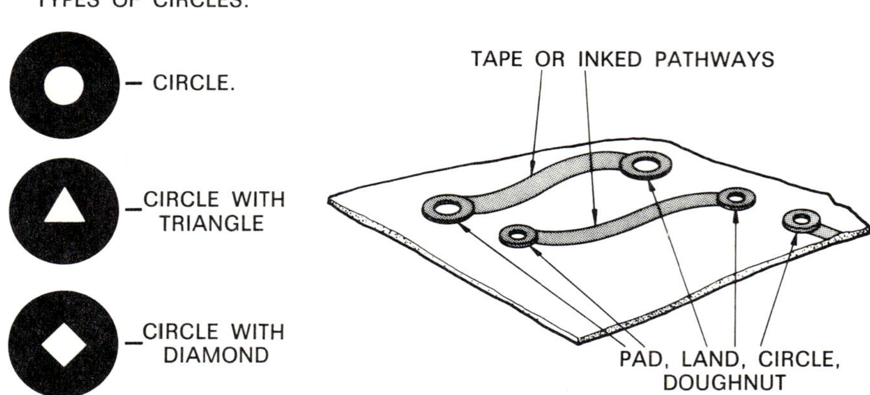

Fig. 9-9 An enlarged view of a pad or land

Conductor crossovers

Since the conductor paths are not laid out in three dimensions, it is necessary to eliminate crossovers almost exclusively. However, it is possible to make either an insulated or noninsulated jumper to eliminate a short circuit brought about by a crossover. See Fig. 9-10.

Conductor sharp corners

The elimination of all sharp corners in the conductor path is essential to good circuit board design.

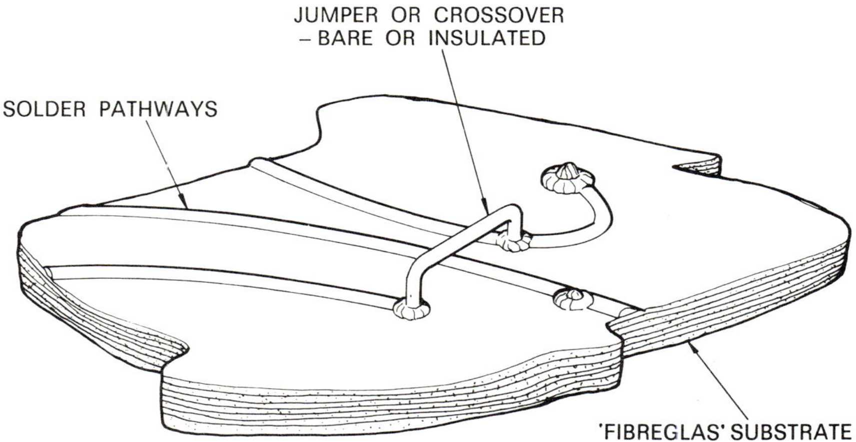

Fig. 9-10 A conductor crossover for a printed circuit

Special drafting materials for printed circuits

A brief mention has already been made in Unit 3 regarding plastic tape coated with pressure sensitive adhesive. This tape is used for much of the circuit path and layout. In addition to this tape which is available in varying widths, press-ons are available in pad, teardrop, tee, twinpad, ell, and universal corner shapes. Fig. 9-11 shows these various types of press-ons.

As an aid to laying out the circuit paths in precise fashion, an accurate undergrid sheet is frequently used.

Also, since spacing of a high dimensional accuracy must be maintained between conductor paths, **polyester** film or glass drafting cloth must be used. The use of these media requires a special 'India' ink that is compatible with the material, otherwise flaking will occur, and portions of the conductor path in the finished printed circuit board may have breaks.

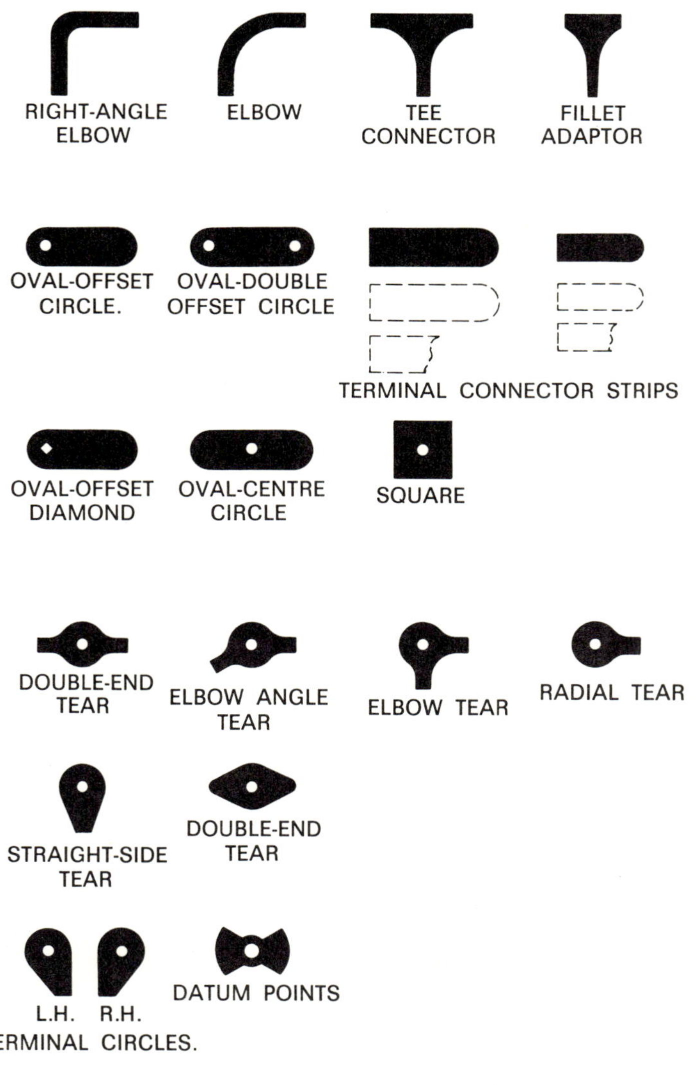

Fig. 9-11 Typical press-ons of various shapes

Flexible, printed circuit boards

The term board as applied to this type of printed circuit is not really an appropriate one. The material from which the board is constructed is either a flexible **polyimide** or **polyester** plastic. In addition, our concept of a board usually implies a square or rectangular shape, whereas the configuration of a flexible board is multitudinous. See Fig. 9-12. Its flexibility is such that it is possible to construct a unit three feet long, roll it into a cylindrical shape, and finally enclose the rolled circuit in a can in which the voids are filled in with plastic foam. The consequence of doing so is to make a shockproof assembly.

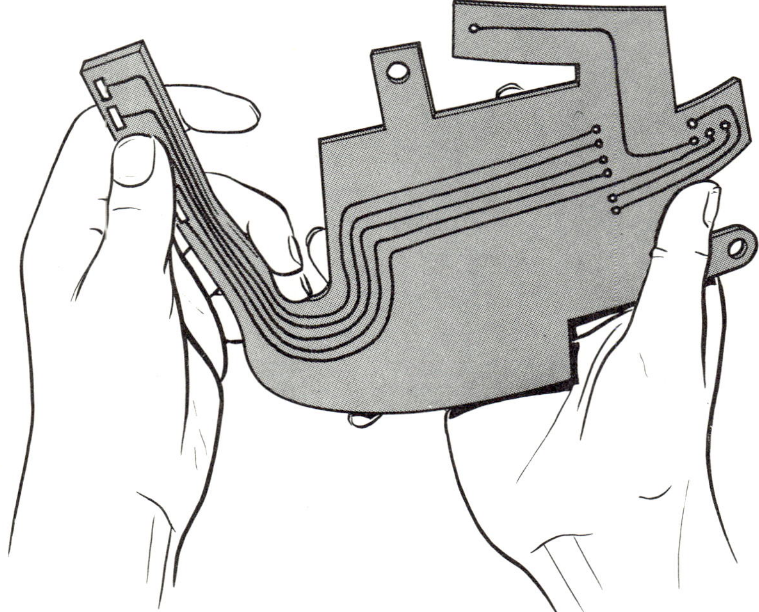

A FLEXIBLE CIRCUIT BOARD HAS THREE DIMENSIONAL ASPECTS. IT CAN BE BENT OR TWISTED INTO ALMOST ANY SHAPE WITHOUT IMPAIRMENT TO THE CIRCUIT

Fig. 9-12 A flexible printed circuit board

Two types of plastics are currently in wide use. For temperatures up to 300°F, **Mylar**, a trade name for polyester, is used. When temperatures above this point are to be encountered **Kapton**, a high temperature polyimide, is used. Its practical top limit is approximately 600°F. Besides these dielectrics, **Teflon** and flexible glass also find application.

In addition to the characteristic of flexibility, low weight, (a flexible board is approximately 0.009 inches thick, compared to 1/16 of an inch for a rigid board), high reliability, and comparative costs to rigid boards, flexible circuit boards provide a desirable means for interconnecting electrical and electronic components. Another important consideration is the large amount of time saved when compared to using standard wiring and cabling procedures. In one specific case a manufacturer was able to reduce the wiring time drastically from 34 hours to 2 hours per unit.

Some of the applications for flexible printed circuits are in the aerospace, automotive, and communications fields. Some of the specific devices in which they find use are telephones, cameras, and computers.

In designing a flexible printed circuit, it is first of all necessary for the draftsman to establish its ultimate shape. The problem here may be either a two or three dimensional one. If we consider either one, it is still necessary to lay out the circuit paths in much the same way as a conventional board, that is to say, maintaining fixed distances and widths of pathways that are appropriate for the electrical power being conveyed. With regard to the ultimate shape of a board (brought about by twisting and/or bending), it is necessary that its overall dimension in the flat is taken into account.

Multilayer printed circuit boards

Another adaptation of printed circuit boards is the multilayer concept. Whereas normal rigid boards have either circuit pathways on one face or both faces, the multilayer board can have up to four layers of circuit pathways, two external and two internal, the internal layers being encapsulated between the top and base laminates. Fig. 9-13 shows a cross section of a typical multilayer printed circuit board.

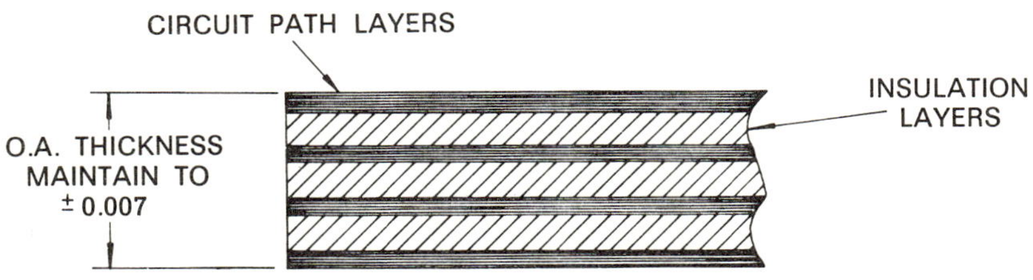

Fig. 9-13 A printed circuit board drawn to a large scale

The original large scale drawings for this type of board are made two to four times full size, and then photographically reduced. In these respects the drafting procedures involved are not unlike a normal one-sided or two-sided board. However, since interconnections are made internally between the laminates by plated-through holes, the registration accuracy of each master drawing is critical and must be of high order. Size and locational dimensions to three places of decimals are a standard drafting feature.

Multilayer circuit boards that are produced by chemical etching can have almost an unlimited number of circuit layers. Boards having twelve or more are not uncommon. The master drawings used with this method are made at least four times the final board size. In addition, it is not possible to make circuit connections internally through the laminates. To accomplish interconnections, external joining using wire wrap is used.

The application of multilayer boards is extensive in computer memory cores, where compactness of complex circuitry is essential.

INTEGRATED CIRCUITS

When circuits are referred to as integrated, it implies that instead of discrete components connected in a functional circuit, the components are formed directly on, or 'grown' in on, minute circuit chips or boards by deposition or other processes. Additional details regarding how these components are formed will be explained under the headings of 'Thin-film' and 'Semi-conductor' circuits.

Thin-film circuits

MATERIALS USED

The deposition of conductive and nonconductive films is made by either of the processes known as **cathode sputtering, vacuum evaporation, vapour plating,** or **anodizing**. Oxides of silicon, titanium, and tantulum are used for capacitor dielectric areas. For resistances, pure tantulum and titanium are used, as well as a nickel chromium alloy, and cadmium sulfide. Fig. 9-14 illustrates the methods of forming both capacitors and resistors.

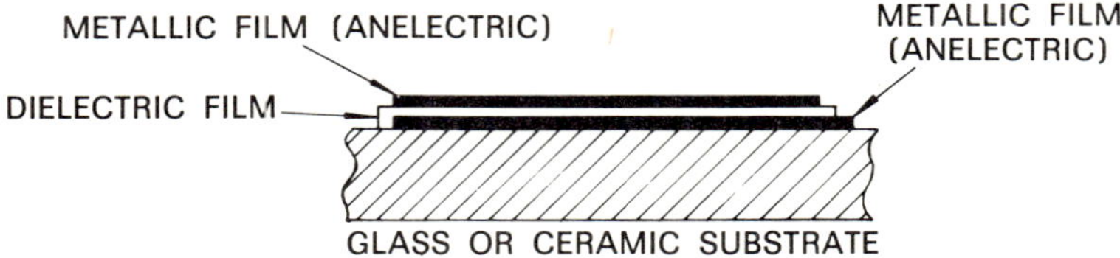

DIAGRAM ILLUSTRATING THE METHOD OF BUILDING UP A THIN-FILM CAPACITOR

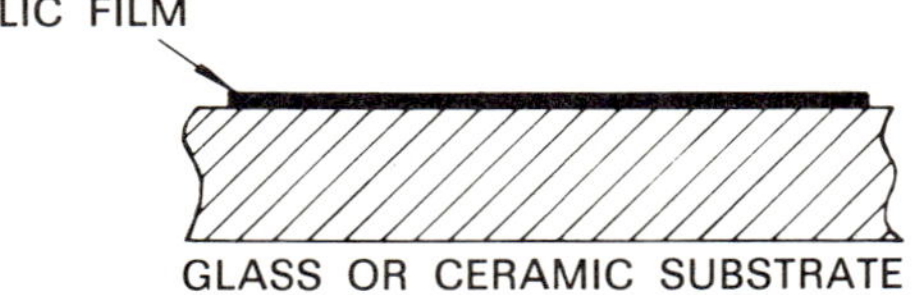

DIAGRAM ILLUSTRATING THE METHOD OF BUILDING UP A THIN-FILM RESISTOR

Fig. 9-14 Large-scale sectional diagrams

MATERIAL THICKNESS

The term **thin film** implies that the thickness of the deposited insulative or conductive material is in the order of a few microns. One micron (1μ) equals approximately four one hundred thousandths (4/100,000) of an inch. Hence the term thin film is aptly appropriate.

STEPS IN DESIGNING A THIN-FILM CIRCUIT

Once the schematic diagram is determined and the electrical values of each resistor and capacitor are known, the linear dimensions of these components are calculated according to the width of the deposited material used. These are conveniently set down in tabular form.

The next step is for the draftsman to make a sketch of suitable size in order to establish the resistor patterns and to lay out the areas for capacitances and contacts. Figures 9-15, and 9-16 show both the initial schematic diagram and the preliminary layout sketch of the thin-film circuit.

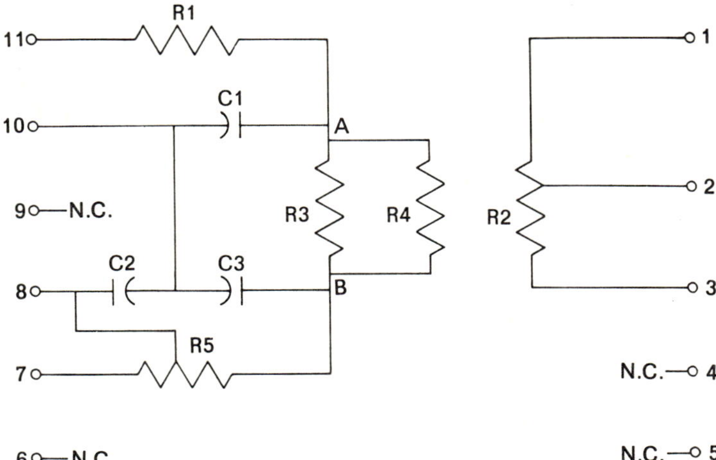

Fig. 9-15 An initial schematic wiring diagram

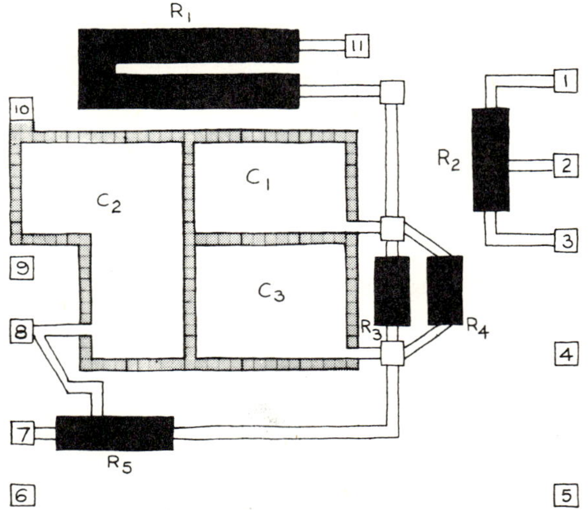

Fig. 9-16 A preliminary layout sketch

The draftsman at this point makes a large scale precision drawing of his preliminary layout of the order of 5:1. Components are indicated and the contact areas are sequentially numbered. The illustration listed as Fig. 9-17 shows the precision drawing for the schematic Fig. 9-15. Using this master precision drawing, a series of masks are made which are photoreduced to the desired size. These are normally used in a **photoresist and etchant process**, a procedure somewhat analogous to producing a photographic negative. The masks are sequentially used in laying down resistive film, capacitive **anelectric** 'conductive' film (metal), and capacitive dielectric film (oxide). Sometimes the mask drawings when photoreduced to the appropriate size are used to make metal masks. These are used directly on top of the substrate. However, fine resolution and detail cannot be obtained by this method, as it can by the photoresist and etchant process. Fig. 9-18 illustrates a series of masks for a thin-film circuit. Construction of the actual thin-film circuit from this point on, involves both photographic and deposition processes. Essentially there is no further drafting involvement. However, it is suffice to say that the quality of drafting preceding these processes must be extremely high.

Thick-film circuits

Entirely different processes and materials are used to produce a thick-film circuit than is required to make a thin-film one. The draftsman, however, is still

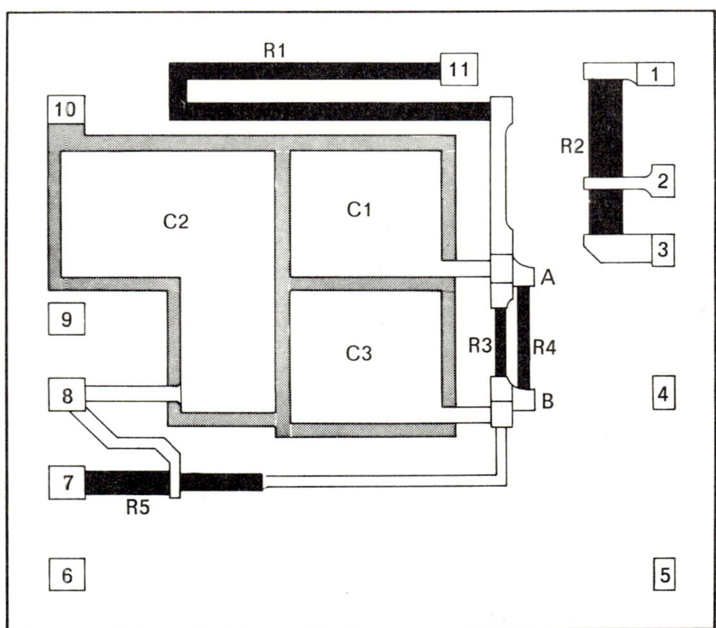

NOTE: C1 IS MEASURED ACROSS 10 AND A
C2 IS MEASURED ACROSS 10 AND 8
C3 IS MEASURED ACROSS 10 AND B

Fig. 9-17 A large-scale precision drawing

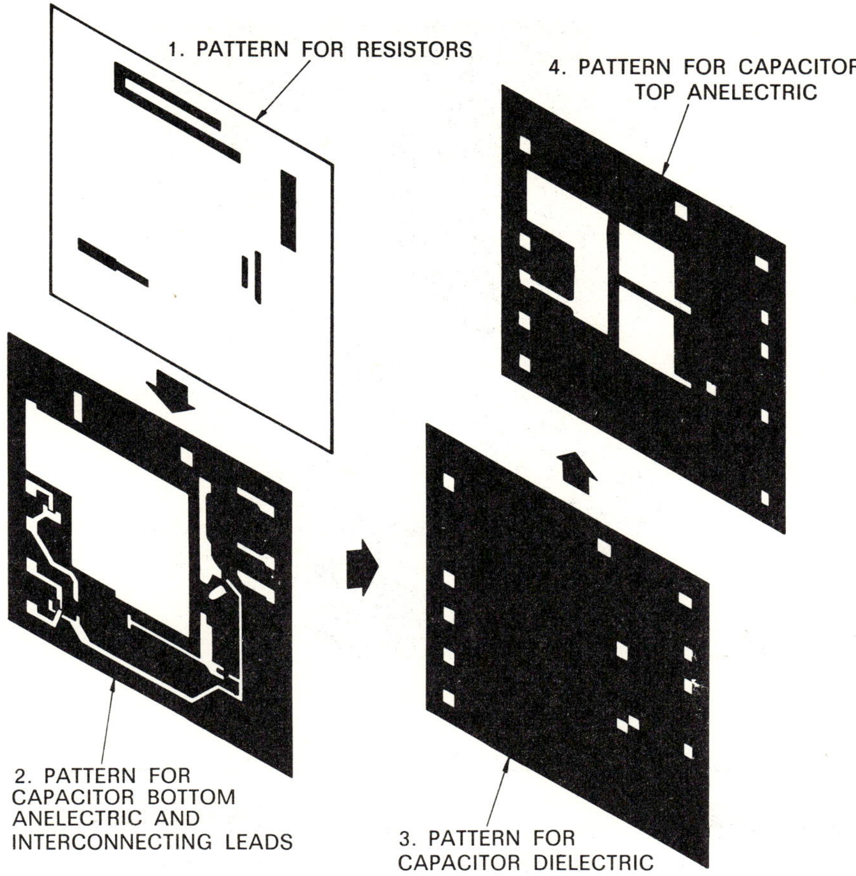

Fig. 9-18 A series of masks for a thin-film circuit. In actual practice, a multipattern concept is used in each stage for reasons of economy

required to make drawings to a large scale because the final product consisting of a functional circuit may vary from 1/4″ to 3″ square. Red Mylar film is used to produce circuit masks and these are scribed to a scale of ten times the actual product size. See Fig. 9-19. The masks when completed are photoreduced, thus producing highly accurate actual product size masks.

These in turn are used in the process which involves the use of a glass plate and a stainless steel screen. The mask image at reduced size is photographically transferred to the plate. This plate is then placed over the stainless steel screen which has been previously coated with a thin layer of light sensitive emulsion. The assembly is then exposed to ultraviolet light which hardens those parts of the stainless steel screen not covered by the design on the glass. The remaining portions having not been exposed remain soft, and are soluble in water which is used to wash them away.

For each Mylar mask subsequently transferred to the plate, a stainless steel screen is made. These screens are used in a printing process in which pastes made from a combination of organic materials, low-melting point glass, and

Fig. 9-19 The creation of a large-scale Mylar film mask for a thick-film circuit

noble metals (i.e. gold, silver, palladium-silver or palladium-gold) are transferred to a white ceramic substrate containing some 96 percent of alumina.

After each film has been transferred by this screening process, the ceramic chip, or board, is dried in an oven at 125° Centigrade. Then the chip, or board, is sintered in a special furnace having four thermal zones, which combine to determine the ultimate electrical properties of it.

A thick-film circuit can be made into a hybrid since it is possible to bond active components such as transistors, diodes or, IC's to it. These devices can be attached or bonded by thermal or eutectic compression.

The advantages of thick-film circuits are a shorter production time, and a simplicity of fabrication: all of which add up to lower manufacturing costs. In addition they have high mechanical and electrical reliability.

Fig. 9-20 shows a block diagram illustrating the processes required to make a thick-film circuit.

Integrated semiconductor circuits

The disadvantage of thin-film circuitry is that transistors and diodes which are active devices cannot be built up. However, when a small silicon wafer is used as a basic building block, elaborate functional circuits of very small size incorporating transistors, diodes, capacitors, and resistors can be made.

Method of construction

The processes required to make a semiconductor circuit are complex and costly. Essentially a **silicon wafer** is polished to a high finish then masked or

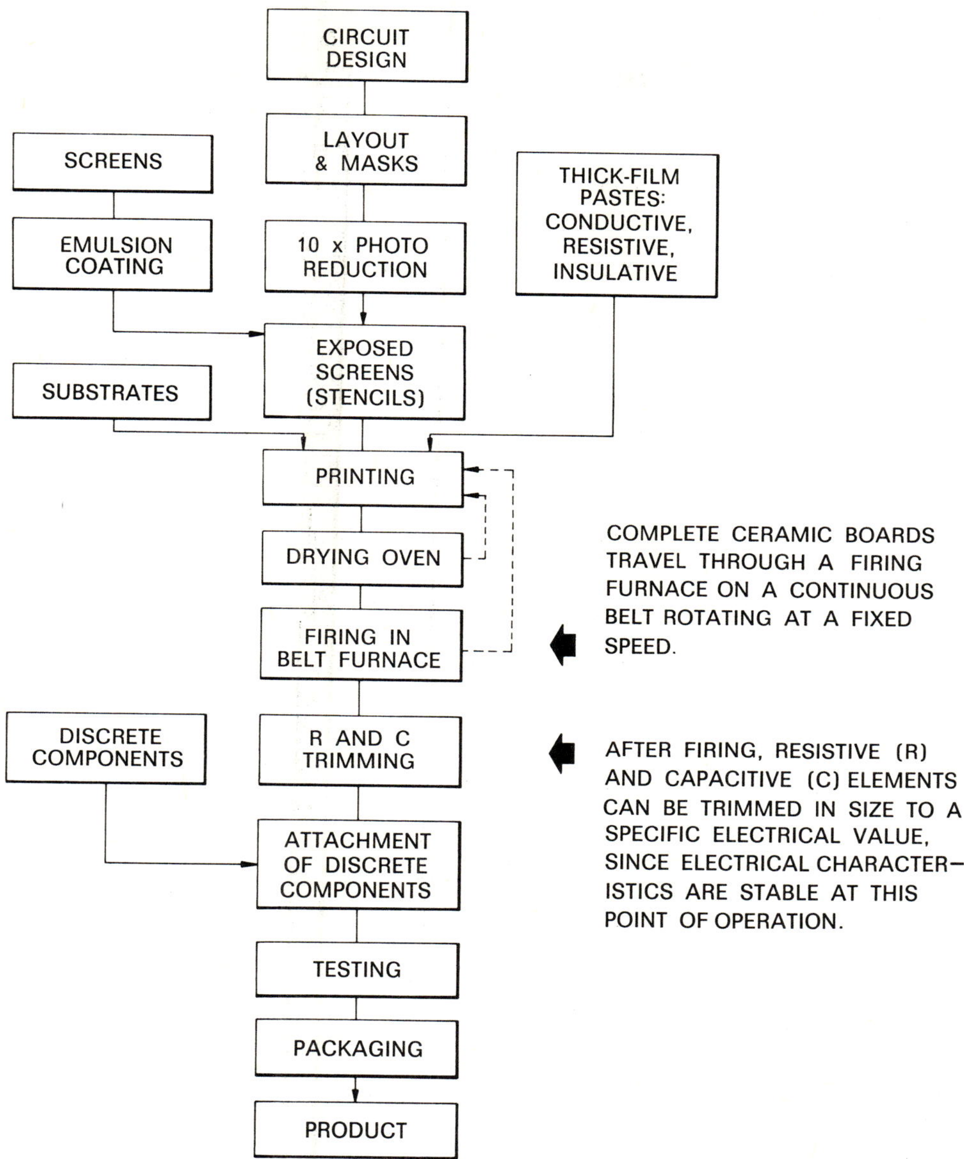

Fig. 9-20 A thick-film circuit process chart

Figs. 9-19 and 9-20 (Courtesy Northern Electric Company Limited, Montreal, P.Q.)

coated over with an oxide. By **diffusing** specific gases or impurities into small openings or apertures on the surface of the polished wafer various electronic components can be formed. These in turn are connected in a functional circuit by the application of a film of conducting material over the wafer. See Fig. 9-21. However, the processes described, while descriptively appearing simple, are in fact sophisticated and elaborate in practice. Eventually specially designed leads are 'welded' to the contact areas while the functioning part of the circuit is

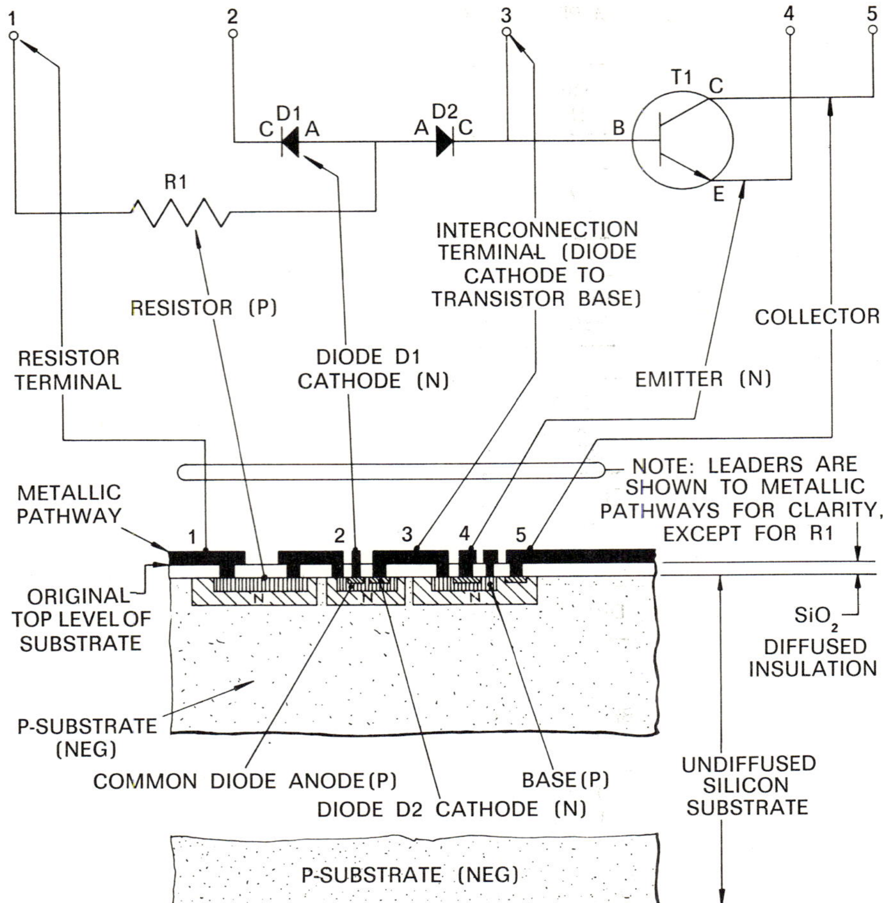

Fig. 9-21 A magnified section of part of a diffused and interconnected integrated circuit

'potted' or encapsulated. Figure 9-22 shows a typical assembled silicon **wafer.**

Obviously it would not be practical to manufacture one chip or wafer at a time. Hence photoduplication is used to provide a number of identical circuit patterns at one time. This multi-pattern is then applied to a larger slice of silicon, the various processes as previously described carried out, and finally the large slice accurately cut to separate the identical circuits. Figures 9-23, and 9-24 illustrate this concept.

Advantages

Silicon microcircuit chips are highly reliable electronic devices. Their extremely small size and lightness in conjunction with their high reliability make them ideal candidates for aerospace electronic equipment, especially in the specific area of rockets and satellites where 'G' (gravity) forces are high on **lift-off** and **reentry**. In addition, where weight penalties of equipment are of prime consideration, and where **radio communication links** and **onboard navigation** computers are required to have 100 percent reliability, if that is possible, their use is futher enhanced.

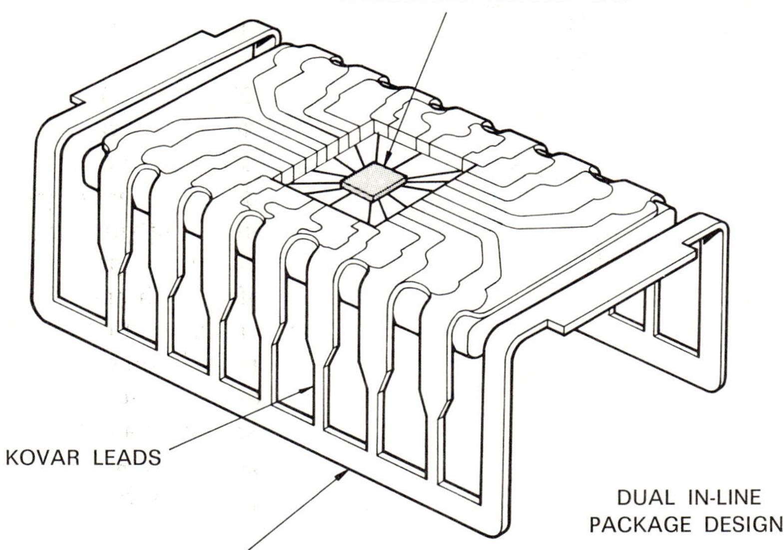

Fig. 9-22 A simplified version of a silicon wafer

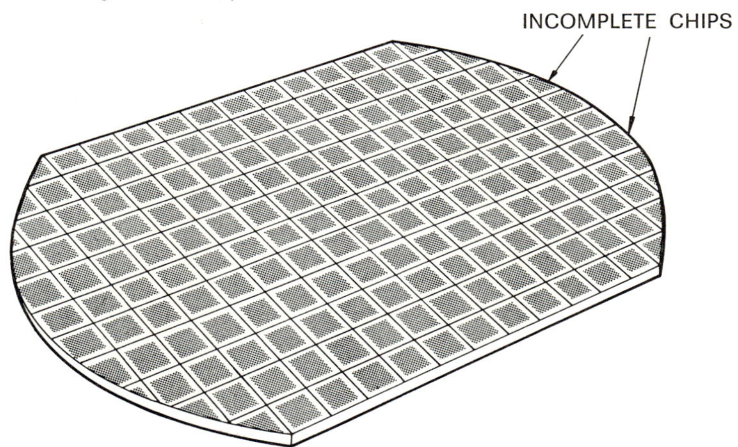

Fig. 9-23 Identical circuit patterns or a multipattern concept for integrated circuitry

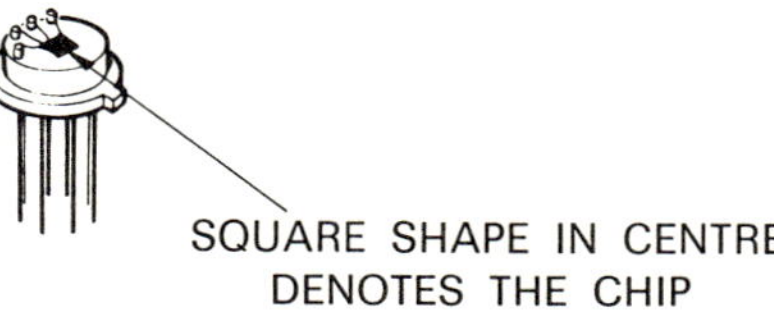

Fig. 9-24 Comparison in size of the chip to the coin

Diagrams required for semiconductor integrated circuit construction

A number of diagrams each made in sequence are necessary to produce a semiconductor integrated circuit chip.

Since the final product may be as small as 1/4" (0.25) by 1/8" (0.125) and may contain the equivalent of twenty, thirty, or more electronic components, all master drawings are drawn many times the ultimate size of the chip. Scales up to the order of four hundred times full size are used for master drawings, depending on the circuit complexity. In order to facilitate drawing even at this large scale, an accurate grid is used, since spacing between components in their finished state may be as close as 5/10,000 (0.0005) of an inch apart. However, the use of electronic computers to establish optium integrated circuit patterns is now an accomplished fact. When properly programmed, the computer gives a printed **readout** showing the X and Y coordinates for a specific circuit pattern. These mathematical values are applied to a machine known as a **coordinagraph**. Its function is to automatically make an accurate large-scale drawing when fed these values. Hence, the use of the draftsman for this phase has been eliminated.

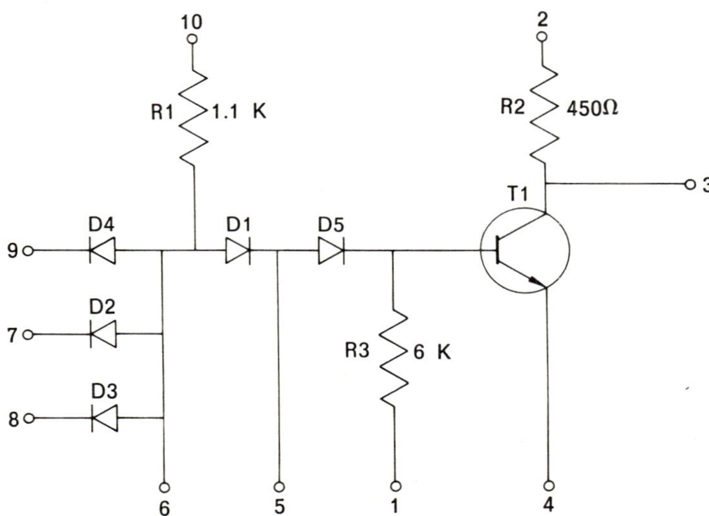

Fig. 9-25 A schematic for an integrated circuit chip

Component pattern diagram

Figure 9-26 shows the initial large-scale precision drawing outlining the complete chip with components located in a specific pattern.

Bar layout diagram

The second drawing required in the production is known as a bar layout drawing and is used to establish the initial diffusion pattern. Figure 9-27 shows a large-scale composite picture of the various electronic components diffused into the silicon chip. The SiO_2 areas are the initial ones created by diffusion, masking, and etching.

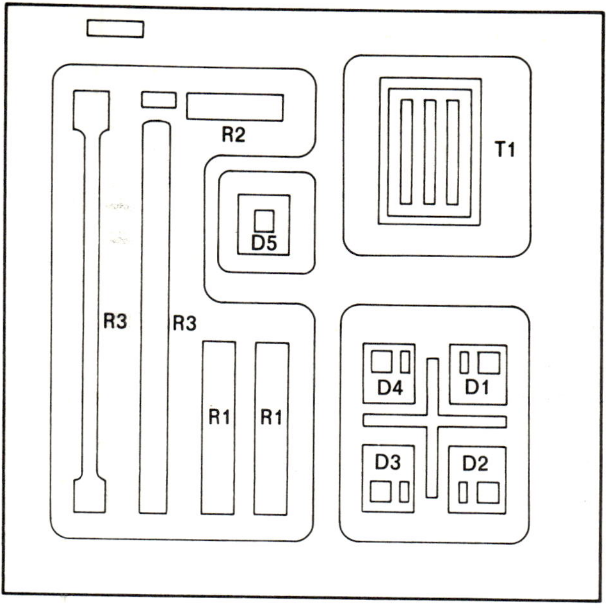

Fig. 9-26 A component pattern diagram

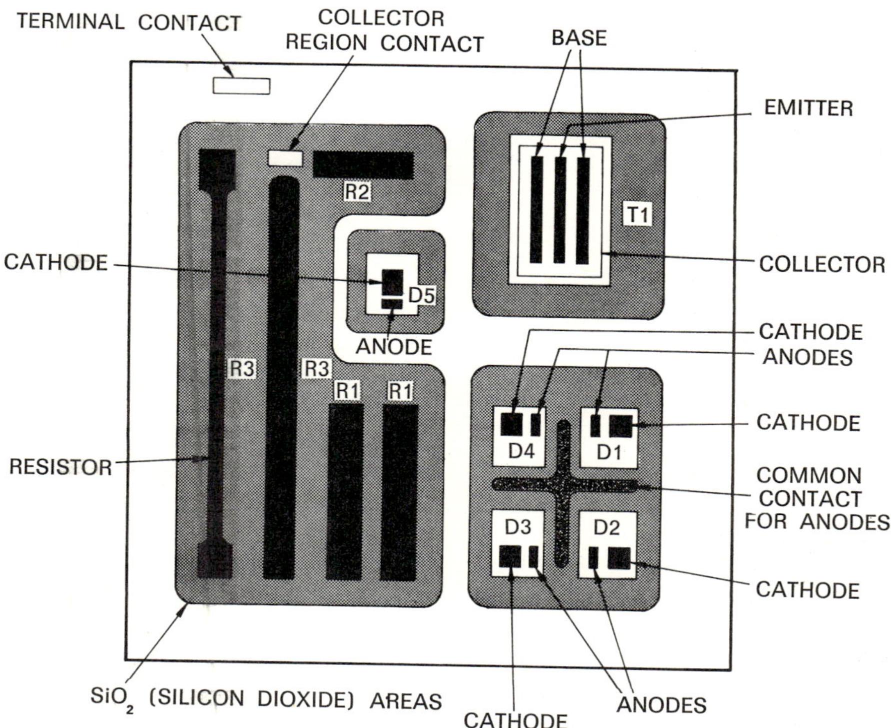

Fig. 9-27 The 'growth' of electronic components into the silicon chip

Connections diagram

The third step involves a drawing which appears very similar to the bar layout diagram, but aluminized connections have been added by an evaporative process.

Bonding diagram

The bonding diagram is identical to the connection diagram except that leads located along the periphery of the rectangular chip are added, in order that the circuitry can be properly terminated. These leads are made from metal with the trade name of **Kovar**.

At this stage, the electronic designer establishes the circuit paths, or maze, to produce a functional circuit.

Intraconnection diagram

The drawing showing the circuit paths is referred to as an intraconnection diagram and is illustrated in Fig. 9-28. From this a mask is made which covers the circuit paths while the blank areas are removed by an aluminum etchant.

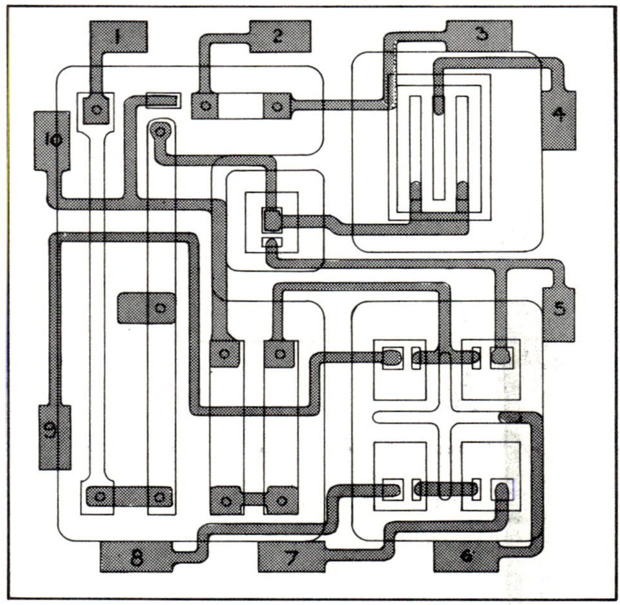

Fig. 9-28 An intraconnection diagram for an integrated circuit chip

Fig. 9-29 shows the complete set of masks that are used in sequence to produce an integrated circuit chip.

Encapsulating

Encapsulation is the process of taking a completed microcircuit and packaging it so that it can be used as a practical unit in a larger electronic system. The package shapes are usually rectangular and flat, or cylindrical.

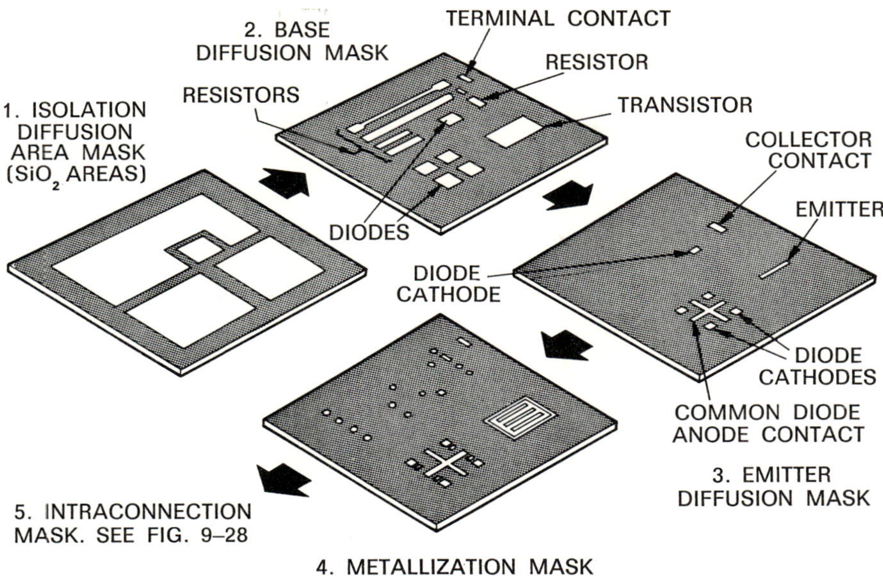

Fig. 9-29 Sequential masking patterns for an integrated circuit chip

See illustrations Figs. 9-30, and 9-31. Encapsulation effectively prevents damage to the delicate microcircuit from mechanical shock and the effects of humidity.

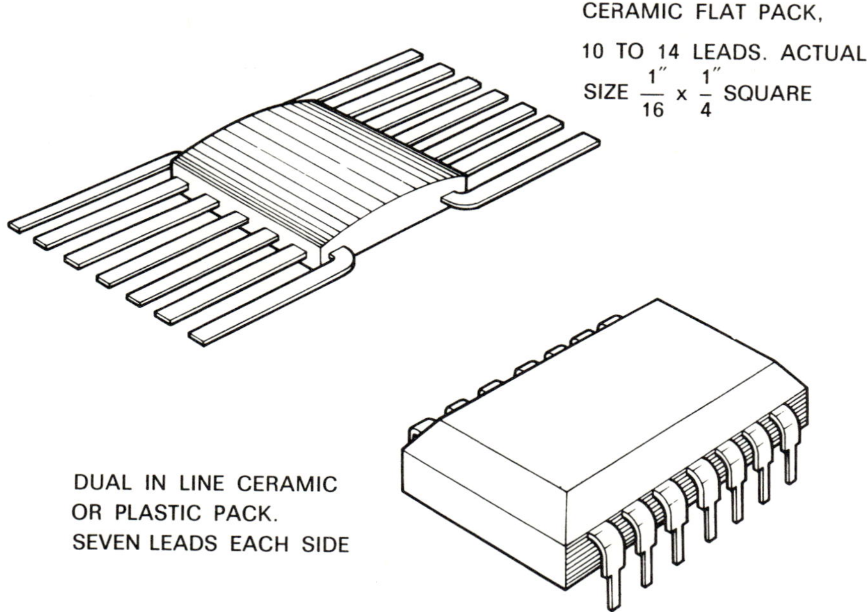

Fig. 9-30 Rectangular shaped chips

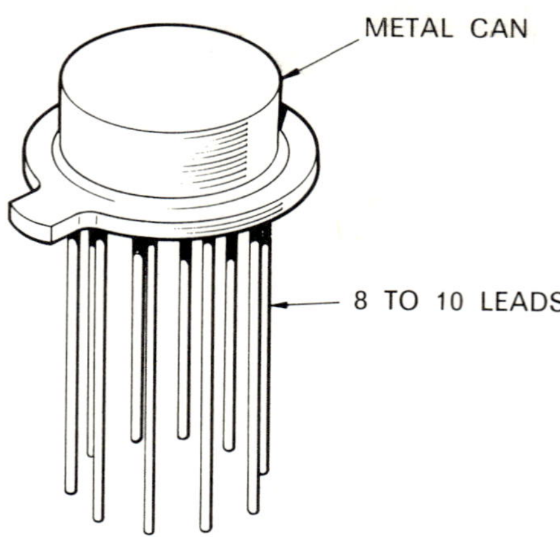

Fig. 9-31 A cylindrical shaped chip

HYBRID CIRCUITS

The term hybrid refers to those integrated circuits composed of elements of both thin-film and semiconductor integrated circuits.

Advantages

Several reasons are given for the requirement of hybrid circuits. If it is the design policy to use thin-film circuits which are solely passive, then active components if they are required must be added. In addition, it is sometimes advantageous from a standpoint of manufacturing or of utilizing their physical characteristics to use both types of circuits conjointly. Figure 9-32 illustrates typical hybrid circuits.

The drawings associated with a hybrid circuit would of course include those already mentioned for both thin-film and semiconductor circuits.

LOGIC CIRCUIT DIAGRAMS

The inclusion of logic circuit diagrams in Unit 9 is because the circuits they represent are manufactured as integrated ones. However, in order to explain their function, block or flow type diagrams are used. It is the intention of the author to discuss only a few of them in a limited way, and leave the remainder for interested students to research on their own.

Logic diagrams are a graphic means of illustrating various circuit switching functions. These functions are related to Boolean Algebra, after George Boole, the mathematician responsible for its development. An important aspect of Boolean Algebra is a mathematical device known as a **truth table**. It so happens that a direct relationship can be made between these truth tables and various switching circuits or functions. Hence, one must understand the truth table, and be able to correlate it with respect to its logic diagram.

Only those draftsmen engaged in computer electronics will likely be con-

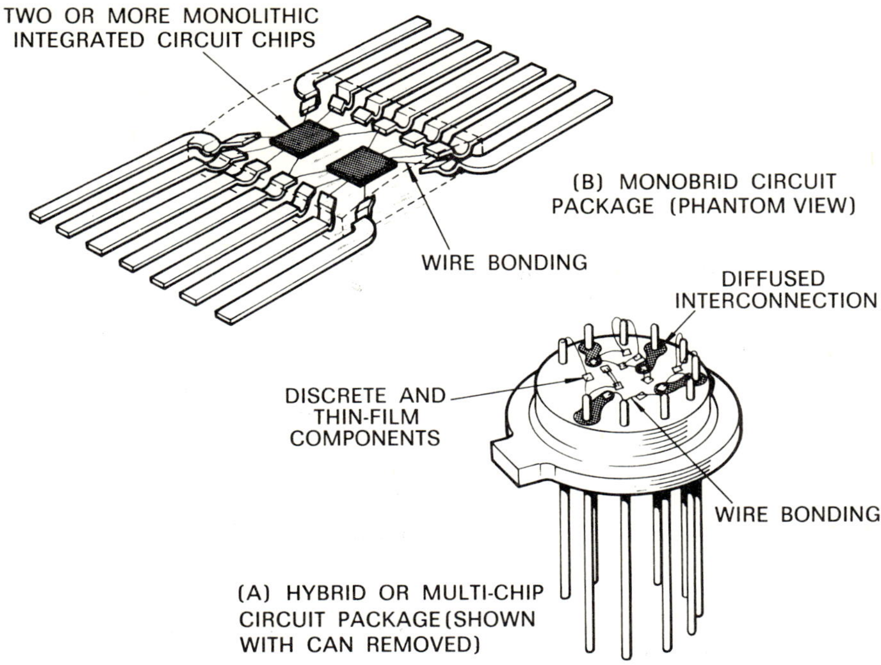

Fig. 9-32 Enlarged typical hybrid circuits

cerned with drawing logic circuit diagrams. Their widest application would appear to be in the realm of manufacturers' handbooks covering this particular subject, and with organizations having extensive computer equipment, and hence involved in 'writing' computer programs.

The AND function

The schematic diagram, Fig. 9-33(a) shows the 'AND' circuit. What the truth table, Fig. 9-33(b) says, is that the only condition whereby an electrical pulse can be received at C, is when the inputs at A and B are triggered at the same instant. The '1-1-1' level has been purposely blocked in to illustrate this point. Figure 9-33(c) illustrates the approved flow diagram for the 'AND' function.

The OR function

The truth table, Fig. 9-34, for the 'OR' function states that three circuit conditions are conducive for providing an output pulse at C. The only condition not so, is when no input exists at A and B at the same instant.

ANALOG AND DIGITAL COMPUTER FLOW DIAGRAMS

Both of these computer systems utilize block, or flow, diagrams to portray graphically the programming necessary for solving specific problems. Beyond these brief comments, it is not the intention of the author to attempt in any way

(a)

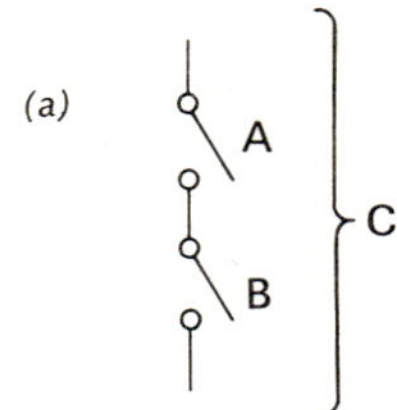

(b)

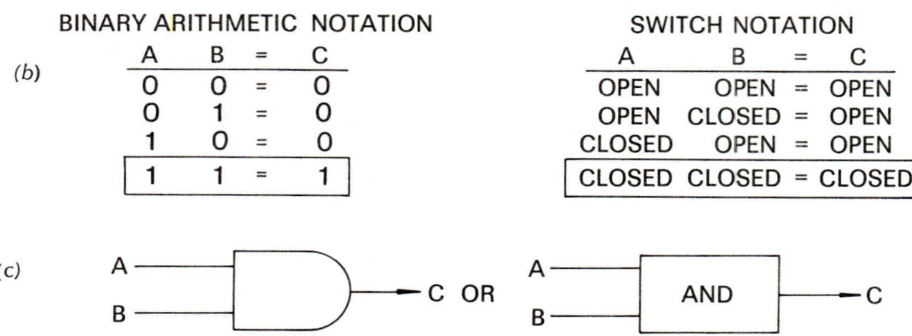

BINARY ARITHMETIC NOTATION

A	B	=	C
0	0	=	0
0	1	=	0
1	0	=	0
1	1	=	1

SWITCH NOTATION

A	B	=	C
OPEN	OPEN	=	OPEN
OPEN	CLOSED	=	OPEN
CLOSED	OPEN	=	OPEN
CLOSED	CLOSED	=	CLOSED

(c)

Fig. 9-33 *The 'AND' function or circuit: (a) schematic diagram, (b) truth table, and (c) flow diagram*

(a)

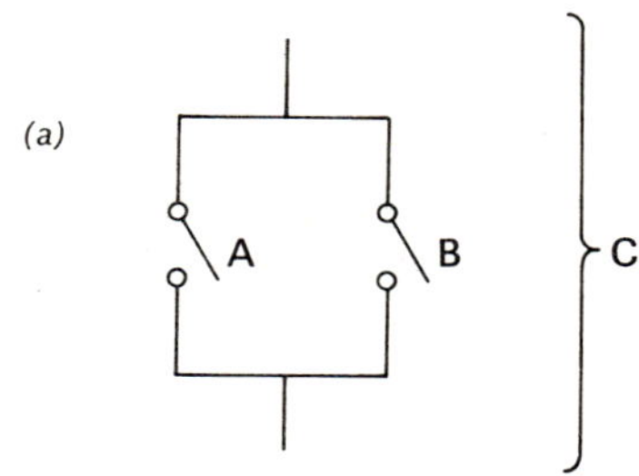

(b)

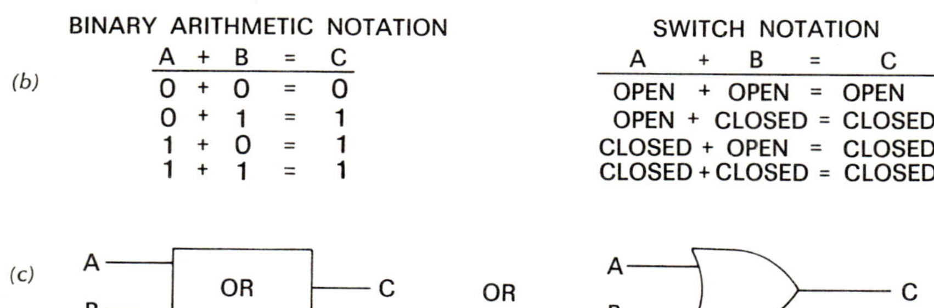

BINARY ARITHMETIC NOTATION

A	+	B	=	C
0	+	0	=	0
0	+	1	=	1
1	+	0	=	1
1	+	1	=	1

SWITCH NOTATION

A	+	B	=	C
OPEN	+	OPEN	=	OPEN
OPEN	+	CLOSED	=	CLOSED
CLOSED	+	OPEN	=	CLOSED
CLOSED	+	CLOSED	=	CLOSED

(c)

Fig. 9-34 *The 'OR' function or circuit: (a) schematic diagram, (b) truth table, and (c) flow diagram*

to delve into the intricacies of this field. Nevertheless, Figs. 9-35, and 9-36 have been included to show representative examples of both analog and digital computer flow diagrams. It should be pointed out, however, that many, although not all, of the block shapes are shown standardized in the ASA Handbook Y32.14, "Graphical Symbols for Logic Diagrams".

In addition, the comprehension of what a truth table indicates may offer a stumbling block to those students having no knowledge of Boolean Algebra, hence the author's position in regard to limiting information on this topic is not duly unfounded.

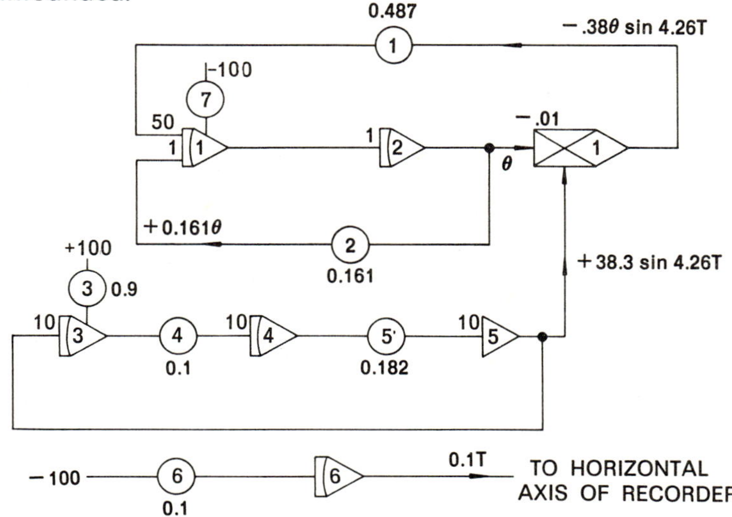

Fig. 9-35 An analog computer flow diagram. (Courtesy Systron-Donner Corporation, Concord, California)

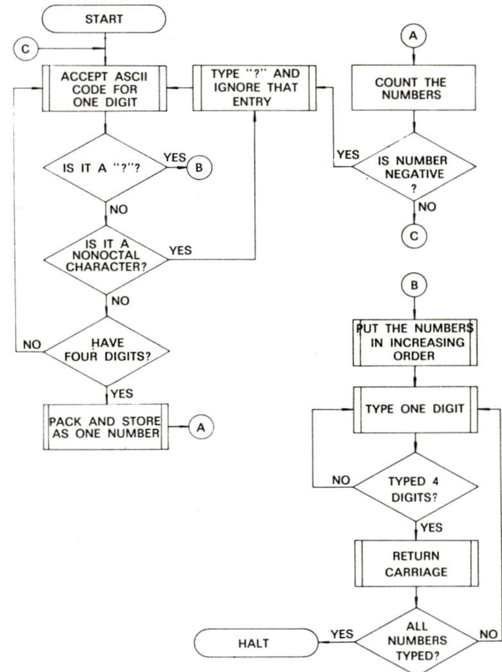

Fig. 9-36 A digital computer program flowchart. (Courtesy Digital Equipment Corporation, Maynard, Mass.)

KNOWLEDGE TESTERS

1. Name the four basic ways of making practical electrical circuits, and briefly explain the nature of each one.
2. What type of electronic components are used in conjunction with printed circuit boards?
3. What process is used to make the circuit pattern on a printed circuit board?
4. Name two different semiconductor devices.
5. List two applications of integrated circuits.
6. What is the difference between an obsolete and an obsolescent electronic device?
7. Describe the correct positioning of the tube envelope symbols and I.F. transformer symbols when drawing a superheterodyne receiver circuit.
8. In Fig. 9-1 are the tube heaters connected in series or parallel?
9. What type of resistor is R4? (Fig. 9-1)
10. Name the type of ground attached to the shielding of transformers T1 and T4. (Fig. 9-1)
11. What type of rectifier tube is V5? Do not give 'Type 35W4' as the answer. (Fig. 9-1)
12. What advantage would the type of antenna shown as L1 offer? (Fig. 9-1)
13. In Fig. 9-2 are the electrostatic shields about the I.F. transformers grounded or nongrounded?
14. What advantage is gained in using the chassis ground system for the receiver shown in Fig. 9-2?
15. Where is the volume control connected in relationship to the diode detector in Fig. 9-2?
16. What is a prototype design?
17. Why do the drawings for a printed circuit board require precise coordinate dimensions for locating the holes of component leads?
18. Why is the layout of electronic components on a printed circuit board not simply a space involvement?
19. Why must the initial large-scale layout for a printed circuit board be of a precise nature?
20. What materials are used for printed circuit boards or substrates?
21. Name three processes that are used in transferring the circuit image to the substrate.
22. How are the blank areas, that is, those other than the circuit paths, removed after the circuit image has been made on the substrate?
23. List the advantages of printed circuits.
24. How does electrical flow and potential affect the spacing, width, and thickness of printed circuit conductor paths?
25. What is the purpose of a pad or land?
26. Since conductor pathways are noninsulated and are printed on a flat plane, how can a crossover be effected?
27. What is the purpose of an undergrid sheet?
28. Why is polyester film, or glass cloth, used as a drafting medium when drawing a printed circuit?
29. List the processes used to form conductive and nonconductive films on a thin-film substrate.
30. Name the substances used to produce the dielectric (insulator) areas for capacitors and resistors on a thin-film circuit.
31. What is the approximate thickness of a thin-film resistive layer?
32. Give the scale that is frequently used when drawing a precision drawing for a thin-film circuit.
33. Name two methods of using masks for thin-film circuits.
34. Is a dielectric substance an insulator or a conductor?
35. What is an important disadvantage of thin-film circuits?
36. Name the basic material used in making an integrated circuit.

37. In simple terms, describe how an IC chip is manufactured.
38. What are the advantages of IC chips?
39. What is the purpose of a coordinagraph?
40. What function does a computer serve in the overall design and manufacture of an IC chip?
41. What scale size is used in producing an IC master drawing?
42. Describe the term 'encapsulation'.
43. Define the terms, 'bonding' and 'intraconnecting' as they pertain to IC drawings.
44. What is a hybrid integrated circuit?
45. In what practical form are logic circuits manufactured?
46. What do logic diagrams illustrate?
47. What is the function of a logic diagram truth table.
48. How many circuit conditions in the AND circuit will provide a pulse at C? Fig. 9-33(a). Name the condition.
49. What is the purpose of the ASA Handbook Y32.14?
50. What is the only condition in the OR circuit that will not provide for an output pulse? Fig. 9-34(a).
51. What plastic materials are used for the substrate of flexible printed circuit boards?
52. List some of the applications for flexible printed circuit boards.
53. Describe a specific design problem that is inherent with flexible printed circuit boards.
54. What are the maximum limits of scales used for multilayer printed circuit boards?
55. Outline the design problem that is critical with multilayer circuit boards.
56. What is the main advantage of the multilayer concept? Give an application of its use.
57. What drafting scale is used in producing the original Mylar masks in thick-film circuit production?
58. How are the circuit designs and passive components transferred to the ceramic substrate of a thick-film circuit?
59. Give examples of three kinds of noble metals used in thick-film circuitry.
60. By what methods are the leads terminating from the circuit of a thick-film chip, or board, bonded?
61. Why are thick-film circuits advantageous to produce and use?

PROJECTS

1. Make a finished drawing of the schematic wiring diagram shown in sketch form. Fig. 9-37. Reroute leads where required to simplify interpretation.
2. Make a drawing for a printed circuit layout of the schematic wiring diagram shown in Fig. 9-38. The finished trim size of the circuit board is to be 2.500" x 1.500", and the circuit pathways are to be on one side only.
3. A shipboard sonar system consists of the following units:

(a) Transducer
(b) Timer
(c) Transmitter
(d) Indicator
(e) Modulator
(f) Receiver

When in operation the purpose of the timer is to produce a reference pulse and trigger the transmitter through the modulator. This transmitted electrical signal, inaudible to the human ear, is converted into sound waves by the transducer. Some of these reflected sound waves upon striking an underwater object are reflected back to the transducer and converted once again into electrical energy. This electrical signal is then fed into the receiver where tuning, detecting, and amplifying take place. Finally the detected signal is displayed in the indicator. The reference pulse created by the timer is also fed to the indicator in order to compare the returned echo with it, and subsequently the distance to the underwater target.

With the foregoing information construct a suitable block diagram. See the example shown in Unit 2.

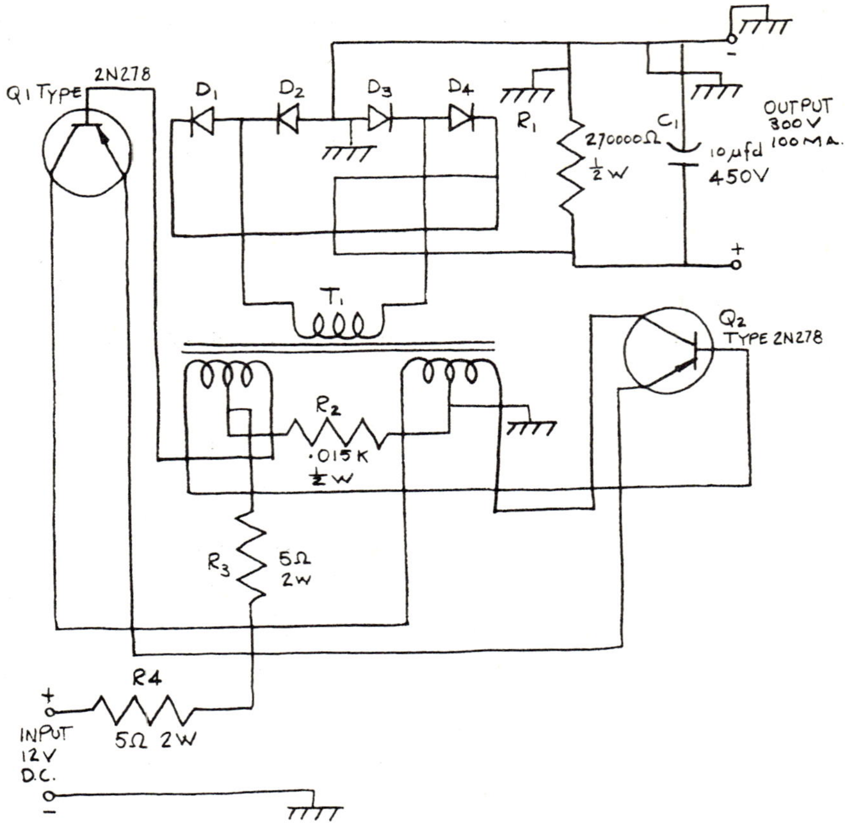

Fig. 9-37 A transistorized D.C. inverter

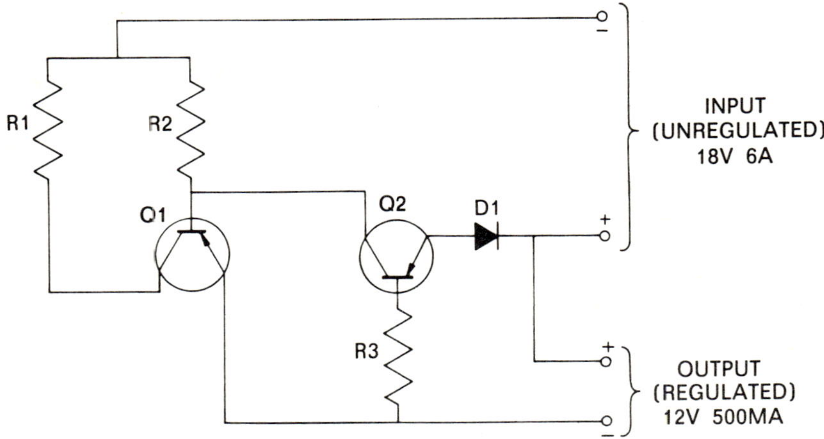

Fig. 9-38 Regulated power supply

4. Make a schematic wiring diagram of the pictorial drawing of the circuit shown in Fig. 9-39.

COMPONENT LIST

CAPACITORS		RESISTORS		TRANSISTORS	
C1	10μfd	R1	22K	Q1	Type 302
C2	220μfd	R2	15K	Q2	Type 302
C3	50μfd	R3	2.2K	Q3	Type 302
C4	6000μfd, 15V	R4	25K	Q4	Type 302
C5	2000μfd, 6V	R5	220Ω	Q5	Type 2N95
C6	4000μfd, 15V	R6	3.3K	Q6	Type 2N35
C7	4000μfd, 15V	R7	3.3K	Q7	Type 2N68
		R8	25K		
		R9	22K		
		R10	8.2K		
		R11	330Ω		
		R12	500Ω		
		R13	560Ω		

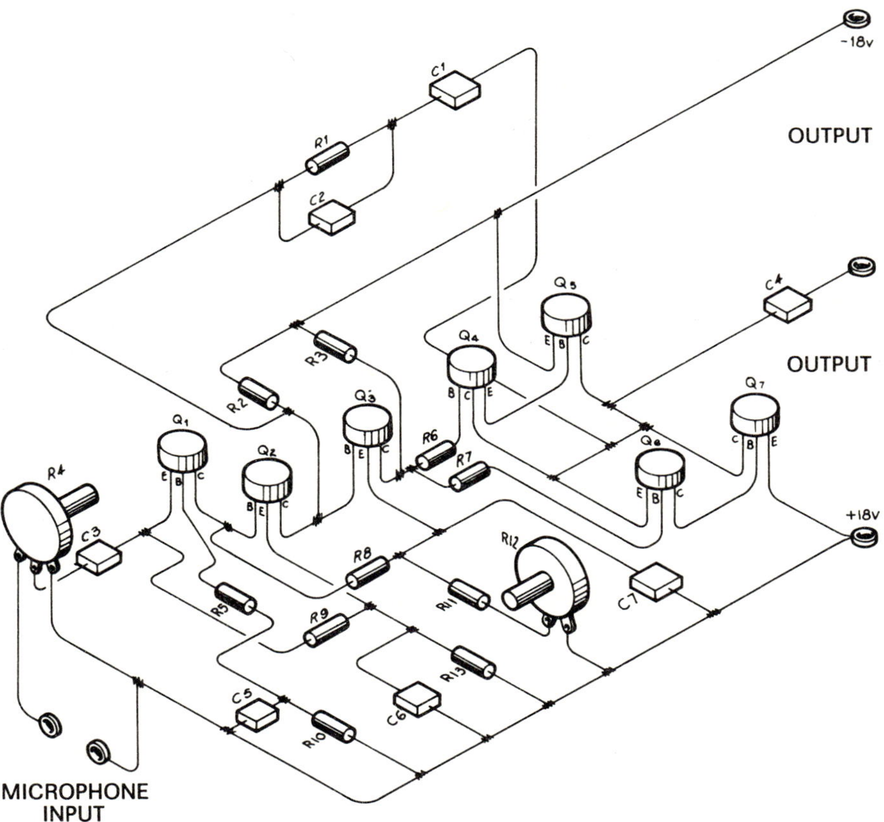

Fig. 9-39 A simplified diagram of a recording amplifier

5. Make a finished drawing of the electronic cabling diagram shown in sketch form in Fig. 9-40.

DRAWING INSTRUCTIONS:

(a) See the cabling diagram example in Unit 2.

(b) The units must be repositioned to provide the optimum pathways.

6. Make an intraconnection drawing for the integrated circuit shown in Fig. 9-41(b). Consult Fig. 9-41(a), the schematic wiring diagram for this circuit.

DRAWING INSTRUCTIONS:

(a) Use Figs. 9-27 and 9-28 as guides.

(b) Make O.A. size of drawing 5.25" x 5.25".

(c) Note electrical value of resistors. Establish size of same by comparison to those shown on Fig. 9-27, (Schematic wiring diagram Fig. 9-25).

(d) Orientation of components need not be as shown, but may be altered for a more convenient layout.

(e) Draw the schematic wiring diagram as shown indicating appropriate number identification terminals. These must conform to those selected for layout of IC chip.

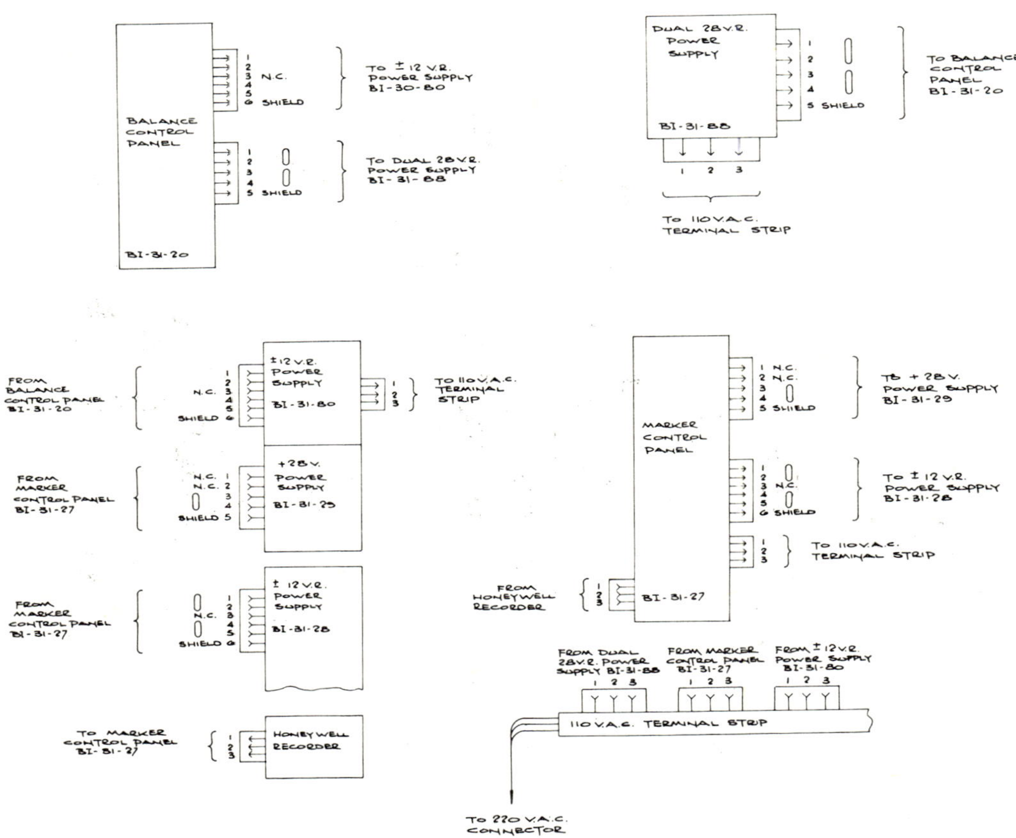

Fig. 9-40 Part of a cabling diagram for an S-band and X-band microwave equipment console

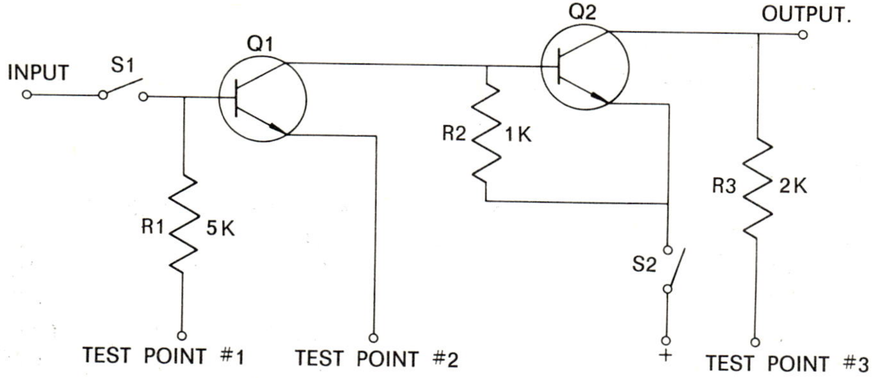

(A) SCHEMATIC WIRING DIAGRAM

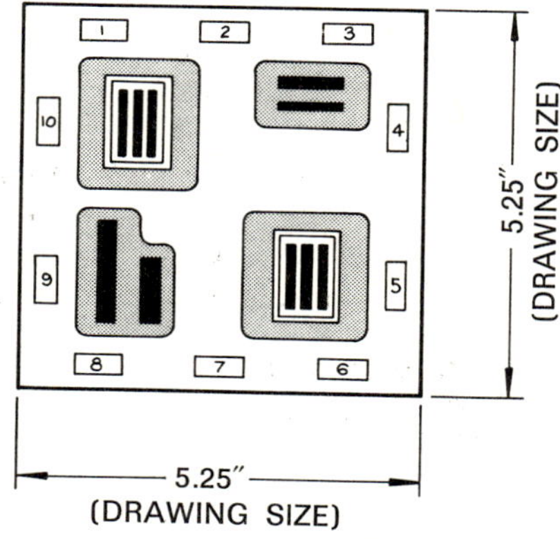

(B) COMPONENT LAYOUT

Fig. 9-41 A transistorized amplifier

7. Make a preliminary layout for a thin-film circuit of the schematic wiring diagram shown in Fig. 9-42.

DRAWING INSTRUCTIONS:

(a) Actual size of board not to exceed 0.75" x 1.50".

(b) Construct drawing four times actual size.

(c) Physical size of resistors to be computed as follows:

(i) Actual width of resistors to be 0.010".

(ii) Actual length of resistors to be determined by the following formula. Length =

$$\frac{\text{(Resistance in ohms}-120)}{40}\text{(width)}$$

(d) Actual contact area between dielectric and anelectric of capacitors C_1 and C_2 to be 0.25" x 0.25".

(e) See illustrations Figs. 9-16 and 9-17 for reference.

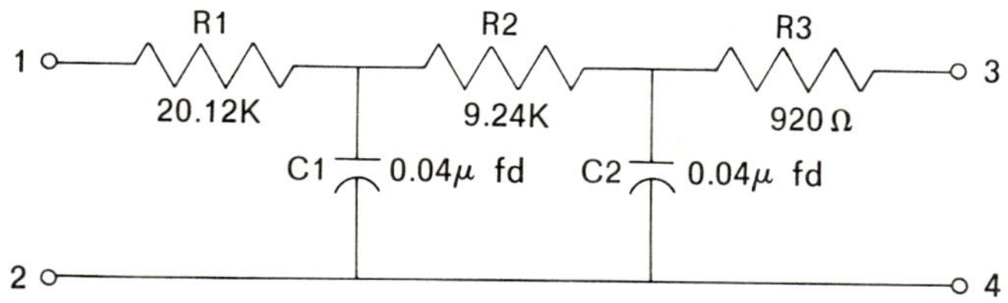

Fig. 9-42 Diagram for Project No. 7

8. Make a schematic wiring diagram of the printed circuit shown in Fig. 9-43.

COMPONENT LIST

CAPACITORS

C1	0.1 mfd 200V.
C2	100 mfd 25V.
C3	10 mfd 15V.
C4	60 mfd 6V.
C5	10 mfd 15V.
C6	10 mfd 15V.
C7	51,000 pfd. 350V.

TRANSISTORS
All transistors
Type 2N652A.

RESISTORS

	1st. BAR	2nd. BAR	3rd. BAR	TOL. BAR
R1	YEL	VIO	BRO	GOLD
R2	BRO	BLA	RED	GOLD
R3	ORA	ORA	ORA	GOLD
R4	BLU	GREY	ORA	GOLD
R5	BRO	GRE	RED	GOLD
R6	BRO	BLA	ORA	GOLD
R7	BLU	GREY	BRO	GOLD
R8	BLU	GREY	RED	GOLD
R9	BRO	BLA	RED	GOLD
R10	BRO	GRE	ORA	GOLD
R11	BLU	SIL	BRO	GOLD
R12	YEL	VIO	BRO	GOLD
R13	YEL	VIO	ORA	GOLD
R14	RED	RED	RED	GOLD
R15	GREY	RED	BLA	SIL

NOTE: Refer to R.M.A. resistor colour code to establish resistor values. See Appendix 5.

9. Obtain the use of a hand-wired radio receiver. Trace the circuit paths and indicate the same in sketch form.
Redraw the sketch as a finished schematic wiring diagram.

10. Make an ink tracing of the finished schematic wiring diagram developed as a solution to Question #1.

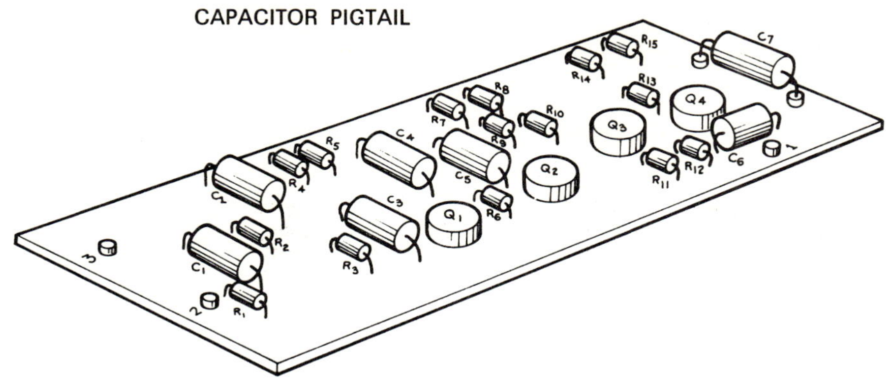

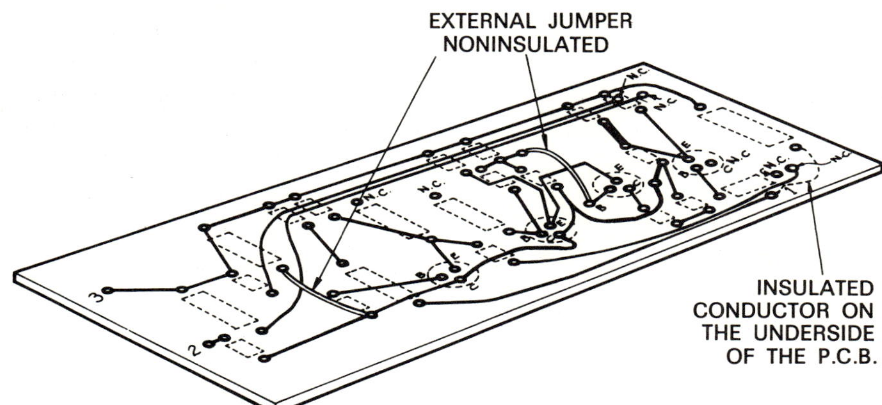

VIEW LOOKING THROUGH COMPONENT SIDE OF BOARD

ABBREVIATIONS:

N.C. — NO CONNECTION

B — BASE

C — COLLECTOR

E — EMITTER

THIS CIRCUIT IS NOT COMPLETELY FUNCTIONAL

Fig. 9-43 Diagram for Project No. 8

UNIT 10

Orthographic Drawings

IN GENERAL, TWO USES FOR ORTHOGRAPHIC DRAWINGS ARE MADE. FIRST of all, manufacturers' production drawings of parts, subassemblies, and assemblies using the projection form already mentioned are widely used. Second, outline drawings containing only overall dimensions and mounting dimensions are found in manufacturers' brochures and pamphlets. This type of drawing also includes vacuum tubes, transistors, and other electronic components.

The procedures used to construct orthographic drawings of electromechanical parts in no way differ from those used to make orthographic drawings of mechanical parts having no electrical significance.

MANUFACTURERS' PRODUCTION DRAWINGS

Production or manufacturing drawings may consist of detail, subassembly, and assembly drawings. A detail drawing normally shows one, two, three, or more than three orthographic views of an individual part. These views contain sufficient dimensions and where applicable, **engineering notes**. If this is so, then the craftsman or technician will be able to fabricate the part properly, provided he adheres to the dimensional and notational instructions shown on the drawing.

A subassembly drawing is one on which a number of parts are shown in their functional relationship with each other. However, the bringing together of these parts does not constitute an entire functional unit. The number of views required may constitute two or more, and the dimensions shown may include both overall and mounting.

An assembly drawing differs from a subassembly drawing only by virtue of the fact that it refers to an entire functional unit.

By using a circuit breaker switch as an example, we can say the following. An individual breaker contact is a part and hence requires a detailed part drawing. The complete breaker contact mechanism is a subassembly drawing. And finally, the entire circuit breaker switch is an assembly of subassemblies, and as such demands an assembly drawing.

In accordance with the author's comments in Unit 1 regarding orthographic drawings, the following illustrations are included for the sole purpose of showing representative examples, since the construction of an electromechanical drawing does not deviate from the use of orthographic projection principles.

Single-view detail drawing

A single view is frequently sufficient to illustrate parts that are cylindrical in shape such as the shaft of a small fractional horsepower motor. See Fig. 10-1. Note that the addition of either the left-side or right-side views or of both of them is of no additional significance. The standard abbreviation 'DIA.' meaning diameter is sufficient information for the reader of the diagram to appreciate that the shaft is cylindrical.

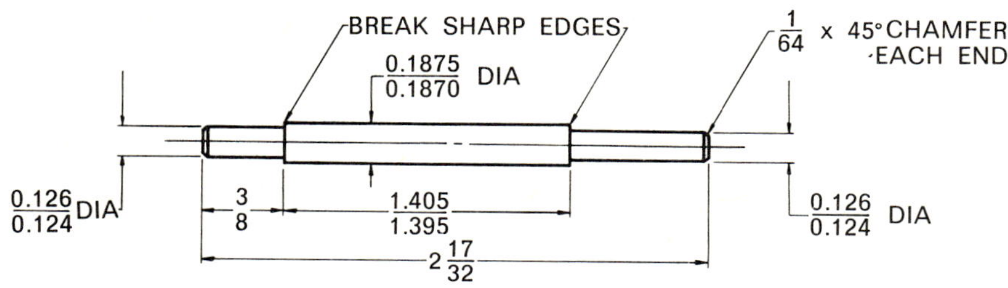

Fig. 10-1 An orthographic front-view drawing

Two-view detail drawing

When a cylindrical or other shaped part has details which cannot be shown with a single view, then both a front view and a combination of another view are required.

Figure 10-2 illustrates a two-view drawing of the **rotor** for a D.C. operated electrical clutch. Note that by using the left-side view, a greater amount of significant detail is shown without the use of hidden lines.

Three-view detail drawing

Frequently three views of a part are required to show its construction ideally. The illustration Fig. 10-3 shows the left-side, front, and right-side views of the rotor of a D.C. operated electrical clutch coupling. The addition of the second side view eliminates concentric circles that otherwise must be shown with broken lines. Also, without a second side view, the interpretation of many of the hidden lines in the front view is obscured.

The drafting concepts llustrated by Figs. 10-2 and 10-3 can be applied equally to parts that are not cylindrical in shape, except that two views are the minimum required unless the part under manufacture happens to be very thin, such as a gasket or shim. In this case a side view is meaningless. However, a note is added to the drawing indicating the shim thickness. See Fig. 10-4. Figures 10-5 and 10-6 illustrate electromechanical parts which are not cylindrical in shape. These illustrations refer to two-view and three-view drawings.

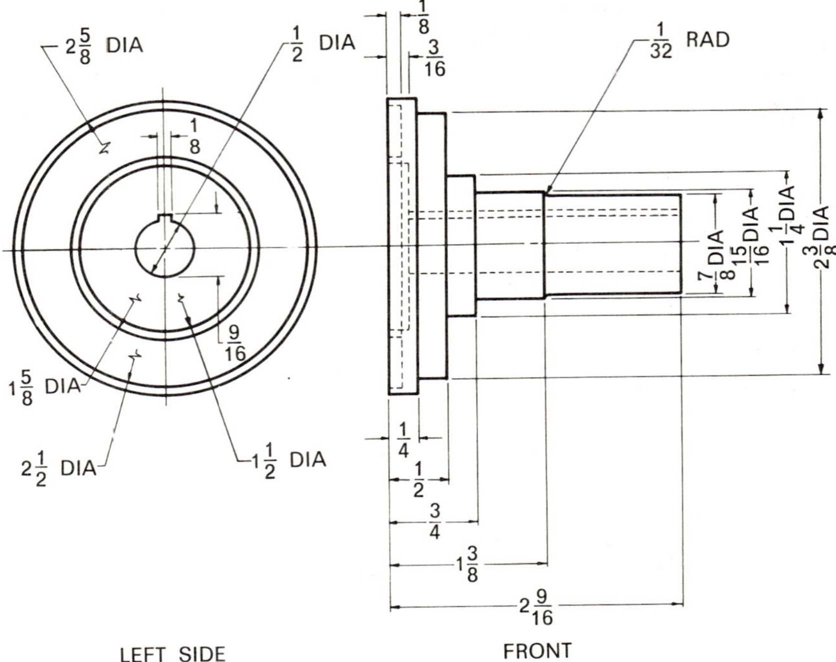

NOTE: ALL DIMENSIONS ARE SHOWN AS NOMINAL FOR REASONS OF SIMPLIFICATION

Fig. 10-2 A two-view orthographic drawing

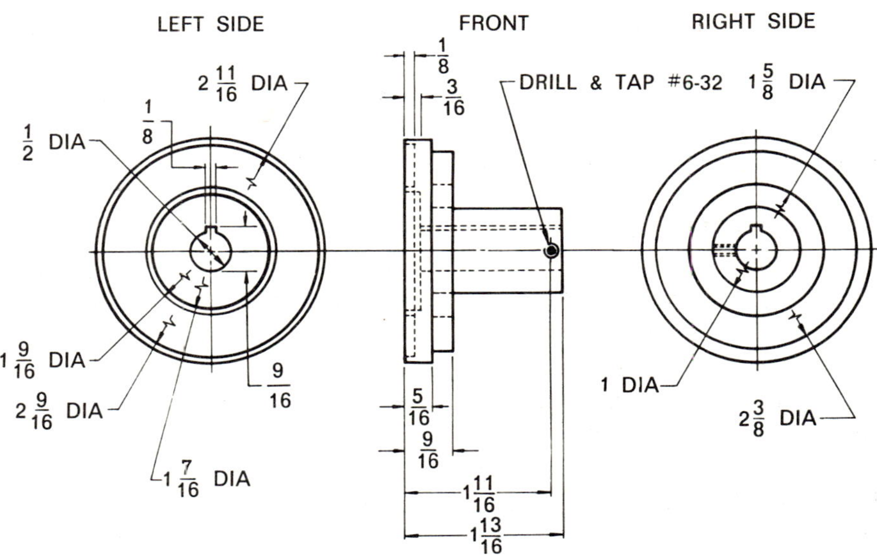

NOTE: ALL DIMENSIONS ARE SHOWN AS NOMINAL FOR REASONS OF SIMPLIFICATION

Fig. 10-3 A three-view orthographic drawing

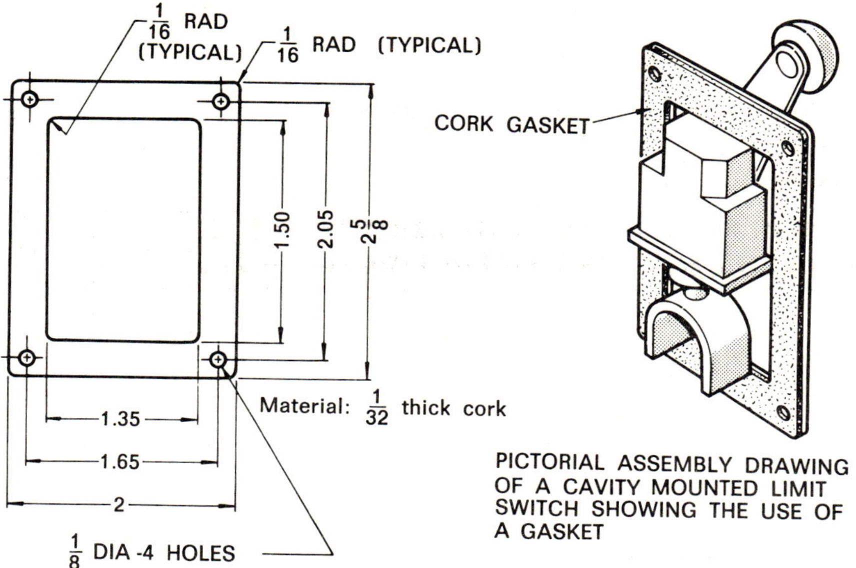

Fig. 10-4 A single-view drawing of a gasket

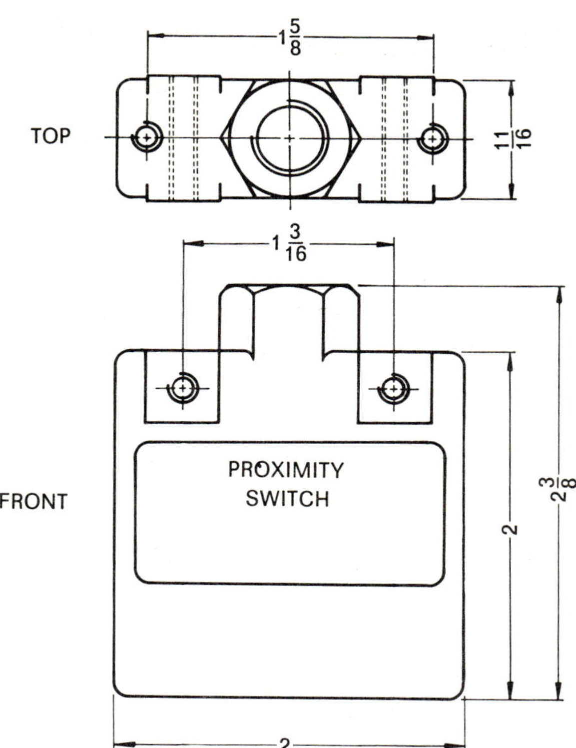

Fig. 10-5 An assembled, permanent magnet proximity switch

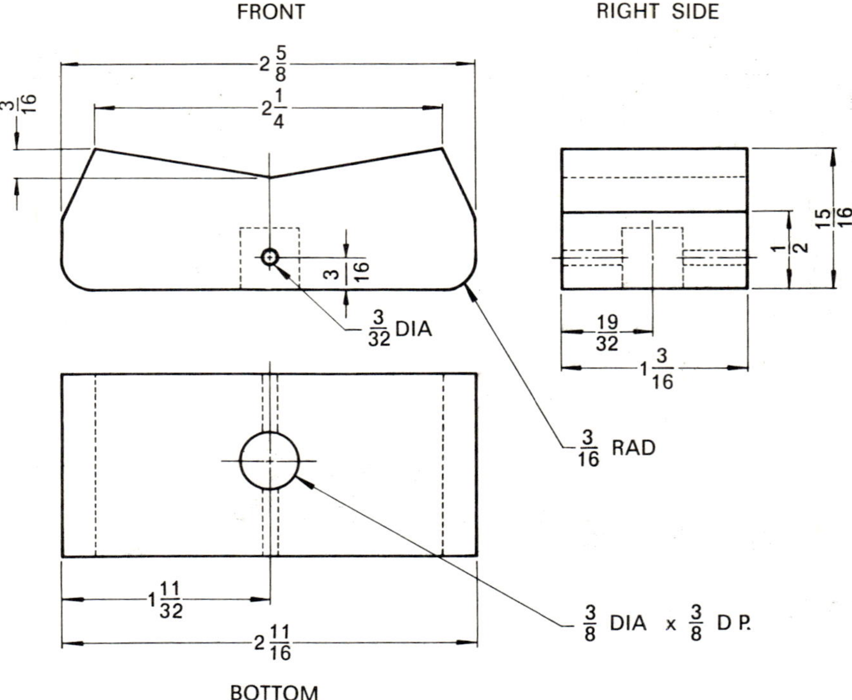

NOTE: ALL DIMENSIONS ARE SHOWN AS NOMINAL FOR REASONS OF SIMPLIFICATION

Fig. 10-6 A manually operated switch, rocker toggle button

Subassembly drawing

Illustration, Fig. 10-7 shows a typical subassembly drawing of a set of circuit breaker contacts. Note the lack of detail dimensions. The inclusion of the identifying balloons and leaders which conform to the 'item numbers' in the parts list is a standard feature of subassembly drawings.

Assembly drawing

The illustration of the circuit breaker switch, Fig. 10-8, is a typical assembly drawing. Note the references to both item and part number for the 'contacts' subassembly, and other subassemblies. However, individual parts such as capscrews and lockwashers are also shown, since they are required to secure subassembly units to others.

Manufacturers' outline drawings

The manufacturers of electromechanical and electronic equipment provide their potential customers with brochures, pamphlets, catalogues, and handbooks which include illustrations known as **outline drawings**.

These drawings are not provided to show how the internal parts of a functional unit operate, but rather to inform engineering staffs of its external configuration, maximum external dimensions, and mounting hole dimensions.

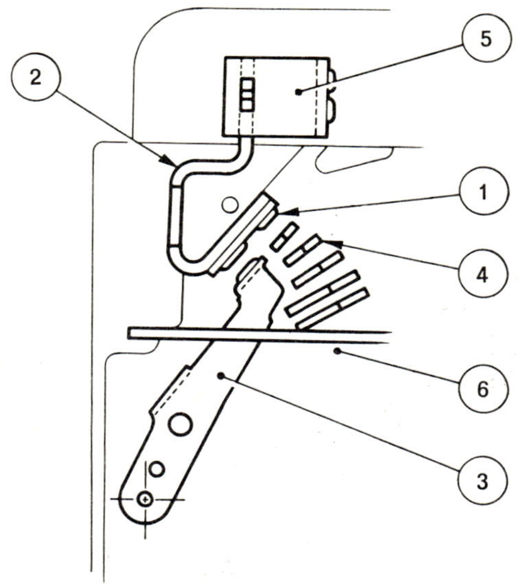

ITEM	PART NO.	QTY.	DESCRIPTION	MAT'L.	SPEC.
6		1	CASE (REFERENCE)		
5		1	LUG, CONNECTION (REFERENCE)		
	VARIES			VARIES	
4	PER	1	MAGNET, BLOWOUT (REFERENCE)	PER	
	MANUFACTURER			MANUFACTURER	
3		1	ARM, HINGED CONTACT		
2		1	ARM, FIXED CONTACT		
1		1	CONTACTS, FIXED		

Fig. 10-7 A subassembly drawing of circuit breaker contacts

Invariably photographs accompany these detail drawings, as well as engineering data frequently set down in chart form. All engineering offices maintain substantial, organized files of brochures and catalogues covering a wide range of equipment. The draftsman must of necessity learn to use these files in order to carry out his task effectively.

Figures 10-9(a) and (b), as well as Fig. 10-10 are representative examples of outline drawings.

In Fig. 10-9(a) a common method is shown whereby one illustration is used for three sets of dimensions. A letter reference is placed on the drawing in lieu of a numerical value, and a simple chart is made showing the three possible dimensions for that letter value.

Two additional comments must be made regarding outline drawings. Since they are to be printed for a wide circulation, and form an integral part of the manufacturers' advertising media, the quality of the drawings must be excellent. Also, lettering should be set by type, or by other mechanical means.

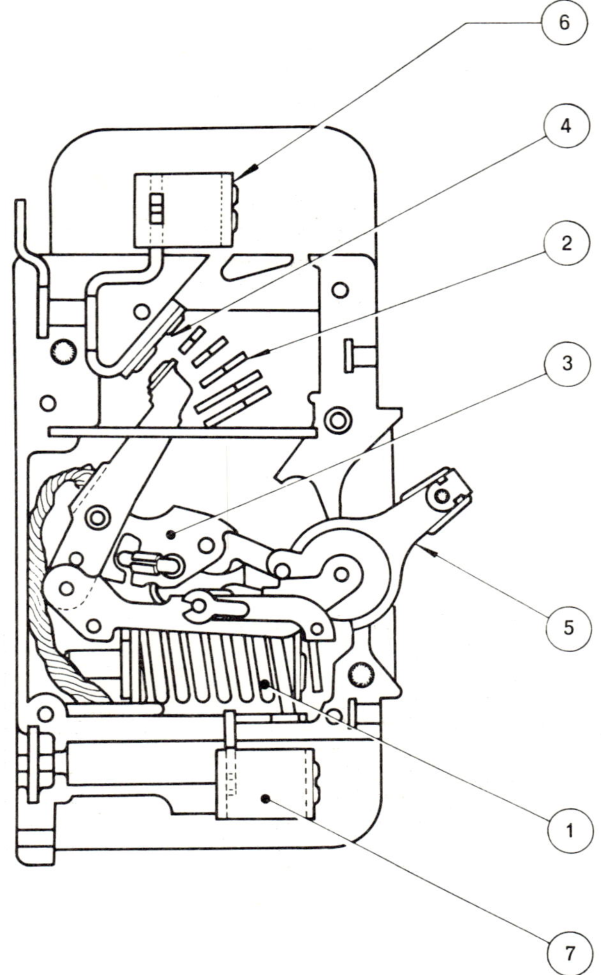

7	CF2-G3-7	1	LUG, LOWER CONNECTION	—	—
6	CF2-G3-6	1	LUG, UPPER CONNECTION	—	—
5	CF2-G3-5	1	HANDLE, TWO-POSITION	—	—
4	CF2-G3-4	1	SUBASSY., CONTACTS	—	—
3	CF2-G3-3	1	SUBASSY., LATCH	—	—
2	CF2-G3-2	1	SUBASSY., ARC BLOWOUT	—	—
1	CF2-G3-1	1	UNIT, HYDRAULIC-MAGNETIC	—	—
ITEM	PART NO.	QTY.	DESCRIPTION	MAT'L.	SPEC.

Fig. 10-8 An assembly drawing of a circuit breaker switch

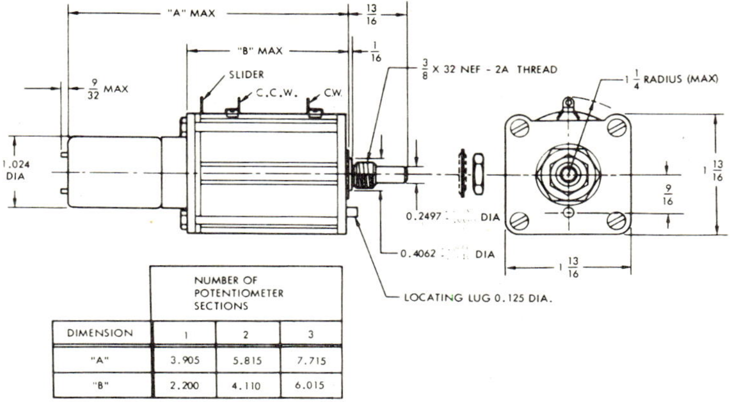

DIMENSION	NUMBER OF POTENTIOMETER SECTIONS 1	2	3
"A"	3.905	5.815	7.715
"B"	2.200	4.110	6.015

MODEL 943

Fig. 10-9 *A manufacturer's outline drawing and accompanying photograph of a Helipot Model 943 potentiometer assembly. (Courtesy Helipot Division, Beckman Instruments, Inc.)*

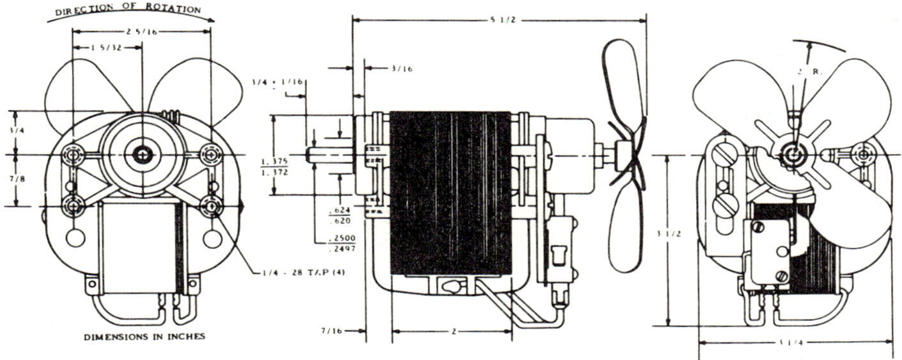

Fig. 10-10 *A manufacturer's outline drawing of a small fan motor assembly. (Courtesy Barber-Coleman Company, Rockford, Illinois)*

KNOWLEDGE TESTERS

1. List the two main areas where orthographic drawings are used for electromechanical parts.
2. What criterion establishes the number of orthographic views required to define a part graphically?
3. How many views are required to describe a simply shaped cylindrical part correctly?
4. In what type of publications are outline drawings found?
5. What are the reasons for showing the overall dimensions on an outline drawing?
6. How does an assembly drawing differ from a subassembly drawing?
7. What method is used to identify both subassemblies and individual parts on an assembly drawing?
8. Why is the left-side view of the fan and motor assembly (Fig. 10-10) required?
9. If only a single view showing the profile of a shim or gasket is required, how is its thickness indicated?
10. What forms of lettering are demanded for outline drawings?
11. Define the term, 'concentric circles'.
12. Using some simple electromechanical objects that you are familiar with, suitably define the concepts of a part, subassembly, assembly.
13. What does the term 'DIA.' mean?
14. Explain the method used whereby a single outline drawing can refer to a number of similar units having some variable dimensions.
15. In what ways does an outline drawing differ from a manufacturer's production drawing?

PROJECTS

1. Make a **working** (manufacturing) drawing, properly dimensioned, of the permanent magnet from a small P.M. speaker shown in Fig. 10-11.
2. Draw front, bottom, and right-side views of the cover, Fig. 10-12, and dimension correctly?
3. Design and draw a gasket to be made from 1/16″ synthetic rubber to fit on the bottom of the cover, Fig. 10-12.
4. Make a two-view dimensioned drawing of the threaded, moulded plastic, electrical cable fitting, Fig. 10-13.
5. (a) Make rear, top, and right-side views of the switch block, Fig. 10-14, in sketch form.
   (b) Draw section A-A and dimension. See Fig. 10-14.
   (c) Draw section B-B and dimension. See Fig. 10-14.
6. Make an assembly drawing of the 'twist-lock' female receptacle, Fig. 10-15.
7. Construct dimensioned detail drawings and a subassembly drawing for the coaxial line receptacle Fig. 10-16. Make three-times full size.
8. Redraw the outline drawing Fig. 10-10, but regard the following instructions:
   (a) Draw only the left-side and front views.
   (b) Draw full size.
   (c) Work out a scale to establish unknown values by measuring a given value. As a suggestion use the fan blade radius.
9. Selecting suitable views, make an outline drawing for a manufacturer's brochure of the step-down transformer, Fig. 10-17.
10. Draw the rear, bottom, and left-side views of the housing shown in Fig. 10-18. Draw three-times full size, and dimension according to good practice.

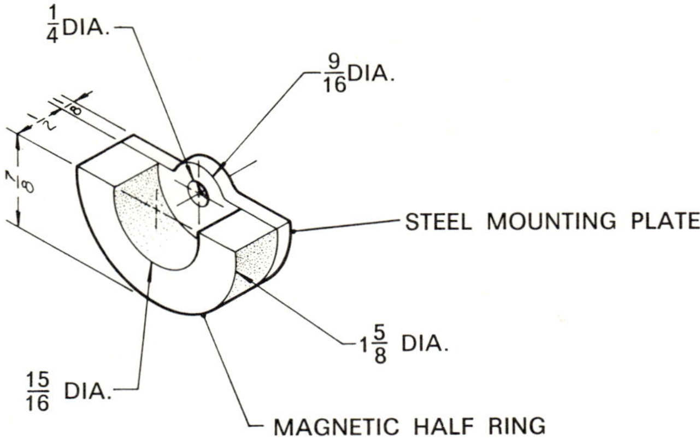

Fig. 10-11 A P.M. loudspeaker magnet

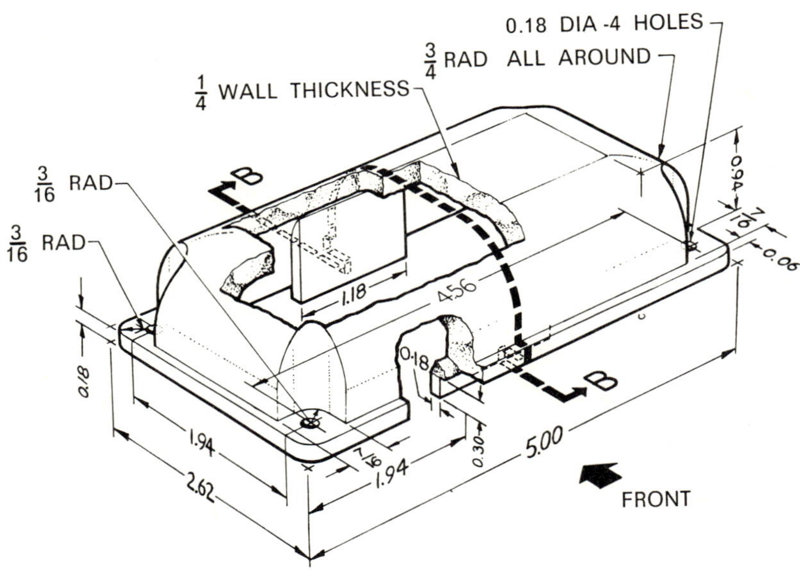

Fig. 10-12 A cover for a moulded plastic, push-button station enclosure

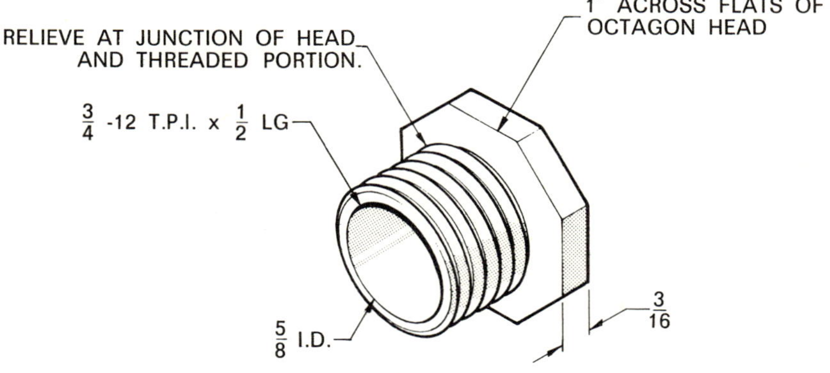

Fig. 10-13 Drawing for Project No. 4

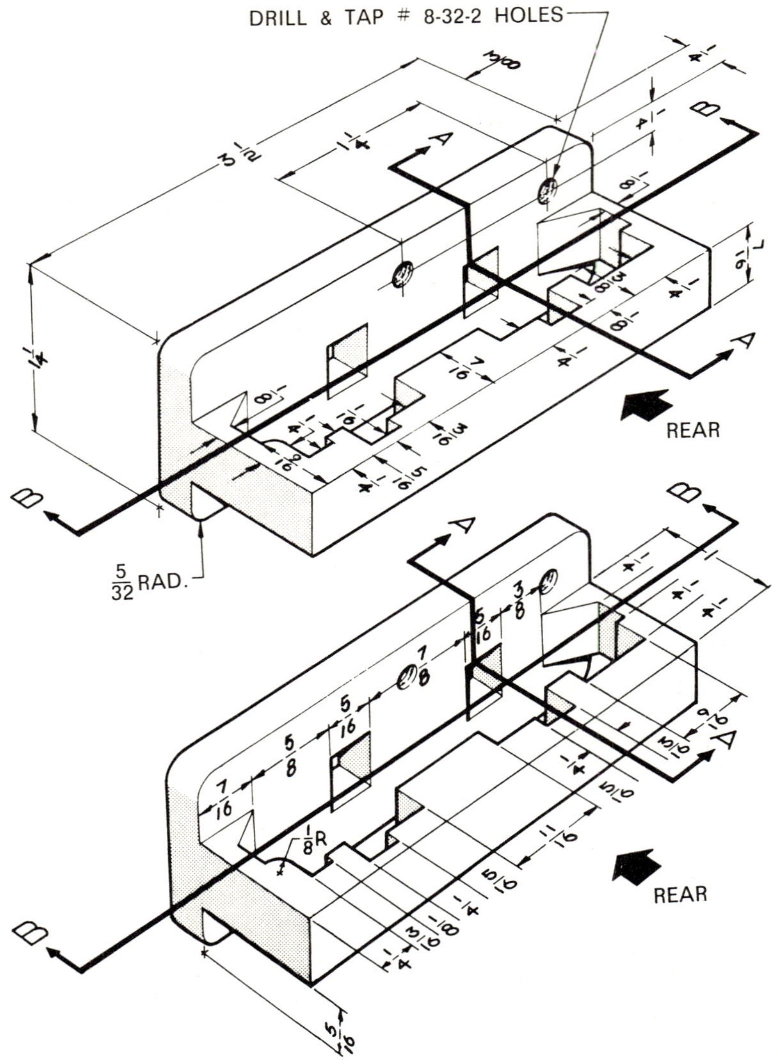

Fig. 10-14 A switch block

Fig. 10-15 A twist-lock female receptacle
Fig. 10-16 A coaxial line receptacle

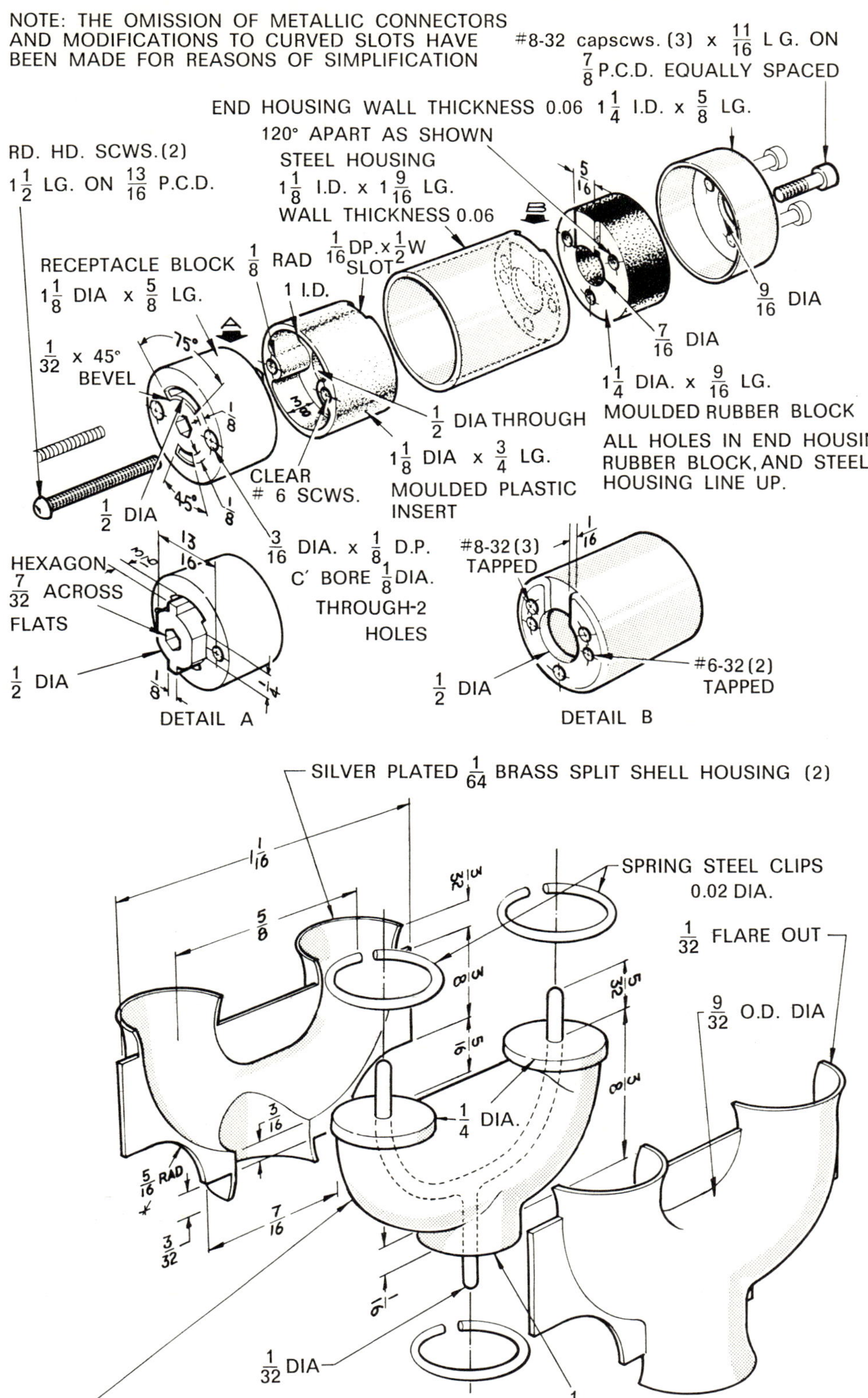
NOTE: THE OMISSION OF METALLIC CONNECTORS AND MODIFICATIONS TO CURVED SLOTS HAVE BEEN MADE FOR REASONS OF SIMPLIFICATION
#8-32 capscws. (3) x $\frac{11}{16}$ LG. ON $\frac{7}{8}$ P.C.D. EQUALLY SPACED
END HOUSING WALL THICKNESS 0.06 $1\frac{1}{4}$ I.D. x $\frac{5}{8}$ LG.
120° APART AS SHOWN
RD. HD. SCWS. (2) $1\frac{1}{2}$ LG. ON $\frac{13}{16}$ P.C.D.
STEEL HOUSING $1\frac{1}{8}$ I.D. x $1\frac{9}{16}$ LG. WALL THICKNESS 0.06
RECEPTACLE BLOCK $1\frac{1}{8}$ DIA x $\frac{5}{8}$ LG.
$\frac{1}{8}$ RAD
$\frac{1}{16}$ DP. x $\frac{1}{2}$ W SLOT
1 I.D.
$\frac{9}{16}$ DIA
$\frac{7}{16}$ DIA
$\frac{1}{32}$ x 45° BEVEL
75°
$1\frac{1}{4}$ DIA. x $\frac{9}{16}$ LG. MOULDED RUBBER BLOCK
$\frac{1}{2}$ DIA THROUGH
ALL HOLES IN END HOUSING, RUBBER BLOCK, AND STEEL HOUSING LINE UP.
$1\frac{1}{8}$ DIA x $\frac{3}{4}$ LG. MOULDED PLASTIC INSERT
CLEAR # 6 SCWS.
$\frac{1}{2}$ DIA
45°
$\frac{3}{16}$ DIA. x $\frac{1}{8}$ D.P. C' BORE $\frac{1}{8}$ DIA. THROUGH-2 HOLES
HEXAGON $\frac{7}{32}$ ACROSS FLATS
$\frac{1}{2}$ DIA
DETAIL A
#8-32 (3) TAPPED
$\frac{1}{2}$ DIA
#6-32 (2) TAPPED
DETAIL B
SILVER PLATED $\frac{1}{64}$ BRASS SPLIT SHELL HOUSING (2)
SPRING STEEL CLIPS 0.02 DIA.
$\frac{1}{32}$ FLARE OUT
$\frac{9}{32}$ O.D. DIA
$\frac{1}{4}$ DIA.
$\frac{5}{16}$ RAD
$\frac{7}{16}$
$\frac{1}{32}$ DIA
$\frac{1}{4}$ DIA.
MOULDED SINGLE CONDUCTOR TO TWIN CONDUCTOR HIGH DIELECTRIC PLASTIC

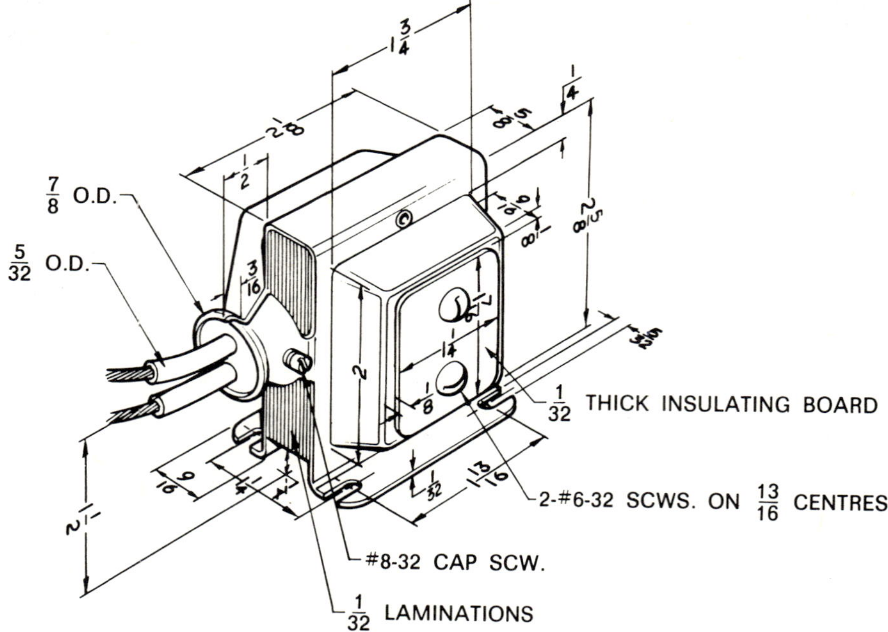

Fig. 10-17 A step-down transformer

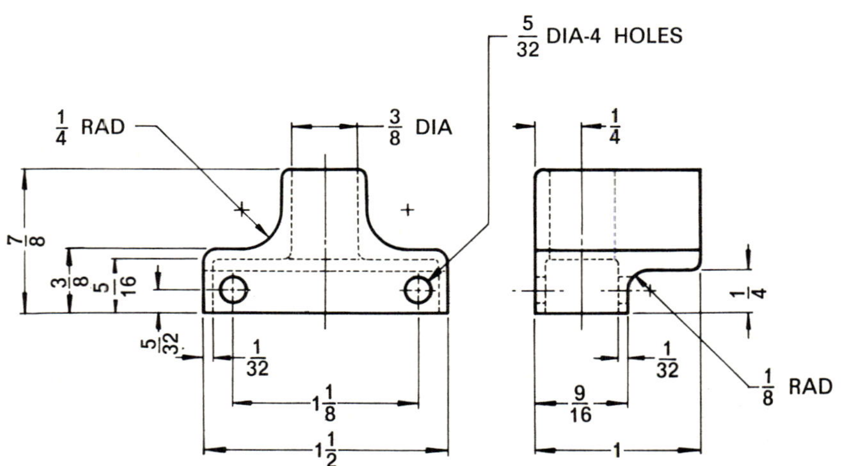

Fig. 10-18 A coaxial line receptacle housing

UNIT 11

Pictorial Drawings

THE TERM 'PICTORIAL DRAWING' REFERS TO AN ILLUSTRATION OR diagram that is three dimensional. By three dimensional we mean that it shows width, height, and depth in one view. Figure 11-1 shows a drawing of a meter case to illustrate these dimensional concepts. Of course, the orientation of the view may be such that any of its surfaces is given prominence. In general, however, the orientation of the view is as shown in Fig. 11-1. Some other possibilities of view orientation are shown in Fig. 11-2, which illustrates a miniaturized audio transformer.

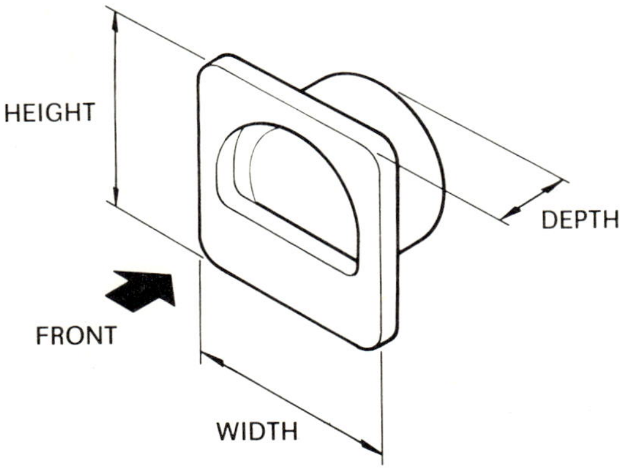

Fig. 11-1 A pictorial drawing of a meter case

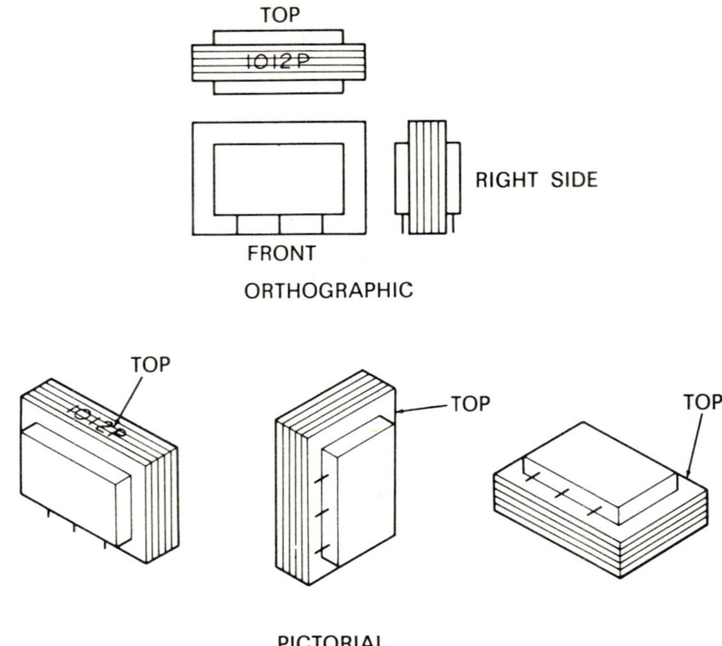

Fig. 11-2 An orthographic and some pictorial drawing possibilities

Purpose of pictorial drawings

Pictorial drawings are used for the following:

1. Electrical handbooks and texts.
2. Servicing and parts catalogues of electrical equipment.
3. Sales brochures of electrical equipment.
4. Electrical engineering pamplets and magazines.
5. As a graphic technique to convey technical information to semiskilled personnel for the assembly of electromechanical devices.

Figure 11-3 shows a pictorial drawing from a sales brochure of an electromechanical part.

The reason that pictorial drawings lend themselves in an adequate way to the areas just mentioned is that they conform closely to the image seen by the human eye. Hence they have a greater impact on one's conception of the configuration of the actual part or assembly.

TYPES OF PICTORIAL DRAWINGS USED

It is not within the scope of this unit to discuss all of the types of pictorial drawings, but to limit the text to those types most frequently used.

Isometric pictorial drawing

From a standpoint of construction simplicity the isometric drawing is perhaps the easiest. It is also acceptable from a point of view of appearance.

RACK MOUNTING	RACK OR FLOOR MOUNTED	ENCLOSURE DIMENSIONS			BLANK MOUNTING PLATE DIMENSIONS		CEMA TYPE 1 GENERAL PURPOSE ENCLOSURE		CEMA TYPE 12 DUST-TIGHT INDUSTRIAL USE		FLOOR MOUNTING
		A	B	C	WIDTH	HEIGHT	CATALOG NUMBER	PRICE	CATALOG NUMBER	PRICE	
	RACK	10	24	6	$7\frac{1}{2}$	$20\frac{1}{4}$	SPECIFIC CATALOG NUMBERS AND ACCOMPANYING PRICES ARE ESTABLISHED BY INDIVIDUAL MANUFACTURERS				
		20	24	6	$16\frac{1}{2}$	$20\frac{1}{8}$					
		20	24	$8\frac{5}{16}$	$16\frac{1}{2}$	$20\frac{1}{8}$					
		20	48	$8\frac{5}{16}$	$14\frac{1}{2}$	$40\frac{3}{4}$					
		20	56	$12\frac{1}{8}$	$13\frac{1}{4}$	$48\frac{1}{2}$					
		30	48	$8\frac{5}{16}$	26	$40\frac{3}{4}$					
		30	48	$12\frac{1}{8}$	26	$40\frac{3}{4}$					
	FLOOR	20	80	$12\frac{1}{8}$	$19\frac{1}{4}$	68					
		30	80	$18\frac{3}{16}$	$29\frac{1}{4}$	68					

EMPTY MODULAR ENCLOSURES

Fig. 11-3 A sales brochure drawing and accompanying information

However, it has certain drawbacks, namely a lack of perspective, and also that receding axes appear to expand outwards or diverge. Figure 11-4 illustrates a typical isometric drawing, and shows the construction of the basic isometric block. Reference must be made to a book on mechanical drafting for the method of locating internal points and lines, and drawing circles and arcs of circles. Figure 11-5 illustrates the disadvantages of an isometric drawing.

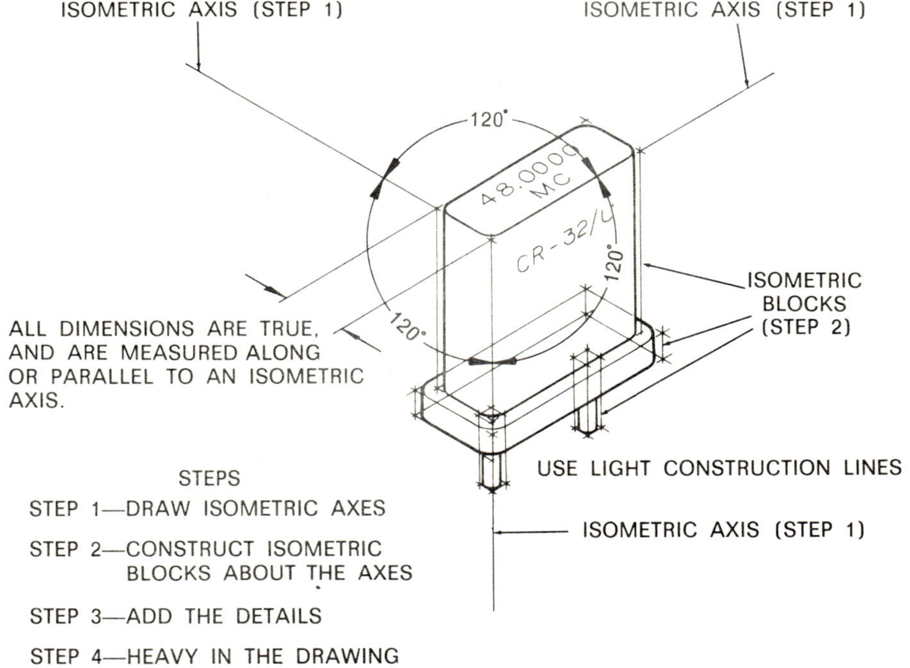

Fig. 11-4 Construction techniques in isometric drawing

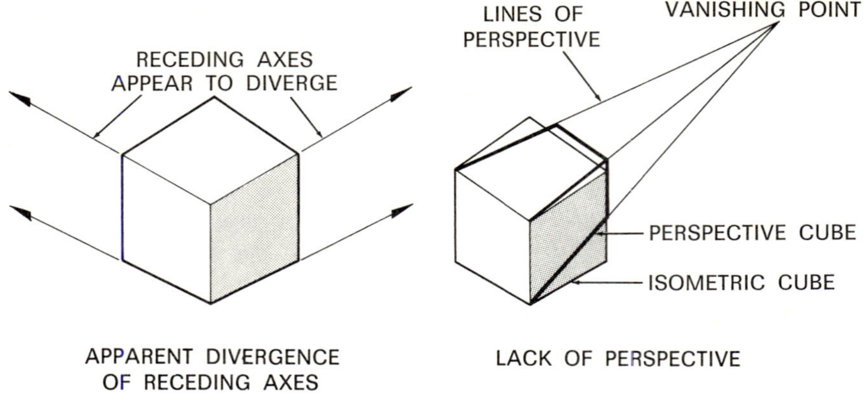

Fig. 11-5 The weaknesses of an isometric drawing

Dimetric drawings

Dimetric pictorial drawings are similar to isometric except that only two of the axes are at the same angle to a horizontal line.

Frequently 'dimetrics' are drawn with two of their axes at 15° to the horizontal, but other combinations can be used.

In addition to this angular difference that is not a characteristic of isometric drawings, special dimetric scales are used for measuring along the axes.

The advantages of dimetric drawings are the reduction of distortion brought about by using the compressed or foreshortened dimetric scales, and the fact that many different views are possible. Figure 11-6 illustrates two possible views of a step-down transformer drawn in dimetric fashion.

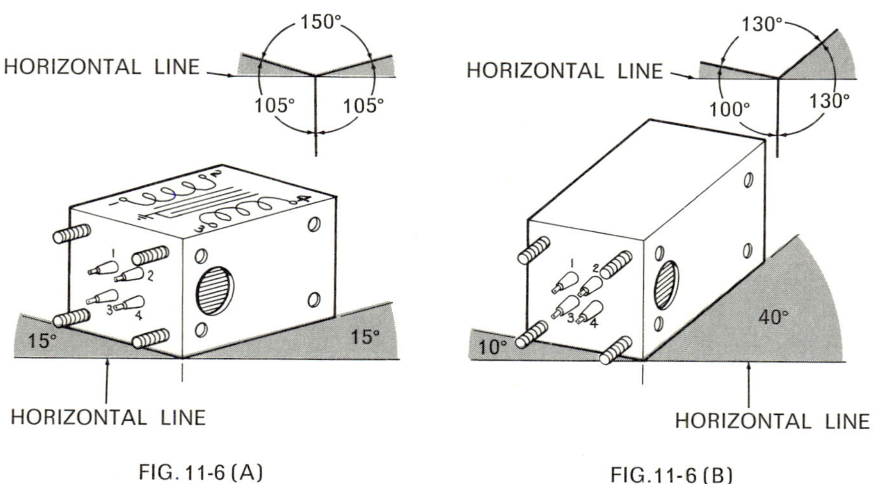

Fig. 11-6 A dimetric drawing (a) with two axis at 15° to the horizontal line and (b) with one axis at 10° and the other at 40° to the horizontal line

To speed up the drawing of a dimetric, commercial grids with foreshortened scales printed on them are available. The grids are placed underneath the tracing paper upon which the original work is being constructed. Complete accuracy is seldom required in making such a pictorial, but the scaled grid does provide the basis for a fully accurate drawing. See Fig. 11-20.

Brokenout or cutaway pictorial drawings

In order to expose certain internal parts of an assembly or subassembly, or simply to expose the inner construction of a part, pictorial drawings are frequently 'cutaway' or 'brokenout'. Figure 11-7 shows an illustration of a proposed communications satellite of Canadian design, in which the assembly has been brokenout to show the internal arrangement of components. In this case, some of the solar panels have been purposely eliminated, and since they are constructed from discrete, modular cells, rectangular in shape, the method of cutting them away is entirely logical. However, since the central structure is a sheet fabrication, the use of a freehand irregular curve is used to break it out, to expose the underlying apogee motor. A further illustration of breakingout a pictorial drawing is Fig. 11-8. Note that in this case, various layers of the assembled motor are cut back to expose underlying ones.

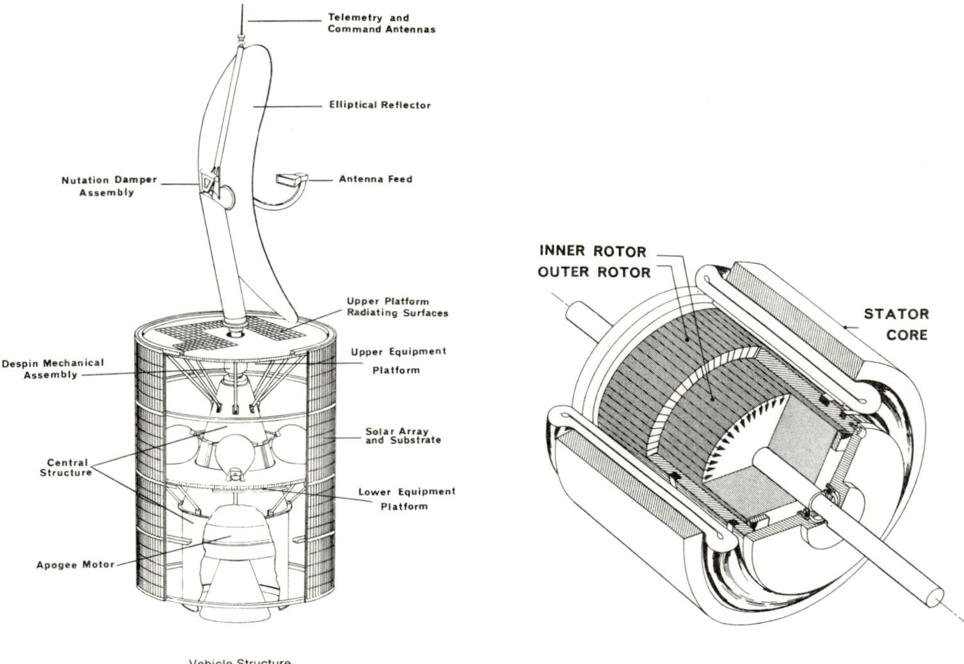

Fig. 11-7 A brokenout or cutaway pictorial drawing. (Courtesy R.C.A. Space Systems, Montreal, P.Q.)

Fig. 11-8 The principle of 'breakingout'. (Courtesy Manufacturer — Reuland Electric Company, Howell, Michigan)

Exploded pictorial drawing

An additional method of showing the internal construction of an assembly, and to show how its component parts are attached to each other is by the use of an exploded pictorial drawing. The plunger operated switch, Fig. 11-9, illustrates a type of exploded pictorial drawing. This particular drawing is not intended to show the internal construction of the switch, but rather to show the relationship of the mounting screws and the bottom mounting plate to the switch. Note the method of offsetting the two screws located on centres 'C'. This is done in the interests of clarity, and to prevent cluttering of the view.

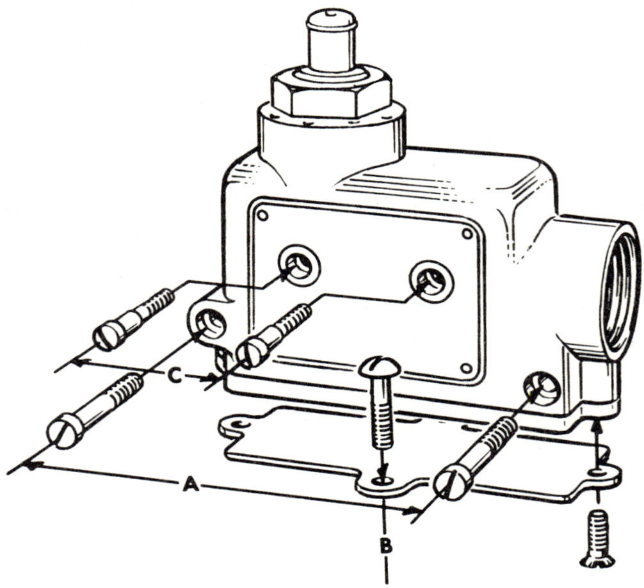

Fig. 11-9 A type of exploded pictorial drawing. (Courtesy Licon Division, Canadian ITW Limited, Don Mills, Ontario)

Figure 11-10 illustrates an additional example of an exploded pictorial drawing. View orientation is of course an important consideration. Some experimentation and trial effort are a necessary prelude to obtaining the optimum results.

If the exploded pictorial is being primarily used to show the fastening of parts, then particular care must be exercised in placing the parts and their associated fastening devices in their correct spatial relationship. This is very important.

Pictorial wiring diagram

Some comments have already been made in Unit 2 regarding this type of drawing. Figure 11-11 illustrates a portion of a circuit only. Since this type of drawing has wide application in the construction of electronic kits, it is necessary for each separate pictorial drawing to show only the leads or conductors that are involved in a particular wiring sequence. This avoids an extremely complicated and difficult to read drawing of a complete wiring installation.

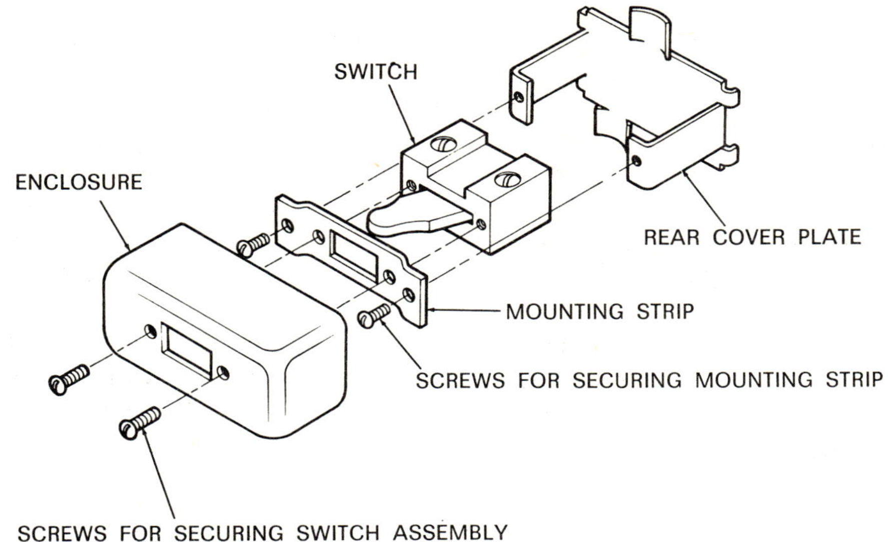

Fig. 11-10 A surface mounted S.P.S.T. switch and enclosure

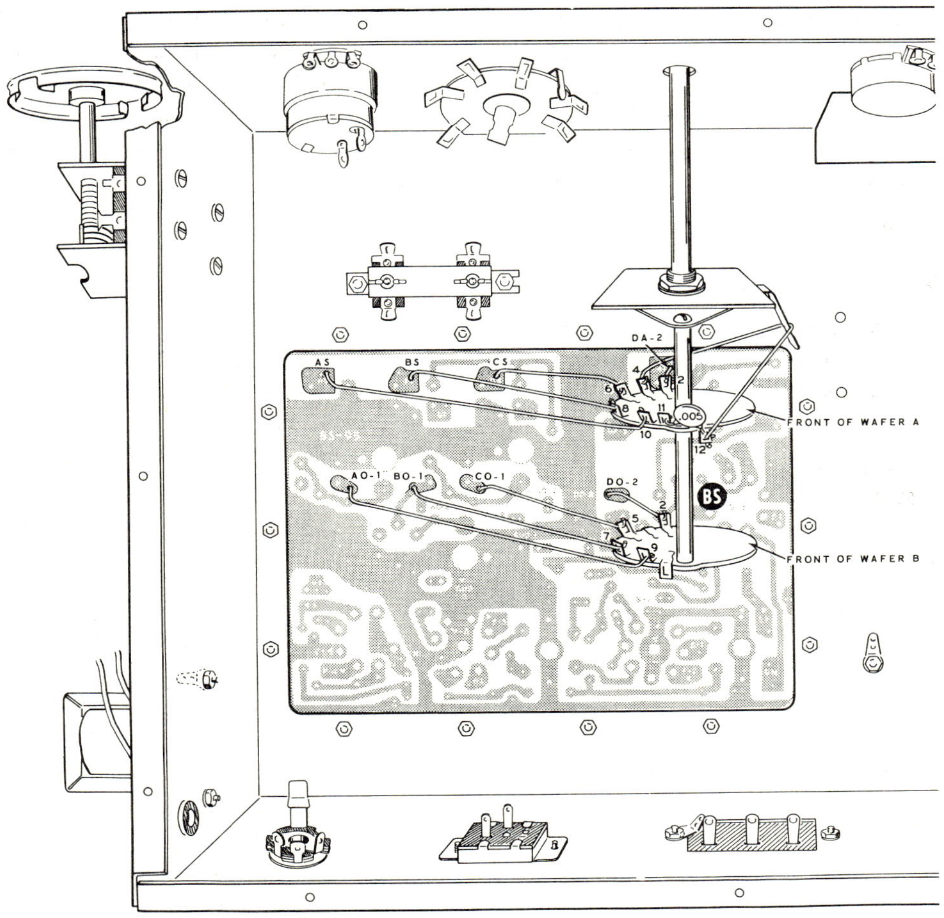

Fig. 11-11 A pictorial wiring diagram. (Courtesy Heath Company)

In the example shown, note that the actual run of the leads is indicated relative to each other and to the associated electronic components. This is very important since the prototypes that are assembled and tested under factory conditions have been done so according to the plans accompanying the kits. Hence, nonadherence to the wiring patterns illustrated is likely to result in a radio or amplifier having poor reception or tonal quality.

Shading is used to a degree on pictorial wiring diagrams. Both freehand techniques and appliqués are evident in Fig. 11-11. What is more important is the concern of emphasising the leads or conductors and their terminating connections.

Quite frequently all of the types of pictorial illustrations mentioned are given tonal effects or texture to improve their appearance. Various shading techniques are used. The following comments are intended to give a brief insight into some of them.

SHADING TECHNIQUES

Dot method of shading

A random placement of dots made freehand using India ink is sometimes used. The dots are concentrated heavily on rounded edges and corners. Thinning takes place in a progressive manner away from these edges and corners. Cylindrical surfaces are also effectively shaded by using the dot technique. Using the dot process is time consuming, nevertheless the process is simple and economical. Figure 11-12 illustrates this method. Another method is by the use of straight lines.

Line method of shading

Parallel lines drawn in ink with a ruling or technical pen and a straight edge are effectively used to produce shading. The spacing of the lines is closer at rounded edges and corners. Cylindrical surfaces are also effectively shaded by using the parallel line technique. In comparison to the dot method, the line technique is faster, and the results are no less pleasing. The pictorial drawing illustrated by Fig. 11-13 shows the line shading method.

Appliqué method of shading

A third method of shading is by the use of preprinted textured appliqués or press-on material. Attachment of the appliqué to the tracing paper or vellum is carried out by gently rubbing the surface of it. This causes the wax emulsion on its reverse side to adhere to the tracing paper. The advantages of such a medium are its uniformity of density and design. Although no drafting skill is required in its application, considerable care is required in trimming the excess appliqué to fit the outline of the part. Also, additional precaution is required after application to prevent lifting, tearing, or moving of the appliqué. Costwise, the expensiveness of the appliqué sheets is probably offset by the reduction in time required for its application. A sample illustration of a pictorial drawing using the appliqué method of shading is shown in Fig. 11-14.

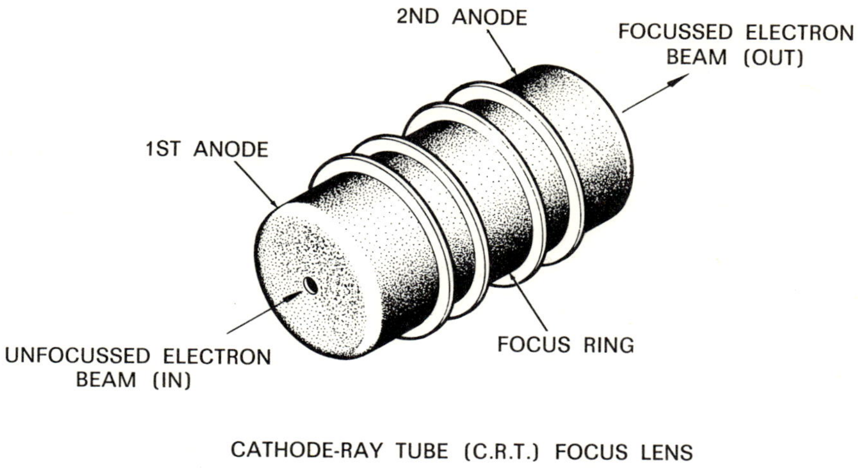

2ND ANODE
FOCUSSED ELECTRON BEAM (OUT)
1ST ANODE
UNFOCUSSED ELECTRON BEAM (IN)
FOCUS RING
CATHODE-RAY TUBE (C.R.T.) FOCUS LENS

Fig. 11-12 The dot technique of shading

6-#26 AWG STRANDED INSULATED LEADS
4-#6-32 UNC-2B THREADED HOLES
16 TEETH 96 PITCH 0.1667 P.D. 20° P.A.
115 VOLT, 50-60 HZ SERVOMOTOR

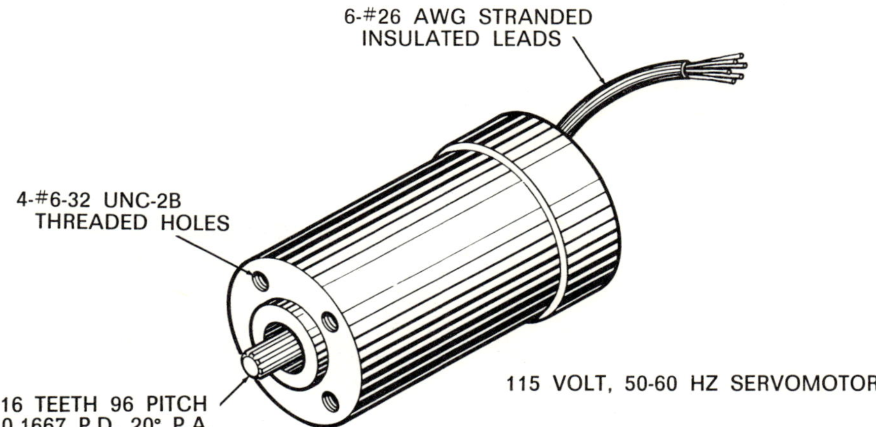

Fig. 11-13 The line method of shading

A DOUBLE-POLE CIRCUIT BREAKER

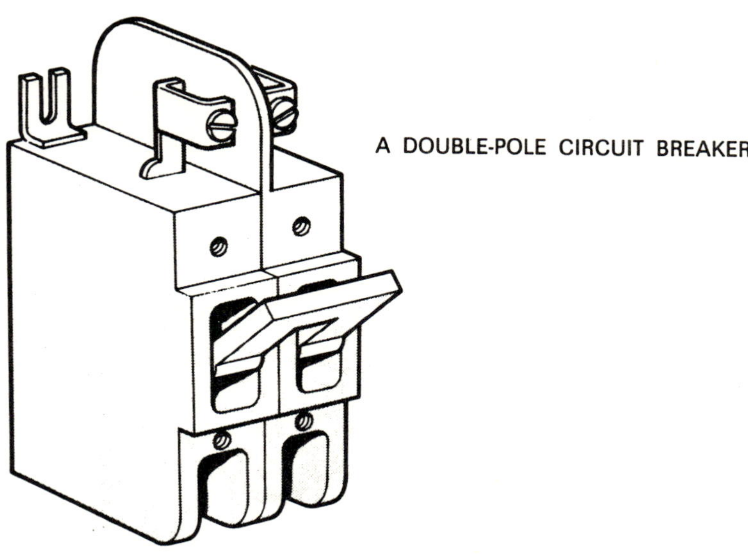

Fig. 11-14 The appliqué method of shading

Air brush shading

The fourth and final method to be presented in this unit is air brush shading. Only a few limited comments are included because this method is generally beyond the scope of most draftsmen. Essentially the equipment involves the use of a small, precision spray brush unit which is operated by air pressure stored in a steel tank. This method has wide application in technical illustration work involving both electrical, and electromechanical parts. Figure 11-15 shows a pictorial drawing shaded by means of the air brush technique. Note the closeness of resemblance of this drawing to an actual photograph.

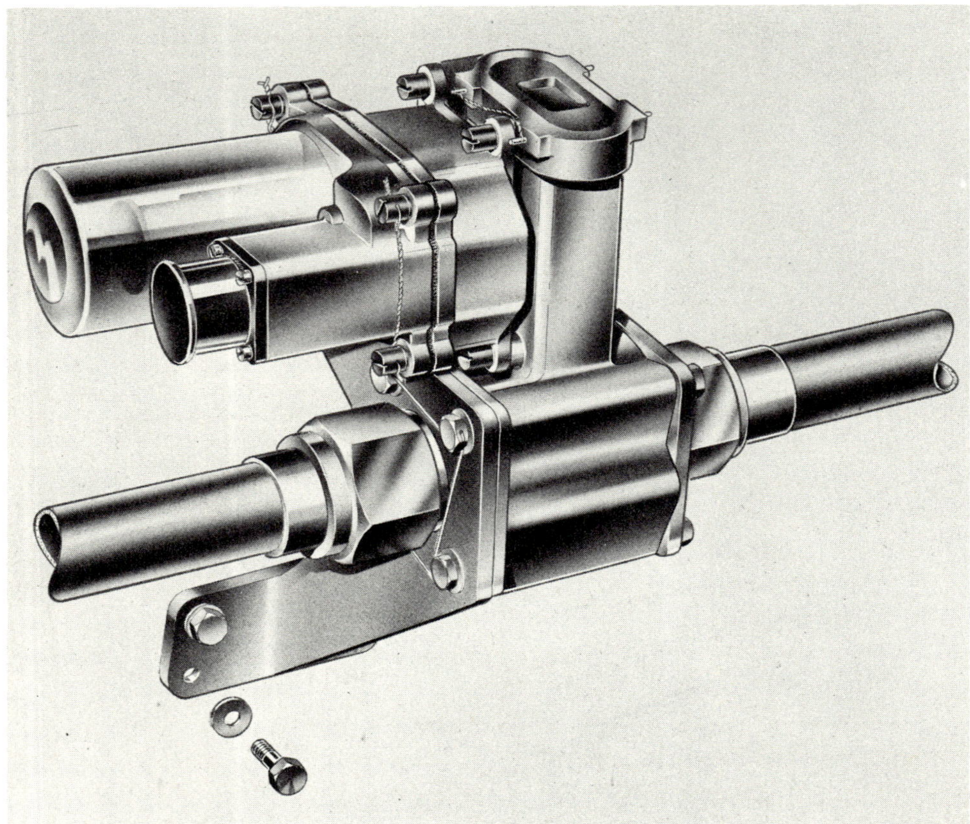

Fig. 11-15 An electronic control for a fuel line

Light and shadow

Regardless of what shading technique is used, from the simplest to the sophisticated, an awareness of light and shadow must be observed. Further information on this subject is to be found in both art and technical illustration texts.

KNOWLEDGE TESTERS

1. Give a definition to describe what is meant by a pictorial drawing of an electromechanical part.
2. List at least four different areas in which pictorial drawings are used to illustrate electrical and electromechanical parts.
3. Why are pictorial drawings an effective medium to illustrate electrical components graphically?
4. How many degrees apart are isometric axes?
5. At what angle are the oblique isometric axes to the horizontal?
6. List two drawbacks of isometric drawings.
7. In what manner are isometric drawings dimensioned?
8. What is meant by
   (a) lines of perspective?
   (b) vanishing point?
9. Compare an isometric to a dimetric drawing.
10. Describe a common way of locating the axes for a dimetric drawing.
11. List the advantages of dimetric drawings.
12. What is a 'dimetric grid'?
13. Give a reason for using brokenout pictorial drawings.
14. What style of line is used in breaking-out the external skin of an assembly?
15. When are exploded pictorial drawings used?
16. What method is sometimes used in displaying parts, especially fasteners, to avoid the cluttering of a view on an exploded pictorial drawing?
17. Why is shading used?
18. List three techniques of shading.
19. List the advantages and disadvantages for the methods or techniques given in answer to Question 18.
20. When is a pictorial wiring diagram used?
21. Why must a draftsman pay strict attention to the location of leads or conductors when drawing a pictorial wiring diagram?

PROJECTS

1. Make an isometric pictorial three-times full size of the miniature transformer shown in Fig. 11-16. Do not dimension or shade, but orient the front to the left-hand side.
2. Redraw Fig. 11-16 (as an isometric pictorial drawing) orienting the front to the right-hand side. Show major overall and mounting dimensions. Use the dot technique of shading where necessary.
3. Make an isometric pictorial drawing three-times full size of the tubular 100 MF. capacitor shown in Fig. 11-17. Shade using the parallel line technique, placing the left circular end of the capacitor to the left-hand side.
4. Make an isometric pictorial drawing of Fig. 11-18 breaking out the waveguide to show that it is hollow and rectangular in cross section. Draw full size. Do not dimension nor shade. Orient the pictorial so that the front is to the left-hand side.
5. Make an isometric pictorial drawing of the aerospace hexagon head electric squib, Fig. 11-19. Draw twice full size. Shade and orient the view to produce the optimum results.
6. Using the special scale provided, Fig. 11-20, make a dimetric drawing of the motor mount Fig. 11-21. Orient the dimetric axes according to the first method suggested in this unit. It will be necessary

to redraw the axes of Fig. 11-20 that are now full size, and extend the length axis to suit the length of the motor mount.

7. Figure 11-22 shows the schematic wiring diagram of an electromagnet circuit. Redraw this schematic as a pictorial wiring diagram. Use Unit 4, Graphic Symbols, to establish the pictorial representation for the symbols.

8. Make a schematic wiring diagram from the pictorial drawing, Fig. 11-23. This problem, while not requiring an isometric pictorial drawing for its solution, has been included to give practise in translating from the pictorial form to the schematic form including the use of graphic symbols.

9. Figure 11-24 shows orthographic views of an electromechanical assembly. Make an exploded isometric drawing of this assembly. Identify individual parts with balloons and leaders, and include a parts list.

10. Make an isometric drawing of Fig. 11-24 and shade using suitable appliqué medium. Draw to a larger scale if required.

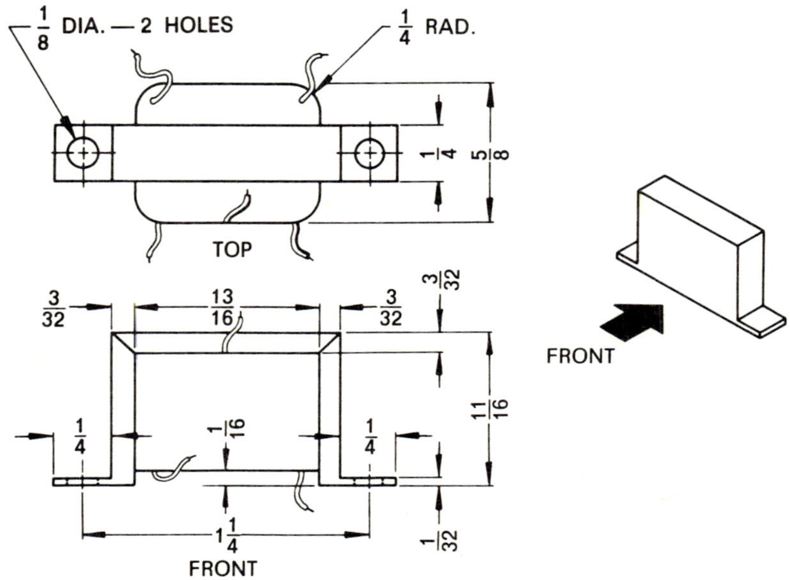

Fig. 11-16 Drawing for Projects No's. 1 and 2

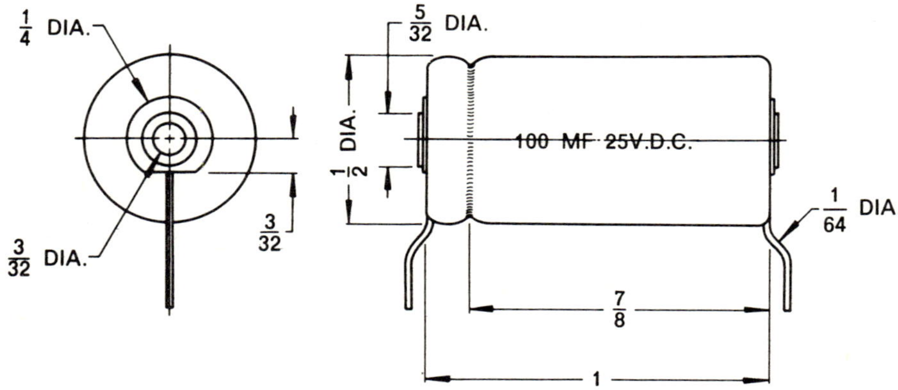

Fig. 11-17 Drawing for Project No. 3

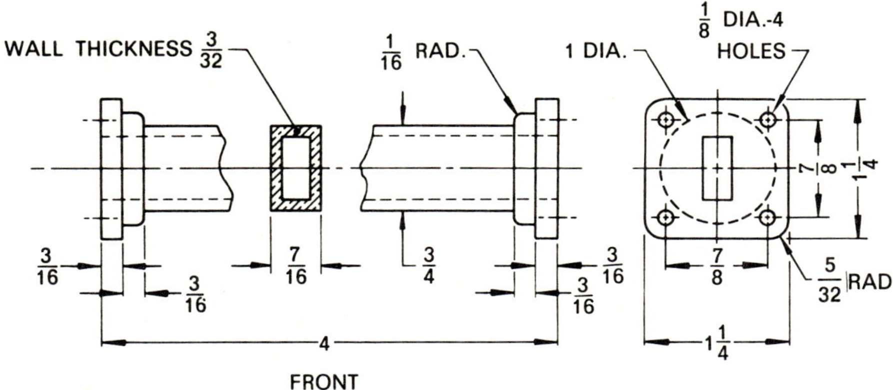

Fig. 11-18 Drawing for Project No. 4

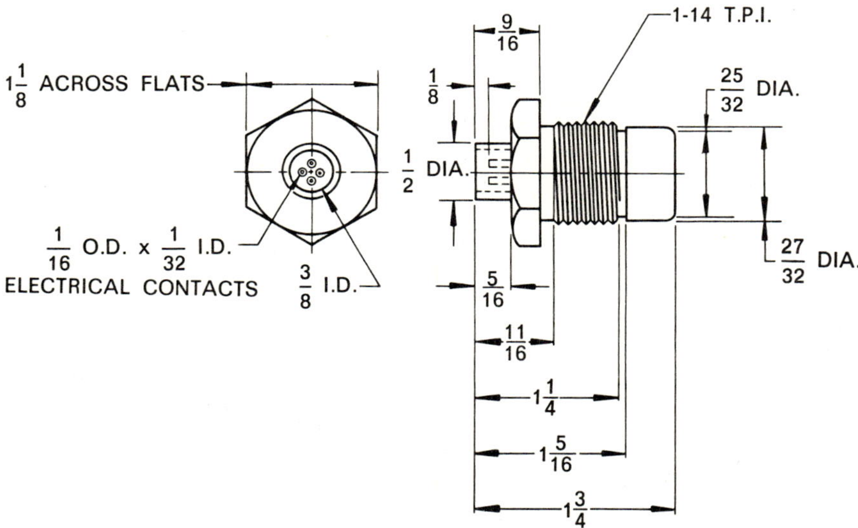

Fig. 11-19 Drawing for Project No. 5

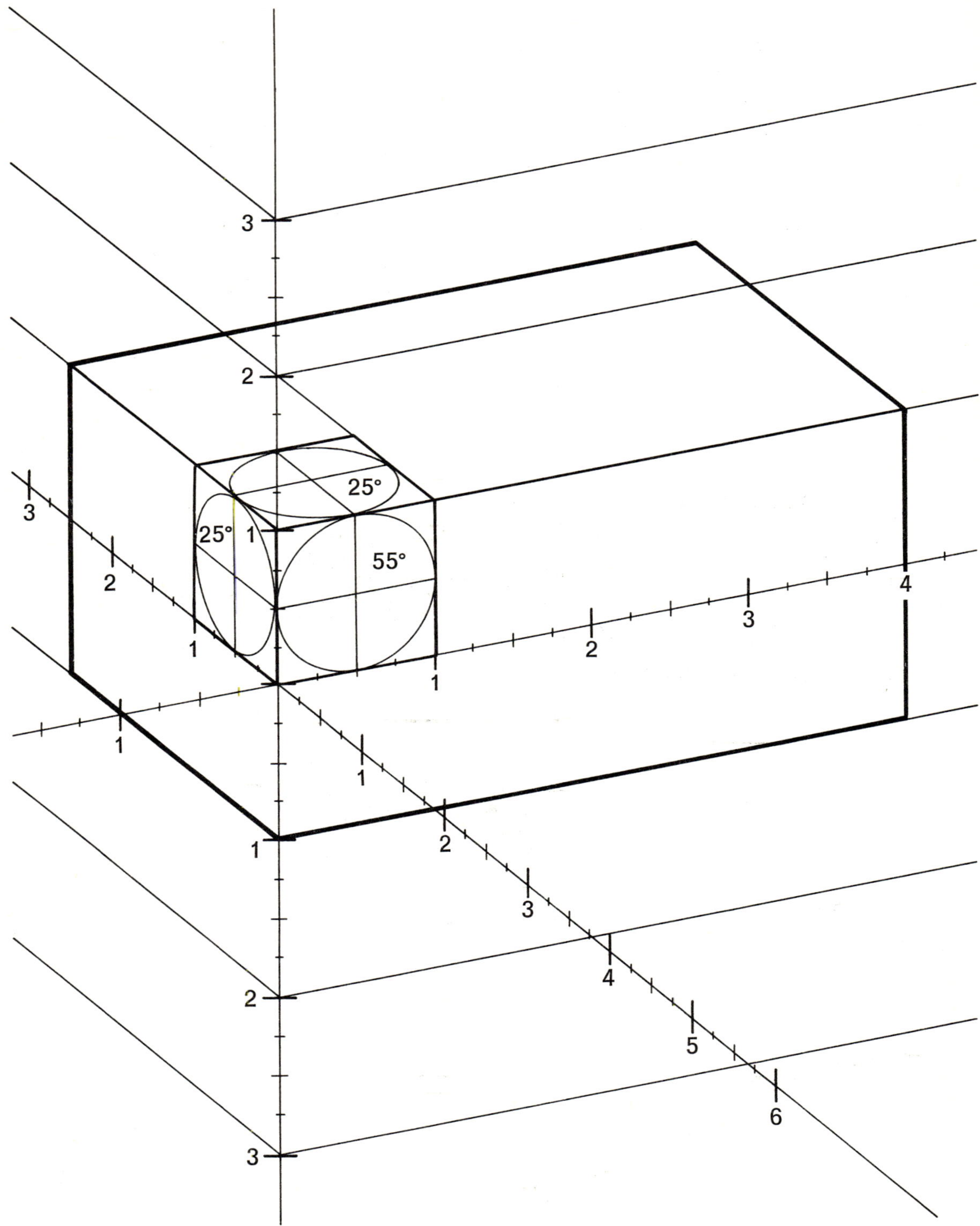

Fig. 11-20 *A dimetric scale for use with Project No. 6. Major units represent inches. Smallest units represent ¼″. If dimensions are smaller than ¼″, determine the dimetric size by approximation*

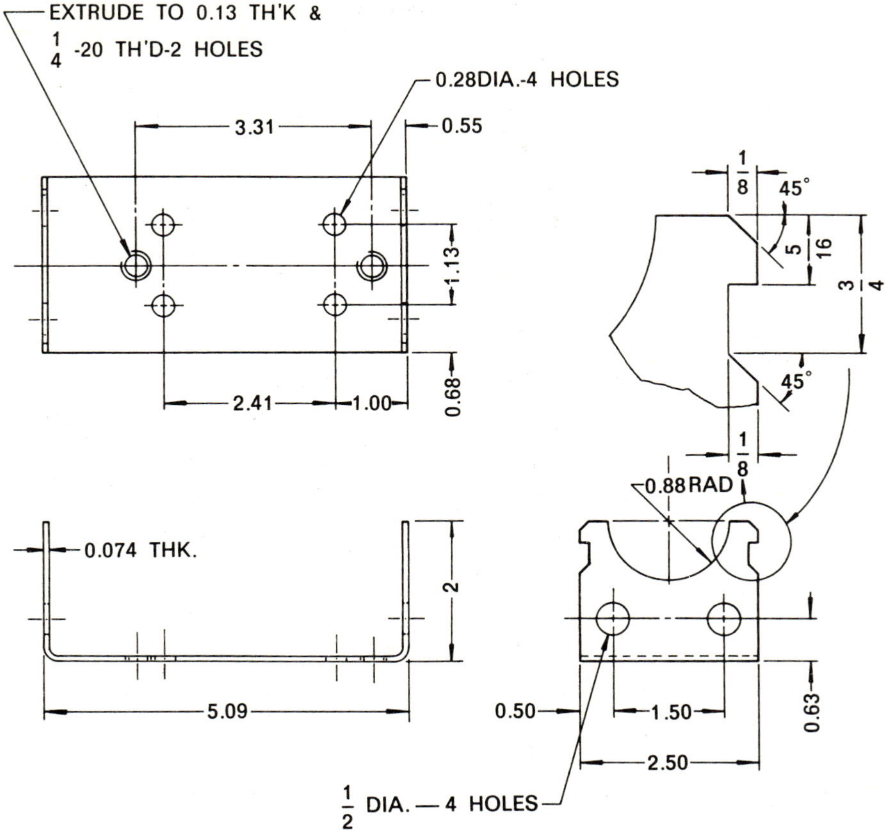

Fig. 11-22 Drawing for Project No. 7

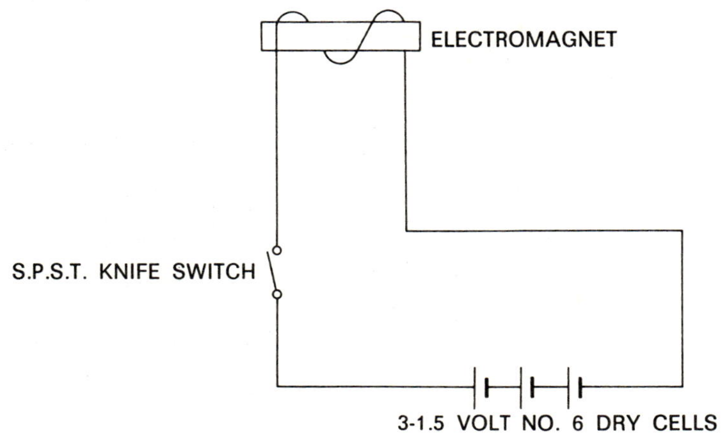

Fig. 11-21 Drawing for Project No. 6

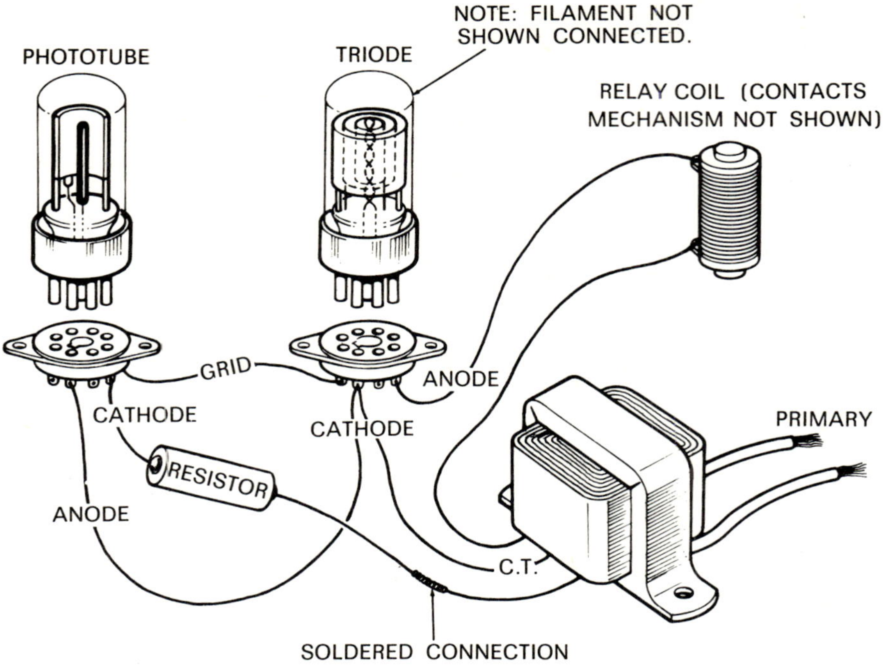

NOTE: FILAMENT NOT SHOWN CONNECTED.
PHOTOTUBE
TRIODE
RELAY COIL (CONTACTS MECHANISM NOT SHOWN)
GRID
ANODE
CATHODE
CATHODE
PRIMARY
RESISTOR
ANODE
C.T.
SOLDERED CONNECTION

PICTORIAL WIRING DIAGRAM OF A TYPICAL PHOTOTUBE CIRCUIT USING ALTERNATING - CURRENT SUPPLY

Fig. 11-23 Drawing for Project No. 8

METHYL ACRYLIC INSULATING PLATE
NOTES:
(1) RELAY CONTACTS AND CONNECTING PARTS HAVE BEEN OMITTED
(2) SCALE THE SIZES OF COMPONENTS DIRECTLY FROM THIS DRAWING.
RIVET
RIGHT-ANGLE MOUNTING BRACKET
SOLDERING LUG
SOLENOID
RIVET
PHENOLIC BASE PLATE

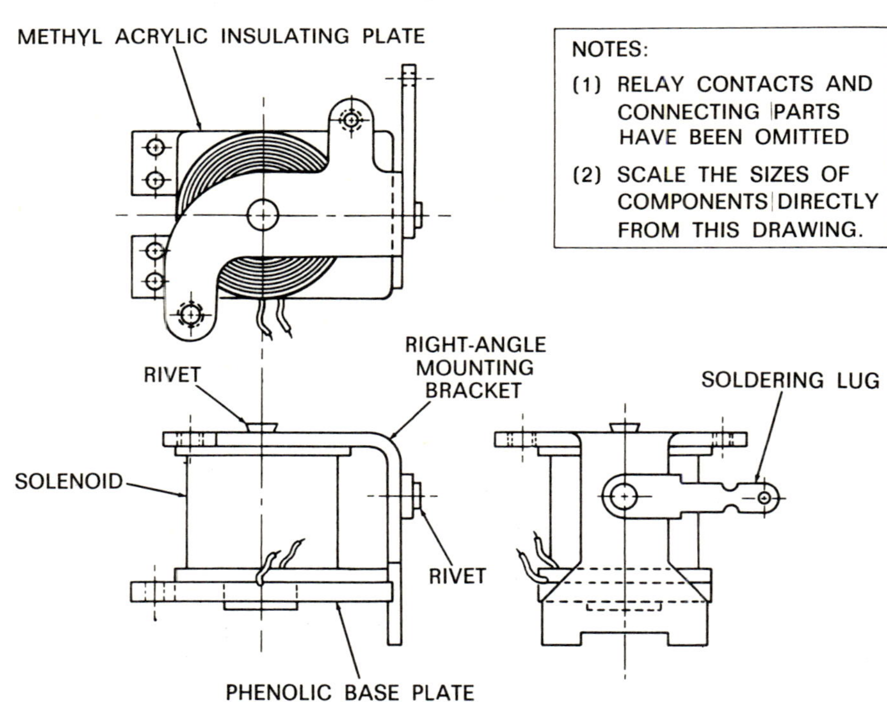

ELECTROMECHANICAL ASSEMBLY DRAWING OF A RELAY

Fig. 11-24 Drawings for Projects No's. 9 and 10

UNIT 12

Sheet Metal Drawings

Fig. 12-1 A sheet metal fabrication. (Courtesy Northern Electric Company Limited, Montreal, P.Q.)

THE USE OF SHEET METAL PARTS AND COMPONENTS IN BOTH THE electrical and electronic industries is extensive. Within the scope of this unit the concern will be largely in the areas of panels, chassis, enclosures, and mounting or supporting devices. These areas do in fact make up a considerable portion of the sheet metal work involved in the aforementioned industries. However, sheet metal fabrication also entails areas of uniqueness and sophistication, for example in communications and aerospace equipment, a **microwave dish** or **parabolic reflector**, Fig. 12-1, could be mentioned.

Drawings of sheet metal parts may involve either the use of orthographic projection or pictorial presentation. The dominant method is the orthographic one. In either case the sheet metal craftsman is required to make a **development drawing**. This drawing is a pattern in the flat before the sheet metal is bent or formed.

DEVELOPMENT METHODS

However, in different industries, the possibility exists that either the draftsman or the sheet metal worker will lay out the actual full-size pattern of the sheet metal part before it is formed to its finished shape. If it is the requirement of the draftsman to make the pattern, then he must understand the processes of development. There are three development methods. One is the **parallel line** method which is used in the development of cylindrical shaped objects. The second is known as **radial line** development. This method is used in the development of conical shaped objects. The third method is called **triangulation** and is used in the development of transition pieces. A transition piece is a sheet metal object which normally has different cross-sectional shapes at either end. Fig. 12-2 shows a viewing hood for a small P.P.I. radar console developed by this method.

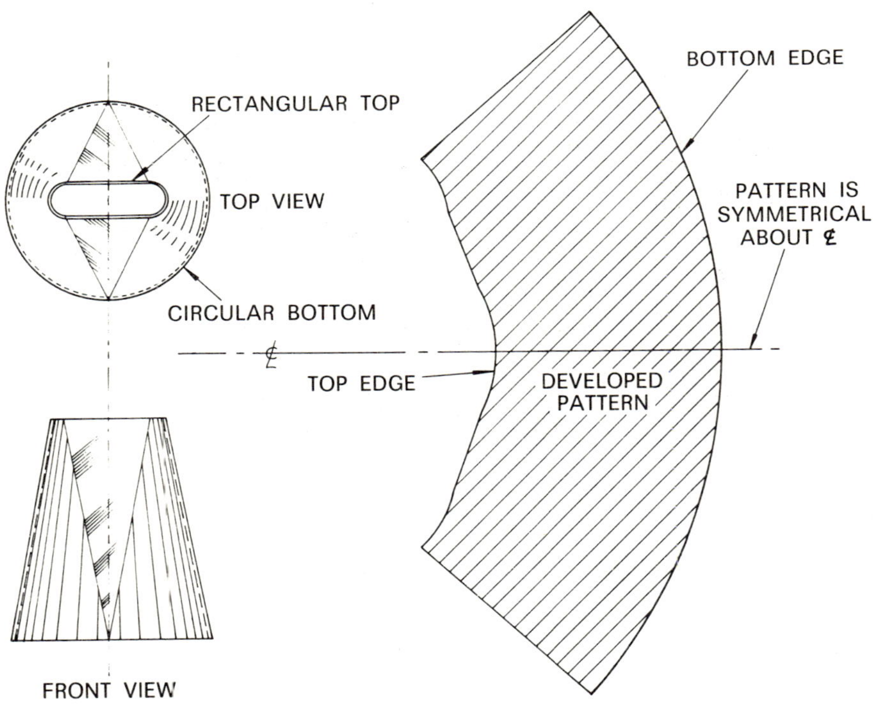

Fig. 12-2 Orthographic views and the developed pattern

Since the methods of sheet metal development are normally included in mechanical drafting texts, only brief references inclusive with illustrative examples are given.

In addition it is necessary for the draftsman to have a knowledge of the materials, fabrication processes, and joining methods.

Parallel line development

Sheet metal objects upon which a series of parallel lines can be drawn are capable of having patterns developed for them. However, in establishing the pattern for a service box or an enclosure cover, it is not necessary to use the parallel line method. Since the orthographic views of the sheet metal part are assumed to be available, the draftsman's ability to visualize what the object will appear like in the flat is the essential requirement. Fig. 12-3 shows the normal orthographic views of a blank radio receiver chassis. The pattern illustrated by Fig. 12-4 shows the same radio receiver chassis previously shown by orthographic views in Fig. 12-3. Note that the **fold** or **bend** lines are shown by means

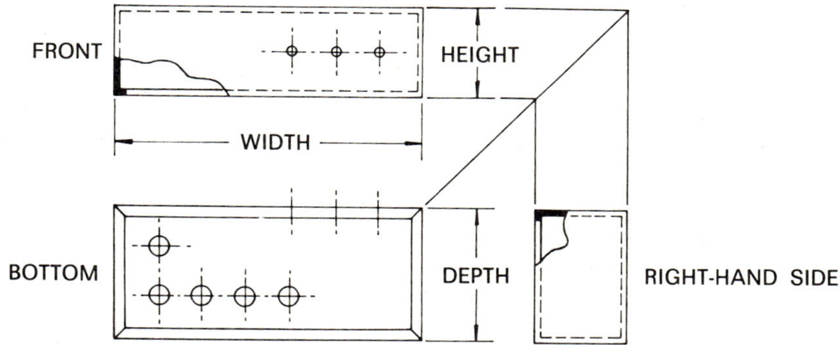

Fig. 12-3 Orthographic views of a radio receiver chassis

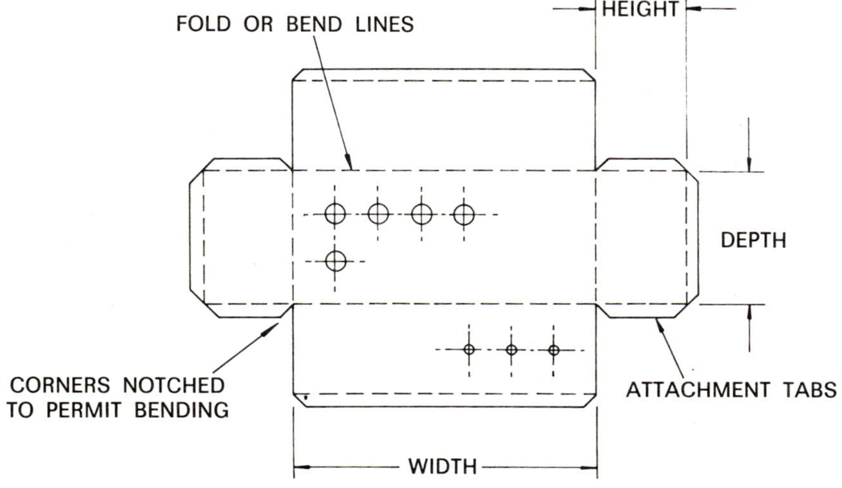

Fig. 12-4 The pattern for the chassis in Fig. 12-3

of broken lines. An important consideration is the notching of inside corners to permit the longer sides to be bent up 90° without interference. Also, **tabs** for spot welding the corners are shown on the top and bottom of the left and right sides.

The pattern for a hollow sheet metal cylinder can be simply developed by the parallel line method. However, the likelihood is that such a device would be **stamped** or **drawn** using dies. Therefore such a pattern would not normally be required. Fig. 12-5(a) shows a two-view orthographic drawing of a split, open-end, vacuum tube R.F. shield, and Fig. 12-6 shows the developed pattern of the same device.

Radial line development

The pattern for a conically shaped sheet metal device is constructed by using a method known as radial line development. For example, the sheet metal lamp shade is for a living room pole lamp. In order to establish the pattern, suitable orthographic views are required. Fig. 12-7 illustrates the top and front views of such a shade, and the developed pattern.

Triangulation

If the lengths of two sides of a right-angled triangle are known, the true length of the third side is the line joining the other two sides. The use of this principle is known as **triangulation**. By using this principle, the true lengths of

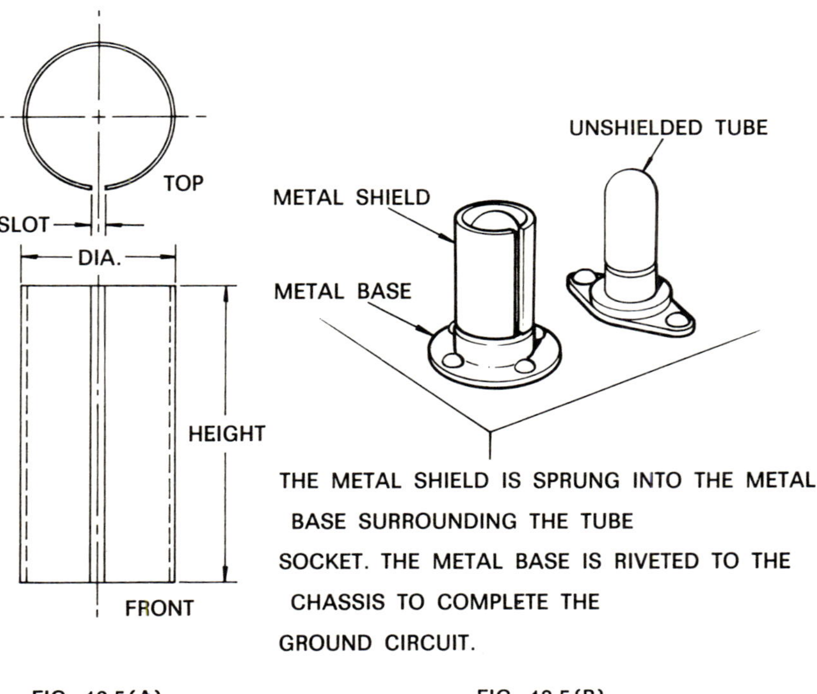

Figs. 12-5(a) and (b) Orthographic and pictorial views

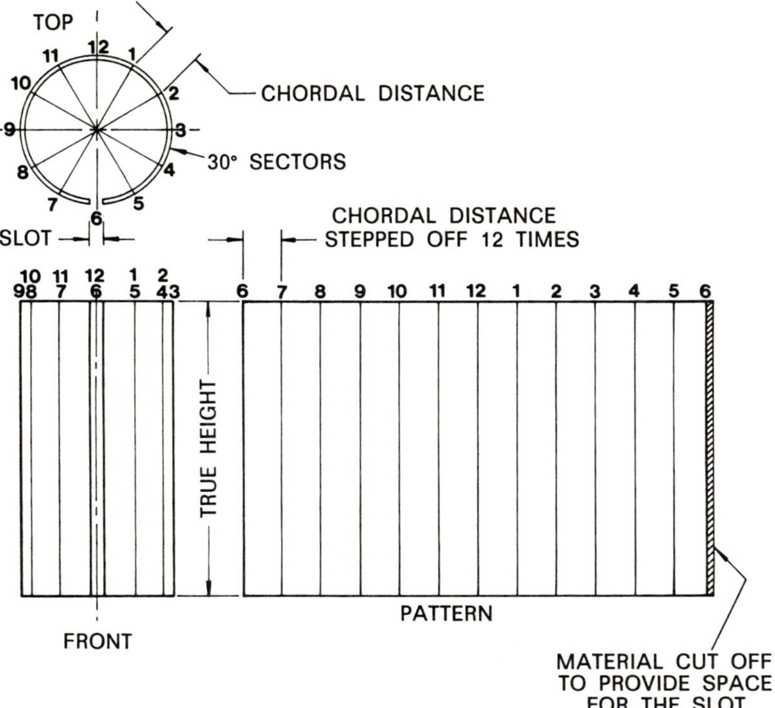

Fig. 12-6 The pattern for the shield in Fig. 12-5

elements or straight lines drawn at equal angular distances on the orthographic views of a sheet metal part can be determined. These in turn can be used to develop the pattern. Fig. 12-8 shows the necessary orthographic views and the pattern developed by triangulation of a viewing hood for a **plan position indicator, P.P.I., radar.**

Triangulation is used more commonly to develop the surface of an object on which a square or rectangular base is reduced to a circular top with a smaller area. An example is a rectangular sheet metal air duct reduced to a circular duct of smaller area. This sheet metal part is used in a computer cooling system, and is known as the transition piece.

Methods of fabrication

A basic knowledge of fabrication methods is required by the draftsman, so that he does not make designs that are impossible, or at the best, exceedingly difficult to manufacture. These methods involve cutting, bending, forming, stamping and punching.

For example, it is pointless for a draftsman to locate a punched hole in such a position that it is deformed when the flat material is bent 90°. See Fig. 12-9. In addition an understanding of the limitations of the shop equipment is an essential prerequisite to making sound designs. Also, the draftsman must be conversant with other operations such as drilling, tapping, spot welding, and brazing. In respect to the materials used, the draftsman must be knowledgeable of their characteristics.

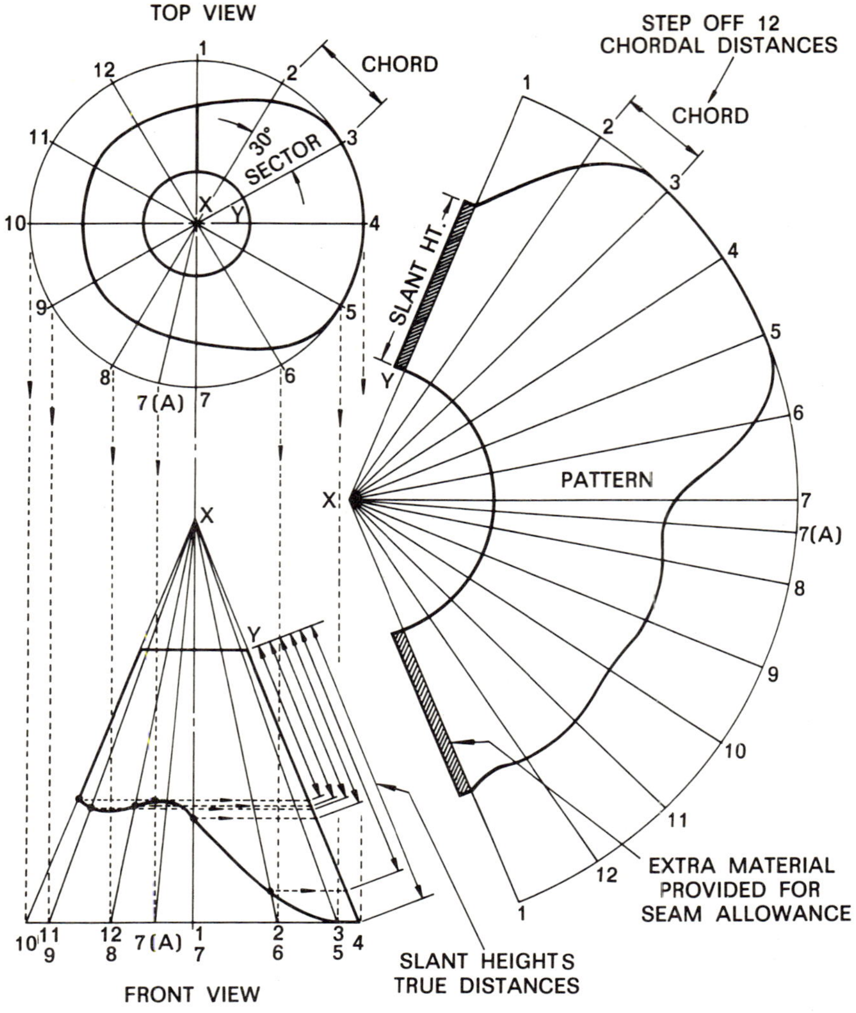

TOP VIEW
CHORD
30° SECTOR
X
Y
STEP OFF 12 CHORDAL DISTANCES
CHORD
SLANT HT.
PATTERN
7(A)
EXTRA MATERIAL PROVIDED FOR SEAM ALLOWANCE
SLANT HEIGHTS TRUE DISTANCES
FRONT VIEW

Fig. 12-7 The method of radial line development

IDEAL LOCATION OF CENTRES FOR HOLES
IMPROPERLY LOCATED CENTRES
IMPROPERLY LOCATED HOLES ARE DEFORMED
WHEN THE MATERIAL IS BENT

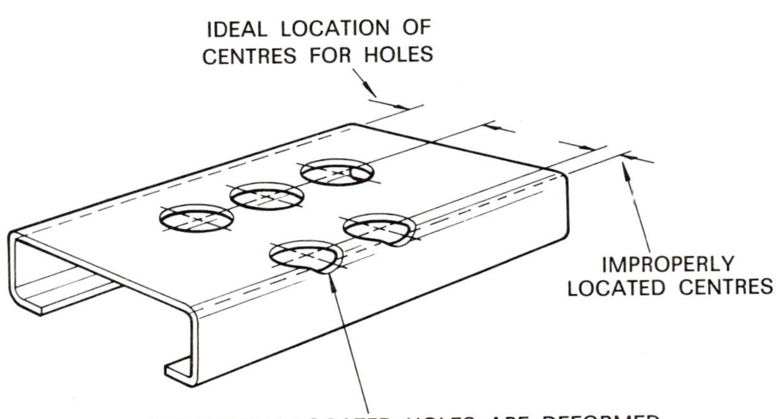

Fig. 12-9 Correct and incorrect design layout

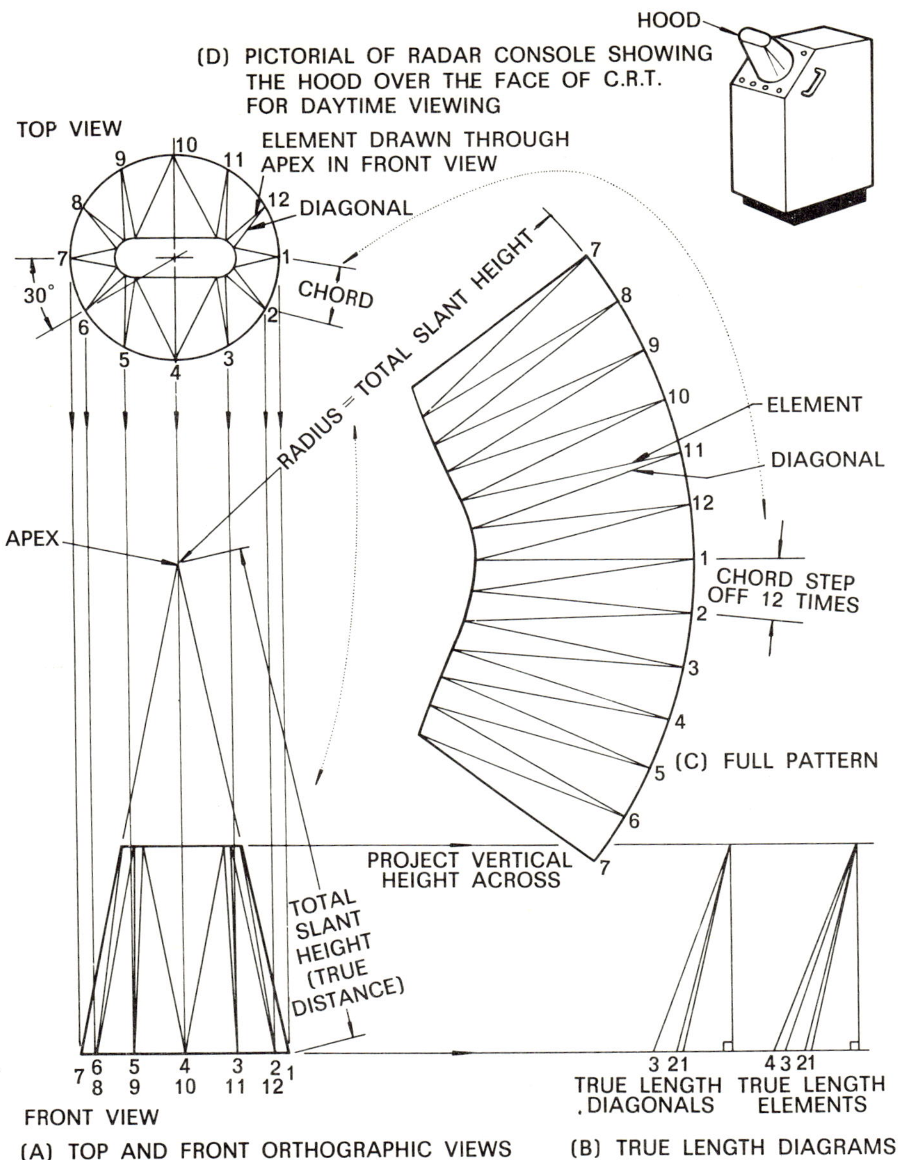

Fig. 12-8 Development by triangulation
NOTE
1. Due to the symmetrical shape of the hood, only three true length diagonals are required.
2. The intersections of the diagonals and the semicircles in the top view are made at 30° intervals.

Materials

Sheet steel of varing thickness is used for enclosures, panels, covers, chassis, and supporting brackets. Its widespread use is dictated by the fact that it is economical, easy to form, has good strength characteristics, and can be 'finished' in many ways to prevent rusting.

Other metallic materials in sheet form also have wide use, in particular aluminum alloy. The advantages of this alloy are its lightness, relative high strength, and its resistance to **oxidation** or rusting.

Hence, when a draftsman makes reference to a specific type of sheet metal with a particular gauge of thickness, he must do so with forethought because the different metallic alloys used have individual characteristics. For example, copper **alloys** are more **ductile** than steel. Certain alloys of aluminum, while exceedingly light and strong cannot be bent sharply without fracturing taking place. Since this is the case, a knowledge of minimum bending radii for different materials of varying thickness is required. A chart outlining bending radii is illustrated in Appendix 9.

An indication has already been given regarding the different sheet metal parts that are fabricated. It is necessary to outline them more specifically and in some detail.

ENCLOSURES AND CHASSIS

Starter Enclosures

Enclosures are manufactured in a number of sizes and styles. Some are square in shape, others rectangular. The enclosure covers are either constructed with sharp 90° corners or are fully rounded. **Conduit knockouts** and prepunched holes for push buttons are standard features. Figs. 12-10 and 12-11 illustrate two styles of enclosures and their accompanying covers.

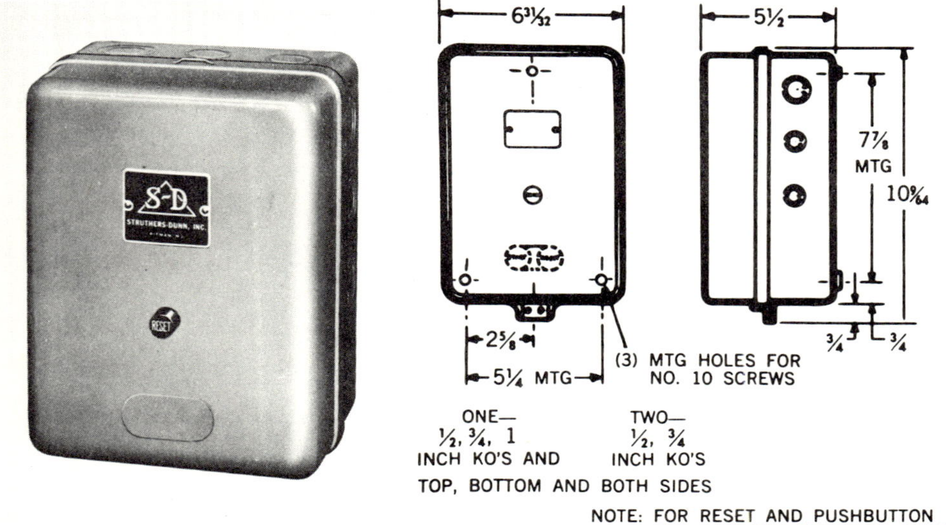

Fig. 12-10(a) *A motor starter enclosure with fully rounded edges and corners, and push-button blanking plate*

Fig. 12-10(b) *An enclosure for a reversing A.C. magnetic starter and contractor*

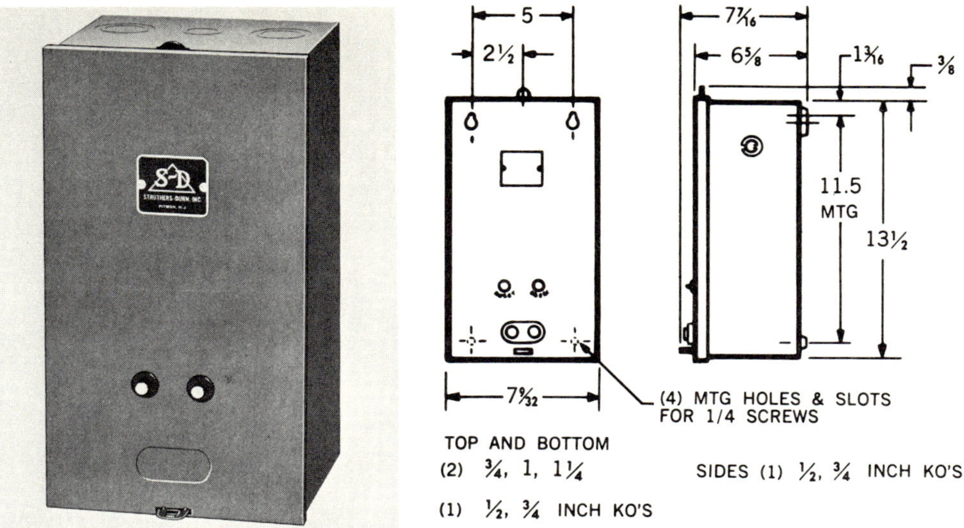

Fig. 12-11(a) A rectangular shaped enclosure with 90° edges and corners. Note the conduit knockouts

Fig. 12-11(b) The orthographic views of the enclosure in Fig. 12-11(a)

Figs. 12-10 and 12-11 (Courtesy Struthers-Dunn, Inc., Pelman, New Jersey)

Service equipment enclosures

Enclosures for service equipment units are designed and manufactured for a range of electrical current ratings. Accordingly the physical size of a 50 ampere unit is considerably smaller than a 100 ampere one. In addition, welded corners and hinged waterproof covers are design features of outdoor units. Certain parts of enclosures, such as an internally threaded top section, are not entirely sheet metal fabrications. The diagram shown in Fig. 12-12 illustrates a typical outdoor service equipment enclosure. Fig. 12-13(a) illustrates a 100 ampere indoor service control equipment unit. Note that the cover is detachable and is designed to fit neatly over the enclosure. It also has a **cutout** for the circuit breaker handles. In contrast to this type of design, other enclosure units are flush mounted. These units have a cover which overhangs the enclosure by approximately a half inch. See Fig. 12-13(b).

Modular enclosures

Some manufacturers produce different enclosure units and wiring troughs that because of their standardized dimensions can be fitted into a vertical rack. This is known as a **modular** system, because each unit's dimensions are identical, or a multiple of the basic. Fig. 12-14 illustrates a typical example.

Electronic enclosures and chassis

Due to the extremely wide variety of sheet metal enclosures and chassis used in the electronic industry it is impossible to discuss them in detail. In addition, standardization of sizes and shapes is not possible, since each manufacturing firm custom designs and fabricates for its own requirements. Light and medium gauge steel and aluminum alloys are used for both enclosures and

Fig. 12-12 *A 50 ampere outdoor type service enclosure. Note the hinged weather cover and the padlock hasp*

Fig. 12-13(a) *A 100 ampere surface mounted indoor service enclosure. Note the open corners and knockouts on the box, and the cutout in the cover for the circuit breaker handle*

Fig. 12-13(b) *A flush mounted indoor service enclosure. Note the overhang on the cover for mounting*

Fig. 12-12 and 12-13 *(Courtesy The Heinemann Electric Co., Trenton, New Jersey)*

Fig. 12-14 A modular rack assembly. (Courtesy Allen-Bradley Canada Limited)

chassis. Radio and television chassis have such a conglomerate of components mounted on them in diverse but specific arrangements that they require ingenuity in layout. Mounting space is usually at a premium and this compounds the difficulty of layout. Also, the electrical peculiarities of one type of component may demand its close separation from another, again causing additional layout problems.

Sometimes an enclosure serves a dual purpose. Besides its designed use it is also used for mounting components on it. In television sets the tuner assembly, Fig. 12-15, is manufactured in this fashion. Note the number and variety of hole shapes punched in this **enclosure/chassis.**

Mounting or supporting brackets and plates

An almost endless number of varieties of mounting and supporting brackets are used. The design of a bracket is indicated by a number of factors. These include the function of the unit or units mounted on it, the shape and mounting arrangement of the unit, and its weight. Aesthetics or the appearance of the bracket may also have a bearing on its design.

Fig. 12-15 A television receiver, tuner unit 'enclosure/chassis'. (Courtesy Standard Kollsman Industries, Inc.)

Frequently, it is expedient by virtue of cost to make two sheet metal brackets, one mounted to another, rather than make a more complex single design. In Fig. 12-16 the motor unit is mounted to a bracket, which in turn is mounted to the end plate or bracket of the assembly. Note how the corners are **relieved** on the motor bracket, thus permitting a sharp, clean 90° bend.

In the interests of cost, corners of brackets are seldom made with a radius. If a sharp corner presents a hazard however, then it is rounded off. The only other reason for doing so is in the interests of appearance. This is apparent in Fig. 12-16.

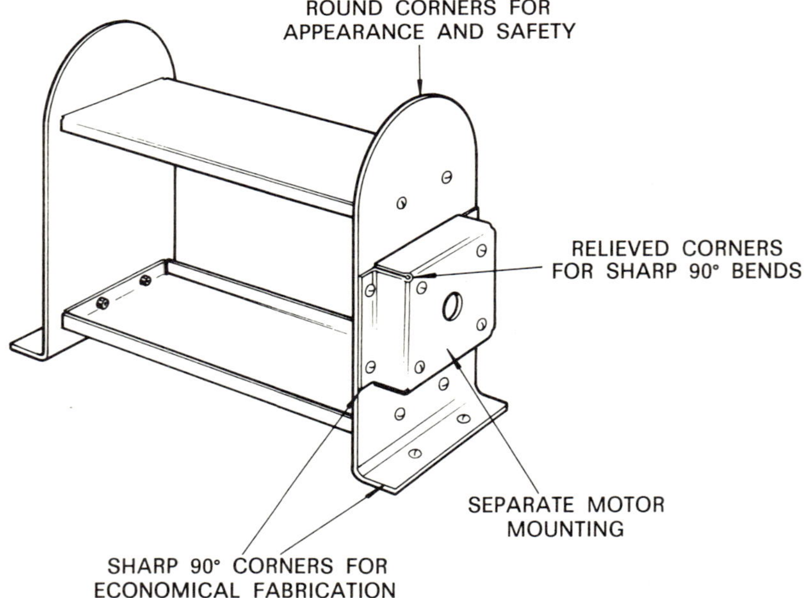

Fig. 12-16 An industrial timer mounting illustrating economical, sound sheet metal practices

Motor mounts fabricated from sheet steel heavier than #16 gauge are frequently used for supporting small fractional horsepower motors. The drawing shown in Fig. 12-17 is typical.

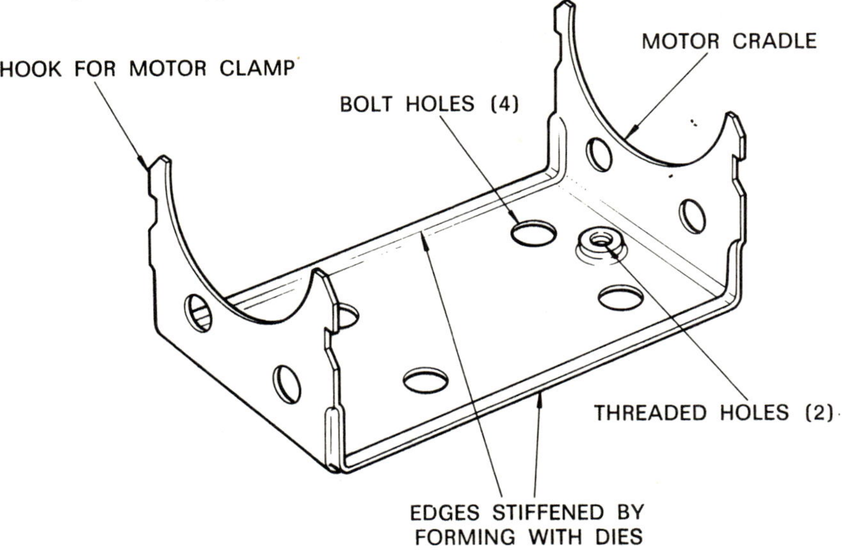

Fig. 12-17 A motor mounting fabricated from No. 14.GA (0.078) U.S.S.G. cold-rolled sheet steel

A method used to provide either mounting feet or a cavity to accept an extended part on a flat base plate is shown in Fig. 12-18. The four depressions and the larger centrally located horizontal crown have been made by **bumping** the sheet metal with suitable dies. In the case of the feet, this eliminates the

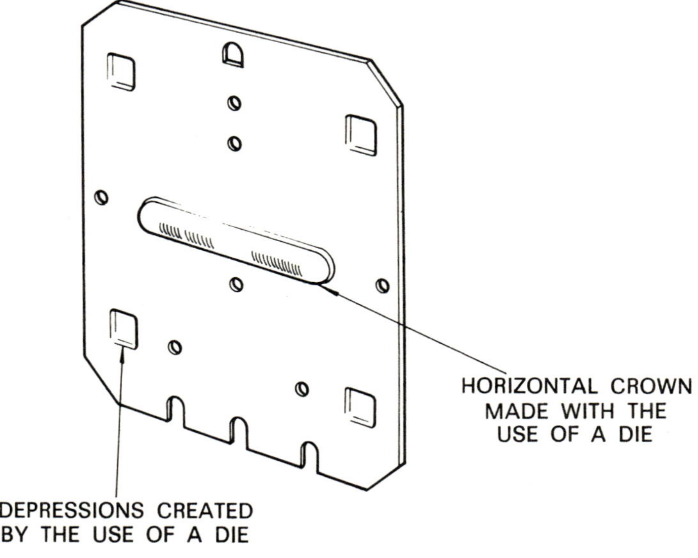

Fig. 12-18 The base plate of a circuit breaker illustrating depressions and crown formed with the use of dies

addition of separate pieces for the same purpose. Flat sheet metal plates are sometimes treated in this same manner to produce **integral stiffening ribs**. See Fig. 12-19.

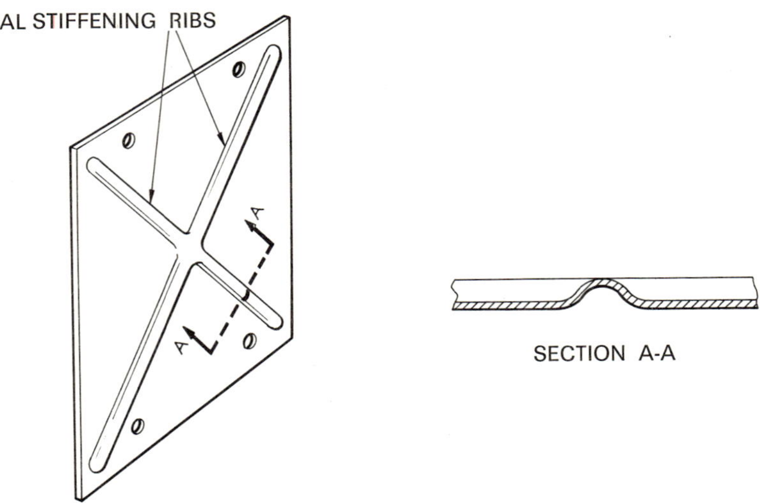

Fig. 12-19 The concept of integral stiffening ribs

KNOWLEDGE TESTERS

1. In simple terms describe what a development drawing is.
2. Name the drafting techniques that are used to develop the patterns for the following parts:
   (a) A hollow, thin-walled sheet metal cylinder.
   (b) A sheet metal funnel.
3. Describe briefly the principle of triangulation.
4. List at least three other areas in addition to developing patterns, in which the draftsman must be knowledgeable.
5. How are fold or bend lines delineated or shown on the pattern drawing?
6. What is the purpose of notching the inside corners on the pattern for a radio chassis?
7. Why are tabs included in the pattern for a chassis?
8. What method can be used to join the corners of a sheet steel chassis?
9. Name a production technique used to produce parts that have the shape of a hollow cylinder.
10. Why must the draftsman have a good knowledge of sheet metal fabrication processes?
11. What is the result of locating a punched hole too close to a bent or folded edge?
12. Give at least four reasons why steel is a suitable material for sheet metal fabrication.
13. What does oxidation mean?
14. Why is aluminum alloy a good material for sheet metal fabrication?
15. Why is aluminum not used in its pure state for sheet metal fabrication? Consult some reference sources to establish why it is used as an alloy.
16. How do the characteristics of different metal alloys of varying thicknesses or gauges relate to their ability to bend properly?
17. Suggest a method to indicate how a starter enclosure having all corners rounded could be fabricated.

18. What standard features are to be found on a starter enclosure? Name their purposes.
19. What criterion establishes the physical size of a service equipment enclosure?
20. Describe the two basic types of service equipment enclosures with reference to their mounting characteristics.
21. Briefly describe the concept of modular design as it pertains to enclosures and wiring troughs.
22. What reasons prevent the standardization of radio and television chassis?
23. Why is it not possible to mount all radio or television components on a chassis extremely close together but not touching each other. Your answer should contain at least three major reasons. Discuss this question with your electronic teacher.
24. Give at least four different features that must be taken into account when designing a mounting bracket.
25. What is the purpose of relieving the corners of a sheet metal part that has sharply bent 90° corners?
26. Why is it desirable to make a complicated mounting bracket from two parts, rather than from one?
27. What process is frequently used to produce depressions and stiffening ribs on sheet metal parts?
28. Besides the obvious saving in costs obtained by using the process known as bumping, what other advantages are gained?

PROJECTS

1. Make a pattern for the radio receiver chassis shown in Fig. 12-20. Dimension the pattern properly.
2. Make patterns of the enclosure, enclosure cover, and cover gusset, Fig. 12-21. Dimension the patterns in the correct fashion. Note that allowances for clearance must be made between the cover and the enclosure.
3. Draw the necessary patterns required for the construction of the education television receiver cabinet, Fig. 12-22. Dimension the pattern properly. Include a removable rear cover that is held in place with four #8 self-tapping screws. Use a drawing scale suitable for the sheet size available.
4. Draw the pattern for the flat elliptically shaped parabolically curved satellite antenna reflector, Fig. 12-23. Show only maximum dimensions of width and height.
5. Draw the pattern for the permanent magnet speaker conically shaped frame, Fig. 12-24. Add the necessary dimensions. A more suitable way to manufacture this part would be by stamping with the use of dies. This method would eliminate the requirement of a welded seam.
6. Draw and dimension the patterns for the cathode-ray tube housing and mounting bracket of the air-borne Loran receiver shown in Fig. 12-25. Include the necessary dimensions.
7. Draw the pattern for the electronic load cell housing shown in Fig. 12-26. Make the pattern one and one-half times full size and include sufficient dimensions for laying out.
8. Draw the pattern of the hood for an electrically operated ventilation fan, Fig. 12-27. Select a scale suitable for the size of drawing sheet available. Include the necessary dimensions.
9. Figure 12-28 shows a two-view orthographic drawing of an electric lamp shade. Draw the patterns of the shade and mounting channel. Use a suitable scale and include the necessary dimensions.
10. Draw the pattern for the sheet metal base of the high-intensity lamp shown in Fig. 12-29. Use a suitable drawing scale and dimension properly. Allow extra material for seam allowances where required. The base is to be chromium plated after fabrication. Select a suitable material for its construction.

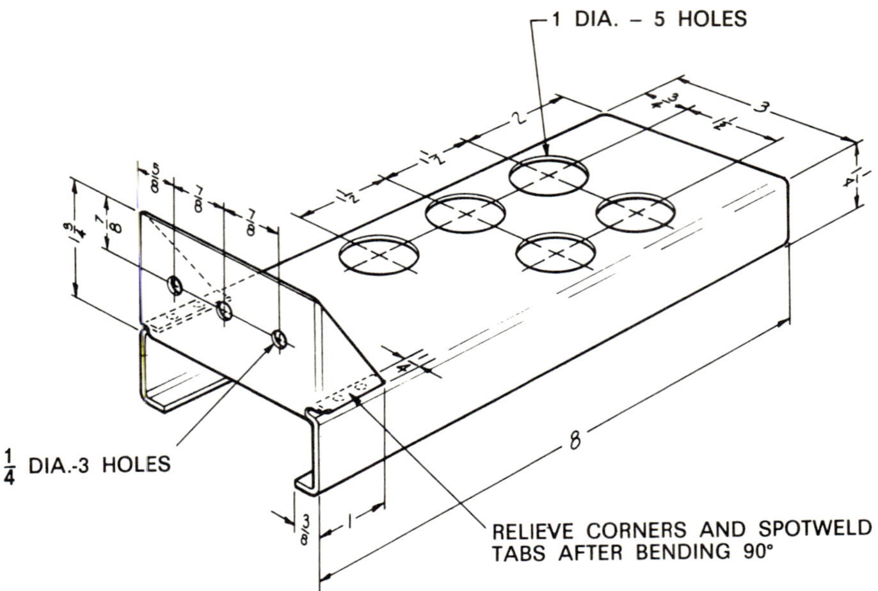

MATERIAL: 0.0478 GA. U.S.S.G.

C.R. SHEET STEEL

NOTE: USE CORRECT BEND RADII TO SUIT

Fig. 12-20 The drawing for Project No. 1

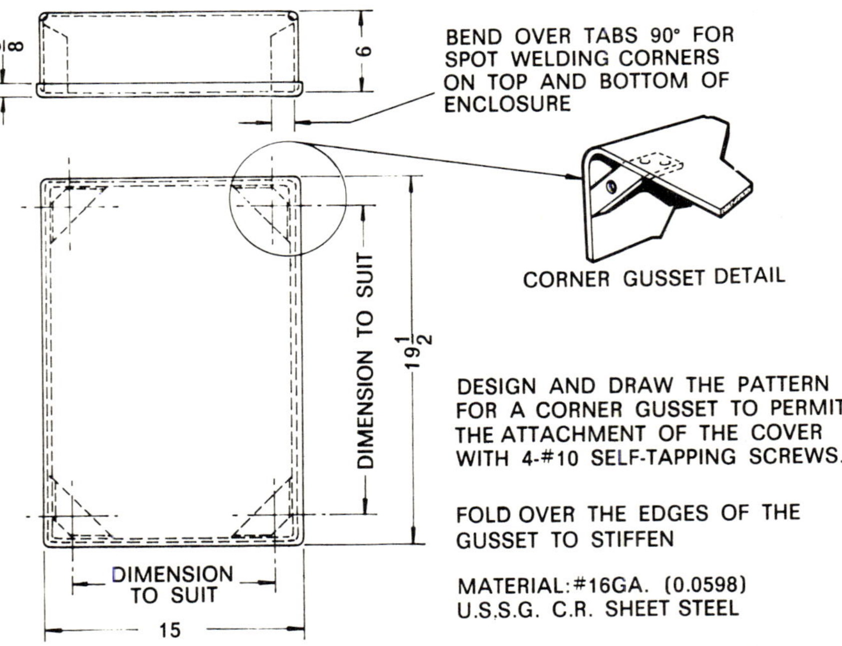

DESIGN AND DRAW THE PATTERN FOR A CORNER GUSSET TO PERMIT THE ATTACHMENT OF THE COVER WITH 4-#10 SELF-TAPPING SCREWS.

FOLD OVER THE EDGES OF THE GUSSET TO STIFFEN

MATERIAL: #16GA. (0.0598)
U.S.S.G. C.R. SHEET STEEL

Fig. 12-21 The drawing for Project No. 2

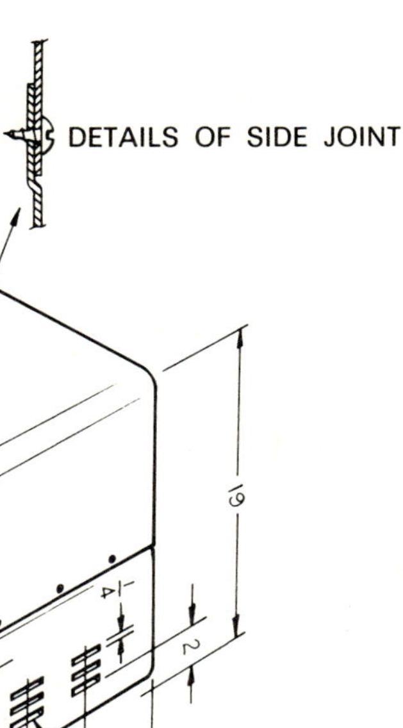

Fig. 12-22 The drawing for Project No. 3

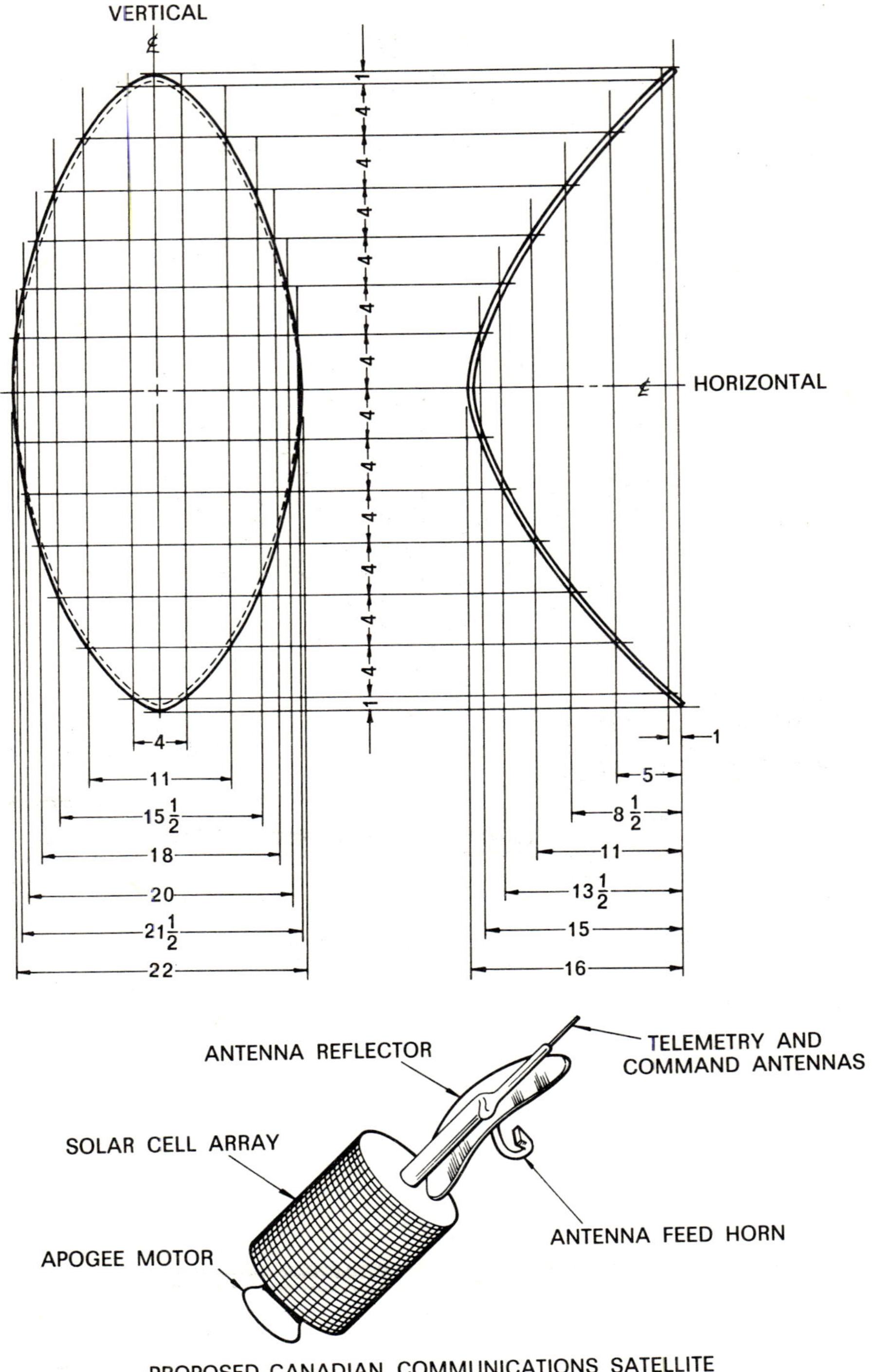

Fig. 12-23 The drawing for Project No. 4

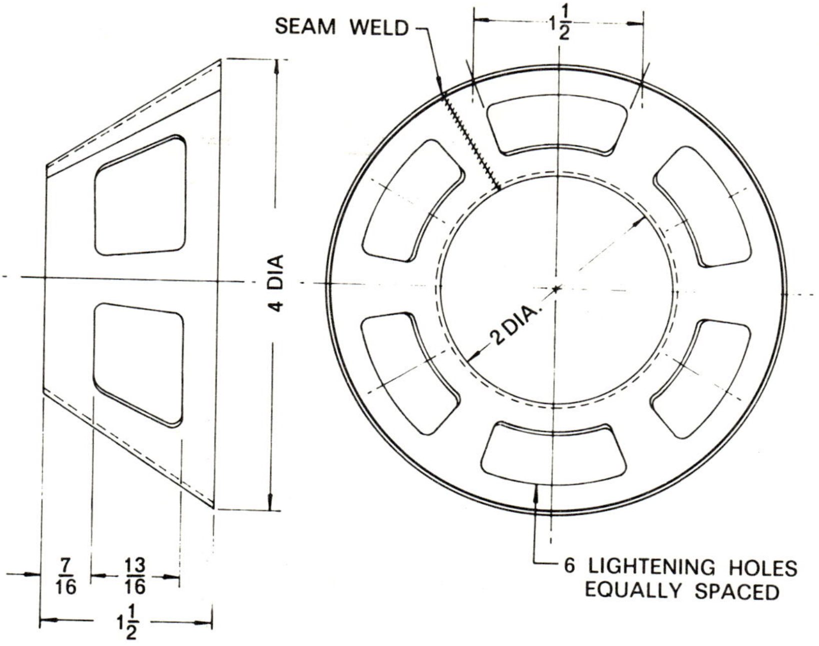

Fig. 12-24 The drawing for Project No. 5

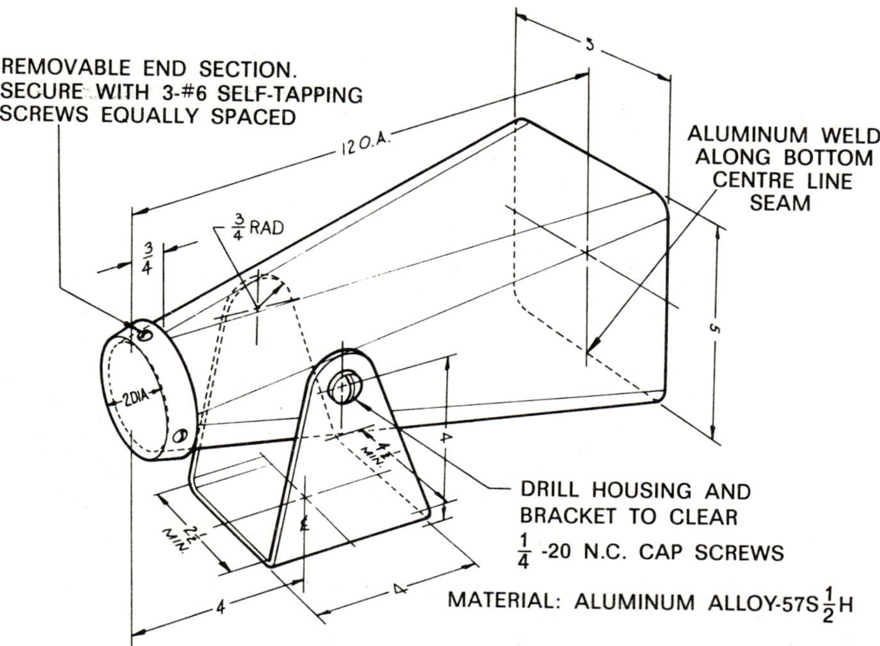

Fig. 12-25 The drawing for Project No. 6

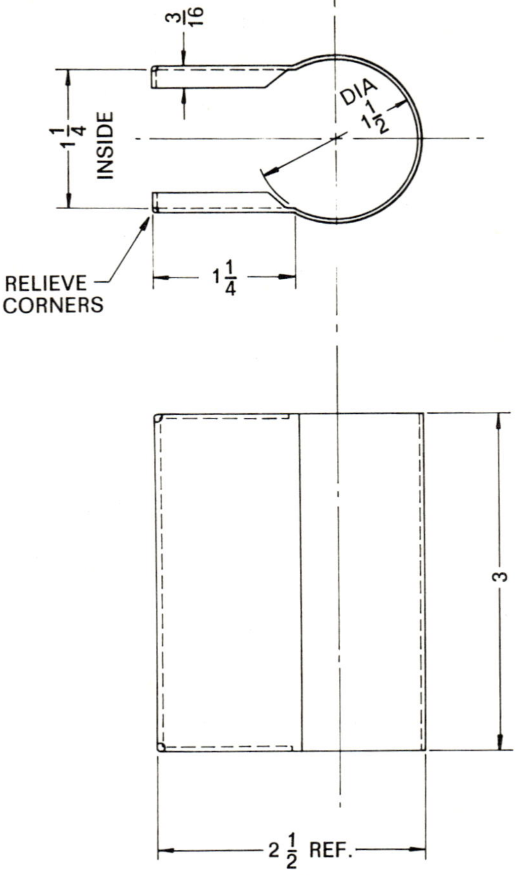

Fig. 12-26 The drawing for Project No. 7

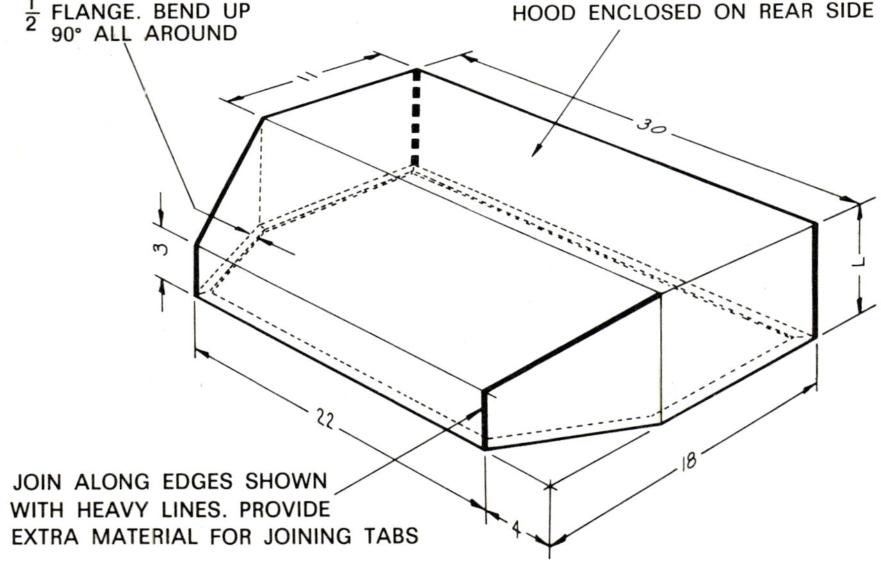

Fig. 12-27 The drawing for Project No. 8

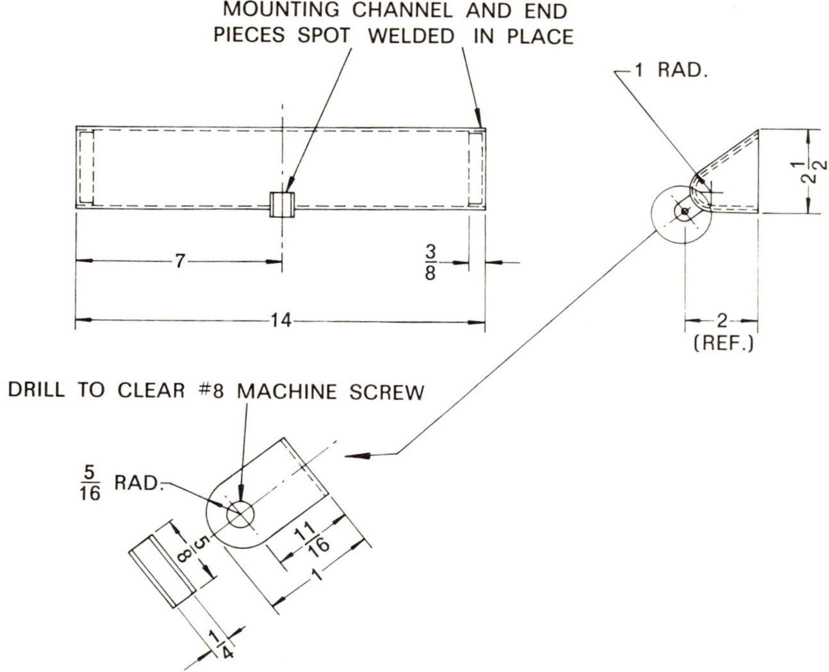

Fig. 12-28 The drawing for Project No. 9

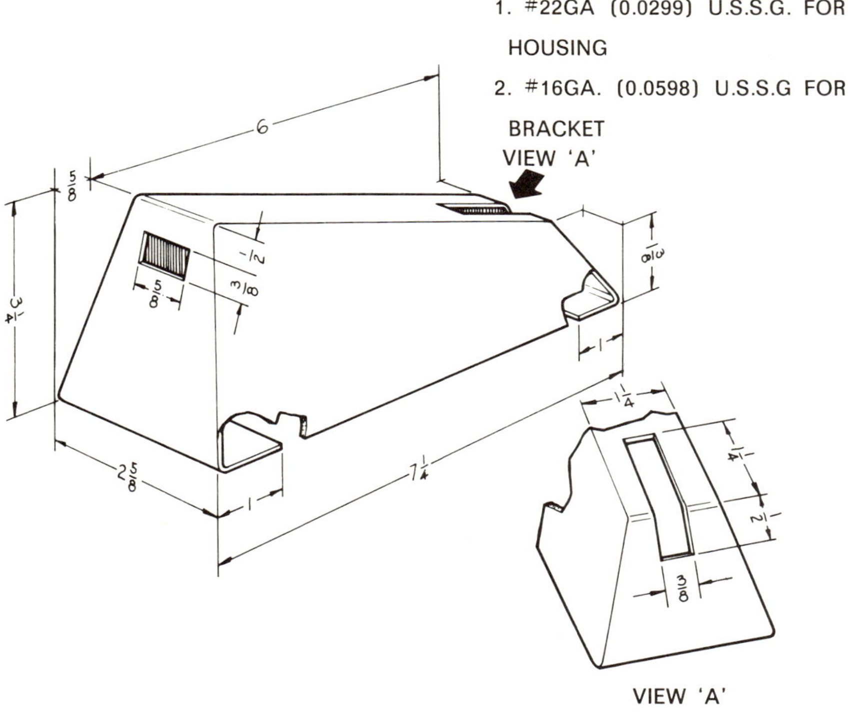

Fig. 12-29 The drawing for Project No. 10

UNIT 13

Graphs and Charts

Purpose

Graphs and charts are devices used to present engineering data in a form that will make the impact on the reader greater than if it is merely put in a written statement. The data is interpreted more effectively by the use of these devices.

In addition graphs provide a simple means of comparing related engineering data and also permit **extrapolation** of unknown values from a plotted curve. That is, curves can be projected into the unknown to provide data having some probability of accuracy.

Characteristics of good graphs

A good chart or graph must have a number of characteristics in order to guarantee its usefulness. They are as follows:

1. The information which they present must be completely factual and without known distortion of the truth.
2. The manner in which the graph is shown must provide ease of reading and interpretation.
3. The appearance of the graph must be aesthetically proper in order that reader's attention is captured and held.

Gathering and assembling data

The accumulation of data is not one of the draftsman's problems. However, it is frequently the task of the technician to record raw data from experiments being carried out in the electrical laboratory or other testing environment. The chances are that the technician or engineer will carry out the preliminary steps and then provide the draftsman with a sketch of the proposed chart or graph. However, the draftsman's competence and expertise may be such that he is entrusted to work directly from the raw data. If the latter is the case, he must have knowledge regarding the shapes of curves that pertain to data.

BASIC CURVE SHAPES

Stepped straight line

1. Data which is discontinuous and has discrete increments must be shown by a series of joined straight lines. See Fig. 13-1.
2. Data without theoretical and mathematical foundation must also be shown as a continuous series of joined straight lines. See Fig. 13-1.

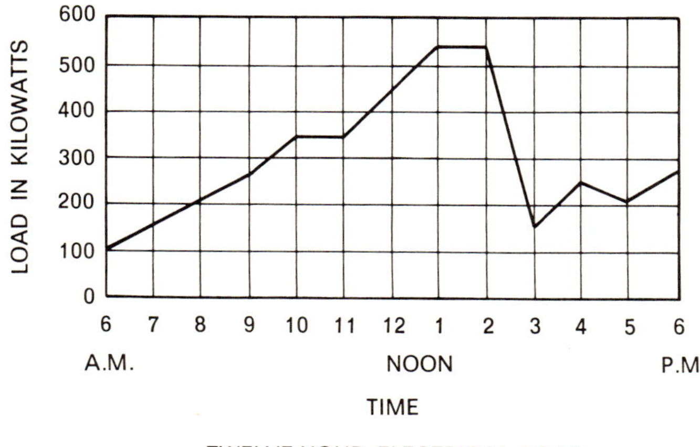

Fig. 13-1 A graph of stepped joined straight lines

Continuous curved line

Data with theoretical foundation and pertaining to physical phenomena is shown with a smooth curved line. See Fig. 13-2.

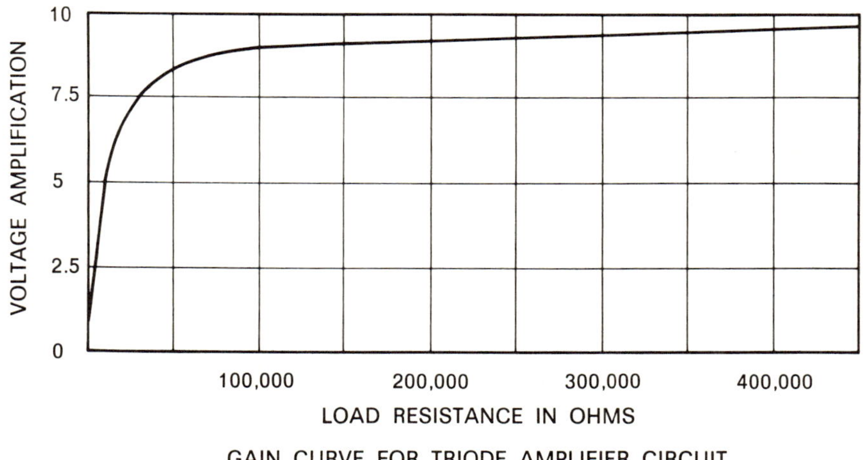

Fig. 13-2 A continuous curved line graph

Family of curves

Frequently graphs are made which show a number of curves on a single drawing with only one set of **coordinates**. For example, receiving tube manuals use this method to show the varying relationship between plate current, plate voltage, and grid voltage for individual types of vacuum tubes. Similar curves are also plotted for transistors. See Fig. 13-3. Some similarity to a **family of curves** is shown in Fig. 13-4 which compares the frequency response between a recent telephone design known as an electret set, and the conventional carbon type.

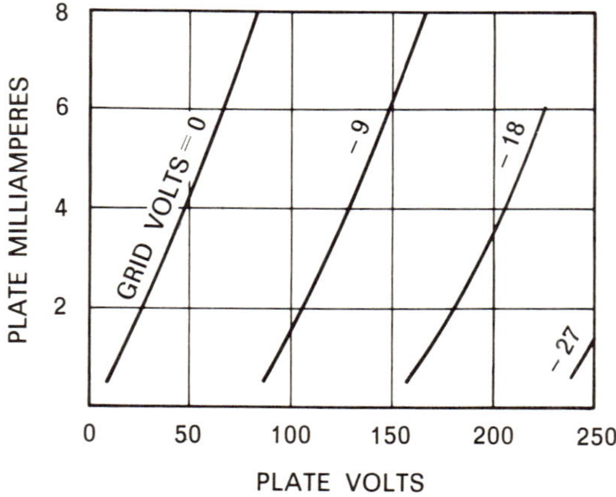

Fig. 13-3 A 'family of curves' graph

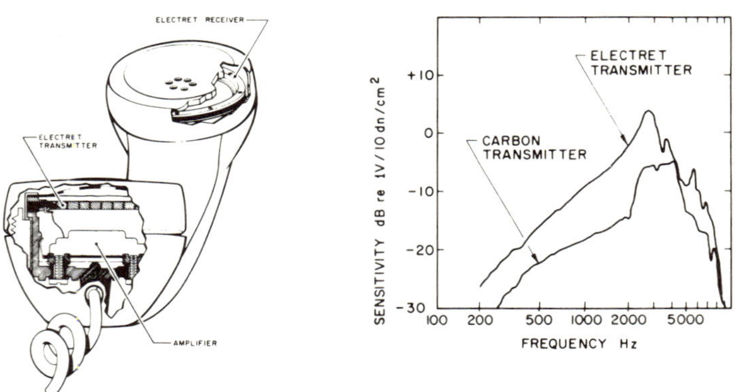

Fig. 13-4 An 'electret' transmitter and receiver and accompanying graph showing a comparison of the frequency response between the conventional carbon transmitter and the electret transmitter. (Courtesy Northern Electric Company Limited, Montreal, P.Q.)

BASIC TYPES OF GRAPHS AND CHARTS

So far we have concerned ourselves solely with linear graphs utilizing both stepped straight lines and smooth curves. There are however, other variations of linear graphs as well as other types not considered as such.

Polar coordinate graph

This type of graph has wide application in graphically portraying the effective radiation pattern from a specific transmitter antenna. This is done by using an instrument known as a **field strength meter** which is moved throughout 360° in **azimuth** at varying but known distances from the antenna. Figure 13-5 shows such a graph plotted for the radiation from an **H.F. monopole** antenna mounted on a 1/30th scale model of a Canadian designed fire fighting aircraft. The small inset drawing indicates the antenna position relative to the aircraft. Note that the pattern produced indicates a more effective radiation **aft** of the aircraft.

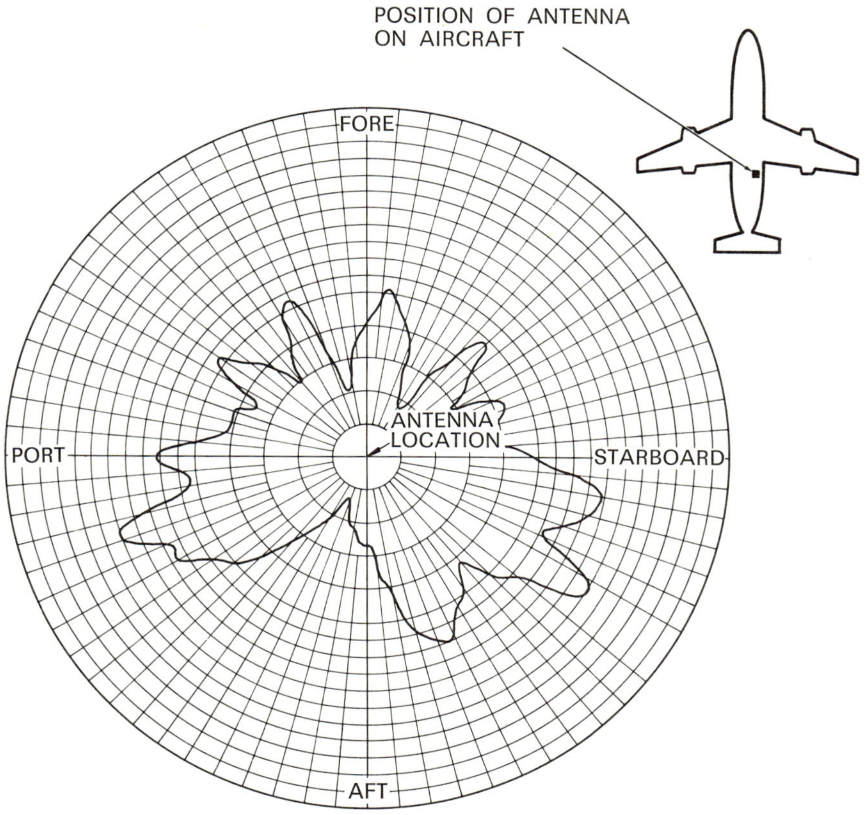

Fig. 13-5 A polar coordinate graph plotted by an automatic plotting device. (Courtesy Canadaire, Montreal, P.Q.)

Bar charts

Bar charts are used in situations that are much less demanding from a technical standpoint than linear graphs. Because of this, and due to their appearance they are interpreted easily by nontechnically oriented personnel.

The elements in the bar chart are set down either vertically or horizontally. This selection is based on the amount of information to be displayed, the pre-established size of the chart, and whether or not a requirement exists for lettering to be read vertically or horizontally. Figures 13-6 and 13-7 show both vertical and horizontal charts. The area within the bars of a chart is often filled in to produce greater attention from the reader's standpoint, and provide an artistically interesting format. If the bars are completely filled in with India ink, or if tape is being used, then the lettering must obviously be placed alongside. However, if appliqué or press-on media is used, a uniformly pleasing texture is obtained. Blank areas can be left in the bars for the application of printed values or other information.

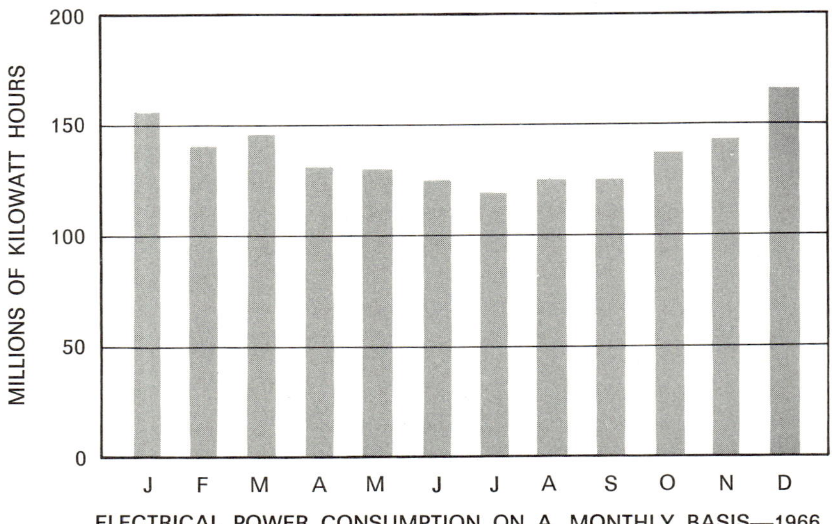

Fig. 13-6 A vertical bar chart

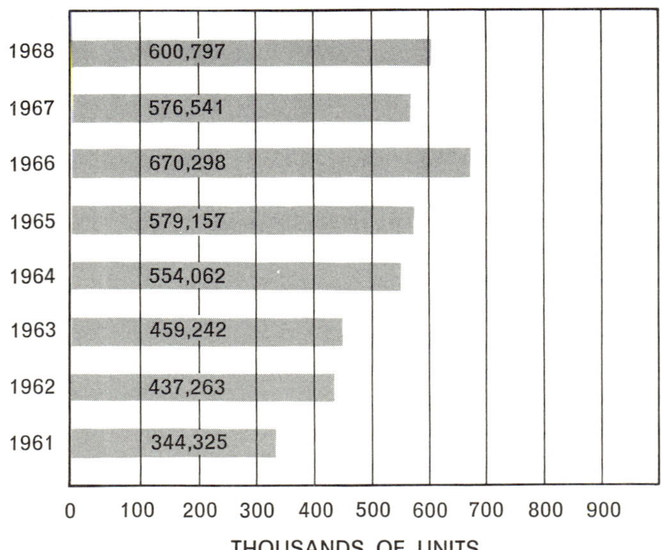

Fig. 13-7 A horizontal bar chart

Pie chart

Another method of graphically portraying data when given as a percentage is a pie chart. The accuracy of such a chart is limited. In addition, its application to the electrical and electronic fields is quite limited. Its chief value is that it is relatively compact in size, and that it is easy to interpret. In constructing such a chart it is common practice to place the largest sector at the twelve o'clock position, and to locate the lesser values sequentially in a clockwise manner. The chart is further enhanced by shading the largest to the smallest sectors from jet black to white. However, this procedure is not mandatory. The diagram shown as Fig. 13-8 is included as representative of a typical pie chart.

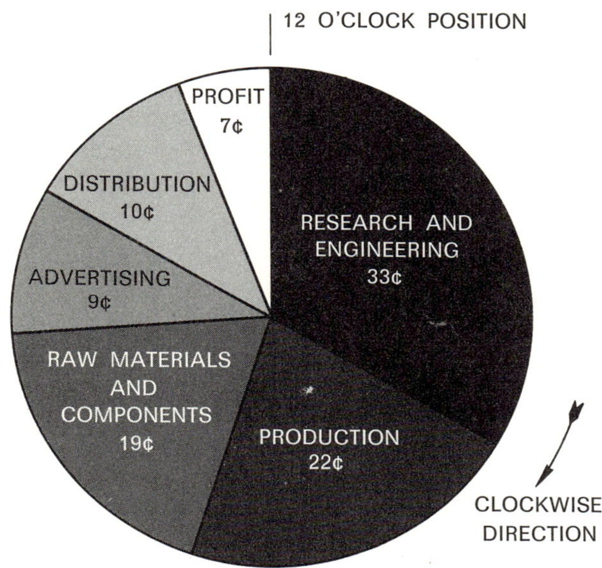

NOTE: THE DISTRIBUTION SHOWN IS A SUGGESTED ONE, AND REPRESENTS THE AUTHOR'S POINT OF VIEW EXCLUSIVELY

Fig. 13-8 A pie chart

Special charts

Besides the conventional charts and graphs already mentioned, data can be displayed in chart form by three dimensional pictorial means. The justification for such a chart must be ample, since the drafting time required to produce one is considerable. Fig. 13-9 illustrates such an example.

In addition, conversion and alignment charts are constructed for electronic work. Also, other special charts are made to illustrate Ohm's Law, reactance, and time constant values. However, the opportunities presented to the draftsman to construct these types of charts are likely to be few. Consequently, no further details are included in the scope of this unit.

Semilogarithmic and logarithmic graphs

Engineering charts are frequently made on either commercially printed semilogarithmic or logarithmic graph paper. The advantages of doing so are that

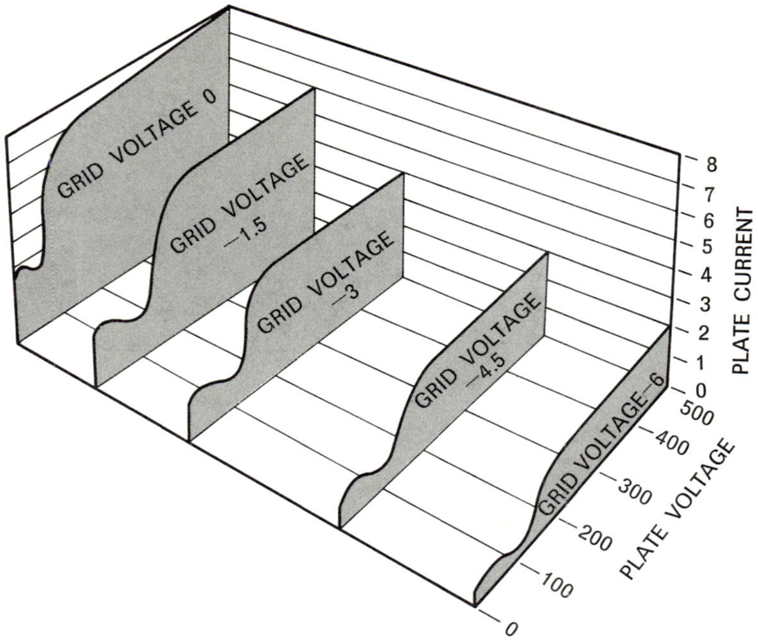

Fig. 13-9 A three-dimensional chart showing the relation between plate voltage and current under varying voltage conditions for a tetrode vacuum tube

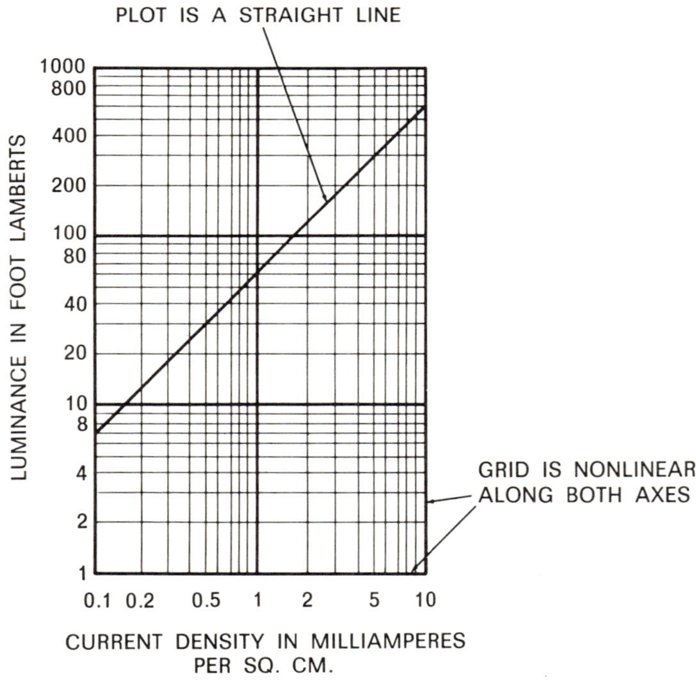

Fig. 13-10 A logarithmic graph showing the curve plotted for a phosphor used in a C.R.T. Notice that luminance increases as the current density of the electron beams increases

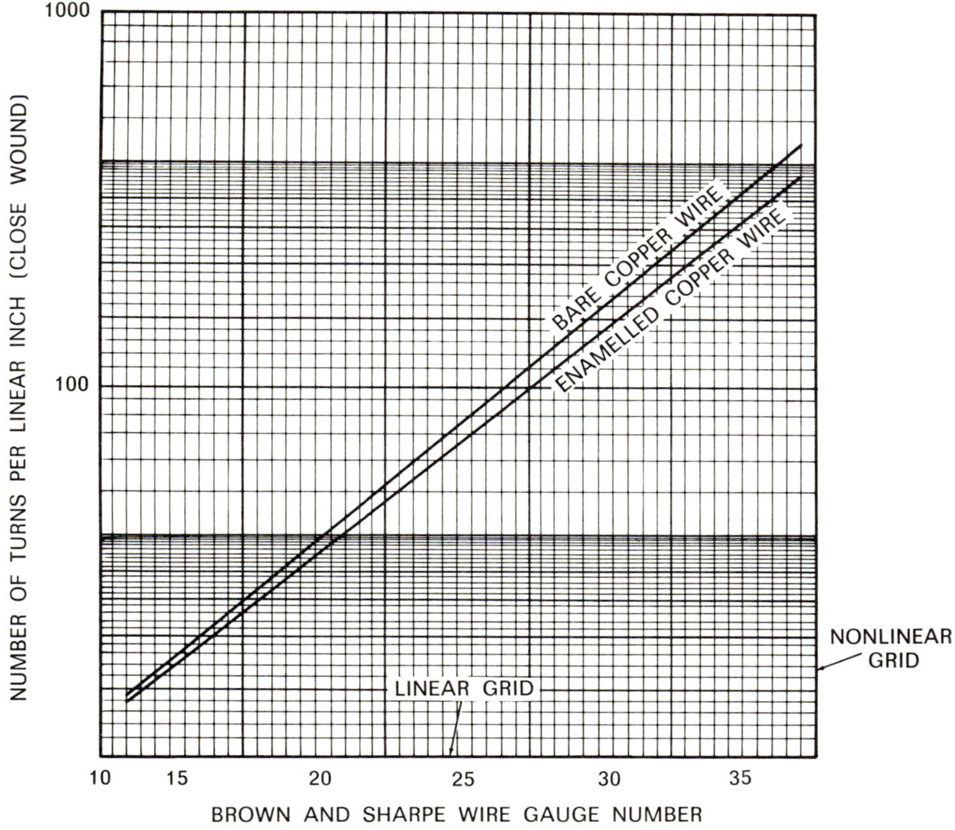

Fig. 13-11 A semilogarithmic graph showing curves plotted to determine the number of turns per inch of wire when using different B & S gauge sizes

graphs which are physically impossible to plot on linear paper because of their length can be done on semilogarithmic paper. As a example, a frequency response curve covering a very wide range of values can be shown. In addition, data plotted can yield a continuous straight line curve which is advantageous in extrapolating other values. Furthermore, if more than one curve is plotted, comparison is relatively simple if the curves are continuous straight line types. Figure 13-10 illustrates a curve plotted on logarithmic graph paper. Note that the grid is nonlinear along both coordinates. The curve shown on the semi-logarithmic graph paper, Fig. 13-11, differs in that the grid along one axis is linear, the other logarithmic.

Steps in making a linear graph

Since the largest requirement will be for the draftsman to make linear graphs consisting of either smooth continuous curves, or curves formed of stepped straight lines, the following information is included in the form of sequential steps.

STEP 1

Assemble the data in tabular form for the sake of convenience. This method will increase both the rapidity and accuracy of the graph construction.

STEP 2

Decide on the style and size of the graph paper. This decision must also include the size of the actual squares or grid.

STEP 3

Work out the scales to be used for both the variable and nonvariable components. The scales selected need not be identical for both axes. As a matter of fact it is often advantageous to vary the scales in order to improve the curve shape. See Figs. 13-12 (a), (b), and (c).

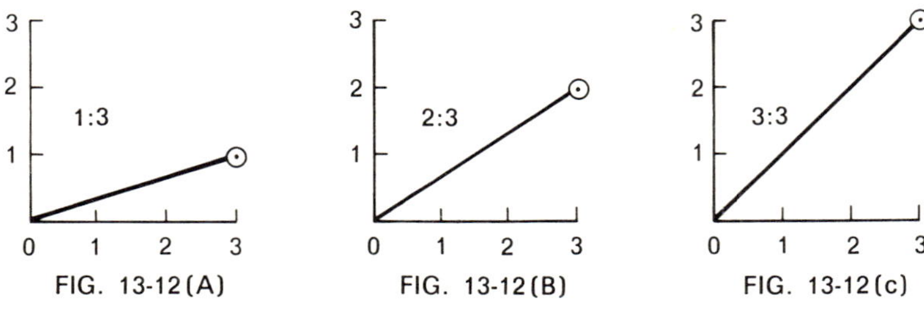

Fig. 13-12(a), (b) and (c) The effects of altering the vertical axis scale while keeping the horizontal axis scale constant

STEP 4

Establish the axes or coordinates for the variable and nonvariable components. Unless this is done at this stage, either axis may turn out to be of insufficient length as the information is plotted.

STEP 5

Decide on the location of the zero point. Generally its location is at the intersection of the coordinates. Sometimes however, it is necessary to move it along the vertical axis above the horizontal one. See Figs. 13-13 (a) and (b).

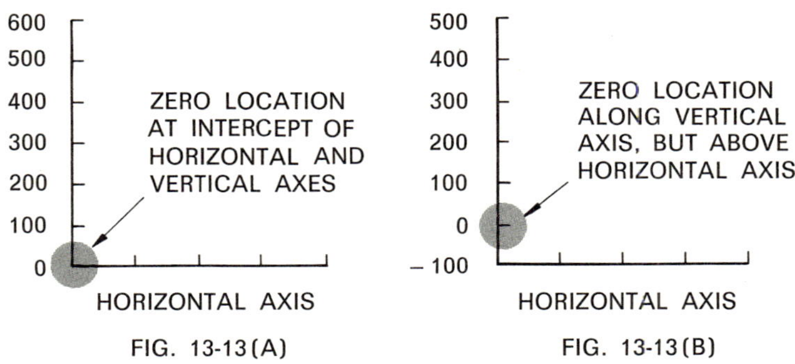

Fig. 13-13 The selection of the zero point location

STEP 6

Plot the values on the blank graph paper taking the information from the tabular form. The points plotted must be located accurately. Use a hard, well pointed pencil. Values of particular importance should be lightly encircled for future reference.

STEP 7

Protect the plotted points required for the completed graph by inking them in with a compass. If the curve to be plotted is a stepped straight line, join the plotted points. The straight lines must not pass through the circles, but they should point towards their centres.

STEP 8

In the case of a continuous smooth curve, it may be necessary to miss some of the plotted points in order to establish a good curve. Figure 13-14 shows such a curve. The broken line indicates the type of curve obtained by joining all points. The solid line indicates the desired curve. When joining the plotted points an irregular curve or drafting **spline** is used. Not less than three points should be joined at one time. Also, all preliminary work must be done extremely light, and of course in pencil.

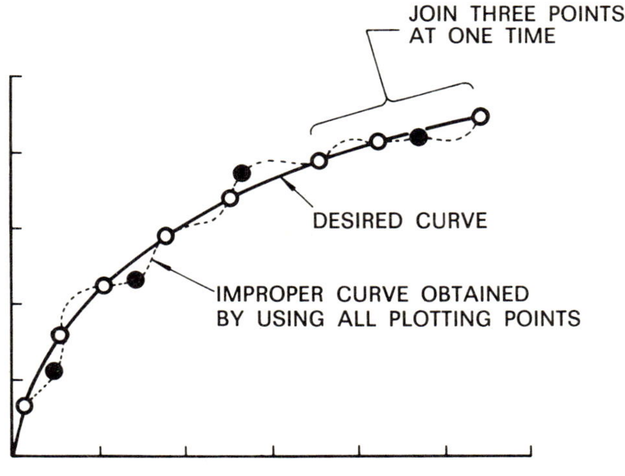

Fig. 13-14 How to draw a continuous smooth curve

STEP 9

Upon completion of the curve, inking can take place. The curve must be inked with a line that is heavier than any of the grid or datum lines. Also, the numerical values should be lettered in. Their location is such that they are centred about their datum or reference line immediately outside the coordinate lines and are read horizontally. Descriptive information is spotted directly below the horizontal numerical values and to the left of the vertical numerical values. In the case of the vertical one, it is read from the bottom of the graph.

Gothic style capitals **both numerals and letters** of appropriate size are ideal for graphs because of their simplicity. See Fig. 13-15.

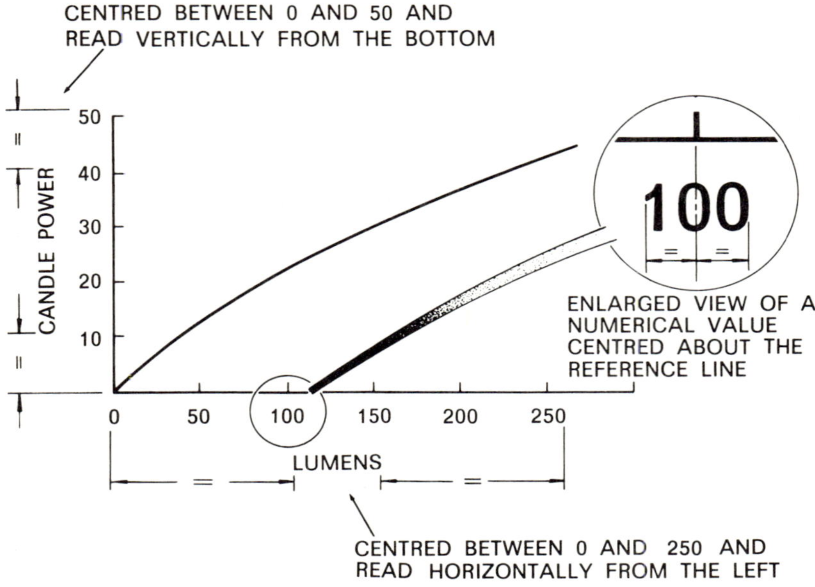

Fig. 13-15 The location of numerical values and descriptive information

STEP 10

Finally the title, the informative references if required, and the border are added to complete the chart. With regard to the title, sufficient thought must be given to establish one which is both brief, but fully and correctly informative.

KNOWLEDGE TESTERS

1. What is the purpose of a graph or chart?
2. List at least three characteristics of a a good graph.
3. When is a continuous smooth curve graph used?
4. What curve type is plotted for data not having either a theoretical or mathematical foundation?
5. Explain what is meant by the term, family of curves.
6. Why are coordinates required for a graph?
7. Obtain a technical dictionary and establish the meaning of the term azimuth.
8. In Fig. 13-5 what value has been used for each sector increment?
9. Why is a polar coordinate graph particularly useful for plotting the radiation pattern for a monopole antenna?
10. By studying Fig. 13-4(b) what con-

clusion can be reached regarding the electret transmitter?

11. When making a bar chart, how is the descriptive information placed for the vertical coordinate?

12. In what fashion are the numerical values placed for a bar chart?

13. When is a bar chart used?

14. List three details that are important when constructing a pie chart.

15. Outline the procedure for drawing a continuous smooth curve.

16. What is the purpose of a drafting spline?

17. Give three important reasons for constructing logarithmic graphs.

18. What is the difference between semilogarithmic and logarithmic paper?

PROJECTS

1. Draw a bar chart comparing the consumption of electrical power for 1960 and 1965 on a monthly basis for Capital City, Canada. Values are in millions of kilowatt hours.

2. Using the same data shown in Problem 1, make a linear graph of the stepped straight line type. Plot the curves for both years on the same graph.

	J	F	M	A	M	J	J	A	S	O	N	D
1960	105	98	102	92	87	83	77	79	84	92	97	110
1965	145	131	135	123	117	113	109	112	116	127	137	149

3. Plot a curve using the polar coordinate system showing the horizontal radiation pattern for a **para-corner reflector** antenna. Note that the pattern is symmetrically shaped, hence values are not given beyond 180°.

In constructing the chart, draw seven concentric circles equidistant apart. The smallest circle should be identified as —15 decibel level; the largest circle +15 decibel level.

Azimuth	0	10	20	30	40	50	60	70	80	90
Decibel level	7.5	7.0	5.0	2.0	—1.0	—4.0	—7.5	—11.0	—12.5	—15.0

Azimuth	100	110	120	130	140	150	160	170	180
Decibel level	—14.0	—16.0	—16.0	—17.5	—16.0	—17.0	—18.0	—19.0	—20.0

4. Plot the family of curves for a 6MD8 tube used in the **matrixing** circuits of colour and black and white television receivers. The graph must show six curves, plotting plate voltage against plate current.

Plate E.M.F. in volts	Grid E.M.F. in volts				
	0	—5	—10	—15	—20
0	0				
50	11				
70		1			
100	25	3			
150	40	11			
160			2		
200		23	4		
225		30		2	
250			12	3	
270			22		
290				7	1
300					2
350				14	14
400					18

5. Draw a pie chart showing the estimated percentage costs of operating the following household electrical items for a one hour period.

(a) Thermostatically controlled frying pan.
(b) Transistor solid-state receiver.
(c) 'Tri-lite' lamp.
(d) Portable room humidifier.
(e) Electrically ignited furnace with fan.
(f) Thermostatically controlled steam iron.
(g) Thermostatically controlled electric blanket.

It is suggested that the student carry out research to establish the wattage ratings of these appliances.

6. Make a continuous curved graph showing the relationship between temperature and resistance for a **thermistor.**

Temperature in degrees Centigrade	—60	—40	—20	0	20	40	60	80	100	120	140
Resistance in Ohms	3000	800	300	90	40	25	8	6	4	3	2

7. Using suitable semilogarithmic graph paper plot the data given for Problem 6.

8. Given the following data, plot a curve to show that the armature current drops off dramatically as the motor speed increases.

Motor speed in %'s of full R.P.M.'s	0	5	25	50	75	100
Armature current in amperes	0	45	30	19	11	5

9. Plot a typical **sine wave** curve for an alternating current given instantaneous current and angular degrees of rotation values. Data given is up to 180° only, since the curve is symmetrical about that datum, except for its negative sign.

Angle in degrees	0	15	30	45	60	75	90	105	120
Instantaneous current in amperes	0	0.366	0.707	1.000	1.225	1.366	1.414	1.366	1.225

Angle in degrees	135	150	165	180
Instantaneous current in amperes	1.000	0.707	0.366	0

10. Obtain data by research work and report it in pie chart form to show the efficiency of producing electrical power by the following methods:—

(a) Hydro electric.
(b) Fossil fuel conversion.
(c) Nuclear power conversion.
(d) Solar cells.
(e) Chemical fuel cells.
(f) Conventional batteries.

UNIT 14

Special Drafting Methods

SOME COMMENTS HAVE ALREADY BEEN MADE REGARDING THE USE OF both materials and machines to lessen drafting time, to provide a high precision of delineation, and to enhance the quality of copies printed.

In addition to these, a variety of methods and machines have been devised in the interests of eliminating many of the tedious and time consuming tasks performed by draftsmen. Many other machines are still in the development and prototype stages. Some of these devices incorporate electronic circuits, especialy the use of computers. Because of their relative high cost, these devices are used by those firms having a large volume of drafting, especially those in the aerospace and electronics manufacturing fields.

MICROFILMING

Since the storage space required for the tens of thousands of drawings of a large firm becomes impossible to cope with, microfilming systems are in widespread use. Essentially what is done, is that the original tracing is photographed with a high resolution camera and reduced to a 35 mm. negative film. The film is mounted in an aperture card, and the original tracing destroyed.

The aperture cards can be conveniently stored in steel filing cabinets and hence can be readily retrieved for reference purposes. Full size prints can be made by photographic methods from the microfilm, and in addition, they can be screened for viewing purposes at an enlarged scale. Other advantages of microfilming are that the aperture cards can be stored in both a fireproof and theftproof vault of much smaller proportions than for original tracings. Since the original master paper tracing is destroyed subsequent to the production of the master film, problems normally inherent with excessive handling of a paper tracing are eliminated.

A special file cabinet known as a Diebold power file is in use to expedite the storage and retrieval of large numbers of microfilm aperture cards. Figure 14-1 illustrates such a device.

Fig. 14-1 A Diebold Power File used to store 105 mm. microfilm masters. (Courtesy of Diebold, Incorporated)

105 mm. - 35 mm. microfilm system

Due to the small size of 35 mm. microfilm aperture cards, a very high degree of draftsmanship is required of the original tracings in order to obtain good results. To overcome this problem, some firms currently use separate 35 mm. and 105 mm. systems of the same films. Obviously this duplication is both costly in terms of processing and storing.

The advantage of the 105 mm. film is that the image quality is much better than 35 mm. film, because of its larger size.

A device known as an optical converter, developed by the Lawrence Radiation Laboratory and the R. A. Morgan Company in the United States, now makes it possible to make 35 mm. aperture cards directly from 105 mm. films. This method provides both advantages in time, and ease of conversion.

The Lockheed Missile and Space Company in Sunnyvale, California uses a 105 mm. - 35 mm. system. To further enhance this system, they utilize paste-ups, stick-ons, cut-outs, and typewritten notes. Furthermore, all changes are made on quality photo blowbacks of 105 mm. film, and drawings are made on nontransparent materials. The composite drawing using such an array of media is 'shot' on 105 mm. microfilm, the paste-up is destroyed, and the 105 mm. master is stored in a file.

From this 105 mm. master, the following can be made:

1. 35 mm. microfilm aperture cards.
2. Blowbacks for engineering department changes.
3. Unlimited copies from 105 mm. offset duplicator masters.
4. 35 mm. microfilm/electrostatic copies.

In terms of space and money saved, Lockheed reports the following:

1. A saving of 9300 square feet of floor space based on 1,000,000 105 mm. films.
2. An annual financial saving of $3,000,000.

Amacus

A further variation of microfilming techniques has been developed for the United States Army at Rock Island, Illinois, by General Precision Systems, Inc. It is known by the acronym, 'Amacus'. This term stands for, automated microfilm aperture card updating system.

This system is capable of revising a 35 mm. microfilm aperture card by fully automatic means in as short a time as five minutes. The operator first of all inserts the card to be revised into a scanner/processor unit. This unit is located in a separate console and is not shown in Fig. 14-2. Here the information on the film portion of the card is converted into digital information by an ultra-high resolution flying spot scanner. It is then stored in a memory system. Revisions in the form of lines, and/or printed data can be made by the use of an electronic light pen or electric input typewriter. When either of these revision procedures are initiated, a fully automated sequence of operations takes place which expose, develop, wash, and dry the newly revised film. Fig. 14-2 shows an operator making line changes to a microfilm by means of an electronic light pen. To his left is the electric input typewriter, and to his right the aperture card printer.

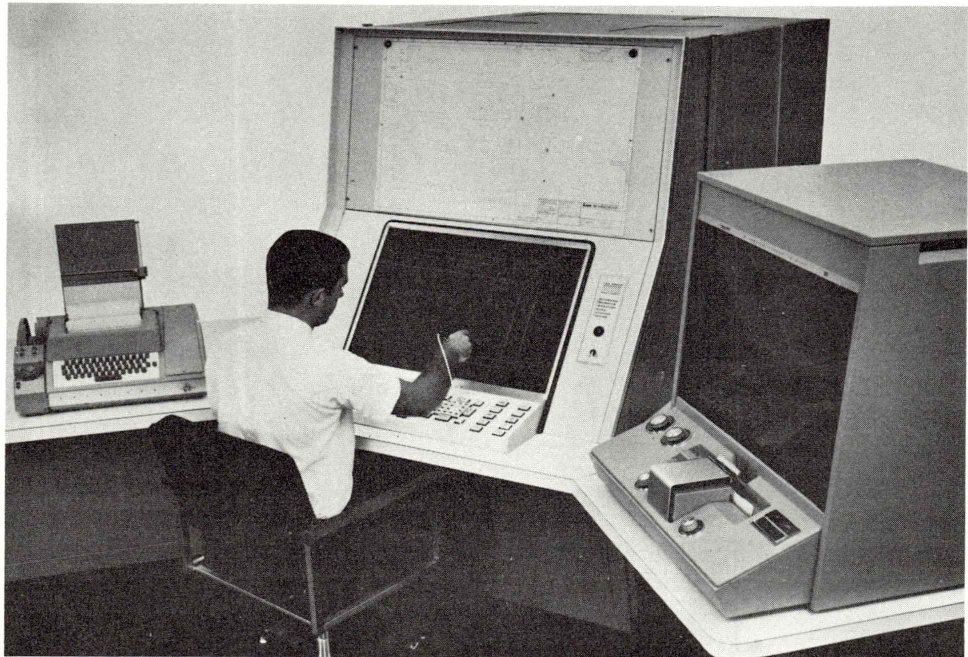

Fig. 14-2 The Amacus microfilming system. The control console consists of a primary electronic display with the operator's control panel above and the light pen positioned directly in front of him, a second optical display with the aperature card printer at his right, and an electric input typewriter at his left. (Courtesy Singer-General Precision Inc., Link Division)

Fig. 14-3 A photodrawing of milk processing equipment. The components are actually scale models. Note the completeness of the plumbing details. (Courtesy Carnation Company)

Photodrafting

A concept that has gained wide popularity because of its flexibility and inherent cost reduction is photodrafting. The Fairchild Instrument Company of Palo Alto, California reports an approximate cost reduction of 50 per cent in producing their final assembly drawings by this method.

In making the photographs of assembled equipment, flat lighting is used to eliminate shadows, and a number of test photographs at different exposures are made to establish the optimum results. From a variety of 'shots' made, the best possible ones are selected for the master photodrawing. Lettering and/or additional drafting is made on the matte surface on the back of the film. Errors on the film can be erased by either mechanical or chemical means.

Some industrial firms, such as the Carnation Milk Company, find photodrafting an exceptional way of documenting precision scale models of major plant alterations. Both isometric and plan views are photographed. To these, engineering dimensions are added if required. Figure 14-3 illustrates an isometric photodrawing of one of their processing installations.

Scissors drafting

Scissors drafting or 'grafting' is a technique that has widespread application in engineering drawing. As the name implies, it literally means to cut and paste the desired pieces of graphic material onto a matrix or base material. In actual practice the materials and method are somewhat more sophisticated. The Alcan Aluminum Limited Company of Canada is actively pursuing this technique in order to liberate their drafting staff from mundane work, so more time will be available to utilize their creative talents.

First of all, copies of their original engineering drawings are produced on a photosensitive translucent film. Then the grafted information, or opaque medium, such as one would find in a catalogue or engineering brochure, is fastened down with translucent adhesive tape at the leading edge only. The advantage of this base film is, of course, that it is virtually shadowproof.

To eliminate any shadow that still might be of annoyance, the Alcan staff apply the grafted film over a sheet of exposed film (without any images on it) when they finally make a photo-master on a sheet of sensitized polyester material.

Fig. 14-4 An illustromat 1200 computer directed drawing machine. The operator traces the orthographic drawing on the table with the styluses to produce the 3-D perspective drawing shown above. (Courtesy Perspective Systems, Inc.)

Beyond the normal drawing procedures that are required to produce the original master, and from which an unlimited number of film sub-masters can be made, tedious erasing and adding of revisions by hand methods is eliminated.

AUTOMATED DRAFTING

Illustromat. A three dimensional drafting machine

A computer-directed machine for making three-dimensional perspective drawings has been designed by Perspective Incorporated, of Seattle, U.S.A. This device, known as an Illustromat, converts two-dimensional drawings into three-dimension ones. The operator traces the two-dimensional views, during which time the analog computer of the Illustromat directs the machine to produce an accurate three-dimensional view. Fig. 14-4 shows an operator in the process of producing a three-dimensional perspective drawing of a machine part that has been produced by an Illustromat 1200.

Enthymat. A three-dimensional drafting machine

Another device being marketed is the Enthymat, or automatic perspective translator. Its designer is Mr. Kermit Bowen of San Diego, California. Unlike the Illustromat, this machine works solely on mechanical principles. The operator controls two master styluses. One controls the height and width, the other depth. The information from these two styluses is directed into a translator mechanism, which in turn operates a reproduction arm. The reproduction arm is designed to make either ink or pencil drawings. Also, its scale can be altered from that of the original through the range of 1 1/2 times original blueprint size to 1/60th of it.

In addition to its ability to produce a perspective drawing from the orthographic views of a part, the Enthymat can produce a true perspective, three-dimensional drawing from a solid object. This feature is especially advantageous to an illustrator when blueprints are not available. Fig. 14-5(a) shows a production model of the Enthymat in which the operator sits at table 'B' upon which is placed an orthographic drawing, such as is shown in Fig. 14-5(b).

During normal operation this table is rotated 90° anti-clockwise from that shown, so that the track mechanism is located parallel to the operator's seated position. After adjusting the machine to produce the desired scale for the required three-dimensional drawing, the operator traces the orthographic views. The result of this action is to produce a three-dimensional true perspective drawing. Fig. 14-5(c) illustrates the true perspective of the casting shown as a five-view orthographic drawing.

Numerical Control Drafting

Automated drafting machines that are controlled by computers are currently engaged by aerospace firms to define external aerodynamic shapes, cross-sections, geometry of component parts, and to prove or verify instructions to tape controlled machine tools. In addition, such items as volumes, and areas

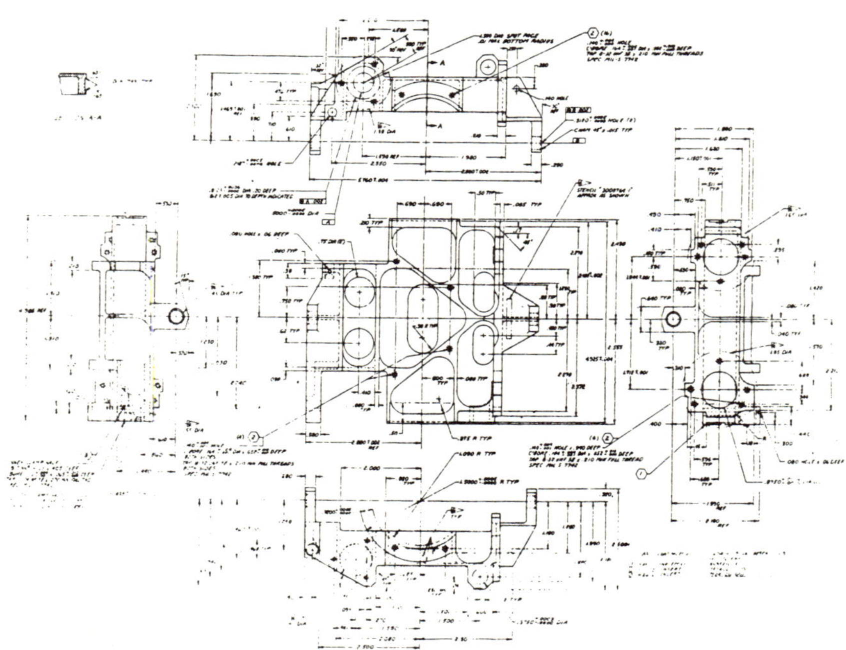

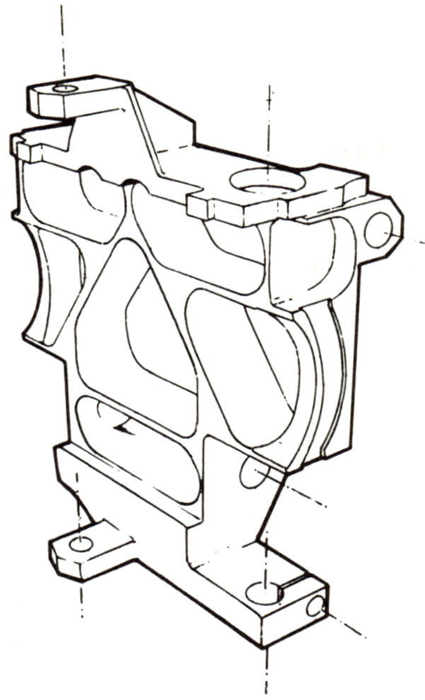

Fig. 14-5 *(a) A production model Enthymat. (b) The orthographic views of a casting. (c) A true perspective drawing made from the orthographic views in (b) by the Enthymat. Invented and patented by Kermit A. Bowen, San Diego, California. (Courtesy of and Manufactured by Enthymat, Inc., San Diego, California)*

of shapes that would be tedious to establish by conventional means, are computer solved. Such a machine is illustrated in Fig. 14-6. It is used by the Grumman Aircraft Engineering Corporation in the U.S.A. The major components of this machine are as follows:

1. Teletype writer.
2. Data processor (computer).
3. Photoelectric tape reader.
4. Carriages and servo mechanical system.
5. Vacuum chuck drafting surface.

When a programmed tape is fed into the machine it is 'read' first by the photoelectric tape reader. Electrical pulses so generated according to the tape format are processed by the data processor to give explicit instructions to the carriages and servo mechanism, and to produce ultimately a drawing of one specific configuration. The vacuum chuck is of course necessary to hold the drafting medium in a completely stationary position. Creepage of the medium would obviously produce drastically poor results.

Last minute changes or revisions can also be made by using the teletype writer. In addition, the positioning of the pen, rate of drafting speed, and other special functions can be controlled by manually operated electrical switches.

Fig. 14-6 Aerodynamic curves being generated to a high degree of accuracy by the servomechanism control of a computer directed drafting machine. Engineering Graphics Magazine. (Courtesy of St. Regis Publications, Inc., New York)

To illustrate the time saving aspect of an N.C. automated drafting machine, Grumman Engineering officials claim that a time factor of approximately 10:1 is involved when producing an identical wing design drawing by hand, as opposed to automated means. Also, the accuracy factor is increased by a ratio of 5:1 by numerical control drafting.

Automated Wiring Diagram Drafting

The production of electrical and electronic wiring diagrams is also brought about by computer controlled equipment. At Control Data Corporation in U.S.A. the following procedure is carried out. Initially the engineer prepares a rough wiring sketch according to predefined sets of symbols. Nonstandard or illegal ones are encircled with coloured pencil. These are later filled in by hand. From the rough sketch, a punched tape is made for the computer. In approximately fifteen minutes the plotter under instructions from the computer completes a finished C-size drawing. In terms of time saved, it takes approximately six times longer to produce an identical drawing by hand. In many cases it is still more economical to correct a tape and generate a new drawing than to make ink revisions.

Some other devices that are available to produce schematics by automated means are the Harris-Intertype Fotosetter, the Keuffel and Esser Photo-Draft System, and the Mergenthaler 'Diagrammer'. The Fotosetter utilizes computerized data in conjunction with a 114 symbol keyboard. The K and E system uses a punched tape made directly from the engineer's rough sketch, and a photographic unit containing a negative master symbol disc having 168 different symbols. The Mergenthaler machine operates on a electromechanical-optical system. Two hundred and fifty-six different symbols can be selected, viewed, and positioned at random for incorporation into a schematic. Fig. 14-7 shows such a device.

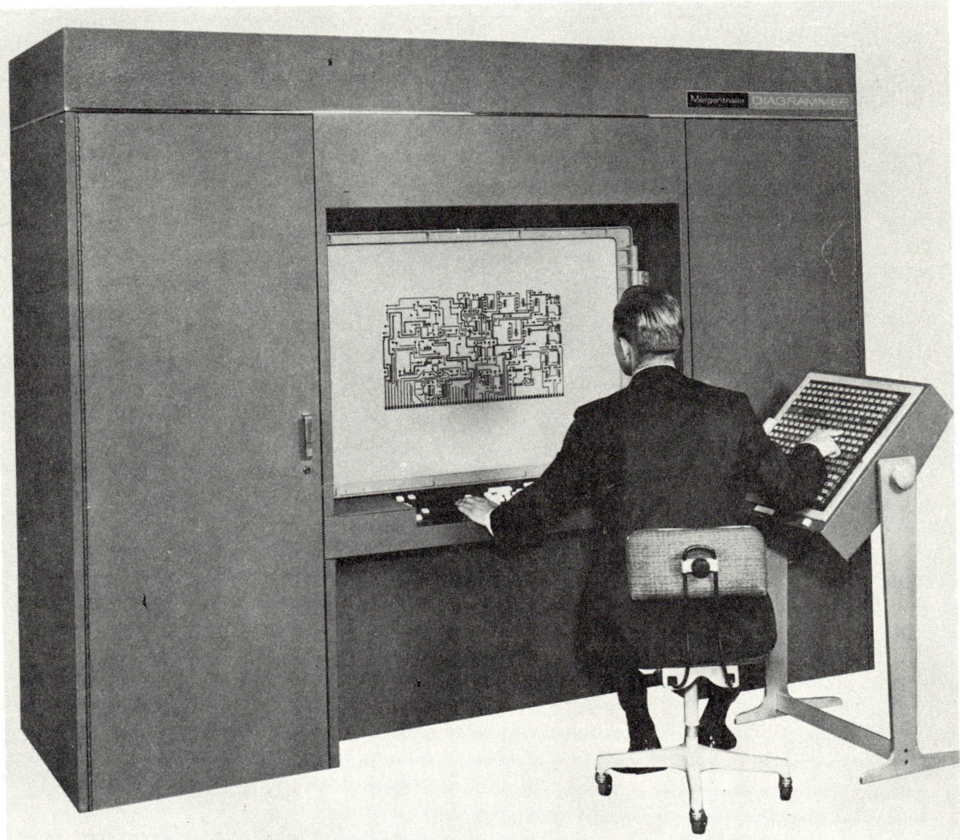

Fig. 14-7(a) A Mergenthaler Diagrammer showing the symbol keyboard at the operator's right and the visual presentation in front of him

Computer controlled graphic displays

A computer controlled graphic display is a sophisticated engineering tool to display numerical information by graphic means, utilizing an advanced electronic system. Graphs or charts of infinite shape can be shown on such a device.

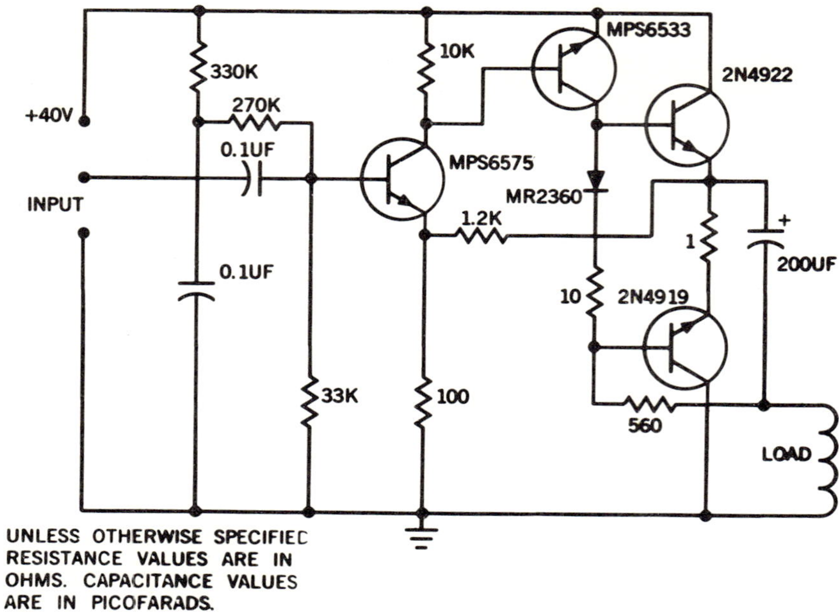

Fig. 14-7(b) *A schematic produced by the Mergenthaler Diagrammer. The symbols shown are those of the company*

Figs. 14-7(a) and (b) *(Courtesy Mergenthaler Linotype Company)*

Another important feature of such a device is its capability of being able to rotate the image being displayed on it through all angles. The IBM Corporation of Kingston, New York, manufactures the IBM 2250, a typical display unit. Figure 14-8 shows this particular model.

KNOWLEDGE TESTERS

1. Give at least three reasons for using microfilming.
2. Briefly describe the process of micro-filming.
3. Why is the 105 mm. microfilming system superior to the 35 mm. one?
4. What is the purpose of the Lawrence optical converter?
5. What is meant by a paste-up?
6. Name the different media that can be produced from the 35 mm./105 mm. system as used by the Lockheed Company?
7. What does the acronymn, Amacus, mean?
8. Describe in simple terms the Amacus system.
9. What is a flying spot scanner?
10. List the special precautions that must be taken when photodrawings are being prepared.
11. Give a brief description of scissors drafting or grafting.
12. What is the essential difference between the design of the Illustromat and the Enthymat?
13. What is the function of either the Illustromat or the Enthymat?
14. List the major units of a numerical controlled automated drafting machine.
15. What special uses are made of N.C. automated drafting machines in the aerospace industries?
16. Describe briefly at least two automated types of drafting machines that are used to produce wiring diagrams.

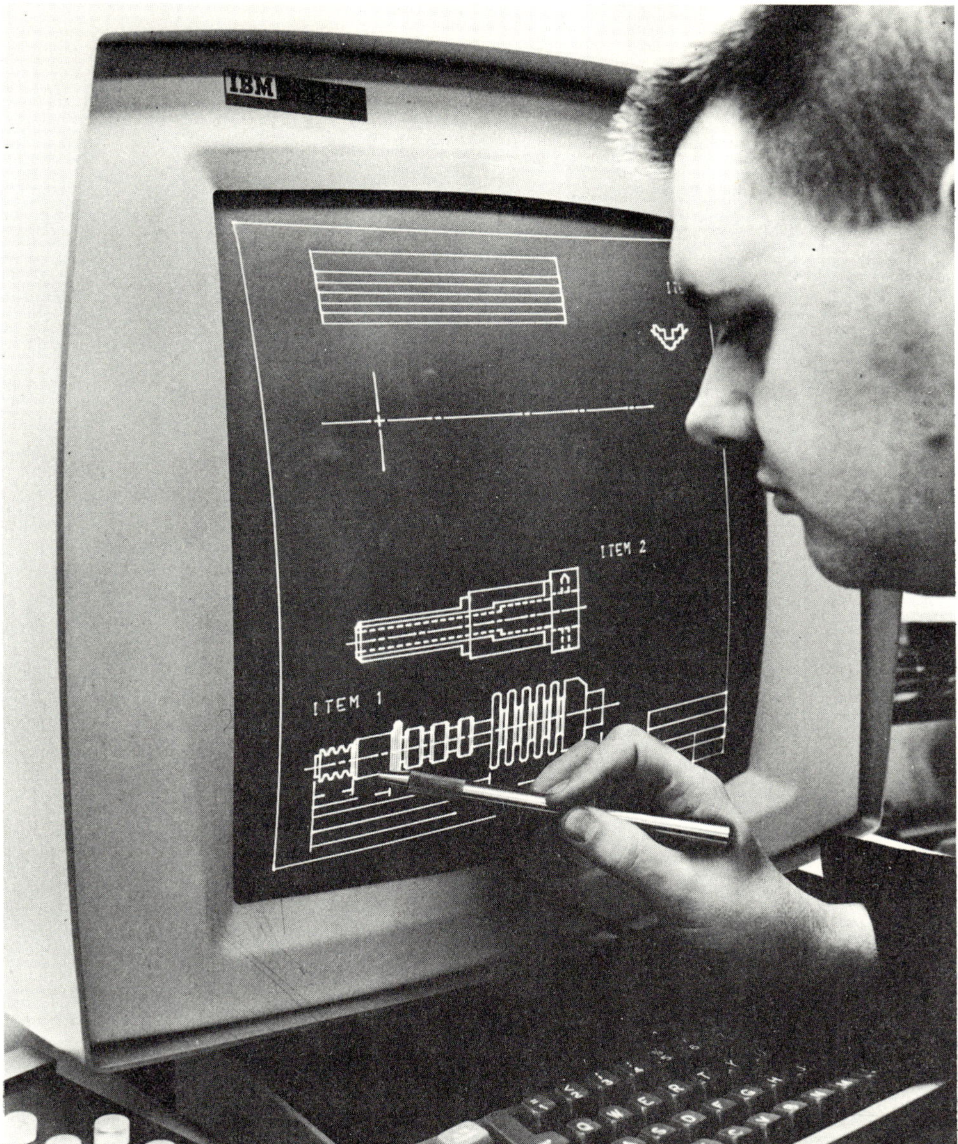

Fig. 14-8 *An engineer working on an IBM 2250 Graphic Data Processing System. The graphic display terminal shown is in communication with a computer, and mechanical designs can be altered by using a light pen. (Courtesy International Business Machine Corporation)*

Appendix

APPENDIX 1.
Capacitor Colour Code

1. A – Standard used
   B – First digit of capacitor value
   C – Second digit of capacitor value
   D – Multiplier, the number of zeroes. List in the order: B, C, and D.
   EXAMPLE
   B dot – green – 5
   C dot – red – 2
   D dot – yellow – 4 zeros
   Value of capacitor is 520,000 $\mu\mu f$
   or
   0.52 μf
   E – Total capacitor value **tolerance**
   F – Characteristics (possibility of 5, i.e. leakage resistance, temperature coefficient, etc.)

NOTE
1. White dot – EIA (New 6 Dot Standard) Mica
   Black dot – MIL STD (Military Standard) Capacitors
   Silver dot – AWS (American War Standard) Paper capacitors
2. Electronic Industries Association

E.I.A. AND MIL 6-DOT STANDARD

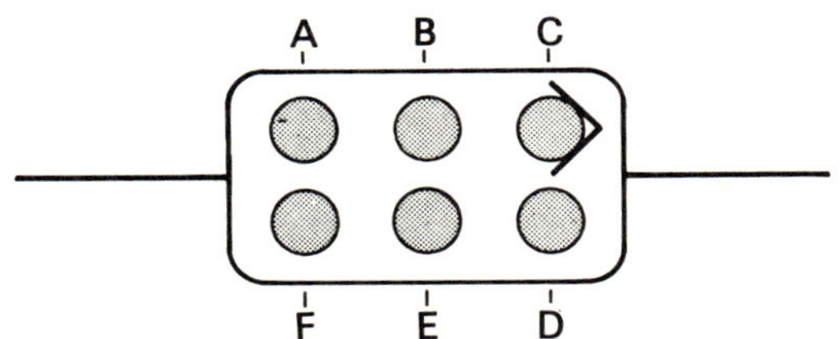

Ceramic Tubular with Axial Leads
A – Temperature coefficient value
B – First digit of capacitor value
C – Second digit of capacitor value
D – Multiplier, the number of zeros. List order: B, C, and D
E – Total capacitor value tolerance

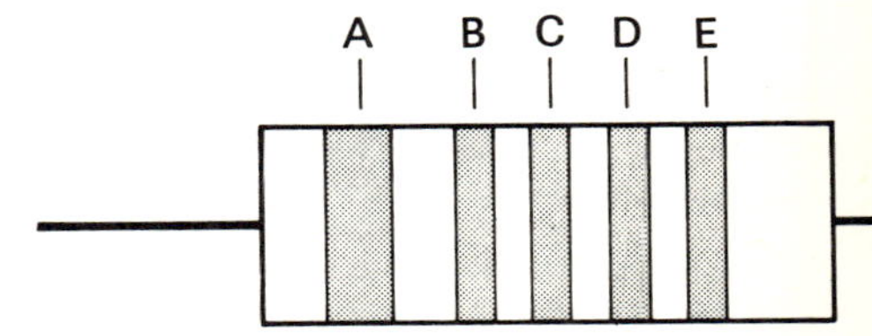

APPENDIX 2.
Capacitor Colour Dot Value

Colour	Value	Multiplier (Paper & Mica)	Multiplier (Ceramic)	Tolerance % (Paper & Mica)	Tolerance % (Ceramic) Above 10pf	Tolerance % (Ceramic) Below 10pf
Black	0	1	1	20	20	2.0
Brown	1	10	10	1	1	
Red	2	10^2	10^2	2	2	
Orange	3	10^3	10^3	3		
Yellow	4	10^4		4		
Green	5	10^5		5	5	0.5
Blue	6	10^6		6		
Violet	7	10^7		7		
Gray	8	10^8	0.01	8		0.25
White	9	10^9	0.1	9	10	1.0
Gold		0.1		5		
Silver		0.01		10		
No colour				20		

APPENDIX 3.
Chassis Wiring Colour Code

Black — Grounds, grounded elements, return paths
Blue — Plates, collectors
Brown — Filament heaters above ground
Gray — A.C. power lines
Green — Control grids, transistor base connection
Orange — Screen grids
Purple — Power supply negative connection (alternate colour)
Red — Power supply B — positive connection
Violet — Power supply negative connection
White — Returns above/below ground, AVC, miscellaneous
Yellow — Cathodes, transistor emitter connection.
Adapted from MIL STD 122

APPENDIX 4.
Loudspeaker Leads Colour Code

Field Coil
Black and red — Start
Slate and red — Tap
Yellow and red — Finish
Voice Coil
Black — Start
Green — Finish

APPENDIX 5.
Resistor Colour Code

Axial Lead

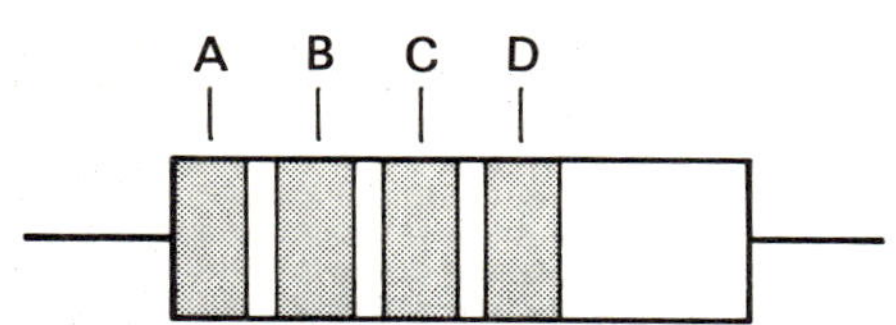

Radial Lead

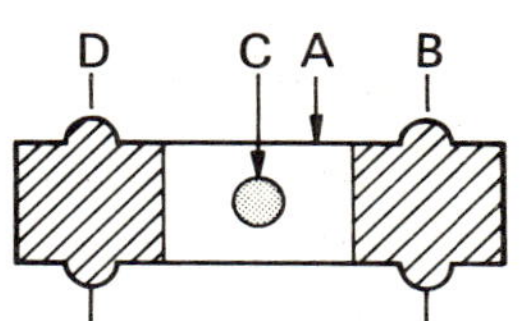

A — First digit of resistor value
B — Second digit of resistor value
C — Multiplier, the number of zeros. List in the order: A, B, and C
D — Total resistor value tolerance

NOTE
Significance of A, B, C, and D, is identical to axial lead resistors, only when leads point down, and dot C faces observer.

Resistor Colour Band Value (Value In OHMS)

Colour	Value	Multiplier	Tolerance %
Black	0	1	
Brown	1	10	
Red	2	10^2	
Orange	3	10^3	
Yellow	4	10^4	
Green	5	10^5	
Blue	6	10^6	
Violet	7	10^7	
Gray	8	10^8	
White	9	10^9	
Gold		0.1	5
Silver		0.01	10
No Colour			20

APPENDIX 6.
Transformer Leads Colour Code

Audio type
Black — Grid return lead
Blue — 1. Primary winding — plate lead (finish)
2. Primary C.T. winding — plate lead (start) May be used if polarity is not important
Brown — Primary C.T. winding — plate lead (start)
Green — 1. Secondary winding — grid lead (finish)
2. Secondary C.T. winding — grid lead (start) May be used if polarity is not important
Red — B+ lead
Yellow — Secondary C.T. winding — grid lead (start)
Intermediate frequency (I.F.) type
Black — 1. Grid return lead
2. Diode return lead
3. Secondary C.T. winding — tap lead
Blue — Plate or anode lead
Green — 1. Grid lead
2. Diode lead
Green with black stripes — Plate or anode lead of second diode (If secondary winding is centre-tapped)
Red — B+ lead

Power type
Black — Primary winding (tapped) — common
Black with red stripes — Primary winding (tapped) — Finish
Black with yellow stripes — Primary winding (tapped) — Tap
Brown — Filament winding No. 2
Brown with yellow stripes — Filament winding No. 2 — Centre tap
Green — Filament winding No. 1
Green with yellow stripes — Filament winding No. 1 — Centre tap
Red — High-voltage plate winding
Red with yellow stripes — High-voltage plate winding — Centre tap
Slate — Filament winding No. 3
Slate with yellow stripes — Filament winding No. 3 — Centre tap
Yellow — Rectifier filament winding
Yellow with blue stripes — Rectifier filament winding — Centre tap
Adapted from E.I.A. colour code standards

APPENDIX 7.

TWIST DRILL SIZES

DRILL SIZE	DECIMAL	DRILL SIZE	DECIMAL	DRILL SIZE	DECIMAL	DRILL SIZE	DECIMAL
80	.0135	42	.0935	**13/64**	**.2031**	X	.3970
79	.0145	**3/32**	**.0938**	6	.2040	Y	.4040
1/64	**.0156**	41	.0960	5	.2055	**13/32**	**.4062**
78	.0160	40	.0980	4	.2090	Z	.4130
77	.0180	39	.0995	3	.2130	27/64	.4219
76	.0200	38	.1015	**7/32**	**.2188**	**7/16**	**.4375**
75	.0210	37	.1040	2	.2210	29/64	.4531
74	.0225	36	.1065	1	.2280	**15/32**	**.4688**
73	.0240	**7/64**	**.1094**	A	.2340	31/64	.4844
72	.0250	35	.1100	**15/64**	**.2344**	**1/2**	**.5000**
71	.0260	34	.1110	B	.2380	33/64	.5156
70	.0280	33	.1130	C	.2420	17/32	.5312
69	.0292	32	.1160	D	.2460	35/64	.5469
68	.0310	31	.1200	**1/4**	**.2500**	**9/16**	**.5625**
1/32	**.0312**	**1/8**	**.1250**	E	.2500	37/64	.5781
67	.0320	30	.1285	F	.2570	19/32	.5938
66	.0330	29	.1360	G	.2610	39/64	.6094
65	.0350	28	.1405	**17/64**	**.2656**	**5/8**	**.6250**
64	.0360	**9/64**	**.1406**	H	.2660	41/64	.6406
63	.0370	27	.1440	I	.2720	21/32	.6562
62	.0380	26	.1470	J	.2770	43/64	.6719
61	.0390	25	.1495	K	.2810	**11/16**	**.6875**
60	.0400	24	.1520	**9/32**	**.2812**	45/64	.7031
59	.0410	23	.1540	L	.2900	23/32	.7188
58	.0420	**5/32**	**.1562**	M	.2950	47/64	.7344
57	.0430	22	.1570	**19/64**	**.2969**	**3/4**	**.7500**
56	.0465	21	.1590	N	.3020	49/64	.7656
3/64	**.0469**	20	.1610	**5/16**	**.3125**	25/32	.7812
55	.0520	19	.1660	O	.3160	51/64	.7969
54	.0550	18	.1695	P	.3230	**13/16**	**.8125**
53	.0595	**11/64**	**.1719**	**21/64**	**.3281**	53/64	.8281
1/16	**.0625**	17	.1730	Q	.3320	27/32	.8438
52	.0635	16	.1770	R	.3390	55/64	.8594
51	.0670	15	.1800	**11/32**	**.3438**	**7/8**	**.8750**
50	.0700	14	.1820	S	.3480	57/64	.8906
49	.0730	13	.1850	T	.3580	29/32	.9062
48	.0760	**3/16**	**.1875**	**23/64**	**.3594**	59/64	.9219
5/64	**.0781**	12	.1890	U	.3680	**15/16**	**.9375**
47	.0785	11	.1910	**3/8**	**.3750**	61/64	.9531
46	.0810	10	.1935	V	.3770	31/32	.9688
45	.0820	9	.1960	W	.3860	63/64	.9844
44	.0860	8	.1990	**25/64**	**.3906**	1	**1.0000**
43	.0890	7	.2010				

B.S.A. TOOLS (CANADA) LIMITED

APPENDIX 8.
Conduit Bending Radius

Conduit Interior Diameter in Inches	Minimum Radius (1) N.M.T.B.A.	(2) N.E.A.	(3) C.E.C.
1/2	4	4	Section 12-140. Radii of bends in raceways and armoured cable. Where armoured cable or raceways of the drawn-in type are bent during installation, the radius of the curve of the inner edge shall be at least 6 times the internal diameter of the armouring or raceway.
3/4	4 1/2	5	
1	5 3/4	6	
1 1/4	7	8	
1 1/2	8 1/4	10	
2	9 1/2	12	
2 1/2	10 1/2	15	
3	13	18	
3 1/2	15	21	
4	16	24	
5	24	30	
6	30	36	

NOTE
1. National Machine Tool Builders Association.
2. National Electrical Code (United States of America).
3. Canadian Electrical Code.
4. The term, raceway, applies to rigid conduit. See C.E.C., Section O, Definitions of special terms.
5. The minimum radius values given DO NOT pertain to conduit carrying lead-sheated cable.

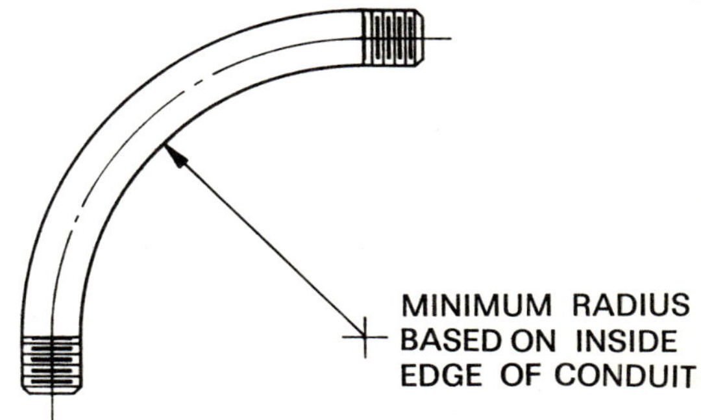

APPENDIX 9.
Minimum Bending Radius For Aluminum And Steel Alloy Sheets

Ninety degree cold bends for either aluminum alloy or steel sheet are dependent on five factors. They are the metallic composition of the alloy, its thickness, and whether or not it is annealed, strain hardened or solution heat treated.

The aluminum alloys listed below are specific alloys of the Aluminum Company of America. Alloy composition can be obtained by referring to its number in the manufacturers handbook. Those alloys that are heat treated have a "T" suffix, those that are hardened an "H" suffix and those that are annealed, an "O" suffix.

The following table gives the minimum bending radius for certain aluminum alloy sheet.

Alloy Designation	1/32" thick	1/16" thick	1/8" thick
Aluminum 1100—0	0	0	0
Aluminum 1100—H18	1/2 — 1 1/2	1 — 2	1 1/2 — 3
Alclad 2014—0	0 — 1	0 — 1	0 — 1
Alclad 2014—T6	3 — 5	3 — 5	4 — 5
Aluminum 2219—31		1/2 — 1 1/2	1 — 1 1/2
Aluminum 2219—T87		2 — 4	3 — 5
Aluminum 2024—0	0 — 1	0 — 1	0 — 1
Aluminum 2024—T86	5 — 7	6 — 8	7 — 10
Aluminum 3003—0	0	0	0
Aluminum 3003—H18	1 — 2	1 1/2 — 3	2 — 4

Cold Rolled Steel
Annealed stainless steel
The ductility of these steel alloys is such that it is possible to form 90° bends with a zero radius. However, when thin sheet is used, a possibility exists that the bending die will cut the metal, rather than allow it to assume its bent contour. Consequently, it is unadvisable to attempt such a procedure.

In most cases a zero radius bend would not be required. A general formula that holds for most requirements is R=T, where R = the minimum bending radius, and T = the material thickness.

A practical rule of thumb frequently carried out is to use a minimum bending radius of 1/32" for 90° bends.

APPENDIX 10.
Wire Gauge Number And Current Carrying Capacity

AWG NUMBER (American wire gauge	nom. dia. bare.)	CURRENT CARRYING CAPACITY (AMPERES)
300 CM*	0.630	240
250 CM	0.575	215
4/0	0.528	195
3/0	0.470	165
2/0	0.419	145
0	0.373	125
1	0.332	110
2	0.292	100
3	0.260	80
4	0.232	70
6	0.184	55
8	0.146	40
10	0.1019	30
12	0.0808	20
14	0.0641	15
16	0.0508	10
18	0.0403	7
20	0.0320	5
22	0.0253	3

These values are based on a cable or raceway installation and would be higher for a single conductor. Also ratings of a higher value are possible with insulative material of a higher dielectric constant.

NOTE
1. Values for conductors smaller than #22 GA. and larger than 300 CM. are obtainable in code handbooks and other electrical texts.
2. *CM — circular mills.

APPENDIX 11.
Wire Thickness And Sheet Metal Gauges

Gauge Number	Steel Wire Gauge (U.S.) Dec. Equiv. in inches	Manufacturers' Standard Gauge for Sheet Steel Dec. equiv. in inches
2/0	0.3310	
1/0	0.3065	
1	0.2830	
2	0.2625	
3	0.2437	0.2391
4	0.2320	0.2242
5	0.2070	0.2092
6	0.1920	0.1943
7	0.1770	0.1793
8	0.1620	0.1644
9	0.1483	0.1495
10	0.1350	0.1345
11	0.1205	0.1196
12	0.1055	0.1046
13	0.0915	0.0897
14	0.0800	0.0747
15	0.0720	0.0673
16	0.0625	0.0598
17	0.0540	0.0538
18	0.0475	0.0478
19	0.0410	0.0418
20	0.0348	0.0359
21	0.0317	0.0329
22	0.0286	0.0299
23	0.0258	0.0269
24	0.0230	0.0239
25	0.0204	0.0209
26	0.0181	0.0179
27	0.0173	0.0164
28	0.0162	0.0149
29	0.0150	0.0135
30	0.0140	0.0120

NOTE

1. The smaller the gauge number, the larger the size of the equivalent wire or sheet.
2. The Steel Wire Gauge is used by practically all of the manufacturers of steel wire in the U.S.A., and while not a legal gauge, it is sanctioned by the U.S. Bureau of Standards.
3. The Manufacturers' Standard Gauge is based upon a weight of 41.82 pounds per square foot per inch thick of material.
4. Aluminum and copper alloy sheets are specified in decimals and fractions of an inch.
5. The use of decimal values in place of gauge numbers is now being used extensively to eliminate confusion.

APPENDIX 12.
Graphical Electrical Symbols For Architectural Plans

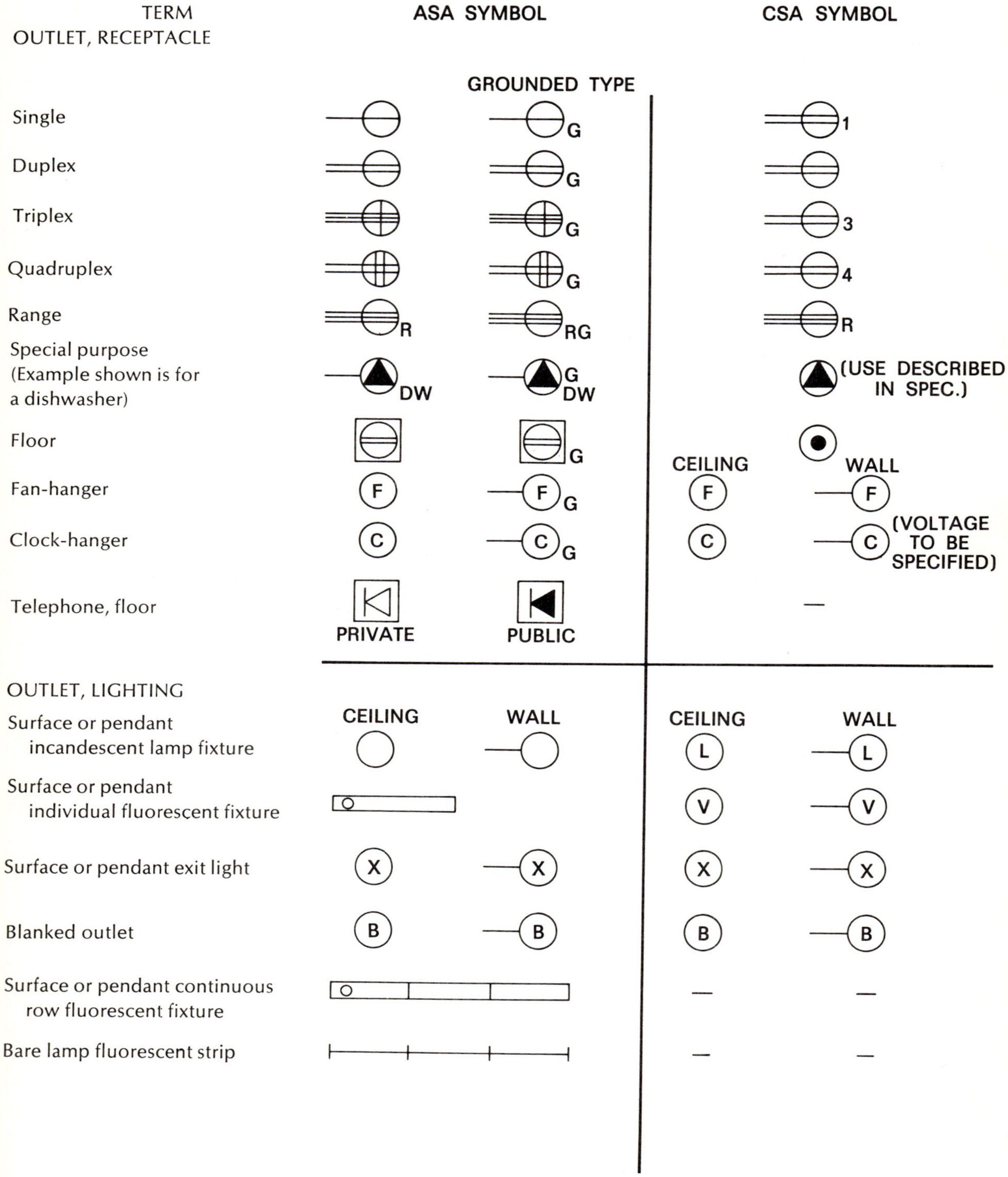

TERM	ASA SYMBOL	ASA SYMBOL (GROUNDED TYPE)	CSA SYMBOL	CSA SYMBOL
OUTLET, RECEPTACLE				
Single		G		1
Duplex		G		
Triplex		G		3
Quadruplex		G		4
Range	R	RG		R
Special purpose (Example shown is for a dishwasher)	DW	G DW		(USE DESCRIBED IN SPEC.)
Floor		G		
			CEILING	WALL
Fan-hanger	F	F G	F	F
Clock-hanger	C	C G	C	C (VOLTAGE TO BE SPECIFIED)
Telephone, floor	PRIVATE	PUBLIC		—
OUTLET, LIGHTING	CEILING	WALL	CEILING	WALL
Surface or pendant incandescent lamp fixture			L	L
Surface or pendant individual fluorescent fixture			V	V
Surface or pendant exit light	X	X	X	X
Blanked outlet	B	B	B	B
Surface or pendant continuous row fluorescent fixture			—	—
Bare lamp fluorescent strip			—	—

TERM	ASA SYMBOL	CSA SYMBOL
Relay operated low-voltage switched outlet	CEILING (L); WALL —(L)	—
OUTLET, SWITCH		
Single pole	S	S
Double pole	S_2	S_2
Three-way	S_3	S_3
Four-way	S_4	S_4
Door	S_D	S_D
Key-operated	S_K	S_K
Switch and pilot lamp	S_P	S_P
Time	S_T	—
Momentary contact	S_{MC}	S_{MC}
Circuit breaker	S_{CB}	S_{CB}
Remote control	—	S_{RC}
Low-voltage switching system	S_L	—
Fused	—	S_F
Ceiling, pull	(S)	(S) CEILING; —(S) WALL
PANELS & RELATED COMPONENTS PANELS		
Flush-mounted lighting	[symbol]	—
Surface-mounted lighting	[symbol]	[symbol]
Power	[symbol]	[symbol]
Controller	[symbol]	[symbol]
Switch	(EXTERNALLY OPERATED DISCONNECT) [symbol]	(ISOLATING) [symbol]
Transformer	—	(T)

TERM	ASA SYMBOL	CSA SYMBOL
OUTLET, SIGNALLING SYSTEM		
Annunciator		
Bell		
Buzzer		
Chime	CH	—
Opener, electric door	D	D
Pushbutton		
Plug, maid's signal	M	M
Box, interconnection		
Transformer, bell-ringing	BT	T
Telephone, interconnecting		
Outlet, radio	R	R
Telephone, outside		
Outlet, television	TV	—
Switchboard, telephone	—	
Alarm, fire	—	F
Device, automatic fire alarm	—	FS
Plug, nurse's signal	—	N
CIRCUIT PATH DESIGNATION		
Branch, concealed in ceiling/wall		
Branch, concealed in floor		
Branch, exposed		
Home run to panelboard	2 1	
(ARROWS INDICATE NUMBER OF CIRCUITS NUMBERS IDENTIFY CIRCUITS)		
Three wire		
Four wire		
Feeder		

CSA symbols based on Canadian Standards Association Standard Z99.3-1946. New CSA Standard is in preparation.

APPENDIX 13.

Partial List Of Electrical/Electronic Components And Devices Reference Designations

(To be used on wiring diagrams involving graphic symbols in conjunction with numbers for identification purposes)

Component or Device	MIL STD 16B	CGSB 33-GP-1
Alarm	DS	DS
Amplifier	A	AR
Annunciator	DS	DS
Antenna	E	ANT
Autotransformer	T	T
Ballast, lamp	—	RT
Battery	BT	BT
Bell	DS	DS
Breaker, circuit	CB	CB
Buzzer	DS	DS
Cable	W	W
Capacitor	C	C
Cell, light sensitive, photo emissive		
Cell, selenium		
Choke	L	L
Coil, relay or tuning	L	—
Conduit	—	W
Connector, fixed	J	J
Connector, movable	P	P
Contactor, electrically operated	K	—
Crystal, diode	CR	CR
Crystal, piezoelectric	Y	Y
Cutout, thermal	S	—
Device, indicating	DS	—
Filter	FL	—
Fuse	F	F
Generator	G	G
Head, erasing, recording, reproducing	PU	—
Heater	HR	HR
Instrument	M	M
Insulator	E	—
Interlock, safety electrical	S	—
Key, switch	S	—
Key, telegraph	—	S
Lamp, pilot, illuminating, signal	DS	LP

Component or Device	MIL STD 16B	CGSB 33-GP-1
Meter	M	M
Microphone	MK	MK
Modulator	A	—
Motor	B	B
Motor-generator	MG	MG
Oscilloscope	M	—
Phototube	V	—
Point, test	TP	—
Potentiometer	R	R
Receiver	HT	RE
Rectifier	CR	CR
Regulator, voltage	VR	—
Relay, electrically operated	K	K
Resistor	R	R
Rheostat	R	R
Shunt	R	—
Solenoid	L	L
Speaker	LS	LS
Strip, terminal	TP	TB
Switch	S	S
Thermistor	RT	—
Thermocouple	TC	TC
Thermostat	S	—
Transducer	MT	—
Transformer	T	T
Transistor	Q	Q
Tube, electron	V	V
Varistor, asymmetrical	CR	—
Varistor, symmetrical	RV	—
Vibrator	—	VP
Waveguide	W	—
Winding	L	—
Winding, inductive	—	L
Winding, resistive heating	—	HR
Winding, resistive voltage dropping	—	R
Wire	W	W

APPENDIX 14.

Partial List of Standard Abbreviations for Electrical/Electronic Terms

(For use on technical drawings, for descriptive purposes)

(Adapted from C.G.S.B. 33-GP-1 and MIL STD 12B)

TERM	ABBREV.
Alternating Current	AC
Alternator	ALTR
Ammeter (AM)	AMM
Ampere	AMP
Amplifier	AMPL
Amplitude Modulation	AM.
Anode	A
Antenna	ANT.
Armature	ARM.
Assembly	ASSY
Attenuator	ATTEN
Audio	AUD
Audio Frequency	AF
Automatic Frequency Control	AFC
Automatic Volume Control	AVC
Autotransformer	AUTO/T
Band Pass	B/P
Band Width	B/W
Battery	BAT.
Beacon	BCN
Beat-frequency Oscillator	BFO
Blowout Coil	BOC
Breaker	BKR
Broadcast	BC
Buzzer	BUZ
Capacitor	CAP.
Cathode	K
Cathode-ray Tube	CRT
Centre Tap	CT
Circuit	CKT
Circular Mil	CM
Commutator	COMM
Continuous Wave	CW
Crystal	XTAL
Crystal Oscillator	C/OSC
Cycle	CY

TERM	ABBREV.
Decibel	DB
Demand Meter	DM
Demodulator	DEMOD
Detector	DET
Diagram	DIAG
Diameter	DIA
Direct Current	DC
Direction Finding	DF
Disconnect	DISC.
Discriminator	DISCR
Distance Measuring Equipment	DME
Double-pole	DP
Double-pole Double-throw	DPDT
Double-pole Single-throw	DPST
Dynamometer (DYNO)	DYNM
Electrolytic	ELECT.
Electronic Industries Assoc.	EIA
Engineering Order	EO
Facsimile	FAX
Farad	F
Federal Communications Commission	FCC
Filament	FIL
Filter	FLT
Fluorescent	FLUOR
Four-conductor	4/C
Four-pole	4/P
Frequency	FREQ
Frequency Modulation	FM
Gas-filled	GF
Generator	GEN
Grid	g
Grommet	GROM
Ground	GND
Heater (HTR)	h
Henry	H

TERM	ABBREV.
High-frequency	HF
High-pass	HP
High-voltage	HV
Inductance	IND
Inductance — capacitance	LC
Inside Diameter	ID
Institute of Electrical and Electronic Engineers	IEEE
Intercommunication	INTERCOM
Intermediate Frequency	IF
Intermediate Continuous Wave	ICW
Junction Box	JB
Kilohertz per Second	KH
Kilowatt	KW
Local Oscillator	LO
Low-frequency	LF
Low-voltage	LV
Lumens per Watt	LPW
Magnetic Amplifier	MAG AMPL
Megahertz per Second	MC
Megohm (MEGO)	MEG
Microampere	μA
Microfarad	μF
Micromho	μMHO
Micromicrofarad	$\mu\mu$F
Microphone	MIKE
Microvolt	μV
Milliampere	MA
Millivolt	MV
Modulated Continuous Wave	MCW
Motor Generator	MG
Mounting	MTG
National Electrical Code	NEC
Negative	NEG
Network	NET.
Oscillator	OSC
Outside Diameter	OD
Pentode	PENT
Permanentmagnet	PM
Phase	PH
Plan Position Indicator	PPI
Polarity	PO
Polyvinylchloride	PVC
Positive	POS
Potentiometer	POT.
Power	PWR
Power Factor	PF
Power Supply (PWR SUP)	PS

TERM	ABBREV.
Primary	PRI
Push-Pull	PP
Quartz	QTZ
Radio Direction Finding	RDF
Radio Frequency	RF
Radio-frequency Choke	RFC
Radio Telephone	RT
Receiver	RCVR
Rectifier	RECT
Regulator	REG
Resistance	RES
Rheostat	RHEO
Root Mean Square	RMS
Rotary Switch	RSW
Schematic	SCHEM
Secondary	SEC
Servomechanism	SERVO
Series	SER
Shield	SHLD
Signal	SIG
Single-conductor	SC
Single-phase	SPH
Single-pole	SP
Single-throw	ST
Solenoid	SOL
Speaker	SPKR
Specification	SPEC
Superheterodyne	SUPERHET
Switchboard	SWBD
Switchgear	SWGR
Synchronous	SYN
Telecommunication	TELECOM
Telemeter	TLM
Teletypewriter	TTY
Terminal	TERM.
Thermistor	TMTR
Thermocouple	TC
Thermostat	THERMO
Three-conductor	3/c
Three-phase	3PH
Three-pole	3P
Three-way	3 way
Time-delay	TD
Toggle	TOG.
Tolerance	TOL
Transceiver	XCVR
Transformer	XMFR
Transistor (TSTR)	XSTR

TERM	ABBREV.	TERM	ABBREV.
Transmit-receive	TR	Very-high Frequency	VHF
Transmitter	XMTR	Vibrator	VIB
Triode	TRI	Video Frequency	VDF
Triple-pole	3P	Volt	V
Triple-pole Double-throw	3PDT	Voltmeter	VM
Two-conductor	2/c	Volt-ohm Milliammeter	VOM
Two-phase	2PH	Volume	VOL
Ultra-High Frequency	UHF	Watt	W
Vacuum tube	VT	Winding	WDG
Variable-frequency Oscillator	VFO	Wire-wound	WW

NOTE
Abbreviations enclosed by brackets are the exclusive designations given by MIL STD 12B.
However, many of the remaining abbreviations are universally used by both American and Canadian authorities, but some may be exclusively either one.
In any event, abbreviations should not be used if ambiguity of meaning becomes obscure. The use of technical terms in their complete spelling, will always eliminate this problem.

Glossary

GLOSSARY OF TERMS

Aerospace — This term implies both aeronautics and astronautics.

Alloy — A metal composed of two or more elements for the purpose of increasing strength, hardness, or other desired characteristics.

Alumina — A term referring to the oxide of aluminum.

Ambient Temperature — The air temperature surrounding a device.

Amplifier — A device which increases the output of an incoming electrical signal.

Amplifier, Intermediate Frequency — In a superheterodyne receiver the I.F. amplifier is that device that amplifies a signal which is the difference between the incoming signal and a local oscillator signal.

Amplifier, Radio Frequency — An amplifier that increases the signal strength of a R.F. signal before it is detected.

Amplifier, Video — An amplifier which increases the gain of the picture or video signal of a television receiver.

Anodizing — A rustproofing process used on aluminum by creating a thin coating of oxide on its surface by the use of an acid bath and direct current.

Aperture Card — In microfilming processes the 35 mm. film is mounted in a standard sized card having an opening, or aperture, to receive the film.

Arcing — An electrical charge that jumps between two metal surfaces and creates a flash.

Armature — The rotating part within a motor, or the moving part or arm in a relay solenoid.

Augur Bunk Feeder — A feeder of this sort consists of a bunk or bunker containing the feed or grain, and which is fed by the screw action of an electrically powered augur.

Azimuth — This is the angular distance measured from 0° to 360° in a horizontal plane.

Baffle — A thin member used as a separator.

Bimental Strip—A metal strip fabricated from a laminate of two metals of different coefficients of expansion.

Bonding — The attachment of an electrical conductor to some contact point by means of brazing, welding, or soldering.

Brazing — The process of joining two metal parts by the addition of molten brazing metal (spelter) without melting the parts.

Bumping — A slang term indicating that the flat surface of a sheet metal plate has been lifted or depressed over a localized area by means of a die.

Bus Bar — A metallic bar of rectangular cross section which serves as a common bonding strip and to which are attached electrical conductors.

Cam — A mechanical device sometimes having a heart shaped configuration. Due to the cam's irregular shape, it is capable of lifting the cam follower as it rotates.

Camera, High-Resolution — When a photograph of high optical fidelity is required, a camera having a high-resolving power is demanded. This means that the camera has an exceptionally good lens system.

Capacitance — The electrical characteristics of a capacitor or condenser, that is, its ability to hold an electrical charge. Capacitance is measured in farads.

Cathode Sputtering — A method of depositing material on the inside walls of a vacuum or gas-filled discharge tube from which the cathode is formed.

Catwhisker — A fine, hard, metallic wire which is used in a point contact diode to make electrical contact on the germanium crystal.

Centrifugal — A force which tends to throw out a rotating body tangentially to its rotating path.

Choke — An electrical device similar to one winding of a transformer that is used in rectifier circuits to smooth out alternating-current ripple.

Coaxial — Two conductors in a cable are coaxial when they have a common axis and the outer cable takes the form of a circle when viewed in cross section.

Code, Morse — Morse code is a method of transmitting intelligence by signals consisting of dots and dashes.

Computer, Analog — A computer that calculates the process of substitution for any given physical quantity in direct proportion to the same laws of behaviour as the original quantity.

Computer, Digital — An electronic computer that calculates by using conventional arithmetic digits.

Concentric — Two circles are concentric when they have a common centre.

Control, Numerical — In modern machine shop practice a method whereby instructions are fed to a cutting machine by means of numerical form on tape or punched cards.

Coordinagraph — An electromechanical device used in making large-scale precision drawings for integrated circuits. When mathematical coordinates are manually, or electronically, fed into this machine, it automatically scribes the I.C. shape.

Coordinate — Mathematical means of establishing the spatial position of a point by the use of horizontal and vertical distances.

Core, Memory — In some computers, a ferroceramic magnetic cylinder or core is used as a storage or memory core.

Corner, Relieved — A corner of a box which has been predrilled before forming has taken place. This eliminates metal stress at that point.

Decibel — A measuring unit used in the comparison of two levels of power in electrical communication circuits.

Delineate — When an engineering drawing is constructed, the lines forming the views are said to delineate them.

Diaphragm — A thin, metal, rubber, or plastic membrane.

Diazo — In the production of whiteprints, a light sensitive chemical known as diazo dyestuff is utilized on the printing paper.

Dielectric Constant — A number rating the insulative value of nonconductors. Air is considered as 1.

Disconnect Switch — A manually operated switch located in a locked enclosure.

Ductile — A metal which is ductile will bend or form easily, without fracturing taking place.

Dynamic — A part or component which is dynamic is a moving one.

Electrostatic — An electrical charge that has been created by frictional methods.

Electrostatic Copy — Opaque copies printed by electrostatic means. This is a rapid, dry print process currently in use in commercial, educational, and engineering fields.

Emission, Secondary — Electrons striking the anode or plate of a vacuum tube and are further emitted.

Emission, Thermionic — The emission of electrons from the glowing filament, or heater (cathode), of a vacuum tube under the attraction of a high-positive potential.

Emulsion — A thin coating of a composite liquid substance which is spread over the surface of a photographic material. It consists of minute particles of one liquid suspended in another.

Encapsulated — When an electronic circuit or device is entirely surrounded by a shock resistant foam or epoxy resin within the confines of a specific three-dimensional shape, it is referred to as being potted or encapsulated.

Envelope — The glass enclosure surrounding the elements in a vacuum tube.

Epoxy — A type of resin that is used as a basic ingredient for paints and adhesives having good moisture resistance and high bonding strength.

Etchant — A fluid which is used to dissolve or etch away a substance.

Eutectic Compression Bonding — In attaching thin metallic leads to thick-film circuits, a bonding method known as eutectic compression bonding is used. This method relies on the principle that when certain alloys of metals are heated, one dissolves into the other, forming an electrical path of high conductivity.

Extrapolate — A method of establishing unknown values along a plot on a graph, or on a projected extension of the plot.

Ferroceramic — A mixture of powdered iron when bonded to a ceramic material.

Filament — The fine metallic element within a tube or lamp which is heated to incandescence by electrical current.

Folded Dipole Antenna — A two-element dipole antenna which has the ends of its open elements folded 180°, then electrically and mechanically joined.

F. M. Tuner — An electronic tuning device that is capable of detecting frequency modulated radio frequency signals. These tuning units can, in some cases, be adapted to radio receivers having the capability of receiving only A.M. (amplitude modulated) signals.

Frequency Standard — An electronic device that provides vibrations or oscillations of a precise number at a constant rate.

Ganged — Two separate mechanical parts joined together to act as a unit.

Gasket — A thin, semiflexible member made from cork, neoprene, copper, or plastic that is used as a seal between two parts conveying a liquid or gas.

Germanium — A rare element possessing the ability to rectify or pass electrical current in one direction, and is used in semiconductor devices.

Hermetically Sealed — When a component is hermetically sealed the air surrounding it is evacuated and maintained at a pressure much lower than atmospheric.

Incandescence — When electrical current is passed through the filament of a lamp, it glows or incandesces.

Inductance — The electrical characteristic of an inductor, or coil. Inductance is measured in henrys.

Inert — A gas or other substance that is inert remains stable and does not react when brought in contact with another gas or substance.

Ionization — A molecule of a substance breaking up or disassociating into two or more oppositely charged ions.

Knockout, Conduit — A prestamped circular shape which can easily be removed from an electrical service box for the installation of a conduit or cable.

Kovar — In the electronics industry, metal-to-glass seals use a metal alloy known as Kovar. This alloy has a temperature coefficient of expansion matching that of many types of glass. It consists of nickel, cobalt, and iron.

Lamination — The attachment of several layers of a material to one another.

Linkage — The arms or links that are sometimes used to actuate a mechanical device remotely. These arms or links are joined together by pivoting connections.

Load Cell — An electromechanical device used in weighing. The electrical current generated by the cell is a function of the load on it, and this current is automatically translated into weight units by means of a special meter.

Lobe — That portion of a cam which comes in contact with the cam follower, and has the greatest radial distance on the cam.

Loran — A radio direction finding method utilizing pulsed radio signals transmitted from two synchronized stations.

Luminaire — A trade term generally used instead of a 'light fixture.'

Manganese — In the construction of dry cells, the oxide of this element is used as a depolarizer.

Matrix — The substance or material upon which another substance is deposited.

Metals, Dissimilar — An example of dissimilar metals is copper and constantan. When heat is applied to a hot junction (welded end) of these metals, an electrical current is produced at the cold junction (open end).

Meter, Field Strength — A meter to measure the radiation pattern given off by the antenna of a transmitter.

Mixer — Either a special multigride tube used in combining a number of input signals, or a circuit used to mix signals from different program sources.

Modulate — An electrical current is modulated when its amplitude, or frequency, is altered according to the variations of another current known as the modulating signal.

Monopole — An antenna having a single element. It is usually oriented in a vertical sense.

Multiconductor — A cable constructed with more than one lead or conductor.

Neon — A rare, inert gas which is used in neon lighting.

Network — A system of resistors and capacitors, or other electronic components connected in a functional circuit.

Nonthermal — A condition in which an increase in heat is not manifest.

Nuclear Power Station — An electrical power generating station utilizing radioactive fuels in a nuclear reactor as a means of providing heat energy. This heat energy is subsequently used to generate steam that drives turbine operated generators.

Offset Duplicator Master — A master printing plate made from thin photosensitized metal, upon which the image of a microfilm has been developed. It is used in the offset printing process to make opaque copies.

Oscillator — An electronic device which vibrates or oscillates at a fixed frequency often under the control of a precisely ground quartz crystal.

Oxidation — The process of rusting of metallic substances.

Palladium — An element of the noble metal type, closely resembling platinum. It is a soft malleable metal with a pronounced tendency to adsorb (the condensing and holding of a gas on a metallic surface) many different types of gasses.

Parabolic — A geometric form, curved in shape and having a focal point located along the central axis of the curve. Reflectors constructed in this shape are used in the transmission of microwaves.

Phase — An electrical term indicating whether or not an electrical current is in step with another electrical current.

Photoblowback — An enlarged photo of a microfilm, brought back to its original size.

Photoengraving — The method of creating an image on a photosensitized printing plate by a photographic process.

Piezoelectric — A phenomena of certain natural occurring crystals, that when a mechanical pressure is applied to them, a minute electrical current proportional to the applied pressure is generated.

Plastic, Methyl Acrylic — The chemical name for a transparent thermoplastic having the trade name of Plexiglas.

Plastic, Polyamide — Plastics bearing the generic name of nylon are a long-chain synthetic polyamide.

Point-to-point — A connection from one specific solder point to another in electronic wiring diagrams.

Polarity — The terminals of a cell or battery have distinct connections or *poles*, which are positive and negative, hence the term polarity.

Polyester — When used in conjunction with a hardener or catalyst, this resin is applied to fibreglas cloth moulded over a form. The resultant is a rigid semitranslucent shape.

Polyphase Device — Any generator that produces electric current or motor that operates on electric current of more than one phase.

Printing, Offset — A method of printing involving a secondary or intermediate cylinder which carries the ink from the plate cylinder to the impression cylinder.

Prototype — A preliminary, full-size working model of a part or assembly. It is used for testing and evaluation purposes.

Pulsating — An electrical current is said to be pulsating if its amplitude is variating.

Raceway — A metallic or plastic trough or conduit, designed to hold electrical conductors or cables running from one point to another. The raceway may be rigid or flexible.

Radar — An acronym meaning radio direction and ranging.

Reflector, Para-corner — A radio transmitter antenna reflector, having two, flat, vertical elements joined together forming a vee, and having its antenna dipole located along the centre of the reflector's horizontal axis.

Region, Intrinsic — The region of a semiconductor that has electrical characteristics similar to a pure crystal.

Registration — Two or more colours superimposed exactly over each other such as are required in printing coloured photographs.

Resistance — The restriction of flow of electrical current through a conductor.

Resonant — The frequency at which a tuned electronic circuit resonates or oscillates.

Response, Frequency — The limit or range of a device used in the production of sound waves in its capacity to reproduce speech or music faithfully.

Rib — A thin mechanical member used for stiffening or shaping.

Rochelle Salts — A type of crystal which is capable of producing a peizoelectric effect.

Rotor — A part or component of a stationary assembly which rotates.

Scanner, Flying Spot — A special electronic tube having the ability to scan information such as would appear on an engineering microfilm, and by electronic means convert this information into corresponding electrical pulses of a digital nature. These electrical pulses could then be stored in a computer memory system.

Selenium — A nonmetallic element which has the ability to permit the passage of electron flow in one direction only, and which is also light sensitive.

Semiconductor — Materials such as selenium and germanium are semiconductors, thus implying their ability to partially conduct electrical current.

Servomechanism — A motorlike device which drives or moves a mechanical apparatus in direct response to electrical signals applied to it.

Sheath — A shroud or covering about a conductor or cable.

Shielding — In order to prevent stray electrical energy from causing problems in electronic circuits, wire braid around cables, and metal cans around components are used. These shields are suitably grounded.

Shim — A thin member frequently constructed from brass shimstock which is used to alter the position of a machined part relative to another by a small, precise amount.

Silicon — An element used in power rectifier diodes because of its ability to pass current in one direction only.

Silicone — The trade name of a resin derived from silicon which has good electrical properties and a high resistance to heat and chemical corrosion.

Silk Screening — A process in which a stencil containing a specific pattern is attached to a porous silk screen through which ink is passed.

Sine Wave — A wave form generated by a single-frequency, alternating current. Its graphic form is a plot of all points traced by the sine of an angle as it is rotated through 360°.

Sintering — The heating of a powdered material into a solid mass without melting taking place.

Slip Rings — Metallic rings joined to the rotating armature windings in a generator, which contact brushes connected to the external circuit.

Slug — A mass of iron or ferroceramic material located within certain types of inductors.

Solid-state — In the electronic industry, solid-state refers to semiconductor devices, as for example, transistors.

Sonar — An acronym meaning sound navigation and ranging. It uses a transmitting and receiving device that sends and receives underwater ultrasonic sound waves.

Squib, Electric — A pyrotechnic device used in aerospace equipment to fire the igniter of a rocket, or operate an explosive bolt.

Substrate — The insulative board upon which the circuit is made in printed circuitry.

Superheterodyne — A type of radio receiver utilizing a local oscillator whose signal is mixed with the incoming signal before detection takes place.

Switch, Arming — An additional switch used when it is necessary to provide a safety aspect to an electrical circuit controlling a vital function.

Switch, Safety Interlock — In situations where high-electrical voltages are used, it is mandatory to provide a safety interlock switch on the door of the enclosure or compound surrounding the components having high voltage.

Synchronous — Two actions are considered to be synchronous when they take place at the same instant.

Tantalum — Because of its inert characteristic, this metallic element finds application in corrosive resistant equipment.

Tap — An independent lead taken off a coil.

Teflon — A trade name for the chemical tetrafluoroethylene. It is a plastic having a high-dielectric constant, and is slippery to the touch. It also has a high resistance to chemical corrosion, and a high-melting point.

Thermal Compression Bonding — A process used in the electronics industry to attach the extremely fine metal leads to micro-circuit chips. By heating the chip and applying mechanical pressure to the lead, attachment or bonding takes place.

Titanium — A metallic element used to make alloys of steel which are very hard and strong. The oxide of it is used as the basis for certain paint pigments.

Tolerance — The amount of variation permitted on a precision dimension.

Torque — When a force is applied radially to a lever arm attached to the axis of a rotating device, a torque or twisting action is set up.

Transformer Vault — The room or area in which large power transformers are safely located. The vault is fireproofed and locked.

Transmitter-receiver — An electronic device that has the capability of both transmitting and receiving.

Tungsten — An element having a high-melting point, and extreme hardness. It is used as a filament in special lamps.

Turbine — A rotating mechanical device having a number of fins of a specialized shape attached to a central shaft. The action of water or gas under pressure against the fins causes the turbine to rotate at high speed. In conjunction with the turbine, an electrical generator is mechanically connected to produce electrical current.

Ultraviolet Light — A type of light which gives off a purplish glow. It has a wave length 320-3900 Angstrom units. (10^{-10} meters). These light waves are longer than X-rays, but shorter than visible light waves.

Vacuum Deposition — A process in which a thin metallic film is vapourized on the surface of a part contained in an evacuated chamber.

Vapour Plating — A process by which a thin film of metal is vapourized on a part in a vacuum chamber.

Variant — A term inferring the variable nature of the state of pressure or temperature of a substance.

Welding, Spot — A form of resistance welding in which the weld areas are highly localized spots. Mechanical pressure and heat are required.

Whiteprint — A copy of a tracing made by a method analogous to photography. The background of the print is white, the lines either blue or black.

Wideband — A term referring to the extent of frequency coverage of a transmitter or receiver.

Bibliography

Alternating Current Fundamentals, J. R. Duff: Delmar Pubs. Inc.

Applications of Electronics, B. Grob & M. S. Kiver: McGraw-Hill Co. of Canada, Ltd.

Applied Electricity, R. F. Brillinger: The Ryerson Press, Toronto

Basic Electricity, K. G. Shoultz: The Macmillan Co. of Can. Ltd.

Basic Electronics, G. Wilcox: Holt, Rinehart & Winston

C.G.S.B. Graphic Symbols, Designations and Abbreviations for use on Electrical and Electronic Schematic Diagrams, National Research Council of Canada

Classification of Electron Tubes, J. Haantjes & H. Carter: The Macmillan Co., New York

Electrical & Electronics Drawing, C. J. Baer: McGraw-Hill Book Co. Inc.

Electrical and Electronic Technology One, H. M. Brouwers: General Publishing Co. Ltd.

Electronic and Electrical Fundamentals, J. E. Remich and staff members: Philco Techrep Division, Philco Corporation

Fundamentals of Integrated Circuits, L. Stern: Hayden Book Co. Inc.

Interior Electric Wiring, K. C. Graham: American Technical Society

Principles of Electrical Theory, K. Schick: McGraw-Hill Co. of Canada, Ltd.

Principles of Electronic Technology, C. B. Weick: McGraw-Hill Co. of Canada, Ltd.

USA Standard, Graphic Symbols for Electrical and Electronics Diagrams, The United States of America Stds. Institute

Index